ELECTRONIC ENGINEERING SYSTEMS SERIES

Series Editor: **J. K. FIDLER**, *University of York*

Asssociate Series Editor: **PHIL MARS**, *University of Durham*

THE ART OF SIMULATION USING PSPICE - ANALOG AND DIGITAL
Bashir Al-Hashimi, Staffordshire University

FUNDAMENTALS OF NONLINEAR DIGITAL FILTERING
Jaakko Astola and Pauli Kuosmanen, Tampere University of Technology

WIDEBAND CIRCUIT DESIGN
Herbert J. Carlin, Cornell University and Pier Paolo Civalleri, Turin Polytechnic

PRINCIPLES AND TECHNIQUES OF ELECTROMAGNETIC COMPATIBILITY
Christos Christopoulos, University of Nottingham

OPTIMAL AND ADAPTIVE SIGNAL PROCESSING
Peter M. Clarkson, Illinois Institute of Technology

KNOWLEDGE-BASED SYSTEMS FOR ENGINEERS AND SCIENTISTS
Adrian A. Hopgood, The Open University

LEARNING ALGORITHMS: THEORY AND APPLICATIONS IN SIGNAL PROCESSING, CONTROL AND COMMUNICATIONS
Phil Mars, J. R. Chen, and Raghu Nambiar
University of Durham

DESIGN AUTOMATION OF INTEGRATED CIRCUITS
Ken G. Nichols, University of Southampton

INTRODUCTION TO INSTRUMENTATION AND MEASUREMENTS
Robert B. Northrop, University of Connecticut

CIRCUIT SIMULATION METHODS AND ALGORITHMS
Jan Ogrodzki, Warsaw University of Technology

Continuous-Time Active Filter Design

T. Deliyannis
Yichuang Sun
J.K. Fidler

CRC Press
Boca Raton   London   New York   Washington, D.C.

Library of Congress Cataloging-in-Publication Data

Deliyannis, Theodore L.
  Continuous-time active filter design / Theodore L. Deliyannis, Yichuang Sun, J. Kel Fidler.
      p.   cm.
  Includes bibliographical references and index.
  ISBN 0-8493-2573-0 (alk. paper)
  1. Electric filters, Active—Design and construction. 2. Continous-time filters—Design and construction. I. Sun, Yichuang. II. Fidler, J. Kel. III. Title.
  TK7872.F5D45 1999
  621.3815′324—dc21   98-44774
                                                                                                                    CIP

   This book contains information obtained from authentic and highly regarded sources. Reprinted material is quoted with permission, and sources are indicated. A wide variety of references are listed. Reasonable efforts have been made to publish reliable data and information, but the author and the publisher cannot assume responsibility for the validity of all materials or for the consequences of their use.
   Neither this book nor any part may be reproduced or transmitted in any form or by any means, electronic or mechanical, including photocopying, microfilming, and recording, or by any information storage or retrieval system, without prior permission in writing from the publisher.
   CRC Press LLC's consent does not extend to copying for general distribution, for promotion, for creating new works, or for resale. Specific permission must be obtained in writing from CRC Press LLC for such copying.
   Direct all inquiries to CRC Press LLC, 2000 Corporate Blvd., N.W., Boca Raton, Florida 33431.

© 1999 by CRC Press LLC

No claim to original U.S. Government works
International Standard Book Number 0-8493-2573-0
Library of Congress Card Number 98-44774
Printed in the United States of America 1 2 3 4 5 6 7 8 9 0
Printed on acid-free paper

Preface

"In this digital age, who needs continuous-time filters?" Such an obvious question, and one which deserves an immediate response. True, we do live in a digital age—digital computers, digital communications, digital broadcasting. But, much though digital technology may bring us advantages over analog systems, at the end of the day a digital system must interface with the real world—the analog world. For example, to gain the advantages that digital signal processing can offer, that processing must take place on *bandlimited* signals, if unwanted aliasing effects are not to be introduced. After the processing, the signals are returned to the real analog world after passing through a *reconstruction* filter. Both bandlimiting and reconstruction filters are analog filters, operating in *continuous time*. This is but one example—but any system that interfaces with the real world will find use for continuous-time filters.

The term *continuous-time* perhaps needs some explanation. There was the time when analog filters were just that—they processed analog signals in real time, in contrast to digital filters which performed filtering operations on digital representations of samples of signals, often not in real time. Then in the 1970s, along came sampled data filters. Sampled data filters did not work with digital representations of the sampled signal, but operated on the samples themselves. Perhaps the best known example of such an approach is that of *switched-capacitor filters,* which as the name suggests, use switches (usually transistor switches) together with capacitors and active devices to provide filter functions. Note that these filters are *discontinuous* in time as a result of the switching which takes place within the circuits; indeed continuous bandlimiting and reconstruction filters are needed as a result. Much research took place in the 1970s and 1980s on switched capacitor filters as a result of the advantages for integrated circuit realization that they promised. There was so much stress on research in this area that development of the more conventional analog filters received relatively little attention. However, when switched capacitor filters failed to provide all the solutions, attention once again turned to the more traditional approaches, and the name *continuous-time filters* was coined to differentiate them from their digital and sampled data counterparts.

This book is about continuous-time filters. The classic LCR filters built with inductors, capacitors, and resistors are such filters, of course, and indeed are still much in use. However, these filters are unsuitable for implementation in the ubiquitous integrated circuit, since no satisfactory way of making inductors on chip has been found. That is why so much attention has been paid to active continuous-time filters over the years. Active filters offer the opportunity to integrate complex filters on-chip, and do not have the problems that the relatively bulky, lossy, and expensive inductors bring—in particular their stray magnetic fields that can provide unwanted coupling in a circuit or system. It is therefore active continuous-time filters on which we shall concentrate here.

As just mentioned, active filters have been around for some time as a means of overcoming the disadvantages associated with passive filters (of which the use of inductors is one). It is a sobering realization that the Sallen and Key circuit (which uses a voltage amplifier, resistors, and capacitors, and is one of the most popular and enduring active-RC filter "architectures") has been around for about 40 years, yet research into active filters still proceeds apace after all that time. Tens of thousands of journal articles and conference papers must have been published and presented over the years. The reasons are manifold, but two particular ones are of note. First, the changes in technology have required new approaches. Thus as cheap, readily available integrated circuit opamps replaced their discrete circuit counterparts (early versions of which used vacuum tube technology, mounted in 19″ racks), it became feasible to consider filter circuits using large numbers of opamps, and new improved architectures emerged. Similarly the development of integrated transconductance amplifiers (the so-called OTA, or operational transconductance amplifiers) led to new filter configurations which reduced the number of resistive components, and allowed with advantage filter solutions to problems using currents as the variables of interest, rather than voltages. In the limit this gives rise to OTA-C filters, using only active devices and capacitors, eminently suitable for integration, but not reducing the significance of active-RC filters which maintain their importance in hybrid realizations. Second, the demands on filter circuits have become ever more stringent as the world of electronics and communications has advanced. For example, greater demands on bandwidth utilization have required much higher performance in filters in terms of their attenuation characteristics, and particularly in the transition region between passband and stopband. This in turn has required filters capable of exhibiting high "Q," but having low sensitivity to component changes, and offering dynamically stable performance – filters are not meant to oscillate! In addition, the continuing increase in the operating frequencies of modern circuits and systems reflects on the need for active filters that can perform at these higher frequencies; an area where the OTA active filter outshines its active-RC counterpart.

What then is the justification for this new book on continuous-time active filters? For the newcomer to the field, the literature can be daunting, in both its volume and complexity, and this book picks a path through the developed field of active filters which highlights the important developments, and concentrates on those architectures that are of practical significance. For the reader who wants to be taken to the frontiers of continuous-time active filter design, these too are to be found here, with a comprehensive treatment of transconductance amplifier-based architectures that will take active filter design into the next millennium. All this material is presented in a context that will enable those readers new to *filter design* (let alone continuous-time active filter design) to get up to speed quickly.

This book will be found interesting by practising engineers and students of electronics, communications or cognate subjects at postgraduate or advanced undergraduate level of study. It is simply structured. Chapters 1 through 3 cover the basic topics required in introducing filter design; Chapters 4 through 7 then focus on opamp-based active-RC filters; finally, Chapters 8 through 12 concentrate on OTA-Capacitor filters (and introduce some other approaches), taking the reader up to the frontiers of modern active continuous-time filter design.

A book such as this requires much work on the part of the authors. In this case it is an achievement of which the authors are particularly proud because it represents the successful collaboration of three engineering academics from quite different cultural background—Greece, China, and the United Kingdom. The catalyst to this collaboration has been Nora Konopka from CRC Press in the U.S., to whom all of us are grateful. In addition, we have many to thank as individuals. Theodore Deliyannis particularly thanks his colleagues I. Haritantis, G. Alexiou, and S. Fotopoulos in Patras, Prof. A. G. Constantinides of Imperial College, London, and the IEE for allowing him to reproduce parts of their common papers published in the *Proceedings of the IEE*. He also expresses his gratitude to Mrs. V. Boile and his postgraduate student K. Giannakopoulos for their help in preparing the manuscript. Finally he thanks his wife Myriam for her encouragement and understanding.

Yichuang Sun acknowledges Prof. Barry Jefferies of the University of Hertfordshire, U.K., for his support, and helpful comments on his work; he is also grateful to Tony Crook for his help in preparing the manuscript. He also expresses his thanks to his wife, Xiaohui, and son, Bo, for their support and patience.

Kel Fidler is particularly grateful to his co-authors Theodore and Yichuang for their incredibly hard work, and their patience and civility at times when things became a little quiet! He also thanks all his friends and colleagues in York for their forbearance and understanding when they observed that, once again, he had taken on too much! In particular he thanks Navin Sullivan, without whom, in many complex ways, these authors would never have come together to write this book.

TLD, Patras
YS, Hatfield
JKF, York

Authors

Professor Theodore L. Deliyannis is Professor of Electronics in the University of Patras, Greece.

Dr. Yichuang Sun is Reader in Electronics in the University of Hertfordshire, U.K.

Professor J. Kel Fidler is Professor of Electronics in the University of York, U.K.

Contents

Chapter 1 Filter Fundamentals

1.1 Introduction 1
1.2 Filter Characterization 1
    1.2.1 Lumped 1
    1.2.2 Linear 2
    1.2.3 Continuous-Time and Discrete-Time 3
    1.2.4 Time-Invariant 3
    1.2.5 Finite 3
    1.2.6 Passive and Active 3
1.3 Types of Filters 4
1.4 Steps in Filter Design 6
1.5 Analysis 7
    1.5.1 Nodal Analysis 7
    1.5.2 Network Parameters 10
        1.5.2.1 One-Port Network 10
        1.5.2.2 Two-Port Network 11
    1.5.3 Two-Port Interconnections 14
        1.5.3.1 Series–Series Connection 14
        1.5.3.2 Parallel–Parallel Connection 15
        1.5.3.3 Series Input–Parallel Output Connection 16
        1.5.3.4 Parallel Input–Series Output Connection 16
        1.5.3.5 Cascade Connection 16
    1.5.4 Network Transfer Functions 17
1.6 Continuous-Time Filter Functions 19
    1.6.1 Pole-Zero Locations 20
    1.6.2 Frequency Response 21
    1.6.3 Transient Response 22
        1.6.3.1 Impulse Response 22
        1.6.3.2 Step Response 23
    1.6.4 Step and Frequency Response 24
1.7 Stability 26
    1.7.1 Short-Circuit and Open-Circuit Stability 27
    1.7.2 Absolute Stability and Potential Instability 27
1.8 Passivity Criteria for One- and Two-Port Networks 29
    1.8.1 One-Ports 29
    1.8.2 Two-Ports 30
    1.8.3 Activity 31
    1.8.4 Passivity and Stability 31
1.9 Reciprocity 32
1.10 Summary 33
References and Further Reading 33

Chapter 2 The Approximation Problem

2.1 Introduction 35

2.2　Filter Specifications and Permitted Functions..35
　　2.2.1　Causality 35
　　2.2.2　Rational Functions..36
　　2.2.3　Stability 36
2.3　Formulation of the Approximation Problem..37
2.4　Approximation of the Ideal Lowpass Filter..38
　　2.4.1　Butterworth or Maximally Flat Approximation 39
　　2.4.2　Chebyshev or Equiripple Approximation 42
　　2.4.3　Inverse Chebyshev Approximation 45
　　2.4.4　Papoulis Approximation 47
　　2.4.5　Elliptic Function or Cauer Approximation 47
　　2.4.6　Selecting the Filter from Its Specifications 49
　　2.4.7　Amplitude Equalization 52
2.5　Filters with Linear Phase: Delays..52
　　2.5.1　Bessel-Thomson Delay Approximation 54
　　2.5.2　Other Delay Functions 58
　　2.5.3　Delay Equalization 59
2.6　Frequency Transformations 59
　　2.6.1　Lowpass-to-Lowpass Transformation 60
　　2.6.2　Lowpass-to-Highpass Transformation 61
　　2.6.3　Lowpass-to-Bandpass Transformation 62
　　2.6.4　Lowpass-to-Bandstop Transformation 63
　　2.6.5　Delay Denormalization..64
2.7　Design Tables for Passive LC Ladder Filters 64
　　2.7.1　Transformation of Elements..65
　　　　2.7.1.1　LC Filters..65
　　　　2.7.1.2　Active RC Filters...68
2.8　Impedance Scaling 70
2.9　Predistortion...71
2.10 Summary 72
References...73

Chapter 3　Active Elements

3.1　Introduction...75
3.2　Ideal Controlled Sources 75
3.3　Impedance Transformation (Generalized Impedance Converters and
　　Inverters)...76
　　3.3.1　Generalized Impedance Converters 78
　　　　3.3.1.1　The Ideal Active Transformer 78
　　　　3.3.1.2　The Ideal Negative Impedance Converter...79
　　　　3.3.1.3　The Positive Impedance Converter...79
　　　　3.3.1.4　The Frequency-Dependent Negative Resistor 80
　　3.3.2　Generalized Impedance Inverters 81
　　　　3.3.2.1　The Gyrator 81
　　　　3.3.2.2　Negative Impedance Inverter 82
3.4　Negative Resistance 82
3.5　Ideal Operational Amplifier..84
　　3.5.1　Operations Using the Ideal Opamp..85
　　　　3.5.1.1　Summation of Voltages..85
　　　　3.5.1.2　Integration 86
　　3.5.2　Realization of Some Active Elements Using Opamps 87

		3.5.2.1	Realization of Controlled Sources	87
		3.5.2.2	Realization of Negative-Impedance Converters	88
		3.5.2.3	Gyrator Realizations	90
		3.5.2.4	GIC Circuit Using Opamps	91
	3.5.3	Characteristics of IC Opamps		93
		3.5.3.1	Open-Loop Voltage Gain of Practical Opamps	93
		3.5.3.2	Input and Output Impedances	94
		3.5.3.3	Input Offset Voltage V_{IO}	95
		3.5.3.4	Input Offset Current I_{IO}	95
		3.5.3.5	Input Voltage Range V_I	96
		3.5.3.6	Power Supply Sensitivity $\Delta V_{IO}/\Delta V_{GG}$	97
		3.5.3.7	Slew Rate SR	97
		3.5.3.8	Short-Circuit Output Current	97
		3.5.3.9	Maximum Peak-to-Peak Output Voltage Swing V_{opp}	97
		3.5.3.10	Input Capacitance C_i	98
		3.5.3.11	Common-Mode Rejection Ratio CMRR	98
		3.5.3.12	Total Power Dissipation	98
		3.5.3.13	Rise Time t_r	98
		3.5.3.14	Overshoot	98
	3.5.4	Effect of the Single-Pole Compensation on the Finite Voltage Gain Controlled Sources		98
3.6	The Ideal Operational Transconductance Amplifier (OTA)			100
	3.6.1	Voltage Amplification		100
	3.6.2	A Voltage-Variable Resistor (VVR)		101
	3.6.3	Voltage Summation		101
	3.6.4	Integration		102
	3.6.5	Gyrator Realization		102
	3.6.6	Practical OTAs		103
	3.6.7	Current Conveyor		104
3.7	Summary			106
References				106

Chapter 4 Realization of First- and Second-Order Functions Using Opamps

4.1	Introduction	107
4.2	Realization of First-Order Functions	107
	4.2.1 Lowpass Circuits	108
	4.2.2 Highpass Circuits	109
	4.2.3 Allpass Circuits	110
4.3	The General Second-Order Filter Function	110
4.4	Sensitivity of Second-Order Filters	111
4.5	Realization of Biquadratic Functions Using SABs	114
	4.5.1 Classification of SABs	115
	4.5.2 A Lowpass SAB	116
	4.5.3 A Highpass SAB	120
	4.5.4 A Bandpass SAB	121
	4.5.5 Lowpass- and Highpass-Notch Biquads	126
	4.5.6 Lowpass Notch ($R_6 = \infty$)	127
	4.5.7 Highpass Notch ($R_7 = \infty$)	129
	4.5.8 An Allpass SAB	129
4.6	Realization of a Quadratic with a Positive Real Zero	132
4.7	Biquads Obtained Using the Twin-T RC Network	134

4.8 Two-Opamp Biquads 136
    4.8.1 Biquads by Inductance Simulation 136
    4.8.2 Two-Opamp Allpass Biquads 138
    4.8.3 Selectivity Enhancement 139
4.9 Three-Opamp Biquads...141
    4.9.1 The Tow-Thomas [25–27] Three-Opamp Biquad......................144
    4.9.2 Excess Phase and Its Compensation in Three-Opamp Biquads 145
    4.9.3 The Åkerberg-Mossberg Three-Opamp Biquad 146
4.10 Summary 147
References 148

Chapter 5 Realization of High-Order Functions

5.1 Introduction..151
5.2 Selection Criteria for High-Order Function Realizations 151
5.3 Multiparameter Sensitivity 153
5.4 High-Order Function Realization Methods..154
5.5 Cascade Connection of Second-Order Sections 155
    5.5.1 Pole-Zero Pairing..156
    5.5.2 Cascade Sequence...158
    5.5.3 Gain Distribution..159
5.6 Multiple-Loop Feedback Filters 162
    5.6.1 The Shifted-Companion-Form (SCF) Design Method 166
    5.6.2 Follow-the-Leader Feedback Design (FLF) 168
5.7 Cascade of Biquartics..171
    5.7.1 The BR Section 171
    5.7.2 Effect of η on ω'_i and Q'_i 173
    5.7.3 Cascading Biquartic Sections 175
    5.7.4 Realization of Biquartic Sections 175
        5.7.4.1 Design Example 176
    5.7.5 Sensitivity of CBR Filters 178
5.8 Summary 180
References 180
Further Reading 181

Chapter 6 Simulation of LC Ladder Filters Using Opamps

6.1 Introduction..183
6.2 Resistively-Terminated Lossless LC Ladder Filters 184
6.3 Methods of LC Ladder Simulation 184
6.4 The Gyrator 185
    6.4.1 Gyrator Imperfections 186
    6.4.2 Use of Gyrators in Filter Synthesis 188
6.5 Generalized Impedance Converter, GIC 190
    6.5.1 Use of GICs in Filter Synthesis 190
6.6 FDNRs: Complex Impedance Scaling 193
6.7 Functional Simulation..195
    6.7.1 Example 198
    6.7.2 Bandpass Filters..199
    6.7.3 Dynamic Range of LF Filters 201
6.8 Summary 202
References 202

Chapter 7 Wave Active Filters

7.1 Introduction 205
7.2 Wave Active Filters 205
7.3 Wave Active Equivalents (WAEs) 208
  7.3.1 Wave Active Equivalent of a Series-Arm Impedance 208
  7.3.2 Wave Active Equivalent of a Shunt-Arm Admittance 209
  7.3.3 WAEs for Equal Port Normalization Resistances 209
  7.3.4 Wave Active Equivalent of the Signal Source 210
  7.3.5 Wave Active Equivalent of the Terminating Resistance 211
  7.3.6 WAEs of Shunt-Arm Admittances 212
  7.3.7 Interconnection Rules 212
  7.3.8 WAEs of Tuned Circuits 214
  7.3.9 WA Simulation Example 216
  7.3.10 Comments on the Wave Active Filter Approach 216
7.4 Economical Wave Active Filters 217
7.5 Sensitivity of WAFs 220
7.6 Operation of WAFs at Higher Frequencies 221
7.7 Complementary Transfer Functions 223
7.8 Wave Simulation of Inductance 224
7.9 Linear Transformation Active Filters (LTA Filters) 224
  7.9.1 Interconnection Rule 227
  7.9.2 General Remarks on the Method 229
7.10 Summary 229
References 229

Chapter 8 Single Operational Transconductance Amplifier (OTA) Filters

8.1 Introduction 231
8.2 Single OTA Filters Derived from Three-Admittance Model 232
  8.2.1 First-Order Filter Structures 232
    8.2.1.1 First-Order Filters with One or Two Passive Components 233
    8.2.1.2 First-Order Filters with Three Passive Components 234
  8.2.2 Lowpass Second-Order Filter with Three Passive Components 235
  8.2.3 Lowpass Second-Order Filters with Four Passive Components 236
  8.2.4 Bandpass Second-Order Filters with Four Passive Components 238
8.3 Second-Order Filters Derived from Four-Admittance Model 241
  8.3.1 Filter Structures and Design 241
    8.3.1.1 Lowpass Filter 241
    8.3.1.2 Bandpass Filter 243
    8.3.1.3 Other Considerations on Structure Generation 244
  8.3.2 Second-Order Filters with the OTA Transposed 245
    8.3.2.1 Highpass Filter 245
    8.3.2.2 Lowpass Filter 247
    8.3.2.3 Bandpass Filter 247
8.4 Tunability of Active Filters Using Single OTA 249
8.5 OTA Nonideality Effects 249
  8.5.1 Direct Analysis Using Practical OTA Macro-Model 249
  8.5.2 Simple Formula Method 253
  8.5.3 Reduction and Elimination of Parasitic Effects 253
8.6 OTA-C Filters Derived from Single OTA Filters 254
  8.6.1 Simulated OTA Resistors and OTA-C Filters 254

        8.6.2 Design Considerations of OY Structures 255
  8.7 Second-Ordre Filters Derived from Five-Admittance Model 258
        8.7.1 Highpass Filter 259
        8.7.2 Bandpass Filter 260
        8.7.3 Lowpass Filter 262
        8.7.4 Comments and Comparison 263
  8.8 Summary 264
References 264

Chapter 9 Two Integer Loop OTA-C Filters

  9.1 Introduction 269
  9.2 OTA-C Building Blocks and First-Order OTA-C Filters 270
  9.3 Two Integrator Loop Configurations and Performance 272
        9.3.1 Configurations 272
        9.3.2 Pole Equations 272
        9.3.3 Design 273
        9.3.4 Sensitivity 273
        9.3.5 Tuning 273
        9.3.6 Biquadratic Specifications 273
  9.4 OTA-C Realizations of Distributed-Feedback (DF) Configuration 274
        9.4.1 DF OTA-C Circuit and Equations 274
        9.4.2 Filter Functions 276
        9.4.3 Design Examples 277
        9.4.4 DF OTA-C Realizations with Special Feedback Coefficients 278
  9.5 OTA-C Filters Based on Summed-Feedback (SF) Configuration 280
        9.5.1 SF OTA-C Realization with Arbitrary k_{12} and k_{11} 281
              9.5.1.1 Design Example of KHN OTA-C Biquad 282
        9.5.2 SF OTA-C Realization with $k_{12} = k_{11} = k$ 282
  9.6 Biquadratic OTA-C Filters Using Lossy Integrators 283
        9.6.1 Tow-Thomas OTA-C Structure 284
        9.6.2 Feedback Lossy Integrator Biquad 284
  9.7 Comparison of Basic OY Filter Structures 285
        9.7.1 Multifunctionality and Number of OTA 285
        9.7.2 Sensitivity 286
        9.7.3 Tunability 286
  9.8 Versatile Filter Functions Based on Node Current Injection 287
        9.8.1 DF Structures with Node Current Injection 288
        9.8.2 SF Structures with Node Current Injection 289
  9.9 Universal Biquads Using Output Summation Approach 291
        9.9.1 DF-Type Universal Biquads 292
        9.9.2 SF Type Universal Biquads 292
        9.9.3 Universal Biquads Based on Node Current Injection and
              Output Summation 293
        9.9.4 Comments on Universal Biquads 294
  9.10 Universal Biquads Based on Canonical and TT Circuits 294
  9.11 Effects and Compensation of OTA Nonidealities ....
        9.11.1 General Model and Equations 295
        9.11.2 Finite Impedance Effects and Compensation 298
        9.11.3 Finite Bandwidth Effects and Compensation 299
        9.11.4 Selection of OTA-C Filter Structures 301
        9.11.5 Selection of Input and Output Methods 302

9.11.6 Dynamic Range Problem 302
9.12 Summary 303
References 304

Chapter 10 OTA-C Filters Based on Ladder Simulation

10.1 Introduction 309
10.2 Component Substitution Method 310
    10.2.1 Direct Inductor Substitution 310
        10.2.1.1 OTA-C Inductors 310
        10.2.1.2 Tolerance Sensitivity of Filter Function 311
        10.2.1.3 Parasitic Effects on Simulated Inductor 312
        10.2.1.4 Parasitic Effects on Filter Function 313
    10.2.2 Application Examples of Inductor Substitution 315
        10.2.2.1 OTA-C Biquad Derived from RLC Resonator Circuit 315
        10.2.2.2 A Lowpass OTA-C Filter 316
    10.2.3 Bruton Transformation and FDNR Simulation 317
10.3 Admittance/Impedance Simulation 320
    10.3.1 General Description of the Method 320
    10.3.2 Application Examples and Comparison 321
    10.3.3 Parial Floating Admittance Concept 324
10.4 Signal Flow Simulation and Leapfrog Structures 325
    10.4.1 Leapfrog Simulation Structures of General Ladder 325
    10.4.2 OTA-C Lowpass LF Filters 328
        10.4.2.1 Example 330
    10.4.3 OTA-C Bandpass LF Filter Design 332
    10.4.4 Partial Floating Admittance Block Diagram and OTA-C Realization 332
    10.4.5 Alternative Leapfrog Structures and OTA-C Realizations 334
10.5 Equivalence of Admittance and Signal Simulation Methods 336
10.6 OTA-C Simulation of LC Ladders Using Matrix Methods 338
10.7 Coupled Biquad OTA Structures 340
10.8 Some General Practical Design Considerations 342
    10.8.1 Selection of Capacitors and OTAs 342
    10.8.2 Tolerance Sensitivity and Parasitic Effects 343
    10.8.3 OTA Finite Impedances and Frequency-Dependent Transconductance 343
10.9 Summary 343
References 344

Chapter 11 Multiple Integrator Loop Feedback OTA-C Filters

11.1 Introduction 349
11.2 General Design Theory of All-Pole Structures 350
    11.2.1 Multiple Loop Feedback OTA-C Model 350
    11.2.2 System Equations and Transfer Function 350
    11.2.3 Feedback Coefficient Matrix and Systematic Structure Generation 353
    11.2.4 Filter Synthesis Prcedure Based on Coefficient Matching 354
11.3 Structure Generation and Design of All-Pole Filters 355
    11.3.1 First- and Second-Order Filters 355
    11.3.2 Third-Order Filters 356
    11.3.3 Fourth-Order Filters 357
    11.3.4 Design Examples of Fourth-Order Filters 359
    11.3.5 General nth-Order Architectures 360

    11.3.5.1 General IFLF Configuration..360
    11.3.5.2 General LF Configureation 361
  11.3.6 Other Types of Realization..362
11.4 Generation and Synthesis of Transmission Zeros...363
  11.4.1 Output Summation of OTA Network..364
  11.4.2 Input Distribution of OTA Network 364
  11.4.3 Universal and Special Third-Order OTA-C Filters..366
    11.4.3.1 IFLF and Output Summation Structure in Fig. 11.10(a) 367
    11.4.3.2 IFLF and Input Distribution Structure in Fig. 11.10(b) 367
    11.4.3.3 LF and Output Summation Structure in Fig. 11.10(c) 367
    11.4.3.4 LF and Input Distribution Structure in Fig. 11.10(d)........................368
    11.4.3.5 Realization of Special Characteristics...368
    11.4.3.6 Design of Elliptic Filters 368
  11.4.4 General nth-Order OTA-C Filters..370
    11.4.4.1 Universal IFLF Architectures...370
    11.4.4.2 Universal LF Architectures..372
11.5 General Formulation of Sensitivity Analysis 373
  11.5.1 General Sensitivity Relations..373
  11.5.2 Sensitivities of Different Filter Structures 375
11.6 Determination of Maximum Signal Magnitude..377
11.7 Effects of OTA Frequency Response Nonidealities 379
11.8 Summary 381
References..382

Chapter 12 Current-Mode Filters and Other Architectures

12.1 Introduction...387
12.2 Current-Mode Filters Based on Single DO-OTA Model 388
  12.2.1 General Model and Filter Architecture Generation......................................388
    12.2.1.1 First-Order Filter Structures..389
    12.2.1.2 Second-Order Filter Architectures 390
  12.2.2 Passive Resistor and Active Resistor 390
  12.2.3 Design of Second-Order Filters 391
  12.2.4 Effects of DO-OTA Nonidealities 394
12.3 Current-Mode Two Integrator Loop DO-OTA-C Filters..396
  12.3.1 Basic Building Blocks and First-Order Filters 396
  12.3.2 Current-Mode DO-OTA-C Configurations with Arbitrary k_{ij}...................397
  12.3.3 Current-Mode DO-OTA-C Biquadratic Architectures with $k_{12} = k_{ij}$ 398
  12.3.4 Current-Mode DO-OTA-C Biquadratic Architectures with $k_{12} = 1$ 399
  12.3.5 DO-OTA Nonideality Effects 401
  12.3.6 Universal Current-Mode DO-OTA-C Filters 401
12.4 Current-Mode DO-OTA-C Ladder Simulation Filters 405
  12.4.1 Leapfrog Simulation Structures of General Ladder 405
  12.4.2 Current-Mode DO-OTA-C Lowpass LF Filters...407
  12.4.3 Current-Mode DO-OTA-C Bandpass LF Filter Design................................409
  12.4.4 Alternative Current-Mode Leapfrog DO-OTA-C Structure.............................410
12.5 Current-Mode Multiple Loop Feedback DO-OTA-C Filters....................................411
  12.5.1 Design of All-Pole Filters...411
  12.5.2 Realization of Transmission Zeros 415
    12.5.2.1 Multiple Loop Feedback with Input Distribution 415
    12.5.2.2 Multiple Loop Feedback with Output Summation 416
    12.5.2.3 Filter Structures and Design Formulas..417

12.6 Other Continuous-Time Filter Structures 419
    12.6.1 Balanced Opamp-RC and OTA-C Structures 419
    12.6.2 MOSFET-C Filters...420
    12.6.3 OTA-C Opamp Filter Design 422
    12.6.4 Active Filters Using Current Conveyors..423
    12.6.5 Log-Domain, Current Amplifier, and Integrated-RLC Filters 425
12.7 Summary 425
References..426

Appendix A   A Sample of Filter Functions..431

Index...437

Chapter 1
Filter Fundamentals

1.1 Introduction

Continuous-time active filters are active networks (circuits) with characteristics that make them useful in today's system design. Their response can be predetermined once their excitation is known, provided that their characteristic function is known or can be derived from their circuit diagram. Thus, it is important for the filter designer to be familiar with the concepts relevant to filter characterization.

These useful concepts are reviewed in this chapter. For motivation, we deal with the filter characterization and the possible responses first. In order to pursue these further, we need to consider certain fundamentals; the analysis of a circuit is explained by means of the nodal method. The analysis of the circuit gives the mathematical expressions, transfer, or other functions that describe its characteristics. We examine these functions in terms of their pole-zero locations in the s-plane and use them to determine the frequency and time responses of the circuit. The concepts of stability, passivity, activity, and reciprocity, which are closely associated with the study and the design of the types of networks examined in this book, are also visited briefly.

1.2 Filter Characterization

The filters examined in this book are networks that process the signal from a source before they deliver it to a load. In terms of a block diagram this is shown in Fig. 1.1.

The filter network is considered here to be lumped, linear, continuous-time, time invariant, finite, passive, or active. These terms are clarified in the following section.

1.2.1 Lumped

In *lumped* networks, we consider the resistance, inductance, or capacitance as symbols or simple elements concentrated within the boundaries of the corresponding physical element, the physical dimensions of which are negligible compared to the wavelength of the fields associated with the signal. This is in contrast to the *distributed* networks, in which the physical elements have dimensions comparable to the wavelength of the fields associated with the signal.

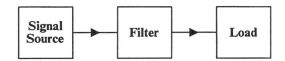

FIGURE 1.1
Block diagram of a filter inserted between the signal source and the load.

1.2.2 Linear

Consider the circuit or system shown in Fig. 1.2(a) in block diagram form, where $r_1(t)$ is the system response to the excitation $e_1(t)$.

The system will be linear (LS) when its response to the excitation $C_1e_1(t)$, where C_1 is a constant, is also multiplied by C_1, i.e., if it is $C_1r_1(t)$, as shown in Fig. 1.2(b). This expresses the principle of proportionality.

For a linear system the principle of superposition holds. This principle is stated as follows: If the responses to the separate excitations $C_1e_1(t)$ and $C_2e_2(t)$ are $C_1r_1(t)$ and $C_2r_2(t)$, respectively, then the response to the excitation $C_1e_1(t) + C_2e_2(t)$ will be $C_1r_1(t) + C_2r_2(t)$, C_1 and C_2 both being constants. Some examples of linear circuits are the following:

- An amplifier working in the linear region of its characteristics is a linear circuit.
- A differentiator is a linear circuit. To show this, let $r(t)$ be the response to the excitation $e(t)$.

$$r(t) = \frac{de(t)}{dt} \tag{1.1}$$

Then, if $e(t)$ is multiplied by a constant C, we will get for the new response $r'(t)$

$$r(t) = \frac{d[Ce(t)]}{dt} = C\frac{de(t)}{dt} = Cr(t) \tag{1.2}$$

- Similarly, for an integrator, the response $r(t)$ to its excitation $e(t)$ is:

$$r(t) = \int_0^t e(\tau)d\tau \tag{1.3}$$

If $e(t)$ is multiplied by the constant C, the new response of the integrator will be:

$$r'(t) = \int_0^t Ce(\tau)d\tau = C\int_0^t e(\tau)d\tau \tag{1.4}$$

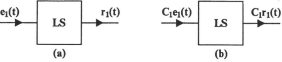

FIGURE 1.2
System response to excitation.

Filter Fundamentals 3

- A time delayer, which introduces the time delay T to the signal, also corresponds to a linear operator, since the response to the excitation $e(t)$ will be

$$r(t) = e(t-T) \tag{1.5}$$

1.2.3 Continuous-Time and Discrete-Time

In a continuous-time filter, both the excitation e and the response r are continuous functions of the continuous time t, i.e.,

$$e = e(t) \quad r = r(t) \tag{1.6}$$

In contrast, in a discrete-time or sampled-data filter the values of the excitation and response are continuous, changing only at discrete instants of time. These are the sampling instants. Only the values of the excitation and response at the sampling instants are of interest. In this case, we have

$$e = e(nT) \quad r = r(nT)$$

where T is the sampling period and n a positive integer.

Details of continuous-time filters are given in Section 1.6, while further reference to discrete-time filters is not within the scope of this book.

1.2.4 Time-Invariant

A time-invariant filter is built up from elements whose values do not change with time during the operation of the filter. In such a filter, if the excitation $e(t)$ is delayed by T, so is its response $r(t)$. This is shown by means of Fig. 1.3.

1.2.5 Finite

The physical dimensions of the filter network are finite; the number of its components is finite.

1.2.6 Passive and Active

A simple definition of a passive filter is given in terms of its elements, i.e., if all of its elements are passive the filter will be passive. Therefore, a passive filter may include among

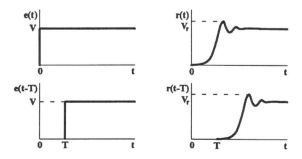

FIGURE 1.3
Defining a time-invariant filter.

its elements resistors, capacitors, inductors, transformers, or ideal gyrators (see Chapter 3). If the elements of the filter include amplifiers or negative resistances, this will be called active.

Another more formal definition is the following: A filter is passive if and only if the following conditions are satisfied:

1. If currents and voltages of any waveform are applied to its terminal, the total energy supplied to the filter is non-negative.
2. No response appears in the circuit before the application of the excitation.

A filter is active if not passive. Condition 2 is necessary in order to avoid the situation in which energy has been stored in some elements and appears before the application of the excitation.

Passivity conditions in terms of network functions are given in Section 1.7.

1.3 Types of Filters

Ideal transmission of a signal from its source to the receiver requires the following two conditions to be satisfied:

1. The spectrum of the signal remains unchanged.
2. The time differences between the various components of the signal remain unchanged.

The latter condition is satisfied if there is no change in the phase of each component during the transmission, or if the phase varies linearly with frequency. Since changes in phase are bound to occur in practice, linearity in phase with frequency is necessary for the Condition 2 to be satisfied.

Thus, the desired transfer function of a transmitting medium should have the following characteristics. Its magnitude should be:

$$|H(j\omega)| = 1$$

and its phase

$$\arg H(j\omega) = -\omega T$$

where T is a constant with the dimensions of time.

The function in Laplace transform notation, which possesses these two characteristics, is the following:

$$H(s) = e^{-sT}$$

However, in real transmission, the signal is usually distorted for various reasons such as interference by other signals, corruption by noise, etc. Then the distorted signal, before reaching the receiver, has to be corrected or processed in order to be restored to its initial form. This can be achieved by means of filters and equalizers.

We distinguish the filters according to their frequency response as lowpass, highpass, bandpass, bandstop, allpass, and arbitrary frequency response (equalizers). The latter are included here, following the general definition of a filter given at the beginning of this chapter.

The basic filter frequency responses are as follows:

1. **The lowpass filter**—The ideal response of a lowpass filter is shown in Fig. 1.4(a). All frequencies below the cutoff frequency ω_c pass through the filter without obstruction. The band of these frequencies is the filter passband. Frequencies above cutoff are prevented from passing through the filter and they constitute the filter stopband.

   However, for reasons explained in Chapter 2, the ideal lowpass filter response cannot be realized by a physical circuit. Instead, the practical lowpass filter response will, in general, be as shown in Fig. 1.4(b). It can be seen that a small error is allowable in the passband, while the transition from the passband to the stopband is not abrupt. The width of this transition band $\omega_s - \omega_c$ determines the filter selectivity. Here ω_s is considered to be the lowest frequency of the stopband, in which the gain remains below a specified value.

2. **The highpass filter**—For reasons similar to those holding for the lowpass filter the ideal highpass filter response is unrealizable. The amplitude response of the practical highpass filter will basically be as shown in Fig. 1.5.

   In the highpass filter the passband is above the cutoff frequency ω_c, while all frequencies below ω_c are attenuated when passing through the filter.

3. **The bandpass filter**—The ideal bandpass is again unrealizable and the amplitude response of the practical bandpass filter is as shown in Fig. 1.6. Here the passband lies between two stopbands, the lower and the upper. Accordingly there are two transition bands.

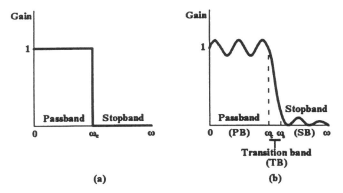

FIGURE 1.4
(a) Ideal and (b) practical lowpass filter amplitude response.

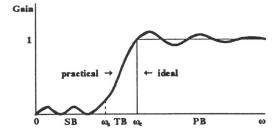

FIGURE 1.5
The basic highpass ideal and practical filter amplitude response.

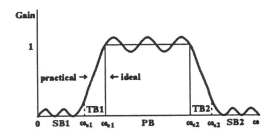

FIGURE 1.6
Amplitude response of the ideal and practical basic bandpass filter.

4. **The bandstop filter**—The amplitude response of the practical band-elimination or bandstop filter is shown in Fig. 1.7, while its ideal response is again unrealizable. It can be seen that the filter possesses two passbands separated by a stopband rejected by the filter. There are also two transition bands.

5. **The allpass filter**—Ideally this filter passes, without any attenuation, all frequencies (0 to ∞), while its characteristic of concern is the phase response. If its phase response is linear, then it can operate as an ideal time delayer. In practice the phase can be linear, within an acceptable error, up to a certain frequency ω_c. For frequencies below ω_c the allpass filter operates as a delayer. It is useful in phase equalization.

   It should be noted that allpass filters are not the only ones that may possess linear phase response. Certain lowpass filters also have similar phase response, as explained in Chapter 2, and they can be used as time delayers.

6. **Amplitude equalizers**—The amplitude equalizer has an amplitude response that does not belong to any of the filter responses considered above. It is used to compensate for the distortion of the frequency spectrum that the signal suffers when passing through a system. Its amplitude response is therefore drawn as complementary to the signal spectrum. In this sense it can be considered arbitrary being suitable for only one distorted signal.

1.4 Steps in Filter Design

Filter design, in effect, involves three separate processes or steps, these being

1. Analysis of circuits
2. Curve approximation
3. Synthesis of the filter

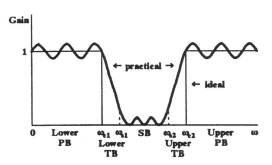

FIGURE 1.7
Amplitude response of the ideal and practical basic bandstop filter.

Filter Fundamentals

These three steps are explained below to clarify matters.

1. **Analysis of circuits**—Conventionally, analysis of a circuit is the procedure to find the characteristics of the filter operation from its diagram and the values of its components. However, analysis of circuits has a more general meaning here, namely to determine general types of operational characteristics for various general types and orders of circuits. These characteristics may be formulated as rational functions of the complex frequency variable s, with constraints depending on the circuit type. These rational functions will be referred to here as the permitted functions.

2. **Curve approximation**—Based on the knowledge of the characteristics and potentialities of the various types of circuits, we may proceed to try to find the solution of a certain design problem. Clearly the filter specifications are not given in the form of rational functions, but as lines or curves that give, for example, maxima and minima of attenuation. These lines determine the so-called *specified curve*. Therefore, the next step in the filter design will necessarily be the determination of the permitted rational function that best approximates the specified curve, i.e., that satisfies the conditions set by the specified curve.

   Usually the complexity (and consequently the cost) of the circuit increases with the order of the permitted function that is selected. It is therefore necessary to determine the simplest permitted function that satisfies the specifications. Once the suitable permitted function has been found, the basic information is available for the determination of the corresponding circuit, i.e., the circuit whose operation characteristics are in agreement with the selected permitted function.

3. **Filter synthesis**—Filter synthesis refers to the process for determining a circuit, i.e., its diagram and the values of its components. Even more ambitiously, we may find all possible circuits that satisfy the specifications and among them select the *best* according to certain criteria (cost, available technologies, power dissipation, etc.).

1.5 Analysis

For the sake of the reader who is not very familiar with the analysis of a general circuit, we include this section to explain the nodal analysis of a circuit and use the results in order to obtain the mathematical relationship(s) connecting its response(s) to the excitation(s). These relationships will, in the general case, give the types of the permitted functions which were mentioned in the previous section.

1.5.1 Nodal Analysis

Nodal analysis is usually used to determine the response of an active circuit to a certain excitation. We will explain the method of nodal analysis by applying it in the case of the circuit in Fig. 1.8.

Let V_i, $i = 1,2,3,4$ be the voltages in the corresponding nodes and apply Kirchhoff's current law (KCL) in each node. We may write for node 1:

$$y_{12}(V_1 - V_2) + y_{13}(V_1 - V_3) + y_{14}(V_1 - V_4) = I_1$$

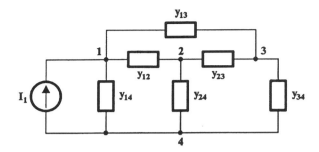

FIGURE 1.8
Circuit illustrating nodal analysis.

or

$$y_{11}V_1 - y_{12}V_2 - y_{13}V_3 - y_{14}V_4 = I_1 \tag{1.7}$$

where $y_{11} = y_{12} + y_{13} + y_{14}$ is the self-admittance of node 1, i.e., the sum of all admittances connected to node 1, while y_{ij}, $i, j = 1,2,3,4$, with $i \neq j$, is the mutual admittance directly connecting node i to node j.

Similarly we may write for the other nodes

node 2: $\qquad -y_{12}V_1 + y_{22}V_2 - y_{23}V_3 - y_{24}V_4 = 0 \qquad (1.8)$

node 3: $\qquad -y_{13}V_1 - y_{23}V_2 + y_{33}V_3 - y_{34}V_4 = 0 \qquad (1.9)$

node 4: $\qquad -y_{14}V_1 - y_{24}V_2 - y_{34}V_3 + y_{44}V_4 = -I_1 \qquad (1.10)$

where y_{ii}, $i = 2,3,4$ is the self-admittance of node 2,3,4, respectively.

It must be noted that these four equations are not independent. For example, if we add the first three we will get Eq. (1.10). To get an independent set of equations we arbitrarily choose one node as the reference node and set its voltage equal to zero. Then, the number of equations required for the calculation of the voltages at the other nodes will be reduced by one. In the case of this example, let $V_4 = 0$ and obtain the following set of three equations:

$$\begin{aligned} y_{11}V_1 - y_{12}V_2 - y_{13}V_3 &= I_1 \\ -y_{12}V_1 + y_{22}V_2 - y_{23}V_3 &= 0 \\ -y_{12}V_1 - y_{23}V_2 + y_{33}V_3 &= 0 \end{aligned} \tag{1.11}$$

where y_{ii}, $i = 1,2,3$ are the self admittances of the nodes including, of course, the mutual admittance connecting the corresponding node to the reference node (node number 4, in this case). Solution of the set of Eq. (1.11) will give the voltages at nodes 1,2,3 referring to the voltage at node number 4.

To complete the analysis, the currents in each admittance should be calculated. This can be easily achieved by applying Ohm's law at each branch. For example, the current I_{ij} in y_{ij} is the following:

$$I_{ij} = y_{ij}(V_i - V_j) \tag{1.12}$$

Filter Fundamentals 9

In the general case of a circuit with N nodes, the $n = N - 1$ independent equations, when the Nth node has been chosen as the reference node, with $V_N = 0$, are as follows:

$$y_{11}V_1 + y_{12}V_2 + \ldots + y_{1n}V_n = I_1$$

$$y_{21}V_1 + y_{22}V_2 + \ldots + y_{2n}V_n = I_2 \quad (1.13)$$

$$\ldots\ldots\ldots\ldots\ldots\ldots\ldots\ldots\ldots\ldots\ldots$$

$$y_{n1}V_1 + y_{n2}V_2 + \ldots + y_{nn}V_n = I_n$$

where, in all y_{ij}, $i \neq j$, the minus sign has been included in the symbol. This set of equations can be written in matrix form.

$$\begin{bmatrix} y_{11} & y_{12} & \cdots & y_{1n} \\ y_{21} & y_{22} & \cdots & y_{2n} \\ \cdots & \cdots & \cdots & \cdots \\ y_{n1} & y_{n2} & \cdots & y_{nn} \end{bmatrix} \begin{bmatrix} V_1 \\ V_2 \\ \cdots \\ V_n \end{bmatrix} = \begin{bmatrix} I_1 \\ I_2 \\ \cdots \\ I_n \end{bmatrix} \quad (1.14)$$

or simply

$$[y][V] = [I] \quad (1.15)$$

where

$$[y] = \begin{bmatrix} y_{11} & y_{12} & \cdots & y_{1n} \\ y_{21} & y_{22} & \cdots & y_{2n} \\ \cdots & \cdots & \cdots & \cdots \\ y_{n1} & y_{n2} & \cdots & y_{nn} \end{bmatrix} \quad (1.16)$$

is an $n \times n$ matrix, and $[V]$, $[I]$ are column matrices. When the admittances are bilateral, i.e., the corresponding currents through them remain the same in magnitude when the applied voltages change their polarity, it is always $y_{ij} = y_{ji}$, and this matrix is symmetric around the main diagonal. All passive RLC networks are characterized by this property. This is true for all reciprocal networks (see Section 1.9).

It is important to realize that matrix $[y]$ can be formed by inspection of the circuit once the nodes have been identified (numbered) and the reference node has been chosen. To this end, one should remember that each self admittance y_{ii} is the sum of all admittances connected to the ith node, while in the symbol for each mutual admittance, y_{ij}, $i \neq j$, the minus sign is included. On the other hand, if matrix $[y]$ is known, it can be used to reconstruct the circuit, following the above observations regarding y_{ii} and y_{ij}, $i \neq j$.

In the above discussion, it was assumed that all independent sources were current sources, and this is most convenient when applying KCL. However, when some excitations are applied via voltage sources, these should be transformed to their equivalent current sources by using Norton's theorem. According to this theorem, a voltage source, Fig. 1.9(a), is equivalent to a current source, Fig. 1.13(b), when

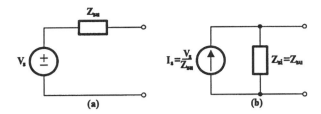

FIGURE 1.9
(a) A voltage source and (b) its equivalent current source.

$$I_s = \frac{V_s}{Z_{su}}$$

and

$$Z_{si} = Z_{su} = Z_s$$

In case V_s is ideal (i.e., $Z_s = 0$), we assume that $Z_s \neq 0$, we carry out the analysis as usual, and in the final expressions we set $Z_s = 0$.

However, when dependent current or voltage sources are present in the circuit, the y matrix is not symmetrical, because then some of the y_{ij} are not the same as the corresponding y_{ji}. In the case of a dependent current source, whether it be current controlled or voltage controlled (see Section 3.2), this is treated as an independent current source in forming the corresponding nodal equation, which is then rearranged in the form of Eq. (1.13). In the case of a dependent voltage source, whether it be voltage controlled or current controlled, this may be transformed to a current source using Norton's theorem, as was explained above.

1.5.2 Network Parameters

The y-matrix that was determined above is useful in determining various network parameters that express the network behavior. We explain this in the cases of one- and two-port networks. These are considered linear, lumped, finite and time-invariant, usually denoted as LLF networks.

1.5.2.1 One-Port Network

A one-port (or two-terminal) network is shown in Fig. 1.10. It is excited by a current source only. V_1 is the response of interest.

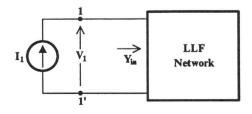

FIGURE 1.10
A one-port network excited by a current source.

Then matrix $[I]$ in Eq. (1.14) will be as follows:

$$[I] = \begin{bmatrix} I_1 \\ 0 \\ \vdots \\ 0 \end{bmatrix} \tag{1.17}$$

From these equations, we obtain V_1 and consequently Y_{in}. Applying Cramer's rule, we will have

$$V_1 = \frac{\Delta_{11}}{\Delta_y} I_1 \tag{1.18}$$

and

$$Y_{in} = \frac{1}{Z_{in}} = \frac{I_1}{V_1} = \frac{\Delta_y}{\Delta_{11}} \tag{1.19}$$

where Δ_y is the determinant of $[y]$, and Δ_{11} is its cofactor when the first row and first column are deleted.

1.5.2.2 Two-Port Network

Consider the two-port network shown in Fig. 1.11. If I_1 and I_2 are the only independent excitations, we may write, as a consequence of linearity, the following equations:

$$V_1 = Z_{11} I_1 + Z_{12} I_2$$
$$V_2 = Z_{21} I_1 + Z_{22} I_2 \tag{1.20}$$

where

$$Z_{11} = \left. \frac{V_1}{I_1} \right|_{I_2 = 0} \quad Z_{12} = \left. \frac{V_1}{I_2} \right|_{I_1 = 0} \quad Z_{21} = \left. \frac{V_2}{I_1} \right|_{I_2 = 0} \quad Z_{22} = \left. \frac{V_2}{I_2} \right|_{I_1 = 0} \tag{1.21}$$

are the so called Z-parameters of the two-port. Here,

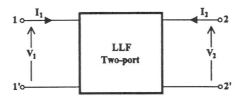

FIGURE 1.11
A two-port LLF network.

$$Z_{11} = \left.\frac{V_1}{I_1}\right|_{I_2=0}$$

means the ratio of V_1 and I_1 when $I_2 = 0$, and similarly for the rest of Eq. (1.21).

These can be obtained from Eq. (1.14) if we set all current excitations, except I_1 and I_2, equal to zero and solve for V_1 and V_2. The result will be

$$Z_{11} = \frac{\Delta_{11}}{\Delta_y} \quad Z_{12} = \frac{\Delta_{12}}{\Delta_y} \quad Z_{21} = \frac{\Delta_{21}}{\Delta_y} \quad Z_{22} = \frac{\Delta_{22}}{\Delta_y} \tag{1.22}$$

where Δ_{ij} is the cofactor of the element y_{ij} of matrix $[y]$. Clearly, all Z_{ij} have the dimensions of impedance.

A corresponding circuit model (equivalent) for the two-port, based on Eq. (1.20), can be drawn as shown in Fig. 1.12. The symbol for $Z_{12}I_2$ and $Z_{21}I_1$ denotes a dependent voltage source.

Alternatively, we may write Eq. (1.20) in the following form:

$$I_1 = Y_{11}V_1 + Y_{12}V_2$$

$$I_2 = Y_{21}V_1 + Y_{22}V_2 \tag{1.23}$$

and determine Y_{ij}, the so called Y-parameters of the two-port as follows:

$$Y_{11} = \left.\frac{I_1}{V_1}\right|_{V_2=0} \quad Y_{12} = \left.\frac{I_1}{V_2}\right|_{V_1=0} \quad Y_{21} = \left.\frac{I_2}{V_1}\right|_{V_2=0} \quad Y_{22} = \left.\frac{I_2}{V_2}\right|_{V_1=0} \tag{1.24}$$

These parameters have the dimensions of admittance and can be obtained from the Z-parameters that were earlier determined from the $[y]$ matrix of the two-port. The conversion formulas between Z- and Y-parameters are given [2, 3] in Table 1.1.

The equivalent circuit of the two-port based on Eq. (1.23) is shown in Fig. 1.13. The symbol for $Y_{12}V_2$ and $Y_{21}V_1$ denotes a dependent current source.

Similarly, we may obtain sets of hybrid parameters of the two-port defined by the following equations:

$$\text{H-parameters:} \quad V_1 = H_{11}I_1 + H_{12}V_2 \quad I_2 = H_{21}I_1 + H_{22}V_2, \text{ or} \tag{1.25}$$

$$\text{G-parameters:} \quad I_1 = G_{11}V_1 + G_{12}I_2 \quad V_2 = G_{21}V_1 + G_{22}I_2 \tag{1.26}$$

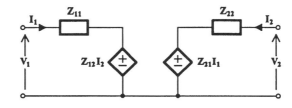

FIGURE 1.12
Equivalent circuit of the two-port using the Z-parameters.

Filter Fundamentals

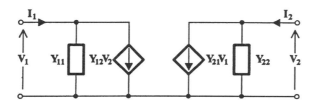

FIGURE 1.13
Equivalent circuit of the two-port based on Eq. (1.23).

TABLE 1.1
Matrix Conversion Table [2]

Z_{11}	Z_{11}	$Y_{22}/\Delta Y$	$\Delta H/H_{22}$	$1/G_{11}$	a_{11}/a_{21}
Z_{12}	Z_{12}	$-Y_{12}/\Delta Y$	H_{12}/H_{22}	$-G_{12}/G_{11}$	$\Delta A/a_{21}$
Z_{21}	Z_{21}	$-Y_{21}/\Delta Y$	$-H_{21}/H_{22}$	G_{21}/G_{11}	$1/a_{21}$
Z_{22}	Z_{22}	$Y_{11}/\Delta Y$	$1/H_{22}$	$\Delta G/G_{11}$	a_{22}/a_{21}
ΔZ	ΔZ	$1/\Delta Y$	H_{11}/H_{22}	G_{22}/G_{11}	a_{12}/a_{21}
Y_{11}	$Z_{22}/\Delta Z$	Y_{11}	$1/H_{11}$	$\Delta G/G_{22}$	a_{22}/a_{12}
Y_{12}	$-Z_{12}/\Delta Z$	Y_{12}	$-H_{12}/H_{11}$	G_{12}/G_{22}	$-\Delta A/a_{12}$
Y_{21}	$-Z_{21}/\Delta Z$	Y_{21}	H_{21}/H_{11}	$-G_{21}/G_{22}$	$-1/a_{12}$
Y_{22}	$Z_{11}/\Delta Z$	Y_{22}	$\Delta H/H_{11}$	$1/G_{22}$	a_{11}/a_{12}
ΔY	$1/\Delta Z$	ΔY	H_{22}/H_{11}	G_{11}/G_{22}	a_{21}/a_{12}
H_{11}	$\Delta Z/Z_{22}$	$1/Y_{11}$	H_{11}	$G_{22}/\Delta G$	a_{12}/a_{22}
H_{12}	Z_{12}/Z_{22}	$-Y_{12}/Y_{11}$	H_{12}	$-G_{12}/\Delta G$	$\Delta A/a_{22}$
H_{21}	$-Z_{21}/Z_{22}$	Y_{21}/Y_{11}	H_{21}	$-G_{21}/\Delta G$	$-1/a_{22}$
H_{22}	$1/Z_{22}$	$\Delta Y/Y_{11}$	H_{22}	$G_{11}/\Delta G$	a_{21}/a_{22}
ΔH	Z_{11}/Z_{22}	Y_{22}/Y_{11}	ΔH	$1/\Delta G$	a_{11}/a_{22}
G_{11}	$1/Z_{11}$	$\Delta Y/Y_{22}$	$H_{22}/\Delta H$	G_{11}	a_{21}/a_{11}
G_{12}	$-Z_{12}/Z_{11}$	Y_{12}/Y_{22}	$-H_{12}/\Delta H$	G_{12}	$-\Delta A/a_{11}$
G_{21}	Z_{21}/Z_{11}	$-Y_{21}/Y_{22}$	$-H_{21}/\Delta H$	G_{21}	$1/a_{11}$
G_{22}	$\Delta Z/Z_{11}$	$1/Y_{22}$	$H_{11}/\Delta H$	G_{22}	a_{12}/a_{11}
ΔG	Z_{22}/Z_{11}	Y_{11}/Y_{22}	$1/\Delta H$	ΔG	a_{22}/a_{11}
a_{11}	Z_{11}/Z_{21}	$-Y_{22}/Y_{21}$	$-\Delta H/H_{21}$	$1/G_{21}$	a_{11}
a_{12}	$\Delta Z/Z_{21}$	$-1/Y_{21}$	$-H_{11}/H_{21}$	G_{22}/G_{21}	a_{12}
a_{21}	$1/Z_{21}$	$-\Delta Y/Y_{21}$	$-H_{22}/H_{21}$	G_{11}/G_{21}	a_{21}
a_{22}	Z_{22}/Z_{21}	$-Y_{11}/Y_{21}$	$-1/H_{21}$	$\Delta G/G_{21}$	a_{22}
ΔA	Z_{12}/Z_{21}	Y_{12}/Y_{21}	$-H_{12}/H_{21}$	$-G_{12}/G_{21}$	ΔA

Since these hybrid parameters are referred to the same two-port, they must be related to the previously defined Z- and Y-parameters. Conversions formulas for these parameters are also given [2, 3] in Table 1.1.

Notice that the hybrid parameters have different dimensions. Thus, H_{11} and G_{22} are impedance functions, H_{22} and G_{11} admittance functions, and H_{12}, H_{21}, G_{12}, and G_{21} are dimensionless. It is for this reason that the parameters are said to be hybrid.

The hybrid equivalent circuits of the two-port based on Eqs. (1.25) and (1.26) are shown in Figs. 1.14(a) and (b), respectively.

Finally, we may write the relationships between port voltages and currents in the following form:

$$V_1 = a_{11}V_2 + a_{12}(-I_2)$$

$$I_1 = a_{21}V_2 + a_{22}(-I_2) \tag{1.27}$$

and thus obtain the a_{ij}, $i, j = 1,2$ parameters, which form the transmission matrix. These parameters relate the input voltage and current to the corresponding output voltage and current and are very useful when studying cascaded two-port networks.

In Eq. (1.27), $-I_2$ is used instead of I_2 to keep in agreement with the initial definition of the a_{ij}, $i, j = 1,2$ parameters, in which I_2 was taken flowing out at port 2 rather than flowing in, as is generally considered.

An equivalent circuit of the two-port based on Eq. (1.27) could also be drawn, but we leave this to the reader as an exercise. Conversion formulae are given [2, 3] in Table 1.1.

1.5.3 Two-Port Interconnections

In certain cases, the analysis of a complex two-port may be simplified, if this can be decomposed into two (or more) subnetworks connected in one of the ways shown below:

1.5.3.1 Series–Series Connection

This connection is shown in Fig. 1.15.

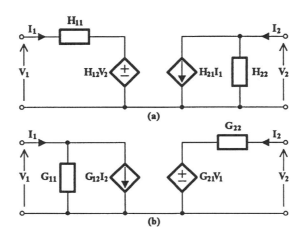

FIGURE 1.14
(a) The *H*-parameter and (b) *G*-parameter equivalent circuits of the two-port.

Filter Fundamentals

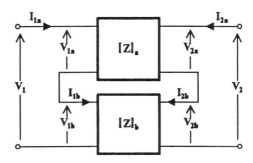

FIGURE 1.15
Series–series connection of two two-ports.

Since

$$I_{1a} = I_{1b} \qquad \text{and} \qquad I_{2a} = I_{2b}$$
$$V_1 = V_{1a} + V_{1b} \qquad\qquad V_2 = V_{2a} + V_{2b}$$

it can be easily shown that the Z-parameter matrix of the overall two-port is the sum of the Z-parameter matrices of the individual two-ports, i.e.,

$$[Z] = [Z]_a + [Z]_b \tag{1.28}$$

This connection of two-ports is sometimes known as *cascode* connection.

1.5.3.2 Parallel–Parallel Connection

In this case, the situation is as shown in Fig. 1.16. Observe that

$$V_1 = V_{1a} = V_{1b} \qquad V_2 = V_{2a} = V_{2b}$$

and

$$I_1 = I_{1a} + I_{1b} \qquad I_2 = I_{2a} + I_{2b}$$

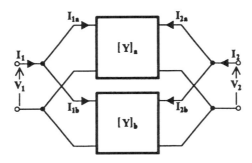

FIGURE 1.16
Parallel–parallel connection.

It can then be easily shown that

$$[Y] = [Y]_a + [Y]_b \qquad (1.29)$$

1.5.3.3 Series Input–Parallel Output Connection
Following similar reasoning, it can be shown that

$$[H] = [H]_a + [H]_b \qquad (1.30)$$

1.5.3.4 Parallel Input–Series Output Connection
Again, it can similarly be shown that

$$[G] = [G]_a + [G]_b \qquad (1.31)$$

1.5.3.5 Cascade Connection
This is shown in Fig. 1.17. Since it is very useful on many occasions, we explain it in detail. It can be seen that

$$I_{2a} = -I_{1b}$$

$$V_{2a} = V_{1b}$$

We may then write

$$\begin{bmatrix} V_{1a} \\ I_{1a} \end{bmatrix} = [A]_a \begin{bmatrix} V_{2a} \\ -I_{2a} \end{bmatrix} = [A]_a [A]_b \begin{bmatrix} V_{2b} \\ -I_{2b} \end{bmatrix}$$

If the behavior of the overall network is described by the relationship

$$\begin{bmatrix} V_{1a} \\ I_{1a} \end{bmatrix} = [A] \begin{bmatrix} V_{2b} \\ -I_{2b} \end{bmatrix}$$

we can easily obtain that

$$[A] = [A]_a [A]_b \qquad (1.32)$$

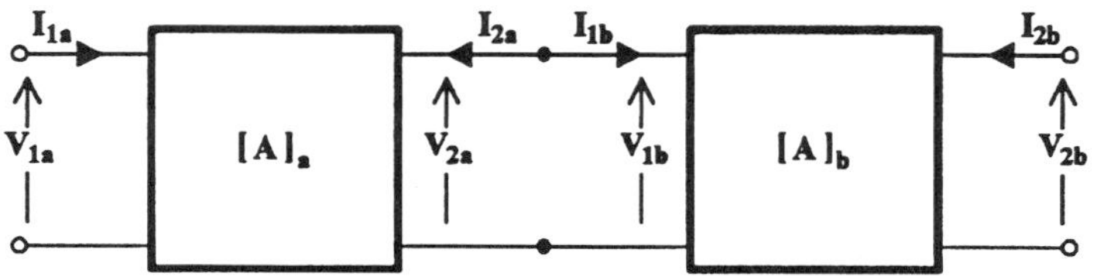

FIGURE 1.17
Cascade connection of two two-ports.

1.5.4 Network Transfer Functions

The five sets of parameters that were introduced in the previous section describe fully the network behavior toward its port terminations. Using these parameters, one can determine various functions, e.g., the input impedance or admittance at one port when another impedance is connected across the other port. Transfer functions are also expressed in terms of these parameters as shown by the following example.

Consider the circuit in Fig. 1.18, where an LLF network is connected between a signal source of voltage E_g and internal resistance R_g and a load resistance R_L. Let the two-port network be described by its Z-parameters.

The voltages and currents at the two-ports of the network are related by Eq. (1.20), which is repeated here for convenience.

$$V_1 = Z_{11}I_1 + Z_{12}I_2$$

$$V_2 = Z_{21}I_1 + Z_{22}I_2 \qquad (1.33)$$

If we wish to determine the voltage transfer ratio V_2/E_g, we may proceed as follows. Observing that

$$V_2 = -I_2 R_L \qquad (1.34)$$

and substituting for I_2 in Eq. (1.33) gives

$$V_1 = Z_{11}I_1 - \frac{Z_{12}}{R_L}V_2 \qquad (1.35)$$

$$V_2 = Z_{21}I_1 - \frac{Z_{22}}{R_L}V_2 \qquad (1.36)$$

Solving Eq. (1.36) for I_1 and substituting in Eq. (1.35), we get, successively,

$$I_1 = \frac{R_L + Z_{22}}{R_L Z_{21}} V_2 \qquad (1.37)$$

and

$$V_1 = \frac{Z_{11}(R_L + Z_{22}) - Z_{12}Z_{21}}{R_L Z_{21}} V_2 \qquad (1.38)$$

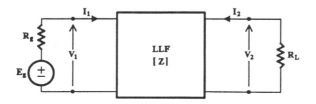

FIGURE 1.18
A two-port connected between source and load.

But
$$V_1 = E_g - I_1 R_g \tag{1.39}$$

Substituting for I_1 from Eq. (1.37) in Eq. (1.39), then equating the sides on the right in Eqs. (1.38) and (1.39) and solving for V_2/E_g, we finally get the following:

$$\frac{V_2}{E_g} = \frac{R_L Z_{21}}{(R_g + Z_{11})(R_L + Z_{22}) - Z_{12} Z_{21}} \tag{1.40}$$

Various other transfer functions for different values of R_g and R_L are given [1] in Table 1.2 using both the Z- and Y-parameters of the two-port where appropriate.

All the network functions that appear on Table 1.2 are the Laplace transforms of corresponding functions of continuous time. Since in this book we are dealing with filters that are characterized by this type of functions only, we review the concept of continuous-time filter functions in the next section in some detail.

TABLE 1.2

Source	Load	Transfer Function
∞	∞	$\dfrac{V_2}{I_1} = Z_{21}$
0	0	$\dfrac{I_2}{V_1} = Y_{21}$
0	∞	$\dfrac{V_2}{V_1} = \dfrac{Z_{21}}{Z_{11}} = -\dfrac{Y_{21}}{Y_{22}}$
∞	0	$\dfrac{I_2}{I_1} = \dfrac{Y_{21}}{Y_{11}} = -\dfrac{Z_{21}}{Z_{22}}$
0	R_L	$\dfrac{I_2}{V_1} = \dfrac{Y_{21}/R_L}{Y_{22} + 1/R_L}$
∞	R_L	$\dfrac{V_2}{I_1} = \dfrac{Z_{21} R_L}{Z_{22} + R_L}$
R_g	∞	$\dfrac{V_2}{E_g} = \dfrac{Z_{21}}{Z_{11} + R_g}$
R_g	0	$\dfrac{I_2}{I_1} = \dfrac{Y_{21}}{Y_{11} + 1/R_g}$
R_g	R_L	$\dfrac{V_2}{V_1} = \dfrac{Z_{21} R_L}{(R_g + Z_{11})(R_L + Z_{22}) - Z_{21} Z_{12}}$

1.6 Continuous-Time Filter Functions

As was mentioned in Section 1.2, the response of a continuous-time filter to the continuous-time excitation $e(t)$ is a continuous-time function $r(t)$ given as follows:

$$r(t) = \int_0^t h(t-\tau)e(\tau)d\tau \tag{1.41}$$

where $h(t)$ is the impulse response (see Section 1.6.3) of the filter.

In the frequency domain this equation is written as follows:

$$R(s) = H(s)E(s) \tag{1.42}$$

where $R(s)$, $H(s)$, $E(s)$ are the Laplace transforms of the time functions $r(t)$, $h(t)$, and $e(t)$ respectively, and s is the complex frequency variable. $H(s)$ is the filter function, transfer or driving-point impedance, or admittance function. These are shown in Fig. 1.19 in block diagram form.

For the filters we are concerned with, $H(s)$ will be a rational function of s, i.e., the ratio of two real and finite polynomials in s. It is written in the following form:

$$H(s) = \frac{N(s)}{D(s)} = \frac{a_n s^n + a_{n-1} s^{n-1} + \ldots + a_1 s + a_o}{s^m + b_{m-1} s^{m-1} + \ldots + b_1 s + b_o} \tag{1.43}$$

where $N(s)$ and $D(s)$ are the numerator and denominator polynomials, respectively, with $m \geq n$, a_i, b_i real and b_i positive (for stability reasons explained below).

If z_i, $i = 1,2,\ldots n$ are the roots of $N(s)$, i.e., the zeros of $H(s)$ and p_i, $i = 1,2,\ldots,m$ are the roots of $D(s)$, i.e., the poles of $H(s)$, then Eq. (1.43) can be written as follows:

$$H(s) = a_n \frac{\prod_{i=1}^{n}(s-z_i)}{\prod_{j=1}^{m}(s-p_j)} \tag{1.44}$$

If the signal is sinusoidal of frequency ω, in Eqs. (1.43) and (1.44) s is substituted by $j\omega$. Function $H(j\omega)$ obtained this way is in fact the continuous-time Fourier transform of $h(t)$. It can then be written in the following form:

$$H(j\omega) = |H(j\omega)|e^{j\varphi(\omega)} \tag{1.45}$$

i.e., in terms of the magnitude and phase of $H(j\omega)$.

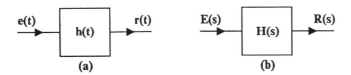

FIGURE 1.19
Block diagram representation of the filter (a) in the time domain and (b) in the frequency domain.

Clearly,

$$\varphi(\omega) = \arg[H(j\omega)] \quad (1.46)$$

It is usual practice to present the magnitude of $H(j\omega)$ in the form

$$A(\omega) = 20 \log|H(j\omega)| \quad (1.47)$$

and thus express it in dB. This gives the filter gain in dB.

However, in most cases, we talk about the filter attenuation or loss, $-A(\omega)$, also in decibels. In some cases, the attenuation is given in nepers obtained as follows:

Attenuation in nepers: $\quad\quad \alpha(\omega) = -\ln|H(j\omega)| \quad (1.48)$

In most filter design cases, $H(s)$ represents the ratio of the Laplace transform of the output voltage to the Laplace transform of the input voltage to the filter being thus dimensionless. However, it may also represent ratio of currents, when it will again be dimensionless, or ratio of output voltage to input current (transimpedance) or output current to input voltage (transadmittance) having the dimensions of impedance or admittance, respectively. Finally, it may represent a driving point function, i.e., the ratio of the voltage to the current in one port of the filter network or vice versa. In these cases, $H(s)$ will represent either an impedance function or an admittance function, again not being dimensionless.

1.6.1 Pole-Zero Locations

The roots of $N(s)$, which are the zeros z_i of $H(s)$ (because for $s = z_i$ $H(s)$ becomes zero), can be real or complex conjugate, since all of the coefficients of $N(s)$ are real. Each of these zeros can be located at a unique point in the complex frequency plane as shown in Fig. 1.20. In case of a multiple zero, all of them are located at the same point in the s-plane.

On the other hand, the roots of $D(s)$, which are the poles p_i of $H(s)$ (because for $s = p_i$, $H(s)$ becomes infinite) can be real or complex conjugate, since $D(s)$ also has real coefficients. However, their real part can only be negative for reasons of stability. Also, for a network to be useful as a filter, its transfer function $H(s)$ should not have poles with real part equal to zero. Thus, the poles of function $H(s)$ should all lie in the left half of the s-plane (LHP) excluding the $j\omega$-axis, while its zeros can lie anywhere in the s-plane, i.e., in the left-half and in the right-half s-plane (RHP).

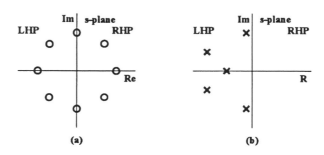

FIGURE 1.20
(a) Possible zero and (b) possible pole locations in the s-plane.

Filter Fundamentals

1.6.2 Frequency Response

Under steady-state conditions (i.e., $s = j\omega$) the magnitude of $H(j\omega)$ and its phase $\arg[H(j\omega)]$, given by Eqs. (1.47) and (1.46), respectively, as $A(\omega)$ and $\varphi(\omega)$, constitute the frequency response of the filter.

To get a good picture of the gain and phase functions of frequency ω, we draw the corresponding plots with the frequency being the independent variable. It is usual in most cases for the scale in the frequency axis to be logarithmic in order to include as many frequencies as possible in the plots. The $A(\omega)$ axis has a linear scale but, in effect, it is also logarithmic, since $A(\omega)$ is expressed in decibels. Finally, the $\varphi(\omega)$ axis is linear, usually expressed in degrees. However, in some cases, instead of working in terms of $\varphi(\omega)$, we work considering the group delay $\tau_g(\omega)$, defined as flows:

Group delay:
$$\tau_g(\omega) = -\frac{d\varphi(\omega)}{d\omega} \tag{1.49}$$

This has the dimensions of time and denotes the time delay that the specific frequency component in the spectrum of the signal experiences, when this passes through the filter.

Since $A(\omega)$ is an even function of ω, its plot against $-\omega$ will be symmetrical around the $A(\omega)$ axis of the plot against ω. On the other hand, $\varphi(\omega)$ is an odd function of ω; therefore, its plots against ω and $-\omega$ will be antisymmetrical around the $\varphi(\omega)$ axis.

To clarify all these terms, let us consider the following example for $H(s)$:

$$H(s) = \frac{s}{s^2 + 0.5s + 1} \tag{1.50}$$

The function has one zero at $s = j0$ and another at $s = j\infty$ (since it takes zero value at $s = j\infty$). It is usual to consider these zeros located on the $j\omega$-axis in the s-plane. The two poles are

$$s_1 = -0.25 + j0.9682$$

$$s_2 = -0.25 - j0.9682$$

They are located in the LHP and are complex conjugate.

To obtain the frequency response, we substitute $j\omega$ for s in Eq. (1.50) to obtain

$$H(j\omega) = \frac{j\omega}{-\omega^2 + j0.5\omega + 1} \tag{1.51}$$

Then

$$|H(j\omega)| = \frac{|\omega|}{\sqrt{(1-\omega^2)^2 + 0.25\omega^2}} \tag{1.52}$$

and

$$\varphi(\omega) = \frac{\pi}{2} - \tan^{-1}\frac{0.5\omega}{1-\omega^2} \tag{1.53}$$

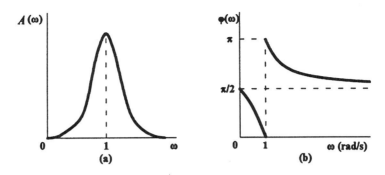

FIGURE 1.21
(a) Amplitude (magnitude) and (b) phase response of the function given by Eq. (1.50).

The plots of the magnitude in dB [i.e., $A(\omega)$] and the phase of $H(j\omega)$ against frequency for positive ω only are shown in Fig. 1.21(a) and (b), respectively.

1.6.3 Transient Response

In filter design, the specifications are usually given in terms of the frequency response. However, in cases of pulse transmission, it is useful to know the response of the filter as a function of time, i.e., its transient response.

In such cases, we usually study the response of the filter to two test functions: the unit impulse or $\delta(t)$ function and the unit step function. The respective impulse response and step response of the filter are briefly reviewed in what follows.

1.6.3.1 Impulse Response

The impulse response of a filter is its transient response when the excitation is the unit impulse or $\delta(t)$ function, which is defined as follows:

$$\int_{-\infty}^{+\infty} \delta(t)\,dt = 1 \tag{1.54}$$

and

$$\delta(t) = 0 \quad \text{for} \quad t \neq 0$$

According to this definition, we may also write

$$\int_{-\infty}^{+\infty} \delta(t-T)\,dt = \int_{T^-}^{T^+} \delta(t)\,dt = 1 \tag{1.55}$$

with $\delta(t)$ being zero for $t \neq T$.

Since the Laplace transform of $\delta(t)$ is

$$L[\delta(t)] = 1 \tag{1.56}$$

substituting for E(s) in Eq. (1.42), repeated here for convenience,

$$R(s) = H(s)E(s) \qquad (1.42)$$

we obtain

$$R(s) = H(s) \cdot 1 = H(s) \qquad (1.57)$$

In other words, the Laplace transform of the filter response to the unit impulse [function $\delta(t)$] is the transfer function $H(s)$.

Taking the inverse Laplace transform of $H(s)$ to be

$$h(t) = L^{-1}[H(s)] \qquad (1.58)$$

we get

$$r(t) = L^{-1}[R(s)] = L^{-1}[H(s)] = h(t) \qquad (1.59)$$

Therefore, the impulse response of a filter is the inverse Laplace transform of its transfer function.

1.6.3.2 Step Response

The step response of a filter is its time response when the excitation is the step function $Ku(t)$, where K is a constant (voltage or current) and $u(t)$ the unit step function defined as follows:

$$u(t) = \begin{cases} 1 & t > 0 \\ 0 & t < 0 \end{cases} \qquad (1.60)$$

with the important property that

$$\delta(t) = \frac{du(t)}{dt} \qquad (1.61)$$

Again, the shifting property holds, i.e.,

$$u(t-T) = \begin{cases} 1 & t > T \\ 0 & t < T \end{cases} \qquad (1.62)$$

Its Laplace transform is

$$L[u(t)] = \frac{1}{s} \qquad (1.63)$$

Substituting for E(s) in Eq. (1.42) gives

$$R(s) = K\frac{H(s)}{s} \tag{1.64}$$

Taking the inverse Laplace transform of $R(s)$ gives the filter step response, i.e.,

$$r(t) = L^{-1}\left[K\frac{H(s)}{s}\right] \tag{1.65}$$

The step response for various filter functions, commonly used, is further examined in the next chapter.

Clearly, since $H(s)$ is the impulse response and $H(s)/s$ the unit step response ($K = 1$) of a filter (or system), the unit step response is the time integral of the impulse response; in other words, the impulse response is the derivative of the step response.

1.6.4 Step and Frequency Response

In Fig. 1.22, a typical step response of a lowpass filter is shown. Characteristic quantities associated with this response are values of the rise time t_r, the delay time τ_o, the settling time t_s, and the overshoot. These quantities are defined as follows:

- The rise time t_r is defined as the time it takes the step response to rise from 10 percent to 90 percent of its final value.
- The delay time τ_o is the time it takes for the step response to rise to 50 percent of its final value.
- The settling time t_s is the time that elapses between the moment of appearance of the first peak and the moment beyond which the step response does not differ by more that 2 percent from its final value.
- The overshoot is the percent of the final value difference between the maximum and the final value of the step response.

The rise time t_r, the settling time t_s, and the overshoot are used as figures of merit in comparing the transient response of various filters. The delay time τ_o is equal to the delay the signal experiences when passing through the filter and is considered again in Section 2.5.

It is interesting to mention here that a very important relationship connects the rise time t_r and the cutoff frequency f_c of a lowpass filter, namely,

$$t_r f_c \cong 0.35 \tag{1.66}$$

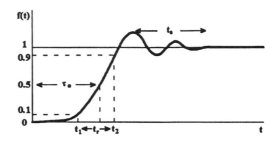

FIGURE 1.22
Typical step response of a lowpass filter.

Filter Fundamentals

or equivalently,

$$t_r \omega_c \cong 2.2 \tag{1.67}$$

This relationship is valid for low overshoot values (<5 percent). For higher overshoot, the constant should increase to a value between 0.35 and 0.5 [1]. Its importance arises from the fact that in order to reduce the rise time, the filter bandwidth has to increase. Its physical implication is that if narrow pulses are to be transmitted through the filter without excessive distortion, their duration should be larger than the rise time of the filter. Equivalently, the filter bandwidth should be larger than the reciprocal of the pulse duration.

Although Eq. (1.66) is generally considered empirical, it can be easily obtained for a filter (or system) with the following transfer function:

$$H(s) = \frac{\omega_c}{s + \omega_c} \tag{1.68}$$

For a unit step input $u(t)$, the response $r(t)$ can be obtained as follows. Since, per (Eq. 1.42),

$$R(s) = H(s)E(s)$$

and

$$E(s) = \frac{1}{s}$$

the response will be

$$R(s) = \frac{\omega_c}{(s + \omega_c)s} \tag{1.69}$$

Taking the inverse Laplace transform of $R(s)$ gives

$$r(t) = 1 - e^{-\omega_c t} \qquad t > 0 \tag{1.70}$$

with $r(\infty) = 1$, the final value of $r(t)$.

Now referring to Fig. 1.22, for $t = t_1$

$$r(t_1) = 0.1 r(\infty) = 0.1 = 1 - e^{-\omega_c t_1}$$

from which we get t_1 to be

$$t_1 \cong \frac{0.1}{\omega_c} \tag{1.71}$$

Next, for $t = t_2$,

$$r(t_2) = 0.9 r(\infty) = 0.9 = 1 - e^{-\omega_c t_2}$$

from which t_2 is found to be

$$t_2 \cong \frac{2.3}{\omega_c} \qquad (1.72)$$

Therefore,

$$t_r = t_2 - t_1 \cong \frac{2.2}{\omega_c} = \frac{2.2}{2\pi f_c} = \frac{0.35}{f_c} \qquad (1.73)$$

and

$$t_r f_c \cong 0.35$$

1.7 Stability

Since the main object of this book is the design of active RC filters, and these filters may be or become unstable under certain conditions, it is appropriate to review the concept of stability here. Filter stability has already been mentioned in Section 1.6 with reference to the pole positions of the filter function.

In practical terms, the output voltage or current of a filter must always follow the input at steady-state, i.e., it should not become uncontrollable. Such an uncontrollable behavior usually leads either to dc saturation of the output voltage or to the generation of a periodic waveform independent of the input signal.

In mathematical terms, stability of a linear network in the time domain, strictly speaking, requires that its impulse response $h(t)$ be absolutely integrable, i.e.,

$$\int_0^\infty |h(t)| dt = M < \infty \qquad (1.74)$$

Consequently, strict stability implies that only terms of the following form are allowed in the expression for the impulse response [3]:

$$At^n e^{-\sigma t} \cos \omega t \quad \text{or} \quad At^n e^{-\sigma t} \sin \omega t$$

where A is a real constant, n is a non-negative integer, σ is a positive real number, and ω is a non-negative real quantity with the dimensions of angular frequency.

For a filter to be useful, it should be strictly stable. Since the impulse response is the inverse Laplace transform of the pertinent network function, strict stability implies that the poles of this function should only be of the form

$$-\sigma \pm j\omega \quad \text{with} \quad \sigma > 0, \omega \geq 0$$

i.e., they should lie in the LH of the s-plane excluding the $j\omega$ axis.

Filter Fundamentals

If the network function has poles on the $j\omega$-axis also, then its impulse response expression will include terms of the form $A\cos\omega t$, $A\sin\omega t$. The network is considered marginally stable in this case, but then it cannot be useful as a filter.

A network is unstable if it is not strictly or marginally stable.

1.7.1 Short-Circuit and Open-Circuit Stability

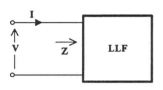

FIGURE 1.23
Two-terminal network.

Consider the two-terminal network shown in Fig. 1.23. The impedance $Z(s)$ appearing across its terminals is the following:

$$Z(s) = \frac{V(s)}{I(s)} \tag{1.75}$$

If the network is driven by a voltage source (excitation), then the response will be the current $I(s)$ given by

$$I(s) = \frac{1}{Z(s)} V(s) \tag{1.76}$$

Clearly, the one-port will be stable if the zeros of $Z(s)$ are all in the LH of the s-plane excluding the $j\omega$-axis. The one-port is then short-circuit stable.

Next, consider the one-port being driven by a current source (excitation). Then the voltage $V(s)$ across its terminals (response) will be given from Eq. (1.77) as follows:

$$V(s) = Z(s)I(s) \tag{1.77}$$

If $Z(s)$ has poles in the LH of the s-plane excluding the $j\omega$-axis, it will be open-circuit stable.

Consequently, the position of the zeros and the poles of $Z(s)$ will determine whether the one-port can be voltage or current driven. If neither the zeros nor the poles of $Z(s)$ lie on the $j\omega$-axis or the RH of the s-plane, the network can be excited by either a voltage source or a current source. For example, if

$$Z(s) = \frac{s^2 + 4s + 13}{s^2 - 6s + 10} \tag{1.78}$$

the corresponding one-port will be open-circuit unstable.

1.7.2 Absolute Stability and Potential Instability

A linear two-terminal network is absolutely stable if it remains strictly stable under any passive termination. It is potentially unstable if there is even one passive termination for which it becomes unstable. Thus, the two-terminal network in Fig. 1.24(a), the impedance of which is $Z(s)$, will be absolutely stable if the impedance $Z_A(s) + Z(s)$ does not possess a zero that lies in the RH of the s-plane including the $j\omega$-axis for any passive impedance $Z_A(s)$, including short-circuit ($Z_A = 0$) and open-circuit ($Z_A = \infty$). Therefore, a short-circuit or open-circuit unstable network is not absolutely stable.

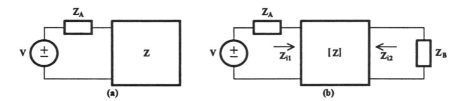

FIGURE 1.24
Defining an absolutely stable (a) one-port and (b) two-port.

In a similar manner, we may define an absolutely stable two-port. Consider the situation in Fig. 1.24(b), where the two-port network with impedance matrix [Z] is terminated at both its input and output ports by the passive impedances $Z_A(s)$ and $Z_B(s)$, respectively.

The two-port will be absolutely stable, if the impedance function

$$Z_1(s) = Z_A(s) + Z_{i1}(s) \tag{1.79}$$

or, equivalently

$$Z_2(s) = Z_B(s) + Z_{i2}(s) \tag{1.80}$$

has no zeros in the RH s-plane including the $j\omega$-axis, for any pair of passive terminations Z_A and Z_B. If this impedance has a zero in the RH or on the $j\omega$-axis, the two-port will be potentially unstable.

The equivalence of Eqs. (1.79) and (1.80) in determining the absolute stability of the two-port can be shown as follows:

$$Z_1(s) = Z_A + Z_{11} - \frac{Z_{12}Z_{21}}{Z_{22} + Z_B} \tag{1.81}$$

and

$$Z_2(s) = Z_B + Z_{22} - \frac{Z_{12}Z_{21}}{Z_{11} + Z_A} \tag{1.82}$$

Assuming that

$$Z_{12}Z_{21} \neq 0$$

the zeros of $Z_1(s)$ will be the zeros of

$$(Z_A + Z_{11})(Z_{22} + Z_B) - Z_{12}Z_{21}$$

plus the poles of $Z_{22} + Z_B$, the latter being always on the LH of the s-plane.

On the other hand, the zeros of $Z_2(s)$ will again be the zeros of

$$(Z_{22} + Z_B)(Z_{11} + Z_A) - Z_{12}Z_{21}$$

plus the poles of $Z_{11} + Z_A$, again the latter being always on the LH of the s-plane. Hence, the equivalence.

1.8 Passivity Criteria for One- and Two-Port Networks

We introduce here, briefly and without any proof, passivity criteria for one- and two-ports in an attempt to relate passivity and activity with stability. Passivity of a network was defined in Section 1.2.

1.8.1 One-Ports

As always, we refer here to LLF time invariant networks. Such a one-port is passive if its input impedance $Z(s)$ or input admittance $Y(s)$ is a positive real function. The impedance function $Z(s)$ is positive real if it satisfies the following two conditions:

1. $Z(s)$ is real when s is real.
2. $Re[Z(s)] \geq 0$ for $R[s] \geq 0$.

Here, Re means the real part of what follows it.

The following three alternative criteria, if satisfied, determine equivalently the positive realness of a network.

1. $Z(s)$ is positive real if and only if [2]
   - It does not have poles with positive real part. Poles on the $j\omega$-axis are simple with real and positive residues and
   - $Re[Z(j\omega)] \geq 0$ for all ω.
2. $Z(s)$ is positive real if [2]
   - It has no poles nor zeros with positive real part,
   - Poles or zeros on the $j\omega$-axis are simple, and
   - $Re[Z(j\omega)] \geq 0$ for all ω.
3. $Z(s)$ is positive real if [3]
   - For $Z(s) = N(s)/D(s)$, the polynomial $N(s) + D(s)$ is Hurwitz, and
   - $Re[Z(j\omega)] \geq 0$ for all ω.

A Hurwitz polynomial has real coefficients, and all its roots lie on the LH of the s-plane excluding the $j\omega$-axis.

Note that, in the above criteria, poles at zero and infinity are considered, in accordance with the convention for zeros at zero and infinity (see Section 1.6.2), to lie on the $j\omega$-axis. Based on these criteria, a preliminary test for positive realness of the impedance function,

$$Z(s) = \frac{N(s)}{D(s)} = \frac{a_m s^m + a_{m-1} s^{m-1} + \ldots + a_1 s + a_0}{b_n s^n + b_{n-1} s^{n-1} + \ldots + b_1 s + b_0} \qquad (1.83)$$

proceeds in the following steps [3]. Check:

1. All a_i, b_i are real and positive.
2. $|n - m| \leq 1$.

3. If $a_o = 0$, then $a_1 \neq 0$, and if $b_o = 0$, then $b_1 \neq 0$.
4. The zeros of $N(s)$ and $D(s)$ on the $j\omega$-axis are simple.
5. There are no missing terms in $N(s)$ and $D(s)$ except when all terms of even powers or of odd powers of s are missing.

If these preliminary conditions are satisfied, we next proceed to test the following, which are necessary and sufficient conditions. Check:

1. $N(s) + D(s)$ is a Hurwitz polynomial.
2. The numerator polynomial of the real part of $Z(j\omega)$ does not have any $j\omega$-axis zeros of odd multiplicity.

The Routh-Hurwitz test can be used to reveal whether a polynomial with positive and real coefficients is Hurwitz. This test involves the expansion in continued fractions of the ratio of the even to the odd part or of the odd to the even part of the polynomial. If the resulting coefficients are all present and positive, the polynomial is Hurwitz. However, with computers at easy reach today, this test may become obsolete.

Next, to test condition b, we observe that

$$Re[Z(j\omega)] = Re\left[\frac{N(j\omega)}{D(j\omega)}\right] = \frac{P(\omega^2)}{|D(j\omega)|^2} \tag{1.84}$$

where $P(\omega^2)$ is the numerator polynomial of the real part of $Z(j\omega)$.

Since $|D(j\omega)|^2$ is positive for all ω, we need only test $P(\omega^2)$. In some cases, this may be easy, but in general we have to examine whether $P(x)$, where $x = \omega^2$ has real roots of odd multiplicity for $0 \leq \omega^2 < \infty$. Although this may be tested by means of Sturm's theorem [2], again, the use of a computer can save effort and avoid errors.

1.8.2 Two-Ports

In the case of a two-port, the criterion for passivity is as follows [2]:

An LLF and time-invariant two-port is passive if and only if:

1. The characteristic polynomial (common denominator of all Z parameters) has no roots in the RH of the s-plane.
2. Any poles of the Z-parameters on the $j\omega$-axis are simple, and their residues at these poles satisfy the following conditions:

$$K_{11} \geq 0 \text{ and real}$$

$$K_{22} \geq 0 \text{ and real}$$

$$K_{11}K_{22} - K_{12}K_{21} \geq 0 \qquad K_{12} = K^*_{21}$$

(K_{ij} residue for parameter Z_{ij})

3. The real (R) and imaginary (X) parts of the Z-parameters (for $s = j\omega$) satisfy the following conditions:

Filter Fundamentals 31

$$R_{11} \geq 0$$

$$R_{22} \geq 0$$

$$4R_{11}R_{22} - (R_{12} + R_{21})^2 - (X_{12} - X_{21})^2 \geq 0$$

The first two conditions may be tested by means of the Routh-Hurwitz criterion and condition c by means of Sturm's theorem. Again, the use of a computer simplifies matters.

1.8.3 Activity

If any of the above conditions for passivity is not satisfied, the network will be active.

The activity of a two-port is directly related to its maximum power gain, which is greater than 1. If we determine this maximum power gain of the two-port in terms of its Z parameters, we finally find that [2]

$$K_{p\,max} = \frac{|Z_{21}|^2}{R_{11}R_{22}[2 + 2\sqrt{1 - r - x^2} - r]} \tag{1.85}$$

where

$$r = \frac{R_{12}R_{21} - X_{12}X_{21}}{R_{11}R_{22}} \qquad x = \frac{R_{12}X_{21} - R_{21}X_{12}}{2R_{11}R_{22}}$$

Substituting for r and x in Eq. (1.85) and taking into account that

$$K_{p\,max} > 1$$

we get that

$$4R_{11}R_{22} - (R_{12} + R_{21})^2 - (X_{12} - X_{21})^2 < 0$$

Thus, condition c for passivity is violated, and the two-port is not passive—it is indeed active.

1.8.4 Passivity and Stability

Based on the above, the following results can be obtained.

One-port networks—The impedance or admittance of a passive LLF time-invariant network, according to the passivity conditions, can have poles and zeros in the LH of the s-plane. If any are on the $j\omega$-axis, they should be simple.

Theoretically, poles and zeros on the $j\omega$-axis occur in the impedance or admittance function of a purely lossless LC one-port. However, in practice, both inductors and capacitors have always some loss associated with their values. The effect of this loss is to move the $j\omega$-axis poles and zeros out of the $j\omega$-axis inside the LHP. In this case, both $N(s)$ and $D(s)$ will be Hurwitz and the passive one-port strictly stable.

On the other hand, any passive termination of the passive one-port will not change the passivity property, since the sum of two positive real functions is also positive real. Therefore, a passive one-port is absolutely stable.

A stable, active one-port, however, is potentially unstable, since a passive termination will always be found that will make it unstable. The condition that $Re[Z(j\omega)] \geq 0$ is not satisfied for all ω in the case of an active one-port.

Two-port networks—A passive two-port terminated at both ports by passive impedances will remain passive. Therefore, it is absolutely stable.

However, contrary to the case of an active one-port, an active two-port may be absolutely stable, because the activity conditions in this case do not violate the passivity conditions. On the other hand, a potentially unstable network will be active.

1.9 Reciprocity

This property of a network is included here for completeness rather than for its usefulness in the development of the material in this book.

A network is reciprocal if [2]

1. (See Fig.1.25) the current occurring in a short between two nodes α and β due to a voltage between the nodes γ and δ is equal to the current occurring in a short between the nodes γ and δ, if the same voltage as before is applied between the nodes α and β, or

2. (See Fig. 1.26) the voltage resulting between the nodes α and β due to a current applied between the nodes γ and δ is the same as that resulting between the nodes γ and δ when the same current as before is applied between nodes α and β.

In terms of the various parameters, the necessary and sufficient condition for a two-port network to be reciprocal is any of the following:

$$Z_{12} = Z_{21}$$

$$Y_{12} = Y_{21}$$

$$H_{12} = -H_{21}$$

$$G_{12} = -G_{21}$$

$$\Delta A = \alpha_{11}\alpha_{22} - \alpha_{12}\alpha_{21} = 1$$

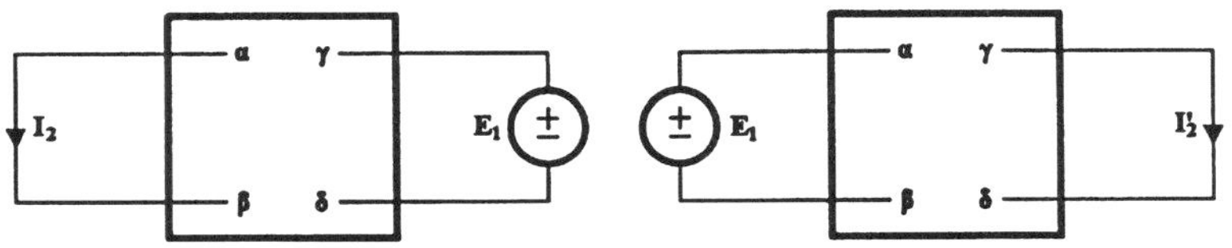

FIGURE 1.25
For reciprocity, $I_2 = I'_2$.

Filter Fundamentals

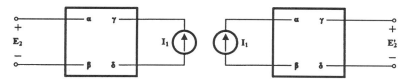

FIGURE 1.26
For reciprocity, $E_2 = E'_2$.

As an example, consider the two-port in Fig. 1.27. For this two-port, we have

$$Z_{12} \equiv \left.\frac{V_1}{I_2}\right|_{I_1 = 0} = R_3$$

and

$$Z_{21} = \left.\frac{V_2}{I_1}\right|_{I_2 = 0} = R_3$$

Therefore, the two-port in Fig. 1.27 is reciprocal.

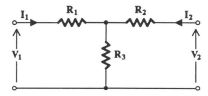

FIGURE 1.27
A reciprocal two-port.

1.10 Summary

Various concepts, necessary for the characterization of the continuous-time active filters, were reviewed in this chapter. These concerned the types of the filters of interest and their responses, ideal and practical.

Next, some useful concepts were reviewed briefly from network theory concerning the nodal analysis method, network parameters and functions, to fundamental properties like stability, passivity, activity, and reciprocity.

The interested reader with no prior knowledge of these concepts could find it useful to get a better insight by consulting the references.

References and Further Reading

[1] E.S. Kuh and D.O. Pederson. 1959. *Principles of Circuit Synthesis*, New York: McGraw-Hill.
[2] Louis De Pian. 1962. *Linear Active Network Theory*, London: Prentice-Hall.

[3] S.K. Mitra. 1969. *Analysis and Synthesis of Linear Active Networks*, New York: John Wiley & Sons, Inc.
[4] W. H. Hayt, Jr., and J.E. Kemmerly. 1993. *Engineering Circuit Analysis*, New York: McGraw-Hill.
[5] R. Spence. 1970. *Linear Active Networks*, New York: John Wiley & Sons, Inc.

Chapter 2

The Approximation Problem

2.1 Introduction

Solution of the approximation problem is a major step in the design procedure of a filter and is equally important in the design of both analog and digital filters. It is through the solution of this problem that the filter designer determines the filter function, the response which satisfies the specifications. Of course, the function obtained this way will satisfy the specifications only approximately and not exactly. However, if the specifications are set within the limitations of the LLF networks, the network realizing the approximating function will fulfil the requirements and thus be suitable for the task for which it is designed.

In this chapter, based on the contents of Chapter 1, we review first the characteristics of the *permitted* functions, and we formulate the approximation problem. Next, we present briefly the best known and most popular functions used in the solution of the approximation problem for the required filter response in the frequency domain. Then, since these functions are lowpass, we introduce suitable frequency transformations in order to obtain highpass, bandpass, or bandstop filters according to the requirements. Finally, we discuss the transformation of elements and the scaling of impedance level.

2.2 Filter Specifications and Permitted Functions

The knowledge gathered from the analysis of LLF networks, in the more general concept of the term *analysis*, which was explained in Section 1.4, can help in the search for the most suitable filter function to meet particular specifications. The results of this analysis impose three important constraints on the *permitted* LLF network functions. These have to be causal, rational, and stable.

Before proceeding to explain how to determine the filter function to meet a set of specifications, we review these constraints briefly.

2.2.1 Causality

In general, causality refers to the fact that there can be no result without cause. In the case of interest here, a causal network will not respond before an excitation has been applied to its terminals. Thus, the unit impulse response is zero for time $t < 0$. The response in Fig. 2.1(a) is not causal; therefore, it cannot be realized. On the other hand, that in Fig. 2.1(b) is causal, therefore realizable. Thus, the ideal lowpass filter is unrealizable, because its impulse response is noncausal.

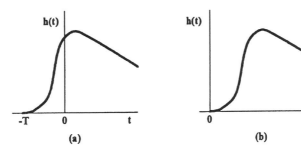

FIGURE 2.1
(a) Noncausal and (b) causal response.

In the frequency domain, causality is determined by means of the Paley-Wiener criterion [1]. Consider the impulse response $h(t)$, which possesses a Fourier Transform $H(j\omega)$ for which

$$\int_{-\infty}^{\infty} |H(j\omega)|^2 d\omega < \infty \qquad (2.1)$$

For $H(j\omega)$ to be causal, the criterion is the following:

$$\int_{-\infty}^{\infty} \frac{|\log|H(j\omega)||}{1+\omega^2} d\omega < \infty \qquad (2.2)$$

Some consequences of this criterion are the following:

1. The magnitude function $|H(j\omega)|$ cannot be zero for a finite frequency band. However, it can be zero at a finite number of distinct frequencies.
2. The magnitude $|H(j\omega)|$ cannot decrease faster than exponentially.
3. Because of this constraint, the ideal filters are unrealizable.

2.2.2 Rational Functions

The LLF network functions are rational, i.e., ratios of two finite polynomials of the Laplace transform variable s. Therefore, it is not possible to realize the function e^{-sT} by such a network, because this function cannot be expressed in the form of a rational function.

2.2.3 Stability

The response of a stable network is bounded if the excitation is bounded. This means that, if $h(t)$ is the impulse response of the network, then

$$\int_0^{\infty} h(t) dt < \infty \qquad (2.3)$$

and $\lim h(t) \to 0$ when $t \to \infty$.

The Approximation Problem

In the frequency domain, stability implies that

1. the network function $H(s)$ does not have poles in the RH of the s-plane,
2. any poles on the $j\omega$-axis are simple, and
3. the degree of the numerator polynomial cannot be higher than the degree of the denominator by more than one.

However, for a filter to be useful, its function $H(s)$ has to be strictly stable, i.e., all its poles must be located in the LH of the s-plane excluding the $j\omega$-axis (poles at zero and infinity are considered to be located on the $j\omega$-axis).

2.3 Formulation of the Approximation Problem

In practice, the specifications of the filter may be given in terms of the cutoff frequency (or frequencies) ω_c, the maximum allowable deviation (error) A_{max} in the passband, the stopband edges (frequencies), and the minimum attenuation A_{min} in the stopband. In the case of equalizers, it may be possible that the required frequency response is specified more closely. In general, from the specifications, we will be able to draw a frequency response magnitude plot, which will correspond to a prespecified curve. For example, for a lowpass filter, this diagram will be of the form shown in Fig. 2.2. The required response will have to lie between the limits set by the diagram.

Theoretically, the approximation problem is stated as follows:

1. Time domain:
    The impulse response $h(t)$ has to be approximated. An approximating function $h^*(t)$ is selected such that some error ε is minimal, where

$$\varepsilon = \int_0^\infty [h(t) - h^*(t)]^2 dt \qquad (2.4)$$

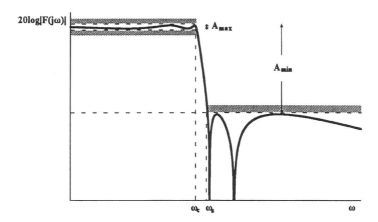

FIGURE 2.2
A possible magnitude response which satisfies the specifications of a lowpass filter.

Of course, $h^*(t)$ should be such that its Laplace transform $H(s)$ is a realizable function by a LLF network.

2. Frequency domain:
In the frequency, domain we often work in terms of lowpass functions, because they are simpler, and because the highpass, bandpass, and bandstop responses can be obtained from lowpass responses by means of suitable frequency transformations.

Our problem here is to find a function $F(s)$ the magnitude and/or phase response of which approximates the prespecified curve according to a predetermined criterion.

The approximation problem has been solved mathematically in various ways. In the case of magnitude approximation, the best known and most popular lowpass functions are the following: the Butterworth or maximally flat, the Chebyshev (Tschebycheff) or equiripple, the monotonic or Papoulis, and the Cauer or elliptic function filters. Of course, with the use of a computer, one may create one's own approximating functions, particularly in the cases of arbitrary responses of filters and equalizers. In such cases, techniques employing linear segments, curve fitting, pole-zero placements, etc., have proved very useful in solving the approximation problem.

In the case of delay approximation, best known functions are the Bessel-Thomson filters, the Padè approximates, both maximally flat, and those of Chebyshev type.

In what follows, we introduce briefly the best known and practically useful approximations to the ideal lowpass filter and to the ideal delay.

2.4 Approximation of the Ideal Lowpass Filter

In practical filter design, the amplitude response is more often specified than the phase response. The amplitude response of the ideal lowpass filter with normalized cutoff frequency at $\omega_c = 1$ is shown in Fig. 2.3. As has already been explained in Section 2.2., this ideal amplitude response cannot be expressed as a rational function of s. It is thus unrealizable. If we accept a small error in the passband and a non-zero transition band, we may seek a rational function $F(s)$, the magnitude of which will approximate the ideal response as closely as possible. A suitable magnitude function can be of the form

$$|F(j\omega)| = M(\omega) = \frac{1}{[1 + \varepsilon^2 w(\omega^2)]^{1/2}} \qquad (2.5)$$

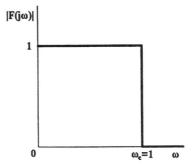

FIGURE 2.3
Ideal lowpass filter amplitude response.

The Approximation Problem

where ε is a constant between zero and one ($0 < \varepsilon \leq 1$), according to the accepted passband error, and $w(\omega^2)$ is a function of ω^2 such that

$$0 \leq w(\omega^2) \leq 1 \qquad 0 \leq \omega \leq 1$$

and which increases very fast with increasing ω, for ω > 1, remaining much greater than one outside the passband.

In general, the numerator of $M(\omega)$ may be a constant other than unity, which will influence the gain (or attenuation) at $\omega = 0$ (at dc).

In the following, we review the most popular functions $w(\omega^2)$ and the corresponding $F(s)$, the magnitude of which approximate the amplitude response of the ideal lowpass filter.

2.4.1 Butterworth or Maximally Flat Approximation

If we let ε = 1 and

$$w(\omega^2) = \omega^{2n}$$

in Eq. (2.5), n being a positive real integer, we will get the following amplitude function:

$$M(\omega) = \frac{1}{[1 + \omega^{2n}]^{1/2}} \tag{2.6}$$

It can be seen that

$$M(0) = 1$$

while $M(\omega)$ decreases monotonically with increasing ω.
At $\omega_c = 1$,

$$M(1) = \frac{1}{\sqrt{2}} = 0.707$$

or

$$20 \log M(1) = -10 \log 2 = -3.01$$

In other words, at $\omega_c = 1$, the amplitude is 3 dB below its value at dc. This is the cutoff frequency of the filter. Clearly, this is independent of n, the order of the filter function, which in fact determines how close to the ideal is the approximating function $M(\omega)$, i.e., how successful the approximation is.

Equation (2.6) for different n gives the amplitude response of the various Butterworth filter functions. The Butterworth approximation is also called the maximally flat approximation, because the first $2n - 1$ derivatives of $M(\omega)$, the maximum number in Eq. (2.2), are zero at $\omega = 0$. The error in the passband is zero at $\omega = 0$ and maximal (3 dB) at cutoff. Between $\omega = 0$ and $\omega = 1$, the error takes intermediate values increasing monotonically from the zero value with increasing ω. For values of $\omega \gg 1$, $M(\omega)$ behaves approximately as

$$M(\omega) = \frac{1}{\omega^n}$$

i.e., it changes asymptotically as

$$20\log M(\omega) = -20n\log\omega \tag{2.7}$$

or, in other words, it falls off by $6n$ dB/octave ($20n$ dB/decade).

We now seek to obtain the network function $F(s)$ whose magnitude with $s = j\omega$ is $M(\omega)$. We proceed as follows. Observing that

$$M^2(\omega) = |F(j\omega)|^2 = F(j\omega)F(-j\omega) = \frac{1}{1+\omega^{2n}} \tag{2.8}$$

we may write

$$F(s)F(-s)\big|_{s=j\omega} = \frac{1}{1+(-1)^n s^{2n}}\bigg|_{s=j\omega}$$

By a process known as analytic continuation, it turns out that we may remove the $s = j\omega$ constraint and write

$$F(s)F(-s) = \frac{1}{1+(-1)^n s^{2n}} \tag{2.9}$$

Now define a function $P(s^2)$ such that

$$P(s^2) = F(s)F(-s) \tag{2.10}$$

when we will also have that

$$M^2(\omega) = P(-\omega^2)$$

Thus, knowing $P(-\omega^2)$ through $M^2(\omega)$, we can obtain $P(s^2)$ by setting s^2 for $-\omega^2$ in Eq. (2.8). Then, expressing $P(s^2)$ in the form of Eq. (2.10), we observe that the poles of $F(s)$ are symmetrical to those of $F(-s)$ about the $j\omega$-axis. Since $F(s)$ has to be a stable function, we identify its poles as those of $P(s^2)$ with negative real part.

The poles of $P(s^2)$ are the roots of the equation

$$1 + (-1)^n s^{2n} = 0 \tag{2.11}$$

It can be shown that the solution of Eq. (2.11) is the following:

$$s_k = \sigma_k + j\omega_k = -\sin\left(\frac{2k-1}{2n}\pi\right) + j\cos\left(\frac{2k-1}{2n}\pi\right) \tag{2.12}$$

for $k = 1, 2, \ldots, 2n$.

The Approximation Problem

The n poles of $F(s)$ are those obtained from Eq. (2.12) with $k = 1, 2,..., n$. All poles have magnitude equal to unity and lie on the circumference of the unit circle equally spaced.

As an example, consider the case for $n = 4$. We have

$$M^2(\omega) = \frac{1}{1 + \omega^8}$$

Then

$$P(s^2) = \frac{1}{1 + s^8}$$

The poles of $P(s^2)$, are found from Eq. (2.12) for $k = 1, 2,..., 8$. These lie on the circumference of the unit circle as shown in Fig. 2.4.

Of these, the first four ($k = 1, 2, 3$, and 4) will be assigned to $F(s)$, since they have to lie on the LH of the s-plane. They are as follows:

$$s_1 = -0.3827 + j0.9239$$

$$s_2 = -0.9239 + j0.3827$$

$$s_3 = -0.9239 - j0.3827$$

$$s_4 = -0.3827 - j0.9239$$

Therefore, the fourth-order Butterworth lowpass function will be

$$F(s) = \frac{1}{(s - s_1)(s - s_2)(s - s_3)(s - s_4)}$$

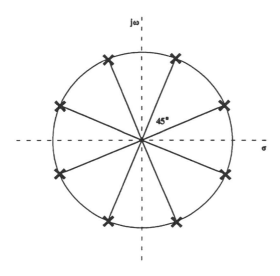

FIGURE 2.4
Poles of the Butterworth filter for n = 4.

Grouping complex conjugate pole terms, it turns out that

$$F(s) = \frac{1}{(s^2 + 0.7654s + 1)(s^2 + 1.8478s + 1)}$$

or multiplying out in full,

$$F(s) = \frac{1}{s^4 + 2.613s^3 + 3.414s^2 + 2.613s + 1}$$

The denominator of this last expression is known as a Butterworth polynomial. The first ten Butterworth polynomials in two forms are given on Table A.1 at the end of the book for easy reference. Finally, in Fig. 2.5 the magnitude response of the third-order Butterworth filter is shown together with the responses of two other filters we are dealing with next for comparison.

2.4.2 Chebyshev or Equiripple Approximation

In this case, Eq. (2.5) takes the following form:

$$|F(j\omega)|^2 = M^2(\omega) = \frac{1}{1 + \varepsilon^2 C_n^2(\omega)} \qquad (2.13)$$

Here again, $0 < \varepsilon \leq 1$, and $C_n(\omega)$ is a Chebyshev polynomial of degree n having the following form:

$$\begin{aligned} C_n(\omega) &= \cos(n \cos^{-1}\omega) & 0 \leq |\omega| \leq 1 \\ &= \cosh(n \cosh^{-1}\omega) & 1 \leq |\omega| \end{aligned} \qquad (2.14)$$

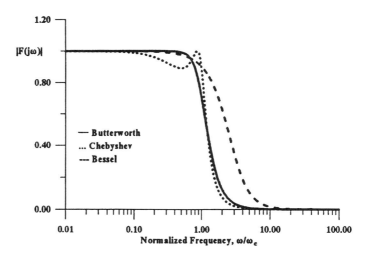

FIGURE 2.5
Magnitude responses of the Butterworth third-order lowpass filter and the corresponding Chebyshev (1 dB ripple) and Bessel filters.

The Approximation Problem

Clearly, $C_n(\omega)$ varies between +1 and –1 in the passband ($0 \leq \omega \leq 1$), while its absolute value increases rapidly with ω above $\omega = 1$. Consequently, $M(\omega)$ varies between 1 and $(1 + \varepsilon^2)^{-1/2}$ in the passband having an oscillatory or ripple error of

$$20\log(1 + \varepsilon^2)^{1/2} = 10\log(1 + \varepsilon^2) \text{ dB}$$

Thus, the accepted error in the passband determines the value of ε.

The Chebyshev polynomials can be obtained by the recursion formula

$$C_{n+1}(\omega) = 2\omega C_n(\omega) - C_{n-1}(\omega) \qquad (2.15)$$

with

$$C_0(\omega) = 1$$

$$C_1(\omega) = \omega$$

A plot of the Chebyshev polynomials with $n = 1, 2$, and 3 is given in Fig. 2.6.

Therefore, at dc ($\omega = 0$), we will have

$$M(0) = \begin{cases} 1 & n \text{ odd} \\ (1 + \varepsilon^2)^{-1/2} & n \text{ even} \end{cases}$$

Outside the passband and for $\omega \gg 1$, $M(\omega)$ behaves approximately like $(\varepsilon\, 2^{n-1}\omega^n)^{-1}$, i.e., the attenuation for $\omega \gg 1$ will be

$$20\log(\varepsilon\, 2^{n-1}\omega^n) = 20\log\varepsilon + 20\log 2^{n-1} + 20\log\omega^n$$

$$= 20\log\varepsilon + 6(n-1) + 20n\,\log\omega \quad \text{dB} \qquad (2.16)$$

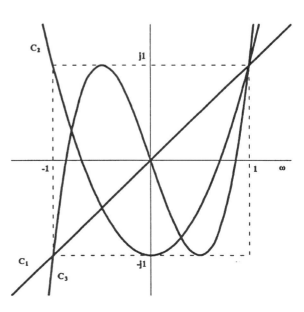

FIGURE 2.6
Plot of Chebyshev polynomials of degrees n = 1, 2, and 3.

compared to $20 n \log \omega$ in the corresponding Butterworth function. Thus, for $\varepsilon > 1$, the Chebyshev approximation has an advantage of $20 \log \varepsilon + 6(n - 1)$ dB over the Butterworth approximation. However when $\varepsilon < 1$, this advantage becomes less significant, since then $\log \varepsilon$ will be negative.

By a similar procedure to the Butterworth case, the poles of the Chebyshev filter functions can be shown to be as follows:

$$s_k = \sigma_k \pm j\omega_k$$

where

$$\sigma_k = \sinh \beta_k \cdot \sin\left(\frac{2k-1}{2n}\pi\right) \tag{2.17}$$

$$\omega_k = \cosh \beta_k \cos\left(\frac{2k-1}{2n}\right) \tag{2.18}$$

with

$$\beta_k = \frac{1}{n}\sinh^{-1}\frac{1}{\varepsilon} \quad k = 1, 2, \ldots, 2n$$

These poles lie on an ellipse defined by the following equation:

$$\frac{\sigma_k^2}{\sinh^2 \beta_k} + \frac{\omega_k^2}{\cosh^2 \beta_k} = 1 \tag{2.19}$$

The major semi-axis of the ellipse falls on the $j\omega$-axis, its length being $\pm \cosh \beta_k$, whereas the length of the minor semiaxis is $\pm \sinh \beta_k$.

The points of intersection of the ellipse and the $j\omega$-axis define the –3 dB frequencies (half-power frequencies), which are thus equal to $\pm \cosh \beta_k$. The corresponding Butterworth frequencies are always $\omega_c = \pm 1$.

We may normalize the poles of the Chebyshev functions in order to have the half-power frequencies (–3 dB frequencies) appearing at $\omega_c = 1$, by dividing s_k by $\cosh \beta_k$. Then, the normalized poles s_k will be as follows:

$$s'_k = \frac{s_k}{\cosh \beta_k} = \sigma'_k \pm j\omega'_k$$

with

$$\sigma'_k = \tanh \beta_k \sin\left(\frac{2k-1}{2n}\pi\right) \tag{2.20}$$

$$\omega'_k = \cos\left(\frac{2k-1}{2n}\pi\right) \tag{2.21}$$

Comparing s'_k with the corresponding poles of the Butterworth functions, it can be seen that they have the same imaginary parts, whereas their real parts differ by the factor $\tanh\beta_k$. The relative locations on the s-plane of the Butterworth and normalized Chebyshev function poles are given in Fig. 2.7 for $n = 3$. For $\varepsilon = 0$, when β_k and $\tanh\beta_k = 1$, the poles of the Butterworth and Chebyshev functions coincide.

The coefficients of the Chebyshev filter functions, as well as their poles, have been tabulated for various ripple values i.e., 0.1, 0.5, 1, ..., 3 dB. A sample of such a tabulation is given on Table A.2 in Appendix A. This table does not give the normalized Chebyshev filter functions. In all these cases, the upper edge of the passband ripple occurs at $\omega_c = 1$.

The amplitude responses for the 1 dB ripple and the 3 dB ripple third-order Chebyshev lowpass functions

$$F_{1dB}(s) = \frac{0.491}{s^3 + 0.988s^2 + 1.238s + 0.491}$$

and

$$F_{3dB}(s) = \frac{0.2506}{s^3 + 0.597s^2 + 0.928s + 0.2506}$$

are shown in Fig. 2.8 for comparison with the response of the corresponding Butterworth filter. It can be seen that the Chebyshev filters have equiripple response in the passband and fall off monotonically outside it.

2.4.3 Inverse Chebyshev Approximation

The Chebyshev polynomials are also used to obtain the so-called Inverse Chebyshev filter functions, the magnitude of which is given as follows:

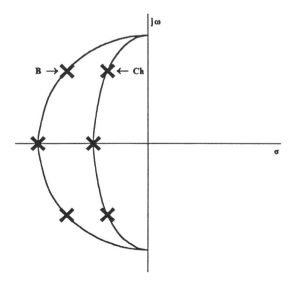

FIGURE 2.7
Relative positions of Butterworth and normalized Chebyshev poles.

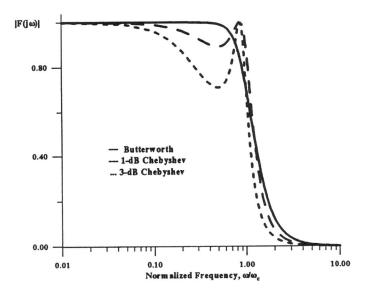

FIGURE 2.8
Comparison of the magnitude response of 1 dB and 3 dB ripple Chebyshev third-order lowpass filters and with the corresponding Butterworth filter.

$$M^2(\omega) = \frac{\varepsilon^2 C_n^2\left(\frac{1}{\omega}\right)}{1 + \varepsilon^2 C_n^2\left(\frac{1}{\omega}\right)} \tag{2.22}$$

The properties of these functions are complementary to those of the Chebyshev functions in the sense that they present maximum flatness in the passband and equiripple behavior in the stopband. Also, their phase response and consequently their group delay is better than that of the Chebyshev filter. In Fig. 2.9, the magnitude response of the third-order Inverse Chebyshev and the corresponding Chebyshev function are shown for comparison.

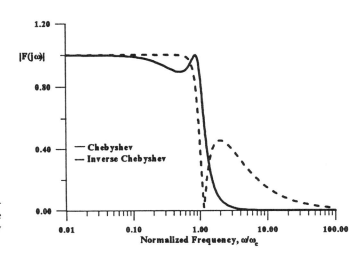

FIGURE 2.9
Magnitude response of the third-order Chebyshev (1 dB ripple) and the corresponding Inverse Chebyshev functions.

2.4.4 Papoulis Approximation

Filter L or Papoulis functions approximating the ideal lowpass filter response are obtained from Eq. (2.5) if we let $\varepsilon = 1$ and

$$w(\omega^2) = L_n(\omega^2) \tag{2.23}$$

with $L_n(\omega^2)$ having the following properties:

a. $L_n(0) = 0$
b. $L_n(1) = 1$
c. $\dfrac{dL_n(\omega^2)}{d\omega} \geq 0$
d. $\left.\dfrac{dL_n(\omega^2)}{d\omega}\right|_{\omega=1} = \max$

Property c secures monotonicity in the amplitude response, whereas property d that the fall off rate at cutoff ($\omega = 1$) is the greatest possible, if monotonicity is assumed.

$L_n(\omega^2)$ polynomials are related to the Legendre $P_k(x)$ polynomials of the first kind. Some of them are given on Table 2.1.

TABLE 2.1
$L_n(\omega^2)$ Polynomials

n	$L_n(\omega^2)$
2	ω^4
3	$3\omega^6 - 3\omega^4 + \omega^2$
4	$6\omega^8 - 8\omega^6 + 3\omega^4$
5	$20\omega^{10} - 40\omega^8 + 28\omega^6 - 8\omega^4 + \omega^2$

The corresponding filters are known as Legendre, Class L, or Papoulis filters [2]. Their poles are found by the procedure that was followed in the case of Butterworth filters.

The main characteristics of these filters are the following:

- Their amplitude response is monotonic.
- The falloff rate at cutoff is the greatest, assuming monotonicity.
- All of their zeros are at infinity.

Because L-filters are less sharp than Chebyshev filters, they are not as popular. However, in cases where the ripple in the passband is undesirable and so the use of Chebyshev filters is excluded, they could be preferable to Butterworth because of their steepest slope at cutoff.

2.4.5 Elliptic Function or Cauer Approximation

The filters examined so far have, except for the Inverse Chebyshev, all of their zeros at infinity. However, in some cases, a higher falloff rate is required in the transition band; in other words, a very high attenuation is required very near the cutoff frequency. This requirement mandates the use of elliptic functions in the approximation, thus obtaining the elliptic functions or, simply, elliptic or Cauer filters.

These filters display equiripple behavior both in the passband and the stopband. The typical magnitude response of a third-order elliptic filter is shown in Fig. 2.10, corresponding to the following general filter function $F(s)$.

$$F(s) = \frac{K(s^2 + \omega_o^2)}{(s + \alpha)(s^2 + \beta s + \gamma)} \quad (2.24)$$

The characteristic quantities that determine the elliptic filter specifications are the maximum passband error, given as maximum attenuation A_{max} in the passband, the minimum attenuation A_{min} in the stopband, the frequency ω_s at which the stopband starts, and the passband edge or cutoff frequency ω_c.

In the case of the elliptic filters, Eq. (2.5) is written in the following form:

$$|F(j\omega)|^2 = \frac{1}{1 + \varepsilon^2 R_n(\omega^2)} \quad (2.25)$$

where R_n, depending on whether n is odd or even, is either

$$R_n(\omega) = \frac{\omega(\omega_1^2 - \omega^2)(\omega_2^2 - \omega^2)\ldots(\omega_k^2 - \omega^2)}{(1 - \omega_1^2\omega^2)(1 - \omega_2^2\omega^2)\ldots(1 - \omega_k^2\omega^2)} \quad n \text{ odd } (n = 2k + 1) \quad (2.26)$$

or

$$R_n(\omega) = \frac{(\omega_1^2 - \omega^2)(\omega_2^2 - \omega^2)\ldots(\omega_k^2 - \omega^2)}{(1 - \omega_1^2\omega^2)(1 - \omega_2^2\omega^2)\ldots(1 - \omega_k^2\omega^2)} \quad n \text{ even } (n = 2k) \quad (2.27)$$

It can be seen from Eqs. (2.26) and (2.27) that

$$R_n\left(\frac{1}{\omega}\right) = \frac{1}{R_n(\omega)} \quad (2.28)$$

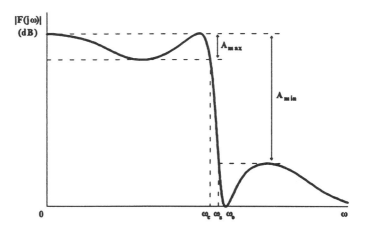

FIGURE 2.10
Typical magnitude response of an elliptic filter of third order.

The Approximation Problem   49

The meaning of this is that the value of $R_n(\omega)$ at a frequency ω' in the band $0 \leq \omega \leq 1$ is the reciprocal of its value at the frequency $1/\omega'$ in the $1 < \omega < \infty$ frequency band. Therefore, if the critical frequencies could be found that lead to equiripple behavior in the passband, automatically the function will have equiripple behavior in the stopband also. Since $|F|^2$ is bounded, the poles of $F(s)$ cannot lie on the $j\omega$-axis. Also, since $|F(j\omega)|^2$ cannot be zero inside the passband, its zeros should lie outside the passband. However, the zeros of $|F(j\omega)|^2$ are the poles of $R_n(\omega)$. Therefore, all the poles of $R_n(\omega)$ should be greater than unity. This means that the zeros of $R_n(\omega)$ should all lie in the band $0 \leq \omega < 1$.

The poles, zeros, and frequencies ω_s have been tabulated for various combinations of values of A_{max} and A_{min}. In such tables, the ripple in the passband is usually given in terms of the reflection coefficient ρ, which is related to A_{max} as follows:

$$A_{max} = -10\log(1-\rho^2) \text{ dB} \tag{2.29}$$

It must be stressed that, with A_{max}, A_{min}, ω_c, and ω_s known, the solution of the approximation problem by means of the elliptic filters requires the lowest-order function and, therefore, it can be realized with the lowest cost. For this reason, the elliptic filters are used most often in practice.

Some elliptic filter functions are given in the Appendix A in Table A.3 for a certain value of maximum attenuation A_{max} in the passband. Clearly, for the selection of the suitable elliptic filter, the specifications should include the values of A_{max}, A_{min}, $\Omega_s(\omega_s/\omega_c)$, and n. In contrast, only the filter order n is required in the case of the Butterworth filter, whereas in the case of the Chebyshev filter the values of n and ε (or A_{max}) should be given. It is very important for the reader to know that, given A_{min}, A_{max}, $\Omega_s(\omega_s/\omega_c)$, the required value of n can be quickly determined for the Butterworth, Chebyshev, and Cauer filters from corresponding nomograms [4].

2.4.6 Selecting the Filter from Its Specifications

In the table giving the Butterworth filter functions, it is assumed that the cutoff frequency is normalized to unity, i.e., $\Omega_c = 1$. The suitable filter function can be read off this table, if its order n has been determined from the specifications. Thus, if the desired rate of fall in the transition band is $6N$ dB/octave, because n is an integer, we select $n = N$ if N is an integer; otherwise, the value of n will be equal to the nearest integer greater than N.

However, in some cases the filter specifications may be given differently. Let us suppose, in the more general case, that the filter specifications require the maximum attenuation in the passband to be A_{max} (<3 dB), occurring at ω_p rad/s (not normalized), and that beyond the frequency ω_s (rad/s) the minimum attenuation should be A_s (in dB). In such cases, the determination of the Butterworth filter order n can proceed as follows.

We suppose that the 3 dB (cutoff) frequency is ω_c rad/s, which of course corresponds to the normalized cutoff frequency Ω_c. Then the normalized frequencies Ω_p and Ω_s will be the following:

$$\Omega_p = \frac{\omega_p}{\omega_c} \qquad \Omega_s = \frac{\omega_s}{\omega_c} \tag{2.30}$$

From Eq. (2.6), with $A = -20\log M(\omega)$, we have

$$A_p = 20\log(1+\Omega_p^{2n})^{1/2} = 10\log(1+\Omega_p^{2n}) \tag{2.31}$$

from which, solving for Ω_p^{2n}, we obtain

$$\Omega_p^{2n} = 10^{0.1A_p} - 1 \tag{2.32}$$

Similarly, we will have for Ω_s

$$\Omega_s^{2n} = 10^{0.1A_s} - 1 \tag{2.33}$$

Dividing Eq. (2.33) by Eq. (2.32) gives

$$\left(\frac{\Omega_s}{\Omega_p}\right)^{2n} = \frac{10^{0.1A_s}-1}{10^{0.1A_p}-1} \tag{2.34}$$

But, because of Eq. (2.30),

$$\frac{\Omega_s}{\Omega_p} = \frac{\omega_s}{\omega_p} \tag{2.35}$$

Then, substituting ω_s/ω_p for Ω_s/Ω_p in Eq. (2.34), we obtain the following required value of n:

$$n = \frac{1}{2}\log\left(\frac{10^{0.1A_s}-1}{10^{0.1A_p}-1}\right)\left[\log\left(\frac{\omega_s}{\omega_p}\right)\right]^{-1} \tag{2.36}$$

It should be pointed out that this value of n will not be necessarily an integer. Then, the order of the required Butterworth filter function will be equal to the nearest integer greater than this value.

Next, we must determine the value of ω_c. Substituting the value of n that was found in Eq. (2.32) [or in Eq. (2.33)], the actual value of Ω_p (or Ω_s) is determined and, using Eq. (2.30), the value of ω_c is obtained. In general, the value of ω_c which is obtained based on the value of Ω_p will be different from that obtained based on Ω_s. However, any one of these values of ω_c will satisfy the filter specifications. The same is true if we use the mean of these two values of ω_c.

To demonstrate, this let the filter specifications be the following:

$$f_p = 3 \text{ kHz}, \quad A_p \leq 1 \text{ dB}, \quad f_s = 6 \text{ kHz}, \quad A_s \geq 20 \text{ dB}$$

Substituting in Eq. (2.36) gives

$$n = 4.3$$

We select n to be the next integer value, i.e., 5.

The Approximation Problem 51

Substituting for n in Eqs. (2.32) and (2.33), we find the following values for Ω_p and Ω_s:

$$\Omega_p = 0.87361$$

$$\Omega_s = 1.5833$$

Then, from Eq. (2.30), for $\Omega_p = 0.87361$, we get one value of ω_c.

$$\omega_{c_p} = 2\pi \times 3.434 \text{ krad/s}$$

and for $\Omega_s = 1.5833$ another

$$\omega_{c_s} = 2\pi \times 3.78955 \text{ krad/s}$$

We may choose to consider as the ω_c value the mean of ω_{c_p} and ω_{c_s}, namely

$$\omega_c = \frac{\omega_{c_p} + \omega_{c_s}}{2} = 2\pi \times 3.61148 \text{ krad/s}$$

Either of ω_{c_p}, ω_{c_s}, or ω_c can be used as the required cutoff frequency ω_c. To show this, we calculate the values of A_p and A_s by means of Eqs. (2.32) and (2.33), which correspond to each of these three values of ω_c. Results are given in Table 2.2.

TABLE 2.2

f_c (kHz)	A_p (dB)	A_s (dB)
3.434	1	24.25
3.78955	0.401	20
3.61148	0.63	22.1

Clearly, the specifications are satisfied in all three cases.

Let us now consider the selection of the Chebyshev filter satisfying certain specifications. Since $C_n(1) = 1$ for any integer n, we will have from Eq. (2.13)

$$20\log[1 + \varepsilon^2 C_n^2(1)]^{1/2} = 10\log(1 + \varepsilon^2) = A_p \quad (2.37)$$

from which

$$\varepsilon = \sqrt{10^{0.1A_p} - 1} \quad (2.38)$$

For the previous example, substituting for $A_p (= A_{max}) = 1$ dB Eq. (2.38) gives

$$\varepsilon = 0.505$$

Considering that at Ω_s the filter function behaves approximately as $(\varepsilon 2^{n-1}\Omega^n)^{-1}$ we obtain

$$A_s \cong 20\log\varepsilon + 6(n-1) + 20n\log\Omega \quad (2.39)$$

In the case of the present example, $\Omega = \Omega_s = 2$ since $\omega_c = \omega_p$. Then, Eq. (2.39) gives

$$20 = 20\log 0.505 + 6(n-1) + 20n\log 2$$

from which, solving for n, we get $n = 2.69$. Therefore, the required order of Chebyshev filter will be $n = 3$, which is lower than the required order $n = 5$ of the Butterworth filter.

The case of the Cauer filter is much simpler, since n can be obtained straight from tables, given the specifications ω_p, A_p, ω_s, and A_s, again using ω_p as ω_c. From such tables [3], or the corresponding nomogram [4] in the case of the previous example, we obtain $n = 3$. Since n is also 3 in the Chebyshev case, we prefer to realize the Chebyshev function, since the cost will be lower, as we shall see in later chapters. It should be mentioned, however, that in practical filter design, in which the fall-off rate is higher than that in this example, the order of the Chebyshev filter is always higher than the order of the corresponding Cauer filter, and the most economical filter will be the Cauer filter.

2.4.7 Amplitude Equalization

In some cases, the amplitude response of the practical filter may not match the amplitude response of the function it realizes. This is because the performance of its components is not ideal. To avoid the subsequent distortion of the signal when passing through the filter, it is necessary that its amplitude response be corrected or, in an other word, equalized.

It is obvious that the transfer function of the equalizer cannot be selected from a predetermined set of functions, since it depends uniquely on the individual filter response that requires equalization. Once the equalizer response has been deduced from the difference between the expected response and the "actual" filter response, the latter obtained by simulation of the filter on the computer using non-ideal components, a function approximating the equalizer response should be found. This can be achieved by applying curve-fitting techniques and an optimization program, while care should be taken in order that the resulting function will be realizable (permitted function). Another approach would start with a certain pole and zero placement and use then the optimization program to adjust their locations until the required response is obtained.

A practical approach [6] suitable in the case of passive filters is to use a cascade of simple networks, e.g., the constant-resistance bridge-T network, the amplitude response of which can be relatively easily adjusted. By properly selecting the component values of the sections, the overall response of the cascaded sections can be adjusted to match the required equalizer response.

2.5 Filters with Linear Phase: Delays

As was explained in Section 1.6.2, if

$$\varphi(\omega) \equiv \arg H(j\omega)$$

we define the group delay τ_g as

The Approximation Problem

$$\tau_g \equiv -\frac{d\varphi(\omega)}{d\omega} \quad (2.40)$$

whereas

$$-\frac{\varphi(\omega)}{\omega} = \tau_p \quad (2.41)$$

is the phase delay.

It can be shown [5] that the definition of the group delay has a physical meaning only when (a) the magnitude function varies slowly with frequency, and (b) the phase varies nearly linearly with frequency over the band of interest.

If $H(s)$ is a rational function, the same will be true for the group delay τ_g, while the phase delay τ_p will not be a rational function.

The function

$$H(s) = e^{-sT}$$

has linear phase, since

$$\varphi(\omega) \equiv \arg e^{-j\omega T} = -\omega T$$

and represents the pure delay T, since

$$-\frac{d\varphi}{d\omega} = T$$

However, e^{-sT} is not a rational function of s.

Thus, although it can be realized by a lossless transmission line terminated at both ends in its characteristic impedance, it cannot be realized by an LLF network.

We can approximate e^{-sT} though by a rational function having either all its zeros at infinity (polyonimic) or in the RH of the s-plane (non-minimum phase function).

This approximation can be achieved either by way of approximating the linear phase $-\omega T$ [Fig. 2.11(a)] or by way of approximating the group delay T [Fig. 2.11(b)], as was indicated in the case of the ideal lowpass filter. Some useful delay approximation functions are briefly reviewed below.

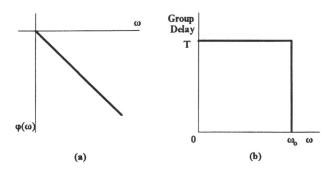

FIGURE 2.11
(a) Ideal linear phase and (b) group delay response.

2.5.1 Bessel-Thomson Delay Approximation

The approximation of the constant group delay T normalized to unity can be obtained by a procedure similar to that followed in the case of magnitude response. Let us select the approximation function $F(s)$ to be polyonimic, i.e., of the form

$$F(s) = \frac{K}{D_n(s)} \tag{2.42}$$

where K is a constant, and $D_n(s)$ a polynomial with positive constant coefficients of order n.

From $F(s)$ we obtain the phase function

$$\varphi(\omega) = \arg F(j\omega)$$

Next, we perform a Taylor expansion of $\varphi(\omega)$ about $\omega = 0$ and take the derivative with respect to ω, which we equate to the negative of the group delay $T = 1$. Equating then coefficients of equal powers of ω, we obtain a number of equations equal to the desired order of approximation n. Clearly, only the constant term of $-d\varphi/d\omega$ is equated to 1. All the other coefficients are set equal to zero. Solution of this set of equations will give the values of the n coefficients of $D_n(s)$. The value of K in Eq. (2.42) is equal to the constant term of $D_n(s)$, normalizing thus the magnitude of $F(j\omega)$ at $\omega = 0$ to unity.

As an example, consider the case $n = 2$. Let

$$D_2(s) = s^2 + \alpha s + \beta$$

Then,

$$\varphi(\omega) = \arg F(j\omega) = -\tan^{-1}\frac{\alpha\omega}{\beta - \omega^2}$$

Taylor's expansion of $\varphi(\omega)$, assuming that $y(\omega) = |\alpha\omega/(\beta - \omega^2)| < 1$, is

$$\varphi(\omega) = -\left[\frac{\alpha\omega}{\beta - \omega^2} - \frac{\alpha^3\omega^3}{3(\beta - \omega^2)^3} + \cdots\right]$$

Taking the derivative of $\varphi(\omega)$ with respect to ω of the first two terms in the series, and ignoring the rest for ω such that $\omega \ll 1$ gives

$$-\frac{d\varphi(\omega)}{d\omega} = \frac{\alpha\beta + \alpha\omega^2}{(\beta - \omega^2)^2} - \frac{1}{3}\left[\frac{3\alpha^3\omega^2(\beta - \omega^2)^3 + 6\alpha^3\omega^4(\beta - \omega^2)^2}{(\beta - \omega^2)^6}\right]$$

or

$$-\frac{d\varphi(\omega)}{d\omega} = \frac{(\alpha\beta + \alpha\omega^2)(\beta - \omega^2)^2 - \alpha^3\omega^2(\beta - \omega^2) - 2\alpha^3\omega^4}{(\beta - \omega^2)^4}$$

Equating this to 1 and multiplying through by $(\beta - \omega^2)^4$ gives the following equation:

$$(\alpha\beta + \alpha\omega^2)(\beta - \omega^2)^2 - (\alpha^3\omega^2(\beta - \omega^2) - 2\alpha^3\omega^4) = (\beta - \omega^2)^4$$

Now we equate the constant term on the one side of this equation to the constant term on the other side and obtain

$$\alpha\beta^3 = \beta^4$$

or

$$\alpha = \beta$$

We do the same for the coefficients of ω^2 and get

$$\alpha\beta^2 + \alpha^3\beta = 4\beta^3$$

But, since $\alpha = \beta$, this equation gives $\beta = 3$. Therefore, the second-order delay function will be the following:

$$F_2(s) = \frac{3}{s^2 + 3s + 3}$$

Following this approach, it is found [5] that the polynomials $D_n(s)$ are related to the Bessel polynomials $G_n(s)$ of degree n by the following relationship:

$$D_n(s) = G_n\left(\frac{1}{s}\right)s^n \tag{2.43}$$

These Bessel polynomials are defined as follows:

$$G_n\left(\frac{1}{s}\right) \equiv \sum_{k=0}^{n} \frac{(n+k)!}{(n-k)!k!(2s)^k} \tag{2.44}$$

It can be shown that all of $D_n(s)$ zeros are located in the LH of the s-plane, and there exists at most one zero on the negative real semi-axis.

The first two polynomials and the recursion formula for obtaining $D_n(s)$ of any degree n are as follows:

$$D_o = 1$$

$$D_1(s) = s + 1$$

$$D_n(s) = (2n-1)D_{n-1}(s) + s^2 D_{n-2}(s) \tag{2.45}$$

The first 5 $D_n(s)$ polynomials and their roots are given in Table 2.3.

The delay functions $F(s)$ obtained this way are called Bessel or Thomson filters, and they approximate the ideal delay according to the maximally flat criterion. Their amplitude

TABLE 2.3
$D_n(s)$ Polynomials and Their Roots

n	$D_n(s)$	Roots of $D_n(s)$
1	$s + 1$	-1
2	$s^2 + 3s + 3$	$-1.5 \pm j0.867$
3	$s^3 + 6s^2 + 15s + 15$	$-2.322, -1.839 \pm j1.754$
4	$s^4 + 10s^3 + 45s^2 + 105s + 105$	$-2.896 \pm j0.867, -2.104 \pm j2.657$
5	$s^5 + 15s^4 + 105s^3 + 420s^2 + 945s + 945$	$-3.647, -3.352 \pm j1.743, -2.325 \pm j3.571$

response is lowpass with a cutoff frequency depending on the value of n and given by the following approximate formula (for $n \geq 3$):

$$\omega_{3dB} = \sqrt{(2n-1)\ln 2} \tag{2.46}$$

This can be easily seen in Fig. 2.12(a), showing the magnitude response of the first three Bessel filters of odd orders.

The corresponding phase response plots are shown in Fig. 2.12(b). It can be seen that the bandwidth with nearly linear phase also increases with increasing n.

In Fig. 2.5 the amplitude, and in Fig. 2.13 the phase response, of the third-order Bessel filter are shown along with the corresponding responses of the Butterworth and the 1 dB ripple Chebyshev filters of the same order. It can be seen that, from the selectivity point of

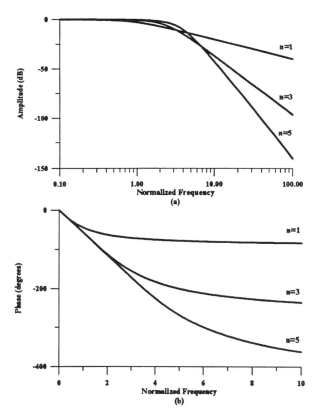

FIGURE 2.12
(a) Magnitude and (b) phase response of the first three Bessel filters of odd orders.

The Approximation Problem

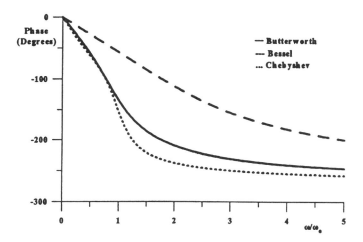

FIGURE 2.13
Phase response of Bessel, Butterworth, and Chebyshev (1 dB ripple) third-order filters.

view, the Bessel filter is at a disadvantage, but its phase response, as far as linearity is concerned, is by far superior—particularly when compared to the Chebyshev phase response.

As a consequence of their superior phase response, the time response of the Bessel-Thomson filters also displays superior performance concerning fidelity to the input waveforms over the other lowpass filters. In other words, they transmit, for example, square pulses with lower distortion than the other filters. This can be easily seen in Fig. 2.14, where the step response of the three filters considered above is shown. Clearly, the rise time, the settling time, and the overshoot are lower for the Bessel filter than for the other two.

It can also be seen that the Bessel response rises to 50 percent of its final value at $t_d = 1s$, which is equal to the unit ($1s$) delay it approximates. This justifies the characterization of t_d as the delay time.

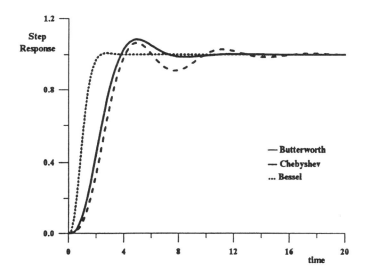

FIGURE 2.14
Step response of the Bessel, Butterworth, and Chebyshev (1 dB ripple) third-order filters.

This observation is nearly true for the corresponding response of the Bessel filters of any order n. However, as n increases, the rise time becomes shorter, and thus the step response comes nearer to the undistorted step. This is so because, as n increases, more frequencies of the infinite spectrum of the input step fall within the bandwidth of nearly linear phase, and thus the approximation comes closer to the input.

The reduced rise time with increasing n can also be proved, if the magnitude response of the Bessel filters is considered. Clearly, from Eq. (2.46) and the plots in Fig. 2.12(a), the 3 dB bandwidth increases with increasing n. Then, from Eq. (1.33), it follows that the rise time should decrease with increasing n.

The bandwidth ω_o of the maximally flat delay is defined [4] as the reciprocal of the delay at $\omega = 0$, i.e.,

$$\omega_o = \frac{1}{T} \tag{2.47}$$

Since the normalized delay is 1, ω_o will also be 1, i.e.

$$\omega_o = 1 \tag{2.48}$$

Clearly, the meaning of Eq. (2.47) is that the product bandwidth times delay is constant. That is, large bandwidth corresponds to short delay and vice versa.

Letting $s = ju$, where

$$u = \omega T = \frac{\omega}{\omega_o}$$

we can create two tables, the first giving the u values (for each n) for certain deviations of the time delay from its ideal value (i.e., its value at $\omega = 0$), and the second giving the values of u (for each n) in which the attenuation is certain decibels below its value at $\omega = 0$. From these tables, the designer can select the value of n and consequently determine the corresponding Bessel-Thomson delay function that suits best the specifications (see Reference 3).

2.5.2 Other Delay Functions

Another class of functions approximating in fact the phase of e^{-sT} according to the maximally flat criterion at $\omega = 0$ are the allpass Padè approximations [6]. All of the zeros of these functions lie on the RHP located symmetrically to the poles with respect to the $j\omega$-axis. The magnitude response of these is unity for all ω and their useful bandwidth is twice that of the Bessel-Thomson delays of corresponding orders. However, in spite of these useful characteristics, their step response displays a very narrow precursor of height about equal to their final value, a highly undesirable characteristic (see Fig. 2.15). To avoid the appearance of this precursor in the step response one may use lowpass Padè delay approximations [7] or other more useful delay functions [8, 9, 10]. Other delay functions which approximate the group delay according to the Chebyshev criterion have also been proposed [11]. These display improved characteristics over the Padè delay functions of corresponding orders.

In general, the function that will be selected for delaying a signal will depend greatly on the type of signal. Thus, if the signal is in the form of a step, a lowpass delay is more suitable than an allpass. In the case of a signal with a certain bandwidth though, an allpass function with linear phase may satisfy the specifications more effectively. In practice, on

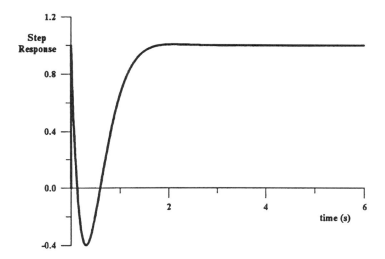

FIGURE 2.15
Step response of the second-order allpass Padè delay function.

many occasions, the combination of a lowpass filter with sharp cutoff and an allpass function filter connected in cascade results in the desirable solution, as explained next.

2.5.3 Delay Equalization

As it was mentioned in the previous section, the step response of the Bessel-Thomson filters, due to their linear phase response, makes them more suitable in pulse transmission than the corresponding Butterworth, Chebyshev or Cauer filters. However, from the selectivity point of view their performance is very poor compared to the other filters.

To achieve both phase linearity and good selectivity in the amplitude response, a practical solution is to use suitable allpass functions of second-order in order to modify the phase of the filter that has the desirable magnitude response. Their use will not affect the filter magnitude response, since they are allpass. Such functions will be of the form

$$F(s) = \frac{s^2 - \beta s + \gamma}{s^2 + \beta s + \gamma} \tag{2.49}$$

They will be selected by means of a computer optimization program, which will determine the most suitable coefficient values β and γ in each case.

This procedure, called phase equalization, proves to be very useful in problems where the required filter should possess high selectivity and at the same time linear phase response, i.e., constant group delay.

2.6 Frequency Transformations

The filter functions that were reviewed in the previous sections refer to lowpass filters. They are given in normalized form, i.e., their passband width is unity. In all these tables, the normalized frequency s_n is implied for sinusoidal excitation $s_n = j\Omega$ with

$$\Omega = \frac{\omega}{\omega_c}$$

where ω is the real frequency variable, and ω_c is the actual cutoff frequency of the desired lowpass filter. Following this convention, the normalized cutoff frequency Ω_c of all filters is

$$\Omega_c = 1$$

We will now show by means of suitable transformations how we can obtain denormalized lowpass, highpass, bandpass, bandstop filters, and delays using data obtained from the tabulated normalized lowpass functions. In all cases, we refer to frequency response, when $s = j\omega$.

2.6.1 Lowpass-to-Lowpass Transformation

In the normalized lowpass function, if we substitute s/ω_c for s_n, we will obtain the denormalized lowpass function with ω_c being its cutoff frequency. For example, for the first-order Butterworth function

$$F(s_n) = \frac{1}{s_n + 1} \tag{2.50}$$

we will get

$$F(s) = \frac{1}{s/\omega_c + 1} = \frac{\omega_c}{s + \omega_c} \tag{2.51}$$

which is also lowpass with its cutoff frequency at ω_c.

Clearly, with this transformation, the normalized frequency band $0 \leq |\Omega| \leq 1$ is transformed to the denormalized frequency band

$$0 \leq |\omega| \leq \omega_c$$

as shown in Fig. 2.16.

As can be seen from Eqs. (2.50) and (2.51) the shape of the frequency response does not change with this transformation. Following this, it is obvious that transmission zeros at Ω_1 will appear at the frequency $\omega_1 = \Omega_1 \omega_c$ in the denormalized magnitude response. This is shown clearly in Fig. 2.17.

Finally, it should be mentioned that the gain of the normalized filter functions is assumed to be normalized, i.e., its maximum value is equal to unity. It is usual that in filter design we are not so interested in the actual value of the magnitude in the corresponding response, but in its relative value (or relative attenuation), which determines the filter selectivity.

FIGURE 2.16
Lowpass-to-lowpass transformation.

The Approximation Problem

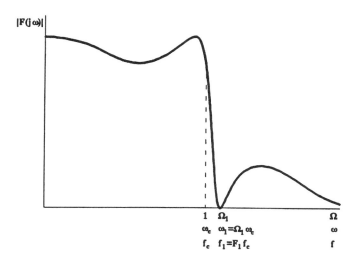

FIGURE 2.17
Lowpass-to-lowpass frequency transformation.

2.6.2 Lowpass-to-Highpass Transformation

Applying the transformation

$$s_n \to \frac{1}{s}$$

a lowpass function is transformed to a highpass.

The frequencies 0 and ∞ of the lowpass function are transformed to ∞ and 0, respectively, while the cutoff frequency of the lowpass, which is 1 in the normalized function, is transformed to itself in the new function. Thus, the passband of the lowpass $0 \le |\Omega| \le 1$ is transformed to the passband $0 \le |\Omega| \le \infty$ of the highpass function as shown in Fig. 2.18(a).

Following the same argument as in the case of the lowpass-to-lowpass transformation, if we substitute ω_c/s for s_n in Eq. (2.50) we will get

$$F(s) = \frac{s}{s + \omega_c} \tag{2.52}$$

which is a highpass function with ω_c its cutoff frequency. The mapping of the lowpass passband $0 \le |\Omega| \le 1$ to the highpass passband $\omega_c \le |\omega| \le \infty$ is shown pictorially in Fig. 2.18(b).

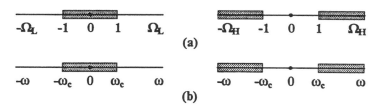

FIGURE 2.18
Lowpass-to-highpass transformation.

2.6.3 Lowpass-to-Bandpass Transformation

Applying the transformation

$$s_n \to s + \frac{1}{s} = \frac{s^2+1}{s}$$

a lowpass function is transformed to a bandpass function with a passband width equal to that of the lowpass, i.e., equal to 1. There are two bandpass cutoff frequencies, Ω_{c1} and Ω_{c2}, such that

$$\Omega_{c1}\Omega_{c2} = 1$$

and

$$\Omega_o^2 = \Omega_{c1}\Omega_{c2} = 1 \tag{2.53}$$

Ω_o is called the normalized center frequency of the bandpass function. The passband mapping is shown in Fig. 2.19(a).

The bandwidth of the passband is

$$\Omega_{c2} - \Omega_{c1} = 1 \tag{2.54}$$

since this has to be equal to that of the lowpass. Solving Eqs. (2.53) and (2.54) for Ω_{c1} and Ω_{c2} gives the following:

$$|\Omega_{c1}| = -\frac{1}{2} + \frac{\sqrt{5}}{2}$$

$$|\Omega_{c2}| = \frac{1}{2} + \frac{\sqrt{5}}{2} \tag{2.55}$$

Using this transformation in the example we considered before, the lowpass Butterworth function [Eq. (2.50)] will be transformed to the function

$$F(s) = \frac{s}{s^2+s+1}$$

which is bandpass, since it becomes zero at $\Omega = 0$ and ∞.

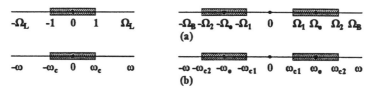

FIGURE 2.19
Lowpass-to-bandpass transformation.

The Approximation Problem

Denormalization of the bandpass function to a center frequency ω_o is obtained by substituting ω/ω_o for Ω in Eqs. (2.53) to (2.55). The bandwidth of the denormalized function will then be

$$B = \omega_{c2} - \omega_{c1} = \omega_o$$

If the bandwidth should be other than ω_o, the transformation has to be modified. Thus, we may obtain the denormalized bandpass function straight from the normalized lowpass function by applying to the latter the following transformation:

$$s_n \rightarrow \frac{\omega_o}{B}\left(\frac{s}{\omega_o} + \frac{\omega_o}{s}\right)$$

Corresponding mapping of the passbands and stopbands are shown in Fig. 2.19(b). In this case,

$$\omega_o^2 = \omega_{c1}\omega_{c2}$$

$$B = \omega_{c2} - \omega_{c1} \tag{2.56}$$

2.6.4 Lowpass-to-Bandstop Transformation

Applying the transformation

$$s_n \rightarrow \frac{1}{s + \frac{1}{s}} = \frac{s}{s^2 + 1}$$

a normalized lowpass function is transformed to a bandstop (bandreject) function, the stopband of which is between the normalized frequencies Ω_{c2} and Ω_{c1} as shown in Fig. 2.20(a).

To obtain the required denormalized bandstop function from the normalized lowpass function, the suitable transformation is

$$s_n \rightarrow \frac{B}{\omega_o\left(\dfrac{s}{\omega_o} + \dfrac{\omega_o}{s}\right)}$$

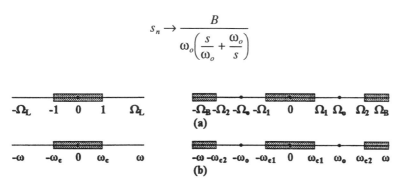

FIGURE 2.20
Lowpass-to-bandstop transformation.

where ω_o is the centre frequency in the stopband with

$$\omega_o^2 = \omega_{c1}\omega_{c2}$$

and B is the bandwidth of the stopband

$$B = \omega_{c2} - \omega_{c1}$$

This is shown in Fig. 2.20(b).

In the case of the example considered above, we will obtain the normalized bandstop function

$$F(s) = \frac{s^2 + 1}{s^2 + s + 1}$$

This is zero at $\Omega = 1$ and 1 at $\Omega = 0$ and ∞, which shows the bandstop behavior of the function.

2.6.5 Delay Denormalization

As we have seen in Section 2.5, the functions that approximate the ideal delay are also given in tables or in the form of recursion formulas. The denormalized function is obtained from the normalized one by substituting $s\tau$ for s_n, where τ is the required delay in seconds.

2.7 Design Tables for Passive LC Ladder Filters

All filter functions introduced in this chapter can be realized by passive networks. There is an abundance of books and papers in the literature describing how this can be done. However, in this book, we are interested only in the realization using doubly terminated LC ladders, since their simulation by active RC networks results in low sensitivity filters (see Chapter 6).

The general structure of such a network is shown in Fig. 2.21. In the case of lowpass polyonimic filters (all zeros at infinity), all Z_is are inductors, and all Y_is are capacitors. In the case of lowpass functions with finite transmission zeros (e.g., Cauer filters), the Z_is will be parallel tuned LC subcircuits, or the Y_is will be series tuned LC combinations.

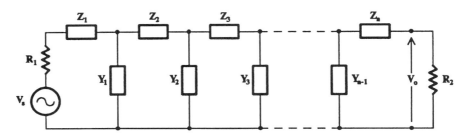

FIGURE 2.21
The general form of a doubly terminated ladder.

Design tables for realizing all the lowpass functions that we have introduced above for various values of A_{max} (or ε), A_{min}, and orders, as well as for various values of the ratio R_2/R_1 are given in many books [4, 12, 13, 14]. Although, from the sensitivity point of view, as we shall see later, the equally terminated ladder ($R_1 = R_2$) is most desirable, in some cases (e.g., even-order Chebyshev and Cauer filters), this is not exactly possible.

These tables are used by the passive filter designer and, of course, by the active RC filter designer who will choose to design a filter by simulating the passive ladder by active elements. Therefore, once the designer has chosen the normalized lowpass filter function from the specifications of his problem, he may use these tables to obtain the passive circuit, which realizes this function. Then, he can obtain the required denormalized filter by applying suitable transformations to the element values as explained below.

As we shall see in later Chapters (e.g., Chapter 6), one powerful method for designing RC active filters is to simulate passive LC ladder filters, either topologically or functionally, using RC active circuits. Thus, the tables of LC passive filters greatly simplify this design and are used accordingly. It is for this very reason that we have introduced these tables for the design of LC ladder filters here.

However, for various reasons, there are no corresponding tables available for the design of RC active filters in general use. For example, an RC active cannot be transformed to a corresponding bandpass or bandstop circuit by the transformation of its elements as in the case of the passive LC circuits, while available tables [15] do not cover all useful active RC circuits and Cauer filter design.

One final point here: the lowpass filter function realized by the circuit in Fig. 2.21 as the voltage ratio V_o/V_s is in fact realized within a constant multiplier that is lower than unity. The reason for this is that at dc ($\omega = 0$), the transfer voltage ratio V_o/V_s reduces to

$$\frac{V_o}{V_s} = \frac{R_2}{R_1 + R_2}$$

which is always less than 1 except for $R_1 = 0$, when it is equal to 1.

2.7.1 Transformation of Elements

In the case of passive filters, as stated above, one can obtain the denormalized highpass, bandpass, or bandstop filters by applying the previously introduced frequency transformations to the impedances of the elements of the normalized lowpass filter.

This approach is not applicable in the case of active RC filters except for the case of obtaining a highpass from the normalized lowpass filter. We examine the element transformation in more detail below:

2.7.1.1 LC Filters

Lowpass-to-lowpass	Transformation $s_n \to \dfrac{s}{\omega_c}$ with ω_c the cutoff frequency	
Element	Impedance	New element value
L_n	$s_n L_n = \dfrac{s}{\omega_c} L_n$	$L = L_n/\omega_c$
C_n	$\dfrac{1}{s_n C_n} = \dfrac{\omega_c}{s C_n}$	$C = C_n/\omega_c$

Lowpass-to-highpass	Transformation $s_n \to \dfrac{\omega_c}{s}$ with ω_c the cutoff frequency		
Element	Impedance		New element value
L_n	$s_n L_n = \dfrac{\omega_c L_n}{s}$		$C = \dfrac{1}{\omega_c L_n}$
C_n	$\dfrac{1}{s_n C_n} = \dfrac{s}{\omega_c C_n}$		$L = \dfrac{1}{\omega_c C_n}$
Lowpass-to-bandpass	Transformation $s_n \to \dfrac{\omega_o}{B}\left(\dfrac{s}{\omega_o} + \dfrac{\omega_o}{s}\right)$		
Element	Impedance	\multicolumn{2}{l}{New element values}
		\multicolumn{2}{l}{L and C in series}
L_n	$s_n L_n = \dfrac{\omega_o}{B}\left(\dfrac{s}{\omega_o} + \dfrac{\omega_o}{s}\right) L_n$	$L = \dfrac{L_n}{B}$	$C = \dfrac{B}{\omega_o^2 L_n}$
		\multicolumn{2}{l}{L and C in parallel}
C_n	$\dfrac{1}{s_n C_n} = \dfrac{1}{C_n \dfrac{\omega_o}{B}\left(\dfrac{s}{\omega_o} + \dfrac{\omega_o}{s}\right)}$	$L = \dfrac{B}{C_n \omega_o^2}$	$C = \dfrac{C_n}{B}$
Lowpass-to-bandstop	Transformation $s_n \to \dfrac{1}{\dfrac{\omega_o}{B}\left(\dfrac{s}{\omega_o} + \dfrac{\omega_o}{s}\right)}$		
Element	Impedance	\multicolumn{2}{l}{New element values}
		\multicolumn{2}{l}{L and C in parallel}
L_n	$s_n L_n = \dfrac{L_n}{\dfrac{\omega_o}{B}\left(\dfrac{s}{\omega_o} + \dfrac{\omega_o}{s}\right)}$	$L = \dfrac{B L_n}{\omega_o^2}$	$C = \dfrac{1}{B L_n}$
		\multicolumn{2}{l}{L and C in series}
C_n	$\dfrac{1}{s_n C_n} = \dfrac{\dfrac{\omega_o}{B}\left[\dfrac{s}{\omega_o} + \dfrac{\omega_o}{s}\right]}{C_n}$	$L = \dfrac{1}{B C_n}$	$C = \dfrac{B C_n}{\omega_o^2}$

All these element transformations are summarized in Table 2.5.

TABLE 2.5

Element Transformations

Elements of lowpass filter	Corresponding elements of the denormalized		
	Highpass	Bandpass	Bandstop
L_n	$C = 1/(L_n \omega_c)$	$L = L_n/B$ $C = B/(\omega_o^2 L_n)$	$L = L_n B/\omega_o^2$ $C = 1/(B L_n)$
C_n	$L = 1/(C_n \omega_c)$	$L = B/(\omega_o^2 C_n)$ $C = C_n/B$	$L = 1/(B C_n)$ $C = C_n B/\omega_o^2$

As an example, consider the design of a bandpass filter having center frequency at 1 krad/s and 100 rad/s bandwidth. Assume that, from additional specifications, a sixth-order Butterworth bandpass filter has been found suitable.

The Approximation Problem

Clearly, the sixth-order bandpass will be obtained from the third-order lowpass Butterworth filter by applying the lowpass-to-bandpass transformation

$$s_n \to \frac{\omega_o}{B}\left(\frac{s}{\omega_o} + \frac{\omega_o}{s}\right)$$

where $\omega_o = 1$ krad/s and $B = 100$ rad/s.

From the tabulated Butterworth filters, Table A.1, we find

$$F(s_n) = \frac{1}{s_n^3 + 2s_n^2 + 2s_n + 1}$$

At this point, we may proceed in one of the following two alternative ways:

1. We may apply the lowpass-to-bandpass transformation to obtain either the normalized or the denormalized bandpass function and proceed to realize it, i.e., to determine the circuit.
2. Alternatively, we may realize the lowpass filter and apply the lowpass-to-denormalized bandpass transformation to the elements of the lowpass making use of the Table 2.5.

The second method has the advantage that the realization of the lowpass filter reduces to choosing the circuit from the available design tables. In cases, when there are no available design tables, the advantage still remains, since the realization of the lowpass filter is simpler than that of the bandpass, because the order of the latter is double that of the lowpass. On the other hand, the second method is not applicable to the synthesis of active RC networks. So, the choice of the most suitable method will depend on the type of the circuit (passive LC or active RC) we choose to design.

Suppose we choose to realize $F(s_n)$ by a passive LC equally terminated ladder. Using the corresponding table, given for example in Reference 12, we find that the suitable circuit is that appearing in Fig. 2.22.

Using Table 2.5, we can easily obtain the element denormalization for $\omega_o = 1$ krad/s and $B = 100$ rad/s. The denormalized w.r.t. frequency bandpass circuit is as shown in Fig. 2.23.

This circuit is still normalized w.r.t. impedance level, since all component values are referred to terminating resistances of 1 Ω. If we want to raise the impedance level to a practical value, e.g., 600 Ω, we should multiply the impedance of each component by 600, when we obtain the component values in parentheses in Fig. 2.23. We treat impedance denormalization in Section 2.7.

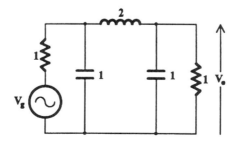

FIGURE 2.22
Butterworth normalized lowpass filter of third-order.

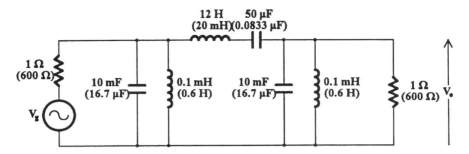

FIGURE 2.23
Frequency denormalized bandpass filter.

2.7.1.2 Active RC Filters

The procedure described above can be applied to obtain the denormalized lowpass or highpass RC active filter once the normalized lowpass function has been selected.

It is not possible to obtain the normalized or denormalized bandpass and bandstop circuits straight from the normalized lowpass, because there is no suitable transformation for this purpose. Clearly with no inductances in these circuits, it is impossible to apply the element transformations for bandpass and bandstop of Table 2.5. In this case, first the normalized bandpass or bandstop function is obtained using the corresponding frequency transformation. Next, the normalized filter is synthesized by a suitable method, as we shall see in later chapters, and then the denormalized bandpass or bandstop filter is obtained by properly scaling the filter time constants.

It has to be emphasized here that frequency transformation must be applied only to time constants, i.e., either to the capacitances or to those resistances that determine the time constants and not to those that determine the gain of the active element.

The following two examples will clarify this, while for a more formal proof the interested reader should refer to References 6 and 16.

Consider first the simple RC circuit in Fig. 2.24(a). The transfer voltage ratio V_2/V_1 is the following:

$$F(s) = \frac{V_2}{V_1} = \frac{1}{RCs + 1} = \frac{1}{RC} \cdot \frac{1}{s + 1/RC} \tag{2.57}$$

If we interchange the position of the elements without changing its topology, as shown in Fig. 2.24(b), the new transfer voltage ratio will be

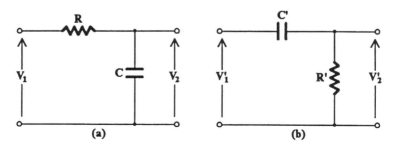

FIGURE 2.24
Simple (a) lowpass and (b) highpass RC circuits.

The Approximation Problem

$$F'(s) = \frac{V'_2}{V'_1} = \frac{sC'R'}{sC'R' + 1} = \frac{s}{s + 1/C'R'} \tag{2.58}$$

It can be seen that, with this interchange of elements, the lowpass circuit in Fig. 2.24(a) has been transformed to the highpass of Fig. 2.24(b). The two filters will have the same cutoff frequency if

$$RC = C'R' \tag{2.59}$$

As a second example, consider the two RC active circuits in Fig. 2.25(a) and (b), using an operational amplifier as the active element (see Chapter 3). Circuit (b) is obtained from circuit (a) by changing all resistors to capacitors and vice versa.

The transfer function V_2/V_1 of circuit (a) is

$$F(s) = \frac{V_2}{V_1} = -\frac{1}{CR_1} \frac{1}{s + \frac{1}{CR_2}} = -\frac{R_2/R_1}{sCR_2 + 1} \tag{2.60}$$

and that of circuit (b) is

$$F'(s) = \frac{V'_2}{V'_1} = -\frac{C_1}{C_2} \frac{s}{s + \frac{1}{C_2R}} = -\frac{sC_1R}{sC_2R + 1} \tag{2.61}$$

Clearly, again the lowpass circuit has been transformed to a highpass simply by changing resistors to capacitors and vice versa in the lowpass circuit. For the two circuits to have the same cutoff frequency, the following relationship should hold:

$$CR_2 = C_2R \tag{2.62}$$

In both these examples, if the required denormalized cutoff frequency is ω_c, the time constants in Eqs. (2.59) and (2.62) have to be divided by ω_c. This means that either the capacitance or the resistance which determine the time constant should be divided by ω_c and not both. Compare this case with LC filters where both L and C are divided by ω_c.

The above is part of the so called RC:CR transformation, by means of which a lowpass RC circuit, passive or active, is transformed to the corresponding highpass, under the con-

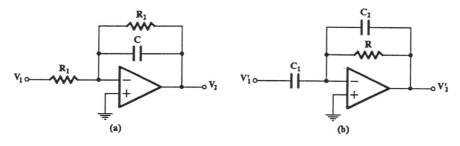

FIGURE 2.25
(a) Lowpass and (b) highpass RC active filters.

dition that the pertinent transfer function is dimensionless, i.e., ratio of voltages or currents. According to this transformation the element R_i is transformed to the element C_i with nominal value $1/R_i$ and vice versa. Resistances, which determine the voltage gain, or the current gain of the active element, do not change, which is not true if the active element is a voltage controlled current source (VCCS) or a current controlled voltage source (CCVS).

As a further example, consider the RC active filter in Fig. 2.26(a), which realizes the second-order Butterworth lowpass function.

$$F(s) = \frac{1}{s^2 + \sqrt{2}s + 1} \tag{2.63}$$

Applying the RC:CR transformation, this circuit is changed to that of Fig. 2.26(b), which is highpass with cutoff frequency $\Omega_c = 1$ equal to the cutoff frequency of the corresponding lowpass in Fig. 2.26(a). If we divide both resistances or both capacitances by ω_c (rad/s), their cutoff frequency becomes ω_c.

2.8 Impedance Scaling

If the desired cutoff frequency of the circuits in Fig. 2.24 is $\omega_c = 1$ rad/s, this can be achieved for

$$C = 1 \text{ F} \quad R = 1 \text{ }\Omega$$

However, the same result can be achieved if

$$C = 1 \text{ mF} \quad R = 1 \text{ k}\Omega$$

or

$$C = 1 \text{ μF} \quad R = 1 \text{ M}\Omega$$

and so on.

When we select the second or third set of values of R and C instead of the first, in actual fact, we have multiplied the impedances of these components by 10^3 or 10^6, respectively,

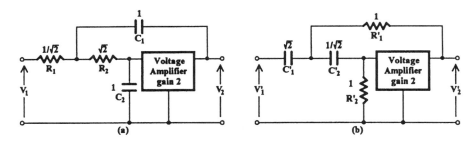

FIGURE 2.26
(a) Second-order RC active lowpass and (b) corresponding RC active highpass filter.

obtaining thus more practical values for the elements R and C. We say that we have raised the impedance level of the circuit by 10^3 and 10^6, respectively.

This denormalization of a circuit to the impedance level R_o requires multiplication by R_o of the values of all resistances, and all inductances in the circuit and division by R_o of all capacitances. In the case when the gain of the active element(s) is determined by the ratio of two resistances, which do not affect the circuit time constants, the impedance level for these resistances may be different from that for the rest of the circuit. Also, if a filter is realized as the cascade connection of low-order functions (i.e., first and second order), which are isolated from each other, again the impedance level in each section can be different from that in the other.

As an example of impedance scaling, we will calculate the values of the components in the circuit in Fig. 2.26(b) for an impedance level of $10^4\ \Omega$ and a cutoff frequency at 10^4 rad/s. In accordance with the above discussion, we multiply all resistances by 10^4 and divide all capacitances by 10^4 in order to perform impedance scaling.

Next we may choose to divide the values of the capacitors by another 10^4 or the values of the resistors by this same scaling factor in order to achieve frequency denormalization. Choosing to frequency scale the capacitances, the following set of component values is obtained:

$$R_1 = R_2 = 10\ k\Omega, \qquad C_1 = \sqrt{2} \times 10^{-8}\ F, \qquad C_2 = \frac{1}{\sqrt{2}} \times 10^{-8}\ F$$

These component values are more suitable for RC active filters (particularly if the active element is an operational amplifier) than if we had chosen to frequency scale the resistances. It should be pointed out, however, that frequency scale could have also been obtained if all resistances had been divided by 10^a and all capacitances by 10^{4-a}, since the time constants that actually matter would have been scaled by $10^a \times 10^{4-a} = 10^4$.

2.9 Predistortion

It is well understood that the components one uses in order to build up a circuit one has designed are not ideal. Thus, the equivalent of a coil is not a pure inductance, but it has some loss associated with it. This loss is modeled by a small resistance r connected in series with its inductance L. Similarly, there is some loss associated with the capacitance of a capacitor, usually negligible in today's capacitors, which is modeled by connecting a conductance g in parallel with the capacitance C.

For the sake of argument, let us suppose that the ratios r/L and g/C are both equal to d. Thus the impedance of the coil and the admittance of the capacitor will be, respectively,

$$Z_L = sL + r = L(s + d)$$

and

$$Y_C = sC + g = C(s + d)$$

The circuit transfer function, as derived by circuit analysis, is found to be a linear function of impedance or admittance ratios. This means that the poles and zeros of the transfer function will move to the left in the s-plane by the same amount, d. Thus, the frequency response of the practical circuit will differ from the expected one, i.e., it will be distorted. In some cases, this effect may not matter, but in cases of highly selective bandpass filters, the distortion in the frequency response will be serious and thus unacceptable.

To counterbalance this type of distortion in the frequency response of the practical circuit to be built out of coils, capacitors, and resistors, the designer can shift the poles and zeros of the transfer function to the right on the s-plane by the same amount d and then proceed to calculate the component values. With the application of this technique, known as predistortion, the poles and zeros of the transfer function of the practical circuit will be placed nearly at the initially wanted positions, moved there, because of the power dissipation of the practical components, coils, and capacitors. Predistortion is in effect a frequency transformation of the initial transfer function $F(s)$ to $F(p)$ where $p = s - d$. This transformation is nonreactive in that it is not applied to the reactive elements of the circuit.

In practice, the losses in coils and capacitors will not be the same. Then, the designer can add resistances in series with the coils and in parallel to the capacitors in order to obtain the same factor d in all components. It should be noted, though, that because of the introduction of dissipative components (the resistors) in the LC circuit, there will be an increase in the flat loss of the circuit. Since this loss can, in some cases, be intolerable, and the number of the additional resistors excessive, the predistortion technique may not always produce an attractive solution to the unavoidable problem of dissipation introduced by the practical reactive components.

2.10 Summary

The problem of determining a filter function satisfying the specifications of a filter has been examined in this chapter to some detail.

Filter specifications may refer mainly either to amplitude (magnitude) requirements or to phase (or delay) requirements. Determining first the order of the lowpass prototype filter function, the designers can then choose the most suitable one among the available in tables Butterworth, Chebyshev, Papoulis, or Cauer functions, if they are interested in the amplitude response. Similarly, if they are interested in the phase response, they may choose a lowpass or an allpass function among the available in tabulated form Bessel-Thomson, Padè, or Chebyshev-type delay functions.

In the case of amplitude or phase equalization, the designers will basically have to work heuristically using the computer as their main tool and a suitable optimization program.

Suitable frequency and element transformations were introduced in order to transform the lowpass prototype filter to the required denormalized lowpass, highpass, bandpass, or bandstop filter. These frequency transformations are also useful in the translation of the denormalized filter specifications to the corresponding lowpass prototype requirements.

Once the most suitable circuit has been chosen (in a way we shall see in later chapters) for the realization of the denormalized function, suitable impedance scaling should be applied to the component values in order to make the circuit more practical within its environment (signal level, source impedance, load impedance, and characteristics of the active element). Impedance scaling was also introduced, while in the final section of this chapter the concept of predistortion was introduced briefly.

Before we examine the selection of suitable circuits for the realization of filter functions we introduce in the next chapter various active elements that will be used in subsequent chapters.

References

[1] F.F. Kuo. 1966. *Network Analysis and Synthesis*, 2/e, New York: John Wiley & Sons.
[2] A. Papoulis. 1959. "On Monotonic Response Filters," *Proc. IRE*, 47, pp. 332–333.
[3] G. Daryanani. 1976. *Principles of Active Network Synthesis and Design*, New York: John Wiley & Sons.
[4] A. Zverev. 1967. *Handbook of Filter Synthesis*, New York: John Wiley & Sons.
[5] E. Kuh and D. Pederson. 1959. *Principles of Circuit Theory*, New York: McGraw-Hill.
[6] R.W. Daniels. 1974. *Approximation Methods for Electronic Filter Design*, New York: McGraw-Hill.
[7] D.F. Tuttle. 1958. *Network Synthesis*, Vol. 1, New York: John Wiley & Sons.
[8] A. Budak. 1965. "Maximally flat phase and controllable magnitude approximation," *Proc. IEEE* CT-12, p. 279.
[9] W.J. King and V.C. Rideout. 1961. "Improved transport delay circuits for analogue computer use," *Proc. 3rd Intl. Meeting of the Association pour le Calcul Analogique*, Opatija, Yugoslavia, Sept. 5.
[10] T. Deliyannis. 1970. "Six new delay functions and their realization using active RC networks," *The Radio and Electronic Engineer*, **39**(3), pp. 139–144.
[11] D. Humphreys. 1964. "Rational function approximation of polynomials to give an equiripple error," *IEEE Trans Circuit Theory* CT-11, pp. 479–486.
[12] L. Weindberg. 1962. *Network Analysis and Synthesis*, New York: McGraw-Hill.
[13] K. Skiwrzinski. 1965. *Design Theory and Data for Electrical Filters*, New York: Van Nostrand.
[14] R. Saal. 1979. *Handbook of Filter Design*, AEG-TELEFUNKEN.
[15] D.E Johnson and J.L. Hilburn. 1975. *Rapid Practical Designs of Active Filters*, New York: John Wiley & Sons.
[16] S. K. Mitra. 1969. *Analysis and Synthesis of Linear Active Networks*, New York: John Wiley & Sons.

Chapter 3

Active Elements

3.1 Introduction

The ideal active elements are devices having one to three ports with properties that make them very useful in network synthesis. Some active elements are more useful than others, in the sense that their realizations are more practical than others.

The most important ideal active elements in network synthesis fall into the following groups:

- Ideal controlled sources
- Generalized impedance converters (GICs)
- Generalized impedance inverters (GIIs)
- Negative resistance
- Current conveyors

Although the GIIs, generally speaking, can be regarded as GICs, they are presented separately here for reasons of clarity.

The first three groups consist of two-port devices, and the fourth is one-port. The fifth is a three-port device. We present each of these groups separately below.

For all these ideal active elements, we give practical realizations using two active devices which are commercially available, namely, the operational amplifier (opamp) and the operational transconductance amplifier (OTA). The opamp and OTA are special cases of two ideal active elements, and their implementations in IC form make them indispensable today, both in discrete and fully integrated analog network design. Because of this exclusive use in active filter design, we introduce them here both as ideal and practical elements, giving emphasis on the imperfections of the practical realizations.

Before proceeding with the development of this chapter, one point should be clarified. Although the transistor, either the bipolar (BJT) or the unipolar (FET), is essentially the basic active element in the realization of all other active elements in practice, we prefer not to consider it as such here but rather to treat it as a type of a nonideal controlled current source.

3.2 Ideal Controlled Sources

An ideal controlled source is a source whose magnitude (voltage or current) is proportional to another quantity (voltage or current) in some part of the network. Table 3.1 lists the four

TABLE 3.1

Ideal Controlled Sources

	Description	Symbol	A Matrix	Reciprocal or Nonreciprocal
1	Voltage controlled voltage source (VCVS)	$gE_1 = E_2$	$\begin{bmatrix} 1/g & 0 \\ 0 & 0 \end{bmatrix}$	Nonreciprocal
2	Current controlled voltage source (CCVS)	$zI_1 = E_2$	$\begin{bmatrix} 0 & 0 \\ 1/z & 0 \end{bmatrix}$	Nonreciprocal
3	Current controlled current source (CCCS)	$hI_1 = -I_2$	$\begin{bmatrix} 0 & 0 \\ 0 & 1/h \end{bmatrix}$	Nonreciprocal
4	Voltage controlled current source (VCCS)	$yE_1 = -I_2$	$\begin{bmatrix} 0 & 1/y \\ 0 & 0 \end{bmatrix}$	Nonreciprocal

types of controlled sources. Their characteristic feature is that the transmission matrix has just one nonzero element. Among them, only the CCVS and the VCCS are basic devices. A VCCS followed by a CCVS gives a VCVS, and reversing the order gives a CCCS.

3.3 Impedance Transformation (Generalized Impedance Converters and Inverters) [1, 2]

Consider a two-port to be terminated at port 2 in an impedance Z_L as shown in Fig. 3.1(a). The input impedance Z_{i1} at port 1 expressed in terms of Z_L and the transmission matrix parameters a_{ij}, $i, j = 1, 2$ is as follows:

$$Z_{i1} \equiv \frac{V_1}{I_1} = \frac{a_{11}Z_L + a_{12}}{a_{21}Z_L + a_{22}} \tag{3.1}$$

Active Elements

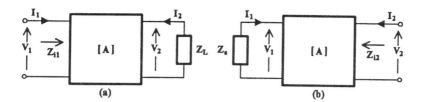

FIGURE 3.1
Two-port device to be terminated at port 2 in an impedance Z_L.

where, in the general case a_{ij}, $i, j = 1, 2$, and Z_L are functions of the complex variable s. Thus, the action of the two-port is to transform the impedance Z_L to another Z_{i1} which, depending on the nature of a_{ij}, can produce very interesting and useful values.

Similarly, if the two-port is terminated at port 1 in an impedance Z_s, the input impedance Z_{i2} at port 2 will be as follows:

$$Z_{i2} \equiv \frac{V_2}{I_2} = \frac{a_{12} + a_{22}Z_s}{a_{11} + a_{21}Z_s} \qquad (3.2)$$

Thus, again the two-port acts as an impedance transformer or converter.

We may consider now the following two specific cases:

- Case a: $a_{11}, a_{22} \neq 0$ while $a_{12} = a_{21} = 0$
- Case b: $a_{11} = a_{22} = 0$ while $a_{12}, a_{21} \neq 0$

Substituting in Eq. (3.1) we get for

Case a: $\qquad Z_{i1} = \dfrac{a_{11}}{a_{22}} Z_L = G_a Z_L \qquad G_a = \dfrac{a_{11}}{a_{22}} \qquad (3.3)$

Case b: $\qquad Z_{i1} = \dfrac{a_{12}}{a_{21}} \dfrac{1}{Z_L} = G_b \dfrac{1}{Z_L} \qquad G_b = \dfrac{a_{12}}{a_{22}} \qquad (3.4)$

Similarly, substituting in Eq. (3.2) we get for

Case a: $\qquad Z_{i2} = \dfrac{a_{22}}{a_{11}} Z_s = \dfrac{1}{G_a} Z_s \qquad (3.5)$

Case b: $\qquad Z_{i2} = \dfrac{a_{12}}{a_{21}} \dfrac{1}{Z_s} = G_b \dfrac{1}{Z_s} \qquad (3.6)$

Clearly, in Case a, when $a_{12} = a_{21} = 0$, the two-port is a generalized impedance converter (GIC), when G_a, the conversion constant, is a function of s. The conversion constant is not the same for port 1 and port 2.

On the other hand, in Case b, when $a_{11} = a_{22} = 0$, the two-port is a generalized impedance inverter (GII). In this case, the inversion constant G_b is the same for port 1 and port 2.

3.3.1 Generalized Impedance Converters [3]

Following the argument presented above, we can define the GIC as the two-port for which the transmission matrix parameters a_{12} and a_{21} are zero for all s, while $a_{11} = k$ and $a_{22} = k/f(s)$, i.e.,

$$[A] = \begin{bmatrix} k & 0 \\ 0 & k/f(s) \end{bmatrix} \qquad (3.7)$$

where k is a positive constant.

Following this definition of the GIC and referring to Fig. 3.1, if the impedances Z_L and Z_s are connected across port 2 and port 1, respectively, then the input impedances Z_{i1} and Z_{i2} at ports 1 and 2 will be

$$Z_{i1} = f(s) Z_L \qquad (3.8a)$$

$$Z_{i2} = \frac{1}{f(s)} Z_s \qquad (3.8b)$$

The conversion function $f(s)$ can take any complex value realizable by an active RC two-port. However, some simple expressions of $f(s)$ have been proven to be of very high practical value in active network synthesis, as shown below.

3.3.1.1 The Ideal Active Transformer

Let $a_{11} = \pm 1/n_1$, and $a_{22} = \pm n_2$, with $n_1 \neq n_2$. Then,

$$\frac{a_{11}}{a_{22}} = \frac{1}{n_1 n_2} \qquad (3.9)$$

and the converter transforms Z_L to

$$Z_i = \frac{1}{n_1 n_2} Z_L \qquad (3.10)$$

If $n_1 = n_2 = n$, the two-port is the ideal transformer (Fig. 3.2) which, of course, is passive and reciprocal.

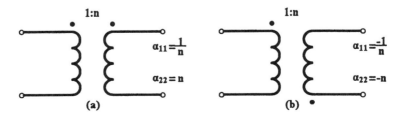

FIGURE 3.2
Ideal transformer, (a) normal and (b) reverse polarity.

Active Elements

3.3.1.2 The Ideal Negative Impedance Converter

Let
$$a_{11} = \mp k \qquad a_{22} = \pm\frac{1}{k} \qquad [f(s)] = 1$$

In either case, we have

$$\frac{a_{11}}{a_{22}} = -k^2 \qquad (3.11)$$

and

$$Z_i = -k^2 Z_L \qquad (3.12)$$

Thus, an ideal negative impedance converter (NIC) is a two-port device that presents across one of its ports the negative of the impedance that is connected across the other port within a constant k^2.

Two types of NICs can be identified according to the signs of a_{11} and a_{22}. When

$$a_{11} = -k \qquad a_{22} = \frac{1}{k}$$

the voltage negative-impedance converter (VNIC) is obtained, since then the polarity of the voltage at port 1 is reversed with respect to the polarity of the voltage at port 2.

On the other hand, when

$$a_{11} = k \qquad a_{22} = -\frac{1}{k}$$

the current negative impedance converter (CNIC) is obtained, since then the direction of the output current with respect to that of the input current is reversed.

With $k = 1$, the VNIC and CNIC of unity gain are obtained. The concept of negative resistance is explained in Section 3.4.

3.3.1.3 The Positive Impedance Converter

Let
$$f(s) = s \qquad \text{and} \qquad k = 1$$

Then, for $Z_L = R$, Eq. (3.8a) gives

$$Z_{i1} = sR \qquad (3.13)$$

Thus, terminating this GIC with a resistor makes the input impedance look like that of a grounded inductor. This is very significant in filter design, as we shall see later (Chapters 4 and 6).

This GIC, when it was first introduced [4], was called the positive-immittance converter or the PIC (*immittance* from **im**pedance and ad**mittance**).

On the other hand, for $Z_L = 1/sC$,

$$Z_{i1} = \frac{1}{C} \tag{3.14}$$

i.e., a grounded resistance of $1/C$ ohms.

Looking now at port 2, if $Z_s = R$, Eq. (3.8b) gives [$f(s) = s$],

$$Z_{i2} = \frac{R}{s} \tag{3.15}$$

i.e., the impedance of a capacitor $1/R$ farads.

The use of the PIC in filter design is explained in Chapter 6.

3.3.1.4 The Frequency-Dependent Negative Resistor [5]

As a last case, consider that

$$f(s) = \frac{1}{s}$$

Then, with $Z_L = R$,

$$Z_{i1} = \frac{R}{s}$$

i.e., the impedance of a grounded capacitor of $1/R$ farads.

However, if $Z_L = 1/sC$, then

$$Z_{i1} = \frac{1}{s^2 C} \tag{3.16}$$

Substituting $j\omega$ for s gives

$$Z_{i1}(\omega) = -\frac{1}{\omega^2 C} \tag{3.17}$$

Clearly, this is a negative resistance dependent on frequency. For this reason it is called the frequency-dependent negative resistor (FDNR) of type D (D-FDNR).

Usually, the impedance of a D-FDNR is written as $1/s^2D$ with the unit of D being farad-second. The symbol of this in a circuit is similar to that of a capacitor but with four parallel lines instead of two. For this reason the D-FDNR is sometimes referred to as **supercapacitor**.

An E-type FDNR can be obtained if a PIC is terminated at port 2 by an inductor. Then, with $Z_L = sL$ and $f(s) = s$, Eq. (3.8a) gives

$$Z_{i1} = s^2 L \tag{3.18}$$

This is sometimes called the **superinductor**, but it is not so useful in active RC filter design as the supercapacitor, as we shall see in Chapter 6.

Active Elements

The GIC is a nonreciprocal two-port, as can be easily derived from its transmission matrix. If it is loaded by the same impedance at both its ports, Z_{i1} and Z_{i2} will be different, as can be seen through Eqs. (3.8a) and (3.8b) for $Z_s = Z_L$. Depending on $f(s)$ and Z_L or Z_s, either of the ports (port one or port two) can be used to represent the component that has been obtained by the impedance conversion.

The GIC is thus a very flexible device, which can be used to simulate the transformer, negative resistance, inductance, and the D-FDNR, all of importance in filter design.

3.3.2 Generalized Impedance Inverters

The generalized impedance inverter (GIV) can be defined as the two-port with transmission parameters $a_{11}, a_{22} = 0$, and $a_{12}, a_{21} \neq 0$ for all s.

If such a two-port is terminated at one port by an impedance Z_L, the impedance Z_i seen in the other port will be

$$Z_i = \frac{a_{12}}{a_{22}} \frac{1}{Z_L} = G_b \frac{1}{Z_L} \qquad (3.19)$$

where $G_b = a_{12}/a_{21}$ can be defined as the inversion constant with units Ω^2.

G_b will be, in general, a function of s. However, in network synthesis, two cases have attracted the interest of the designers: the positive impedance inverter or gyrator and the negative impedance inverter. These are now considered.

3.3.2.1 The Gyrator

The gyrator, or positive impedance inverter, is a very attractive two-port, because it can be used to simulate inductance. Its symbol and transmission matrix are shown in Fig. 3.3.

This definition through its transmission matrix, with $g_1 \neq g_2$ and positive, refers to the active gyrator. However, if $g_1 = g_2 = g$, the gyrator is a passive two-port.

Clearly, the gyrator is a nonreciprocal two-port, since

$$a_{11}a_{22} - a_{12}a_{21} \neq 1$$

Its importance in network synthesis stems from the fact that, if it is terminated at port 2 by a capacitance C_L, the impedance seen in port 1, according to Eq. (3.19), is

$$Z_i = G_b s C_L = s \frac{C_L}{g_1 g_2} = s L_{eq} \qquad (3.20)$$

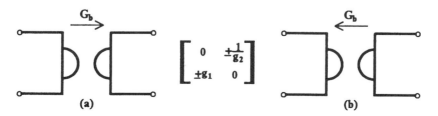

FIGURE 3.3
Symbols of gyrator: (a) $a_{12}, a_{21} > 0$, and (b) $a_{12}, a_{21} < 0$ with $G_b = (g_1 g_2)^{-1}$.

i.e., the impedance of an equivalent inductance

$$L_{eq} = \frac{C_L}{g_1 g_2} \quad (3.21)$$

The use of the gyrator in network synthesis is explained in Chapter 6, where this device is studied more rigorously.

3.3.2.2 Negative Impedance Inverter

A negative impedance inverter (NIV) is a two-port device whose input impedance at one-port is the negative reciprocal of the terminating impedance at the other port. This can be obtained, if G_b in Eq. (3.19) is equal to -1, i.e.,

$$G_b = -1$$

Then,

$$Z_i = -\frac{1}{Z_L} \quad (3.22)$$

Since, in this case, either

$$a_{12} = -1 \quad a_{21} = 1$$

or

$$a_{12} = 1 \quad a_{21} = -1$$

with $a_{11}, a_{22} = 0$, the NIV is a reciprocal two-port.

3.4 Negative Resistance

The concept of negative resistance is exciting from both theoretical and practical points of view. A negative resistance is a two-terminal device defined by the relationship between the voltage and the current in it, i.e.,

$$V = -RI \quad R > 0 \quad (3.23)$$

Its physical meaning can be explained by the fact that it absorbs negative power; therefore, it acts as an energy source.

The defining Eq. (3.23) is valid in practice for a limited range of voltages and currents, over which it can behave linearly.

In practice, negative resistance can be seen at one port of an NIC or NIV when the other port is terminated in a positive resistance. Its presence can be detected by the simple experimental setup shown in Fig. 3.4. A positive resistance of value equal to the magnitude of

Active Elements

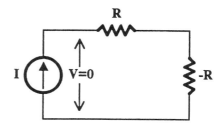

FIGURE 3.4
Demonstrating the action of a negative resistance.

the negative one is connected in series with −R. If a current is sent through this combination, the voltage measured across it is zero, in spite of the fact that the voltage drops across R and −R are nonzero.

The concept of negative resistance can be also explained through the V-I characteristics of the tunnel diode and the unijunction transistor. In these cases, incremental negative resistance appears in the part of characteristics with negative slope. There are two types of such characteristics shown in Fig. 3.5. The S-type corresponding to the V-I characteristic of the unijunction transistor and the N-type corresponding to the V-I characteristic of the tunnel diode (or the tetrode electronic tube).

The negative resistance obtained by the means explained above is supposed to be independent of frequency, and indeed this is true in practice for a range of frequencies. It is, however, possible to obtain negative resistance dependent on frequency, and this has been exploited usefully in network synthesis.

Consider the case of a GIV with an inversion "constant" ks to be terminated at port 2 by a capacitor C_L. The input impedance seen at port 1 will be, from Eq. (3.4),

$$Z_i = G_b \frac{1}{Z_L} = s^2 k C_L$$

Substituting $j\omega$ for s in this equation, we obtain

$$Z_i = -\omega^2 k C_L$$

which is, in fact, a negative resistance dependent on ω^2. This is the frequency-dependent negative resistance type E (E-FDNR) that we saw in Section 3.3.1.

The second FDNR type, type D, can be obtained if a GIC with a conversion function $f(s) = 1/s$ is terminated at port 2 by a capacitor C_L. Then, the input impedance at port 1 using Eq. (3.3), will be

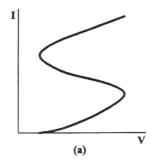
(a)

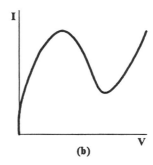
(b)

FIGURE 3.5
(a) S-type and (b) N-type V-I characteristics.

$$Z_i = G_b \frac{1}{Z_L} = s^2 k C_L$$

which for $s = j\omega$ gives

$$Z_i = -\frac{1}{\omega^2 C_L}$$

As mentioned earlier, the D-type FDNR will be realized and used in filter synthesis in Chapter 6.

3.5 Ideal Operational Amplifier

The operational amplifier, or opamp, is the most versatile active element. All active elements that have been used in active network synthesis in the past can be realized using the opamp.

The ideal opamp is an ideal differential voltage controlled voltage source (DVCVS) with infinite gain. It has infinite input impedance and zero output impedance. Its symbol and equivalent circuit are shown in Fig. 3.6. The ground connection in Fig. 3.6(a) is not generally shown.

By definition,

$$v_o = A(v_1 - v_2) \tag{3.24}$$

with v_1 applied to the noninverting input and v_2 to the inverting input.

Assuming finite v_o, infinite A calls for

$$v_1 - v_2 = 0 \tag{3.25}$$

or equivalently

$$v_1 = v_2 \tag{3.26}$$

This equality holds approximately quite satisfactorily in practice also, since the input voltage difference $v_1 - v_2$ is A times ($A \approx 10^5$) smaller than v_o, the maximum value of which can be, say, up to 10 V for IC opamps, depending on the power supply voltage.

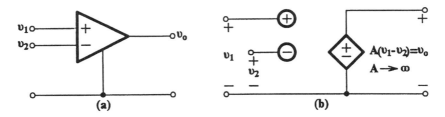

FIGURE 3.6
The ideal operational amplifier (a) symbol and (b) equivalent circuit.

Active Elements

In many cases, the noninverting input is grounded, which leads to the inverting input being at nearly earth potential, i.e.,

$$v_2 \cong 0$$

In such cases, the inverting input node of the opamp is called the virtual earth (VE) or virtual ground point.

3.5.1 Operations Using the Ideal Opamp

The infinite voltage gain of the ideal opamp, coupled with its infinite input resistance and zero output resistance, make it suitable for performing some useful mathematical operations on voltages. The most important of these operations are explained below.

3.5.1.1 Summation of Voltages

The circuit arrangement for such an operation is shown in Fig. 3.7, where the opamp is used in its single input mode.

Assuming a virtual ground at the inverting input (i.e., $V = 0$), we can write for this node

$$\frac{V_1}{R_1} + \frac{V_2}{R_2} + \ldots + \frac{V_n}{R_n} + \frac{V_o}{R_f} = 0$$

which is obtained through Kirchhoff's current law. This leads to

$$V_o = -\left(\frac{R_f}{R_1}V_1 + \frac{R_f}{R_2}V_2 + \ldots + \frac{R_f}{R_n}V_n\right) \tag{3.27}$$

If $R_f = R_1 = R_2 = \ldots = R_n$, then

$$V_o = -(V_1 + V_2 + \ldots + V_n) \tag{3.28}$$

Thus, the negative of the sum of voltages can be obtained. If the difference of two voltages is required, the arrangement in Fig. 3.8 can be used.

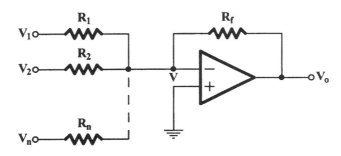

FIGURE 3.7
The opamp as a summer.

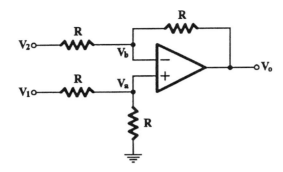

FIGURE 3.8
Circuit giving the difference of two voltages.

Clearly, we can write the following:

$$V_a = \frac{1}{2}V_1 \tag{3.29}$$

and

$$V_b = \frac{1}{2}(V_2 + V_o) \tag{3.30}$$

Since

$$V_a \cong V_b$$

using Eqs. (3.29) and (3.30), we get

$$V_o = V_1 - V_2 \tag{3.31}$$

In the case that some of the voltages in Eq. (3.28) have to be added with the opposite sign, a second opamp should be used to sum those voltages first in the manner shown in Fig. 3.7. Then this sum should be fed through the appropriate resistor to the input node of the main opamp.

3.5.1.2 Integration

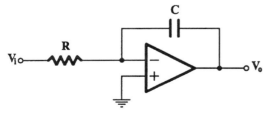

FIGURE 3.9
Integrator.

The arrangement to obtain the integration of a voltage is shown in Fig. 3.9 (if V_i and V_o are the Laplace transforms of voltages v_i and v_o, respectively.) Assuming zero initial conditions [i.e., $v_o(0) = 0$], we will have

$$\frac{v_i}{R} + C\frac{dv_o}{dt} = 0$$

from which we obtain the following:

$$v_o(t) = -\frac{1}{RC}\int_0^t v_i \, dt \qquad (3.32)$$

Initial conditions can be introduced by charging the capacitor to the appropriate voltage before starting the integration.

In the complex frequency domain, Eq. (3.32) is written as

$$V_o = -\frac{1}{RCs} V_i \qquad (3.33)$$

where V_i and V_o are the Laplace transforms of voltages v_i and v_o, respectively.

As an example, if v_i is the unit step voltage $u(t)$, from Eq. (3.33), we obtain

$$v_o(t) = -\frac{1}{RC} \cdot t \qquad (3.34)$$

It is seen that the slope of the ramp thus obtained is determined by the time constant RC.

If the position of the passive components in Fig. 3.9 is interchanged, the circuit of a differentiator results. However, in practice, such a circuit will not work properly because of excessive noise. Differentiation using an opamp in this configuration is never used. However, with a resistor of some low value connected in series with C, the noise can be reduced, but only approximate differentiation will be obtained.

3.5.2 Realization of Some Active Elements Using Opamps

The opamp can be "programmed" to realize other active elements that are useful in the synthesis of active networks. We include here some examples of such circuits, whereas others such as the gyrator, PIC, GIC, FDNR, and FDNC will be presented in Chapter 6, where they are also used in filter synthesis.

3.5.2.1 Realization of Controlled Sources

Clearly, the opamp, being a voltage-controlled voltage source in itself, is most suitable for realizing other controlled voltage sources of finite gain. In Fig. 3.10(a), the realization of a finite-gain K VCVS is shown, while in Fig. 3.10(b) that of a finite-gain CCVS is shown.

For the arrangement in Fig. 3.10(a), if V is the voltage at the inverting input of the opamp, we have

$$V = \frac{R_1}{R_1 + R_2} V_o$$

Since $V_i \cong V$ (because $A \cong \infty$), we easily deduce that

$$K \equiv \frac{V_o}{V_i} = 1 + \frac{R_2}{R_1} \qquad (3.35)$$

Similarly, for the circuit in Fig. 3.10(b), since the inverting input of the opamp is at virtual earth, we get

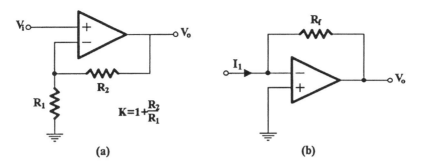

FIGURE 3.10
Realization of (a) a finite-gain VCVS and (b) a finite-gain CCVS.

$$I_1 + \frac{V_o}{R_f} = 0$$

or

$$V_o = -R_f I_1 \quad (3.36)$$

Clearly, the circuit in Fig. 3.10(b) acts as a current-to-voltage converter.

The opamp can also be used to translate voltage-to-current or current-to-current. An example of a voltage-to-current converter is shown in Fig. 3.11. The derivation of this follows the observation that

$$V_s = R_1 I$$

and consequently the current through Z is independent of Z.

3.5.2.2 Realization of Negative-Impedance Converters

As explained in Section 3.3.1, there are two types of negative-impedance converters: the current NIC and the voltage NIC. If their conversion ratio is unity, they possess the following A matrices:

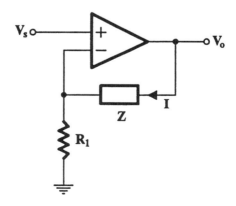

FIGURE 3.11
A voltage-to-current converter.

INIC: $$[A] = \begin{bmatrix} 1 & 0 \\ 0 & -1 \end{bmatrix} \qquad (3.37)$$

VNIC: $$[A] = \begin{bmatrix} -1 & 0 \\ 0 & 1 \end{bmatrix} \qquad (3.38)$$

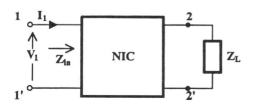

FIGURE 3.12
NIC terminated by impedance Z_L.

A NIC of either type terminated in an impedance Z_L at port 2,2' (Fig. 3.12) has an input impedance Z_{in} at port 1,1' given by

$$Z_{in} \equiv \frac{V_1}{I_1} = -Z_L \qquad (3.39)$$

If Z_L is purely resistive, the resulting negative resistance can be used to compensate for a positive or dissipative resistance of equal magnitude, thus reducing power dissipation, e.g., the copper loss in a wire transmission system. The concept of negative resistance and its types are explained in more detail in Section 3.4.

The realization of both types of NIC using an opamp is shown in Fig. 3.13(a) for the INIC and Fig. 3.13(b) for the VNIC [6]. To prove this, consider the circuit in Fig. 3.13(a) first. Clearly,

$$V_1 = V_2 \qquad I_1 = \frac{V_1 - V_0}{R}$$

Therefore,

$$I_2 = \frac{V_2 - V_0}{R} = \frac{V_1 - V_0}{R}$$

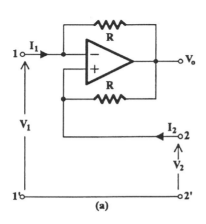 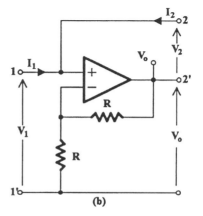

FIGURE 3.13
Opamp realization of (a) the INIC and (b) the VNIC.

or

$$I_1 = -(-I_2)$$

which is the case for an INIC.

Coming now to Fig. 3.13(b), it can be observed that

$$V_0 = 2V_1$$

$$V_2 = V_1 - V_0 = V_1 - 2V_1 = -V_1$$

$$I_1 = -I_2$$

which conforms to the transmission matrix of a VNIC.

3.5.2.3 Gyrator Realizations

A number of gyrator realizations using opamps have appeared in the literature. Some of these have been successfully used in practice [7, 8, 9]. The Orchard-Wilson gyrator [7] is a single-opamp active one ($g_1 \neq g_2$), whereas Riordan's [8] employs two opamps. Useful gyrator circuits have also been suggested by Antoniou [9].

In Fig. 3.14, the Riordan arrangement for inductance simulation is shown. Straightforward analysis, assuming ideal opamps, gives that the input impedance Z_{in} is

$$Z_{in} = \frac{V_1}{I_1} = sCR^2$$

Thus,

$$L_{eq} = CR^2$$

One of Antoniou's gyrator circuits is shown in Fig. 3.15. This is a four-terminal circuit, and it cannot be used when a three-terminal one is required. It is a very useful circuit though, because from this a useful generalized-immittance-converter circuit is obtained as shown below.

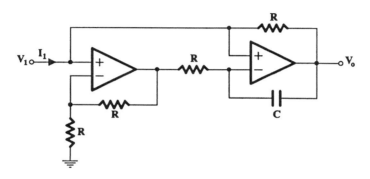

FIGURE 3.14
Riordan circuit for inductance simulation.

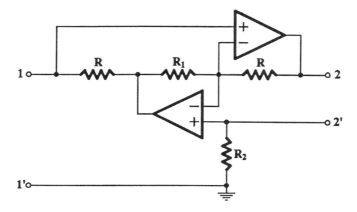

FIGURE 3.15
One of Antoniou's gyrator circuits.

Assuming $R_1 = R_2 = R$ and identical open-loop gains of the opamps, it can be shown [9] that this circuit is unconditionally (absolutely) stable, while that in Fig. 3.14 is conditionally stable.

3.5.2.4 GIC Circuit Using Opamps

The Antoniou gyrator circuit [9] that appears in Fig. 3.15 is redrawn, in its general form, in Fig. 3.16 (within the broken lines). All Y_i are admittances.

It can be seen that the voltages at nodes a, b, and c, assuming ideal opamps, are equal. Thus,

$$V_1 = V_b = V_2$$

To determine I_1 in terms of $-I_2$, we can write the following successively:

$$I_1 = Y_2 (V_1 - V_3) \tag{3.40a}$$

$$Y_3 (V_3 - V_1) = Y_4 (V_1 - V_4) \tag{3.40b}$$

$$Y_5 (V_4 - V_1) = -I_2 \tag{3.40c}$$

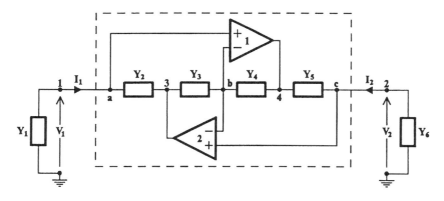

FIGURE 3.16
General GIC circuit using opamps.

From Eq. (3.40c), V_4 is obtained, which is then inserted in Eq. (3.40b) to give V_3. Next, the value of V_3 is inserted in Eq. (3.40a) to give

$$I_1 = \frac{Y_2 Y_4}{Y_3 Y_5}(-I_2) \tag{3.41}$$

Thus, the transmission matrix of the GIC is as follows:

$$[A] = \begin{bmatrix} 1 & 0 \\ 0 & \dfrac{Y_2 Y_4}{Y_3 Y_5} \end{bmatrix} \tag{3.42}$$

Consequently, the conversion function $f(s)$ and the constant take the following values:

$$k = 1 \qquad f(s) = \frac{Y_3 Y_5}{Y_2 Y_4} \tag{3.43}$$

Assuming that $Y_2 = Y_3 = Y_4 = Y_6 = R^{-1}$ and $Y_5 = sC$, the input impedance $Z_{i,1}$ at port 1 will be

$$Z_{i,1} \equiv \frac{V_1}{I_1} = f(s) \cdot \frac{1}{Y_6} = sCR^2 \tag{3.44}$$

The same result can be obtained if $Y_5 = R^{-1}$ and $Y_3 = sC$, whereas if $Y_6 = sC$, then $Z_{i,1} = R$.

If the admittance Y_1 is connected across port 1, the input impedance $Z_{i,2}$ at port 2 will be

$$Z_{i,2} \equiv \frac{V_2}{I_2} = \frac{1}{f(s)} \frac{1}{Y_1}$$

Then again, for $Y_1 = Y_2 = Y_3 = Y_4 = R^{-1}$ and $Y_5 = sC$,

$$Z_{i,2} = \frac{1}{sC} \tag{3.45}$$

while, if Y_1 is also equal to sC, then

$$Z_{i,2} = \frac{1}{s^2 C^2 R} \tag{3.46}$$

giving a supercapacitor or D-FDNR. The same results are obtained for $Z_{i,2}$ if $Y_5 = R^{-1}$ and $Y_3 = sC$.

One important point that should be noted is that, if $Y_1 = sC$ and $Y_6 = R^{-1}$ are both connected to the GIC circuit as shown in Fig. 3.16, the overall circuit will be a resonator, simulating a parallel LC circuit. Another observation concerns the connections of the opamps to the nodes of the Y-subnetwork. Inspection of Eqs. (3.40a) through (3.40c) reveals that it is immaterial which input terminal of opamp 1 is connected to node a and which to node b. The same is true for opamp 2. We will make use of this circuit in Chapters 4 and 6 to simulate inductance.

3.5.3 Characteristics of IC Opamps

Practical opamps have characteristics that differ from those of the ideal element used in previous sections. Apart from their open-loop voltage gain, which is noninfinite, their input impedance and output admittance are not infinite either. There are also some additional parameters associated with the operation of the practical opamp [10] which degrade its performance, and the designer should always keep them in mind. In spite of these, though, the nonideal behavior of the practical opamp does not prevent it from being the most versatile linear active element in use today.

3.5.3.1 Open-Loop Voltage Gain of Practical Opamps

The dc and very low frequency open-loop voltage gain of most IC bipolar opamps is of the order of 10^5 (100 dB), and for MOS opamps at least one order of magnitude lower. In most practical cases, the error introduced in circuits incorporating opamps is not very significant, and the operation at dc can still be considered ideal. This, however, is not true at frequencies above a few hertz.

For reasons of stability of the circuits in which the opamp is embedded, its magnitude response is shaped such that the falloff rate is 6 dB/octave, as shown in Fig. 3.17(a). The associated phase response is shown in Fig. 3.17(b). This behavior can be described mathematically as follows:

$$A(s) = \frac{A_o}{1 + s\tau} \quad (3.47)$$

where A_o is the dc gain and τ a time constant that creates a pole at $-1/\tau$. The cutoff frequency is

$$\omega_c = 2\pi f_c = \frac{1}{\tau}$$

with f_c equal to about 10 Hz for general-purpose bipolar IC opamps such as, e.g., the 741. The frequency f_T at which the magnitude of $A(j\omega)$ becomes unity is the most important characteristic of each opamp, since it actually denotes its gain-bandwidth (GB) product. We can explain this as follows.

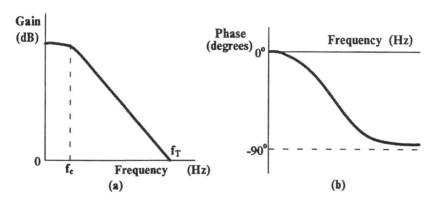

FIGURE 3.17
(a) Magnitude and (b) phase response of the opamp (741 type) open-loop voltage gain.

With $\omega_c = 1/\tau$, Eq. (3.47) gives

$$A(s) = \frac{A_o}{1 + s/\omega_c}$$

and so

$$|A(j\omega)| = \frac{|A_o|}{\left[1 + \left(\frac{\omega}{\omega_c}\right)^2\right]^{1/2}}$$

Thus, when $|A(j\omega)| = 1$,

$$\left[1 + \left(\frac{\omega_T}{\omega_c}\right)^2\right]^{1/2} = |A_o|$$

and since $(\omega_T/\omega_c)^2 \gg 1$, we have

$$\frac{\omega_T}{\omega_c} \cong |A_o|$$

whence

$$\omega_T = |A_o|\omega_c$$

i.e., the gain-bandwidth product—the product of the dc gain and the 3-dB bandwidth. Note that if $\omega_T = 2\pi f_T$, then $f_T = |A_o| f_c$ also.

Clearly, the high gain of the opamp is not available at frequencies higher than about 10 Hz. It should also be mentioned that, as it can be easily obtained from Eq. (3.47), the maximum gain that can be obtained at the frequency f_x using the opamp is

$$|A(j\omega_x)| = \frac{f_T}{f_x} \qquad (3.48)$$

This model will be taken into consideration whenever we examine the performance of various circuits using opamps throughout this book.

The effect of this single-pole model of the opamp on the performance of the VCVS realized using the opamp is examined in Section 3.5.4. In some cases, when we are interested in frequencies well above f_c, this single-pole model can be simplified by writing Eq. (3.47) in the following approximate form:

$$A(s) \cong \frac{\omega_T}{s} \qquad (3.49)$$

3.5.3.2 *Input and Output Impedances*
The input impedance of the opamp can be defined when measured either between each input terminal and the ground, or, as differential, i.e., between the two input terminals.

Although a function of frequency, it is usually considered purely resistive, R_i. The value of R_i is around 150 kΩ for bipolar opamps, while for opamps using FETs as the input stage or MOSFETs throughout, it is very high indeed. However, even in the case of bipolar opamps, since they are always used with negative feedback, the error introduced due to its presence is insignificant and therefore can be neglected in practice. Similarly, the output impedance of a bipolar general-purpose IC opamp is of the order of 100 Ω, which when the opamp is used with negative voltage feedback introduces an insignificant error, usually ignored in practice.

To be sure that these impedances will not affect the performance of the circuit using bipolar IC opamps, the impedance level of the associated circuit should be chosen greater than 1 kΩ and smaller than 100 kΩ, with 10 kΩ being the most appropriate choice. The upper limit is set by other imperfections of the opamp, which are explained below.

3.5.3.3 Input Offset Voltage V_{IO}

If both inputs of the real opamp are grounded, the output voltage will not be zero in practice, as would be expected. This is a defect that causes the output voltage to be offset with respect to ground potential. For large ac input signals, the output voltage waveform will then be unsymmetrically clipped; that is, the opamp will display a different degree of non-linear behavior for positive and negative excursions of the input signals. The input offset voltage V_{IO} is that voltage which must be applied between the input terminals to balance the opamp. In many opamps, this defect may be "trimmed" to zero by means of an external potentiometer connected to terminals provided for this reason.

3.5.3.4 Input Offset Current I_{IO}

This is defined as the difference between the currents entering the input terminals when the output voltage is zero. These currents are actually the base bias currents of the transistors at the input stage of the opamp (for bipolar opamps), and their effect is the appearance of an undesired dc voltage at the output. This defect of the opamp can be modeled by connecting two current generators at the input terminals of the ideal opamp. This is shown in Fig. 3.18 for the case of the circuit in Fig. 3.11, which is used to provide $1 + R_f/R_1$ voltage gain. R_2 is inserted to reduce the effect of the input bias currents as we show below and has no effect on the signal. If the input voltage V_s is zero, and assuming linear operation of the opamp, we may observe the following.

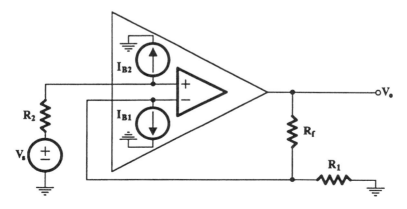

FIGURE 3.18
Current sources I_{B1}, I_{B2} represent the presence of input offset currents.

The action of I_{B1} causes the output voltage to be ($I_{B2} = 0$)

$$V_{01} = R_f I_{B1} \qquad (3.50)$$

The action of I_{B2}, assuming $I_{B1} = 0$, will result in the output voltage

$$V_{02} = -\left(1 + \frac{R_f}{R_1}\right) I_{B2} R_2 \qquad (3.51)$$

Then applying superposition when both I_{B1} and I_{B2} are present, we get the output voltage

$$V_0 = V_{01} + V_{02} = I_{B1} R_f - \left(1 + \frac{R_f}{R_1}\right) I_{B2} R_2 \qquad (3.52)$$

For this voltage to be zero when $I_{B1} = I_{B2}$, which is the optimistic case, the following relationship between the resistor values should hold:

$$R_2 = \frac{R_1 R_f}{R_1 + R_f} \qquad (3.53)$$

However, even under this condition, when $I_{B1} \neq I_{B2}$, the output voltage will be

$$V_0 = (I_{B1} - I_{B2}) R_f = I_{IO} R_f \qquad (3.54)$$

i.e., nonzero. Note though that without R_2, $V_o = I_{B1} R_f$ and, since $I_{IO} \ll I_{B1}$ in practice, the output voltage arising from the input bias currents is reduced by including R_2.

3.5.3.5 Input Voltage Range V_I

Assuming that the imperfections of the opamp due to input offset voltage and input offset current have been corrected, the voltage transfer characteristic of the amplifier will be as shown in Fig. 3.19, where V_i represents the differential input voltage. It can be seen that the opamp behaves linearly only in the region of V_i.

$$-V_2 \leq V_i \leq V_1$$

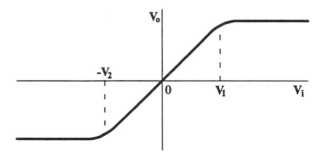

FIGURE 3.19
The saturation characteristics of the opamp.

i.e., only for this range of V_i can one get the benefit of the full voltage gain of the opamp. Beyond this voltage range, the amplifier goes to saturation.

Although the nonlinear behavior of the opamp is a cause of concern in the design of active RC filters, one can get advantage of the saturation characteristic to build analog voltage comparators, which are very useful in practice (for example, as zero crossing detectors).

3.5.3.6 Power Supply Sensitivity $\Delta V_{IO}/\Delta V_{GG}$

This is the ratio of the change of the input offset voltage ΔV_{IO} to the change in the power supply ΔV_{GG} that caused it. The change in the power supply is considered symmetrical.

3.5.3.7 Slew Rate SR

The rate of change of the output voltage cannot be infinite due to the various internal time constants of the opamp circuitry. The slew rate (SR) is defined as the maximum rate of change of the output voltage for a unit step input excitation. This is normally measured for unity gain at the zero voltage point of the output waveform.

The SR sets a serious limitation to the amplitude of the signal at high frequencies. This can be shown in the case of a sinewave as follows. Let

$$v_o = V_m \sin \omega t$$

Then,

$$\frac{dv_o}{dt} = V_m \omega \cos \omega t$$

which becomes maximal at the zero crossing points, i.e., when $\omega t = 0, \pi, 2\pi,\ldots$. Thus, at $\omega t = 0$

$$\left.\frac{dv_o}{dt}\right|_{\omega t = 0} = V_m \omega \qquad (3.55)$$

Since this cannot be larger than the SR, i.e.,

$$SR \geq V_m \omega \qquad (3.56)$$

it is clear, that for linear operation at a high frequency ω, the amplitude of the output voltage cannot be greater than SR/ω. Thus, at high frequencies, the opamp cannot work properly at its full input voltage swing, as it does at low frequencies.

3.5.3.8 Short-Circuit Output Current

This denotes the maximum available output current from the opamp, when its output terminal is short circuited with the ground or with one of its power supply rails.

3.5.3.9 Maximum Peak-to-Peak Output Voltage Swing V_{opp}

This is the maximum undistorted peak-to-peak output voltage, when the dc output voltage is zero.

3.5.3.10 Input Capacitance C_i

This is the capacitance between the input terminals with one of them grounded.

3.5.3.11 Common-Mode Rejection Ratio CMRR

Ideally, the opamp should reject completely all common-mode signals (i.e., the same signals applied to both inputs) and amplify the differential-mode ones. However, for reasons of circuit imperfections, the amplifier gain is not exactly the same for both of its inputs. The result of this is that common-mode signals are not rejected completely. A measure of this imperfection is the common-mode rejection ratio (CMRR). Expressed in dB, the CMRR is the ratio of open-loop differential gain to the corresponding common-mode gain of the opamp. Its value at low frequencies is typically better than 80 dB, but it decreases at higher frequencies.

3.5.3.12 Total Power Dissipation

This is the total dc power that the opamp absorbs from its power supplies, minus the power that the amplifier delivers to its load.

3.5.3.13 Rise Time t_r

This is the time required for the output voltage of the amplifier to increase from 10 to 90 percent of its final value for a step input voltage. It can be shown simply that $t_r \times f_c \cong 0.35$ (see Section 1.4.1).

3.5.3.14 Overshoot

This is the maximum deviation of the output voltage above its final value for a step input excitation.

3.5.4 Effect of the Single-Pole Compensation on the Finite Voltage Gain Controlled Sources

Consider the two circuits in Fig. 3.20. For an ideal opamp, the voltage gains of these circuits are the following:

Fig. 3.20(a) (noninverting) $\qquad \alpha_n = \dfrac{R_1 + R_2}{R_1}$

Fig. 3.20(b) (inverting) $\qquad -\alpha_i = -\dfrac{R_f}{R_1}$

Assuming that the open-loop gain A of the opamp is finite, the voltage gain $G_N(s)$ of the circuit in Fig. 3.20(a) is written as follows:

$$G_N \equiv \frac{V_o}{V_i} = \frac{A}{1 + \beta A} \qquad (3.57)$$

Active Elements

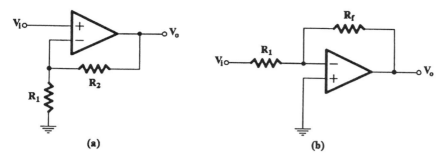

FIGURE 3.20
(a) Noninverting and (b) inverting voltage amplifiers using the opamp.

where β is the feedback ratio given by

$$\beta = \frac{R_1}{R_1 + R_2} \tag{3.58}$$

If A follows the single-pole model, i.e.,

$$A(s) = \frac{A_o \omega_c}{s + \omega_c} \tag{3.59}$$

substituting for A in Eq. (3.58), we get

$$G_N = \frac{A_o \omega_c}{s + \omega_c + \beta A_o \omega_c} \cong \frac{A_o \omega_c}{s + \beta A_o \omega_c} = \frac{A_o \omega_c}{s + \dfrac{A_o \omega_c}{\alpha_N}} \tag{3.60}$$

where we assumed that $\beta A_o \gg 1$, which is quite reasonable in practice.

Applying the same procedure in the case of Fig. 3.20(b), we can obtain, for the gain G_I, the following:

$$G_I \equiv \frac{V_o}{V_i} = -\frac{R_f}{R_1 + R_f} \frac{A}{1 + \beta A} \tag{3.61}$$

where

$$\beta = \frac{R_1}{R_1 + R_f}$$

Then, substituting for A from (3.59), and after some arithmetic manipulations, we obtain

$$G_I = -\alpha_I \frac{\dfrac{A_o \omega_c}{1 + \alpha_I}}{s + \dfrac{A_o \omega_c}{1 + \alpha_I}} \tag{3.62}$$

It can be seen from Eqs. (3.60) and (3.62) that both G_N and G_I have a single-pole behavior. This is to be expected, since $A(s)$ behaves similarly. However, the unexpected is that for equal nominal gains at low frequencies, i.e.,

$$\alpha_N = \alpha_I$$

the useful bandwidth of G_N is larger than that of G_I. In particular, when

$$\alpha_N = \alpha_I = 1$$

the bandwidth of the noninverting amplifier is double the bandwidth of the inverting one.

3.6 The Ideal Operational Transconductance Amplifier (OTA)

The ideal OTA is a differential-input voltage-controlled current source (DVCCS). Its symbol is shown in Fig. 3.21(a), and its operation is defined by the following equation:

$$I_o = g_m(V_1 - V_2) \tag{3.63}$$

The transconductance g_m can be controlled externally by the current I_B. Both voltages V_1 and V_2 are with reference to ground.

The equivalent circuit of the ideal OTA is shown in Fig. 3.21(b). Some simple applications of the OTA are described below [11].

3.6.1 Voltage Amplification

Inverting and noninverting voltage amplification can be achieved using an OTA as shown in Fig. 3.22(a) and 3.22(b), respectively. Any desired gain can be achieved by a proper choice of g_m and R_L. It should be noted that the output voltage V_o is obtained from a source with output impedance equal to R_L. Zero output impedance can be achieved only if such circuits are followed by a buffer or voltage follower.

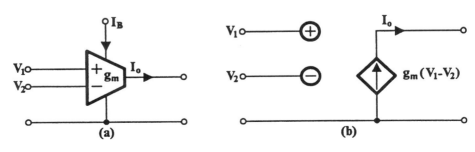

FIGURE 3.21
Ideal operational transconductance amplifier, (a) symbol and (b) equivalent circuit.

Active Elements

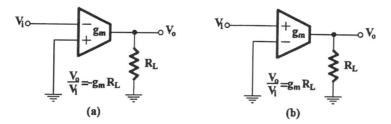

FIGURE 3.22
(a) Inverting and (b) noninverting voltage gain using an ideal OTA.

3.6.2 A Voltage-Variable Resistor (VVR)

A grounded voltage-variable resistor can be easily obtained using the ideal OTA as shown in Fig. 3.23. Since $I_o = -I_i$, we will have the following:

$$Z_i = \frac{V_i}{I_i} = \frac{V_i}{-I_o} = \frac{V_i}{g_m V_i} = \frac{1}{g_m} \tag{3.64}$$

Using two such arrangements cross-connected in parallel, a floating VVR can be obtained. On the other hand, if in Fig. 3.23 the input terminals are interchanged, the input resistance will be $-1/g_m$. Thus, using OTAs, both positive and negative resistors become available without actually having to build them on the chip. These, coupled with capacitors, lead to the creation of the so-called active-C filters discussed later in this book.

3.6.3 Voltage Summation

Voltage summation can be obtained using OTAs, which in effect translate voltages to currents. These are easily summed as shown in Fig. 3.24 for two voltages V_1 and V_2.

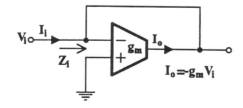

FIGURE 3.23
Grounded voltage-variable resistor.

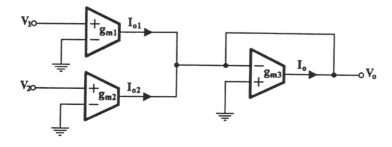

FIGURE 3.24
Voltage summation.

It is clear that

$$I_{01} + I_{02} + I_0 = 0$$

or

$$g_{m1}V_1 + g_{m2}V_2 - g_{m3}V_0 = 0$$

Solving for V_o, we get

$$V_o = \frac{g_{m1}}{g_{m3}}V_1 + \frac{g_{m2}}{g_{m3}}V_2 \qquad (3.65)$$

By changing the grounded input of one of the input OTAs, voltage subtraction can be achieved. These operations are useful for the realization of transfer functions.

3.6.4 Integration

The operation of integration can be achieved very conveniently using the OTA as is shown in Fig. 3.25. Clearly,

$$V_o = \frac{I_o}{sC} = \frac{g_m}{sC}(V_1 - V_2) \qquad (3.66)$$

It follows that both inverting and noninverting integration is easily achieved. Of course, in all cases, the output impedance of the circuit is nonzero.

If a resistor is connected in parallel with C in Fig. 3.25, the integration will become lossy. On the other hand, connecting the circuit in Fig. 3.23 at the output of that in Fig. 3.25, the integration becomes both lossy and adjustable.

3.6.5 Gyrator Realization

The defining equations of the gyrator can be written in the Y matrix form as follows:

$$\begin{bmatrix} I_1 \\ I_2 \end{bmatrix} = \begin{bmatrix} 0 & g_1 \\ -g_2 & 0 \end{bmatrix} \begin{bmatrix} V_1 \\ V_2 \end{bmatrix} \qquad (3.67)$$

These equations can be interpreted in the form of an equivalent circuit comprising two voltage controlled current sources connected as shown in Fig. 3.26. Thus OTAs, being voltage controlled current sources, are most suitable for the realization of the gyrator—more

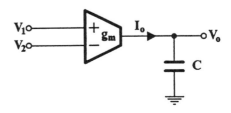

FIGURE 3.25
Integration of the difference of voltages V_1, V_2.

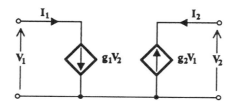

FIGURE 3.26
Gyrator realization using two VCCSs.

suitable than opamps, which are voltage controlled voltage sources. Such a circuit using OTAs is shown in Fig. 3.27. Clearly,

$$I_2 = g_{m1}V_1 \quad \text{and} \quad I_1 = -I_3 = g_{m2}V_2$$

Then the A matrix, not considering Z_L as part of the circuit, will be

$$[A] = \begin{bmatrix} 0 & \dfrac{1}{g_{m1}} \\ g_{m2} & 0 \end{bmatrix} \tag{3.68}$$

Thus, the circuit consisting of the two OTAs realizes, in general, the ideal active gyrator ($g_{m1} \neq g_{m2}$). In case $g_{m1} = g_{m2}$, the gyrator behaves as a passive circuit.

With Z_L connected as shown in Fig. 3.27, the input impedance Z_i is as follows:

$$Z_i \equiv \frac{V_1}{I_1} = \frac{1}{g_{m1}g_{m2}Z_L} \tag{3.69}$$

If Z_L represents the impedance of a capacitor C_L, the equivalent inductance L_{eq} will be

$$L_{eq} = \frac{C_L}{g_{m1}g_{m2}} \tag{3.70}$$

3.6.6 Practical OTAs

The versatility of the OTA as an active element, as demonstrated above, makes it very useful in VLSI circuits. Also, discrete IC OTAs in bipolar and MOS technology are available.

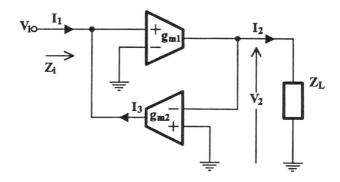

FIGURE 3.27
Gyrator realization using two OTAs.

These practical ICs have certain advantages over opamps as well as disadvantages. Their advantages include higher bandwidths and simpler circuitry. The former make them more useful than the opamps in the design of active filters operating at high frequencies (up to the megahertz region). On the other hand, their simpler circuitry, coupled with the controllability of their g_m, leads to versatility in integration and tuning.

However, they have some drawbacks. Currently available IC OTAs have a performance that is limited by certain imperfections, some of which are similar to those explained in the case of practical IC opamps. Some additional ones, though, need more attention. One important imperfection is the limited range of input voltage (< 20 mV) for linear operation [12, 13]. This problem can be solved by using a potential divider at the input terminals in order to reduce the differential input voltage. This divider, however, reduces the effective input impedance of the OTA.

Other important imperfections include the finite input and output impedances of the OTA, as well as the frequency dependence of transconductance g_m [12, 13].

The input impedance can be modeled by connecting a resistance R_{ic} in parallel with a capacitance C_{ic} from each input terminal of the ideal OTA to ground and a capacitance C_{id} in parallel with a resistance R_{id} between the input terminals. When one of the input terminals is grounded the input impedance is simplified being the parallel combination of the resistances R_{ic}, R_{id} and the capacitance $C_{ic} + C_{id}$.

The OTA output impedance is modeled by the parallel combination of a resistance R_o and a capacitance C_o connected between the OTA output terminal and the ground.

Finally, the frequency dependence of the OTA transconductance can be approximately described by a single pole model given by

$$g_m(s) = \frac{g_{mo}}{1+s\tau} \quad (3.71)$$

where g_{mo} is the value of g_m at dc, and $\tau = 1/\omega_b$, ω_b being the OTA finite bandwidth.

Also, the phase model is often used, which is described as follows:

$$g_m(j\omega) = g_{mo}e^{-j\varphi} \quad (3.72)$$

In this equation, $\varphi = \omega\tau$ is the phase delay with $\tau = 1/\omega_b$ giving the time delay.

Both Eqs. (3.71) and (3.72) can be further approximately written as

$$g_m(s) \approx g_{mo}(1-s\tau) \quad (3.73)$$

These OTA g_m models will be used alternatively in later chapters where the OTA is used in filter design as the active element.

In spite of all these imperfections, though, careful design can minimize their effect on the available bandwidth, which remains much higher than that of an opamp. This makes OTAs very useful for the design of active filters at high frequencies, as shown later in this book.

3.6.7 Current Conveyor [14]

The current conveyor (CC) is a three-port active element classified as a current mode device. We introduce the ideal element here, briefly. The definition of the ideal CC type 1 (CCI) is given by means of the following mathematical description with reference to its symbol, shown in Fig. 3.28:

Active Elements

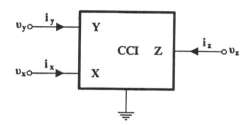

FIGURE 3.28
Symbol of current conveyor, CCI.

$$\begin{bmatrix} i_y \\ v_x \\ i_z \end{bmatrix} = \begin{bmatrix} 0 & 1 & 0 \\ 1 & 0 & 0 \\ 0 & \pm 1 & 0 \end{bmatrix} \begin{bmatrix} v_y \\ i_x \\ v_z \end{bmatrix} \quad (3.74)$$

This is the earlier version of the current conveyor, which was followed later by the type-II current conveyor, CCII, mathematically defined by the following equation:

$$\begin{bmatrix} i_y \\ v_x \\ i_z \end{bmatrix} = \begin{bmatrix} 0 & 0 & 0 \\ 1 & 0 & 0 \\ 0 & \pm 1 & 0 \end{bmatrix} \begin{bmatrix} v_y \\ i_x \\ v_z \end{bmatrix} \quad (3.75)$$

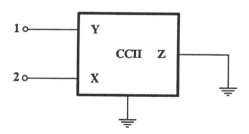

FIGURE 3.29
CCII realizing an ideal VCVS.

The CCII is a more versatile device than the CCI. By means of this, all previously introduced active one- and two-port active elements can be realized. As a first example, consider the situation shown in Fig. 3.29.

If terminal 1 and earth constitute the input port, and terminals 2 and earth the output port, from Eq. (3.75) we can easily obtain the following transmission matrix:

$$[T] = \begin{bmatrix} 1 & 0 \\ 0 & 0 \end{bmatrix}$$

which clearly is that of an ideal voltage-controlled voltage source.

As a second example, consider the situation in Fig. 3.30. With $Y(Z)$ and earth representing the input terminals, and X and earth the output terminals, we can easily obtain from Eq. (3.75) the following:

$$\begin{bmatrix} v_1 \\ i_1 \end{bmatrix} = \begin{bmatrix} v_2 \\ -i_z \end{bmatrix} = \begin{bmatrix} 1 & 0 \\ 0 & -1 \end{bmatrix} \begin{bmatrix} v_x \\ -i_x \end{bmatrix} \quad (3.76)$$

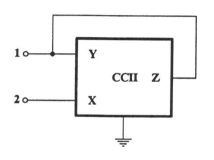

FIGURE 3.30
CCII+ realizing an NIC.

Clearly, this equation describes the ideal unity gain INIC, as was shown in Section 3.3.1.

We will consider this active device again in Chapter 12. The reader interested in its applications to filter design may also be referred for example to [15].

3.7 Summary

In continuous-time active filter design, we employ active elements and passive components. The active elements are mostly two-port active building blocks like controlled sources, NICs, GICs, PICs, gyrators, etc. They all can be realized using opamps and/or OTAs. However, opamps and OTAs, themselves being special types of controlled sources, are also used as active elements on their own right.

Because of the importance of these two amplifiers in active filter design, understanding of their imperfections is absolutely necessary. This can help the designer to avoid problems that will surely arise in practical circuits if they are not taken into consideration.

The active elements, which were introduced in this chapter, will be employed in the design of active filters in all subsequent chapters.

References

[1] T. Deliyannis. 1966. "Synthesis procedures and sensitivity studies for two types of active RC networks," PhD thesis, University of London.

[2] S. K. Mitra. 1969. *Analysis and Synthesis of Linear Active Networks*, New York: John Wiley & Sons.

[3] A. Antoniou. 1970. "Novel RC-active network synthesis using generalized-immittance converters," *IEEE Trans on Circuit Theory*, CT-17, pp. 212–217.

[4] J. Gorski-Popiel. 1967. "RC-active synthesis using positive-immittance converters," *Electron. Lett.* 3, pp. 381–382.

[5] L. T. Bruton. 1980. *RC-Active Circuits, Theory and Design*, Englewood Cliffs, NJ: Prentice-Hall.

[6] A. Antoniou. 1965. "Negative-impedance converters using operational amplifiers," *Electron. Lett.* **1**, p. 88.

[7] H. J. Orchard and A.N. Wilson Jr. 1974. "New active-gyrator circuit," *Electron. Lett.* 27.

[8] R.H.S. Riordan 1967. "Simulated inductors using differential amplifiers," *Electron. Lett.* 3, pp. 50–51.

[9] A. Antoniou. 1969. "Realization of gyrators using operational amplifiers and their use in RC-active network synthesis," *Proc. IEE* **116**, pp. 1838–1850.

[10] Various manufacturers, specifications books on operational amplifiers.

[11] R.L. Geiger and E. Sanchez-Sinencio. 1985. "Active filter design using operational transconductance amplifiers: A tutorial," *IEEE Circuits and Devices* 1, pp. 20–32.

[12] R. Schaumann, M. S. Gausi and K. R. Laker. 1990. *Design of Analog Filters: Passive, Active RC and Switched Capacitor*, Prentice-Hall Intl. Editions.

[13] K.R. Laker and W. M. C. Sansen. 1994. *Design of Analog Integrated Circuits and Systems*, McGraw-Hill Intl. Editions.

[14] K.C. Smith and A. Sedra. 1970. "A second generation current conveyor and its applications," *IEEE Trans. Circuit Theory* CT-17, pp. 132–134.

[15] C. Toumazou, F. J. Lidgey, and D.G. Haigh, eds. 1990. *Analogue IC Design: The Current-Mode Approach*, London: Peter Peregrinus.

Chapter 4

Realization of First- and Second-Order Functions Using Opamps

4.1 Introduction

All active elements, which were introduced in the previous chapter, are useful in the realization of filter functions, although some of them are more useful than others. Among these, the opamp is the most versatile active element in use today up to frequencies of the order of 100 kHz. As was shown in the previous chapter, it can be used to realize other active elements, e.g., all controlled-sources, GICs, etc., but it can also be used as a high (in theory infinite) gain amplifier on its own right. Both of these aspects of the opamp employment in filter design are exploited in this chapter.

As we shall see in the next chapter, one useful method of designing high-order filters is to cascade first- and second-order stages. Usually, one first-order stage will be required only when the filter function is of odd order, the remaining stages being of second-order. In another method of high-order filter design, multiple feedback is applied in the cascade connection of low-order stages. If the filter function is lowpass or highpass only first-order stages will be required whereas, if it is bandpass, bandreject, or allpass, all cascaded stages will be of second-order. It is therefore fully justified to study first- and second-order stages using the opamp(s) as their active element(s).

An abundance of circuits realizing, in particular, second-order functions have been proposed over the past 30 to 35 years, some of which are more "suitable" than others. Criteria of suitability are usually set by the filter designer according to the problem at hand; however, some are clearly objective. Among these, the following three will be of main concern to us in this chapter, namely, (a) the possibility that the circuit can realize the specific second-order function, (b) its sensitivity to component value variations (defined in Section 4.4), and (c) its cost (number and tolerance of its components both passive and active). Based on these criteria we have chosen to include in this chapter only a small number of second-order circuits. This does not necessarily imply that these circuits are the best in all cases, but the filter designer should not ignore their existence and usefulness.

4.2 Realization of First-Order Functions

Alongside its use in the integrator circuit in Fig. 3.9, the opamp can be also used in the realization of other first-order transfer functions, which are useful in filter design. Such functions are lowpass, highpass, and allpass. Although the lowpass and highpass functions can be realized using RC circuits only, the presence of the opamp in the circuit can provide it with gain and isolation from the circuit that follows it. Thus, the presentation of these circuits here is justifiable.

4.2.1 Lowpass Circuits

A first-order lowpass circuit is shown in Fig. 4.1(a). Its transfer voltage ratio can easily be shown to be

$$H(s) \equiv \frac{V_o}{V_i} = -\frac{1/CR_1}{s + 1/CR_2} \quad (4.1)$$

Thus, the circuit can realize the transfer function

$$F(s) = \frac{a}{s+b} \quad (4.2)$$

with

$$a = -\frac{1}{CR_1}, b = \frac{1}{CR_2}$$

The dc gain is

$$\frac{a}{b} = -\frac{R_2}{R_1}$$

and can be adjusted to any desired practical value.

An alternative circuit is that shown in Fig. 4.1(b), which does not introduce a phase inversion (sign reversal) as does that in Fig. 4.1(a). Its transfer function is

$$H(s) \equiv \frac{V_o}{V_i} = \frac{K/CR}{s + 1/CR} \quad (4.3)$$

where

$$K = 1 + \frac{R_2}{R_1}$$

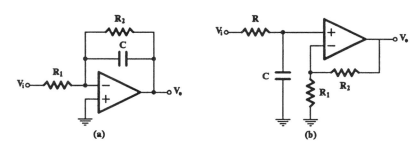

FIGURE 4.1
(a) A first-order lowpass circuit and (b) an alternative circuit.

Equating similar coefficients in Eqs. (4.2) and (4.3) we obtain the following design relationships:

$$a = \frac{K}{CR}, \quad b = \frac{1}{CR}, \quad \text{dc gain } \frac{a}{b} = K \quad (4.4)$$

4.2.2 Highpass Circuits

Both circuits in Fig. 4.2 can realize the first-order highpass function

$$F(s) = \frac{as}{s+b} \quad (4.5)$$

The transfer function of the circuit in Fig. 4.2(a) is

$$H(s) = -\frac{R_2}{R_1} \frac{s}{s + 1/CR_1} \quad (4.6)$$

Therefore, following coefficient matching, we get from Eqs. (4.5) and (4.6) the design equations as follows:

$$a = -\frac{R_2}{R_1}, \quad b = \frac{1}{CR_1} \quad (4.7)$$

Clearly, the value of one component will have to be selected arbitrarily.
Similarly, the transfer function of the circuit in Fig. 4.2(b) is

$$H(s) = \frac{Ks}{s + 1/(RC)} \quad (4.8)$$

and, after coefficient matching, we obtain

$$a = k = 1 + \frac{R_b}{R_a}, \quad b = \frac{1}{CR} \quad (4.9)$$

Here, two components should have their values arbitrarily selected.

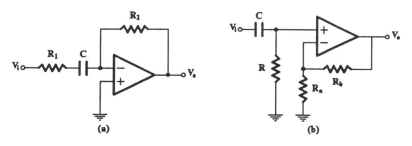

FIGURE 4.2
Two alternative first-order highpass circuits.

4.2.3 Allpass Circuits

Two circuits [1] suitable for the realization of the first-order allpass function,

$$F(s) = \frac{s-a}{s+a} \tag{4.10}$$

are shown in Fig. 4.3.

The transfer function of the circuit in Fig. 4.3(a) is the following:

$$H(s) = \frac{V_o}{V_i} = -\frac{s - 1/CR}{s + 1/CR} \tag{4.11}$$

The same holds for the circuit in Fig. 4.3(b), but without the negative sign in front. For both circuits, coefficient matching gives the following design relationship:

$$a = \frac{1}{CR} \tag{4.12}$$

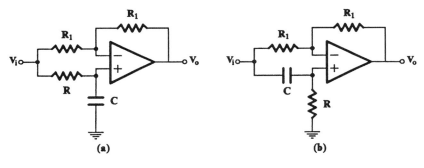

FIGURE 4.3
Two alternative first-order allpass circuits.

4.3 The General Second-Order Filter Function

The second-order filter function, in its general form, is the following:

$$F(s) = \frac{a_2 s^2 + a_1 s + a_0}{s^2 + \beta s + \gamma} \tag{4.13}$$

Realization of this function using active RC networks is of interest only in the case that

$$\sqrt{\gamma} > 0.5\beta$$

i.e., when the poles of $F(s)$ are complex conjugate. Otherwise, when the poles are negative real, the realization can be achieved using passive RC networks only.

$F(s)$ can be written alternatively in general "biquadratic" form as follows:

$$F(s) = K \frac{s^2 + \frac{\omega_{oz}}{Q_z}s + \omega_{oz}^2}{s^2 + \frac{\omega_{op}}{Q_p}s + \omega_{op}^2} \quad (4.14)$$

In Eq. (4.14), ω_{oz} and ω_{op} are the undamped natural frequencies of the zeros and poles respectively, while Q_z and Q_p are the corresponding quality factors, or Q factors. The zero or pole frequency is the magnitude of the zero or pole, respectively, while their quality is a measure of how near the $j\omega$-axis is the corresponding zero or pole in the s-plane.

Comparing Eq. (4.14) to Eq. (4.13), we get

$$\omega_{op} = \sqrt{\gamma} \quad (4.15)$$

$$Q_p = \frac{\sqrt{\gamma}}{\beta} \quad (4.16)$$

It is common to use ω_o instead of ω_{op} and Q instead of Q_p, and we will adopt these symbols here, too. In the case of the second-order bandpass filter, ω_o coincides with the filter center frequency (when the magnitude response is plotted on a log frequency scale), while the Q factor determines the relative width of the frequency response with respect to the center frequency.

Depending on the positions of the zeros on the s-plane, the second-order filter frequency responses of interest, obtained from Eq. (4.13), are as shown in Fig. 4.4. We will examine below the most useful practical active RC circuits realizing these functions, taking into consideration mainly the criterion of low sensitivity, which we introduce first.

4.4 Sensitivity of Second-Order Filters

The term *sensitivity* is used to express the degree of influence of a variation in the value of one (or more) component on the performance of the circuit in which it is embedded. Variations in the component values may be due to one or more of the following reasons:

- changes in the environmental conditions, e.g., temperature
- component aging
- component substitution due to failure
- component tolerances during the production of the circuit

The less sensitive the circuit is to component variations, the more stable its characteristics will be and, thus, the more likely it will be to remain within its specifications, regardless of these changes. Therefore, sensitivity is a major factor in determining how useful a circuit can be in practice.

Sensitivity measures of greatest interest in the case of second-order filters are introduced below. Other sensitivity measures, useful mainly in the case of high-order circuits, are presented in the next chapter.

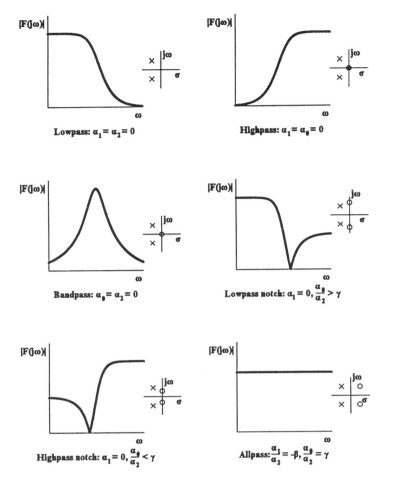

FIGURE 4.4
Frequency responses of various useful second-order filter functions.

Let us suppose that we are interested in the variation of the filter characteristics due to a change in the value of the circuit element x. Writing

$$H(s) = H(s, x) \tag{4.17}$$

we define the relative sensitivity of $H(s)$ to x as

$$S_x^H \equiv \frac{\partial \ln H}{\partial \ln x} \tag{4.18}$$

which can also be written as follows:

$$S_x^H \equiv \frac{\partial H/H}{\partial x/x} = \frac{x}{H}\frac{\partial H}{\partial x} \tag{4.19}$$

Clearly, S_x^H is a complex quantity. However, if H is replaced by $|H(j\omega)|$, we can obtain the sensitivity of the frequency response to variations in the element value and plot it against

Realization of First- and Second-Order Functions Using Opamps

frequency. Note that the sensitivity $S_x^{|H(j\omega)|}$ is of considerable practical use in estimating the effect of element changes. Thus, if x changes by p percent, to a first approximation, $|H(j\omega)|$ will change by $S_x^{|H(j\omega)|} \cdot p$ percent.

It is also useful to note that at any frequency $s = j\omega$.

$$S_x^H = S_x^{|H|} + jQ_x^\varphi$$

where $Q_x^\varphi = x\dfrac{d\varphi}{dx}$ is the semi-relative phase sensitivity.

The sensitivity of H, as it was defined above, can be calculated for a filter function of any order; i.e., this definition is not restricted to the case of second-order functions. More suitable sensitivity measures for second-order filter functions are the pole, ω_o, and Q-factor sensitivities.

The pole p semi-relative sensitivity is defined by

$$S_x^p \equiv \frac{dp}{dx/x} \tag{4.20}$$

and expresses the pole displacement due to the variation in the value of the element x. It is also a complex quantity in general.

The Q factor and ω_o sensitivity measures are particularly useful in the case of 2nd-order bandpass filters. These are defined as follows:

$$S_x^Q \equiv \frac{dQ/Q}{dx/x} = \frac{x}{Q}\frac{dQ}{dx} \tag{4.21}$$

$$S_x^{\omega_o} \equiv \frac{d\omega_o/\omega_o}{dx/x} = \frac{x}{\omega_o}\frac{d\omega_o}{dx} \tag{4.22}$$

The element x in the previous considerations can be passive or active. In the case where it is active, it can be the open-loop gain of an opamp, the g_m of an OTA, or the gain of a controlled source of finite gain, obtained for example from an opamp using resistive feedback. In the latter case, its actual gain variation, due to a variation in the open-loop gain of the opamp, will greatly depend on the nominal value of x.

To show this let us consider the case depicted in Fig. 4.5, where an opamp is used to obtain the gain k of a VCVS. With

$$\beta = \frac{R_1}{R_1 + R_2} \tag{4.23}$$

we can easily find that

$$k \equiv \frac{V_o}{V_i} = \frac{A}{1 + \beta A} \tag{4.24}$$

and subsequently

$$dk = \frac{dA}{(1 + \beta A)^2} \tag{4.25}$$

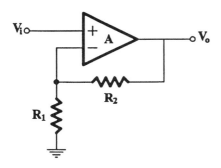

FIGURE 4.5
Opamp used to obtain gain k.

Thus, for given values of A and dA, the actual variation dk of k will depend on β. But,

$$\beta = \frac{1}{1 + \frac{R_2}{R_1}} = \frac{1}{k_n}$$

where k_n is the desired or nominal value of k. Substituting in Eq. (4.25), we get

$$dk = \frac{dA}{(1 + A/k_n)^2} \qquad (4.26)$$

We conclude then that, independently of the sensitivity of Q or ω_o to k, their actual variations will greatly depend on the nominal value of k, being small for small values of k_n.

This observation has led to the introduction [1] of the *gain sensitivity product* as the sensitivity measure that takes into consideration the importance of the value of k, when examining the actual Q and ω_o variations. Under these circumstances, Eqs. (4.21) and (4.22) will be written as follows:

$$GS_k^Q = S_k^Q \cdot k \qquad (4.27)$$

$$GS_k^{\omega_o} = S_k^{\omega_o} \cdot k \qquad (4.28)$$

Since the open-loop gain of an opamp, and to a less extent k, are functions of frequency, one should take this into consideration in calculating the variations of Q and ω_o. It has been shown [2] that the percentage ω_o variations causes a much larger variation in the transfer function of an active RC filter (about 2Q times larger) than that caused by the same percentage Q variation. Therefore, it is important, when comparing second-order filter realizations, to determine the percentage change in ω_o

$$\delta\omega \equiv \frac{\delta\omega_o}{\omega_o} \qquad (4.29)$$

which is caused by the finite bandwidth of the amplifier.

As has been mentioned, the sensitivity measures examined above are suitable for studying the sensitivity of active RC filters of second order. We have also introduced sensitivity measures concerning the overall transfer function variation when all circuit components, passive and active, are varying simultaneously. However, these do not give any important information concerning the sensitivity of second-order active RC filters in addition to the information given by the sensitivity measures introduced above. Such *multiparameter* sensitivity measures are suitable for studying the sensitivity of higher-order active RC filters and are introduced in Chapter 5, where the design of such filters is considered.

4.5 Realization of Biquadratic Functions Using SABs

An active RC circuit realizing a biquadratic transfer function is called a *biquad*. A single-amplifier biquad (SAB) is a biquad using one amplifier.

4.5.1 Classification of SABs

Most SABs can be classified [3] in one of the two general structures shown in Fig. 4.6. In Fig. 4.6(a) the enhanced positive feedback (EPF) configuration is shown, while in Fig. 4.6(b) the enhanced negative feedback (ENF) configuration is shown. The passive RC network is a complex zero-producing section, and it can be of second- or third-order. The input signal is fed to the circuit by inserting the signal voltage source between the terminal b and ground in Fig. 4.6(a) and, similarly, between the terminal c and ground in Fig. 4.6(b). It can be also fed as a current to one of the nodes of the RC passive network, either internal or external (not to terminals b and c). In some cases, when complex zeros should be realized, the input signal is also fed through a resistor connected to the common node of R_a and R_b (R_a' and R_b').

A SAB is called *canonic* if the RC network employs two capacitors only, since this is the minimum number to provide a biquadratic transfer function. However, it can employ three capacitors, as it is the case with the RC subnetwork being a twin-T.

The term *enhanced* stems from the fact that the resistive feedback by means of R_a, R_b or R_a', R_b' affects the position of the complex poles of the overall function, moving them on a circle towards the $j\omega$-axis.

An important property may be possessed by the two configurations as they appear in Fig. 4.6 if the following conditions are satisfied:

1. The RC networks are the same but with the respective connections of terminals b and c to earth and the opamp output interchanged, as is suggested in the figure.
2. If R_b is written as

$$R_b = (n-1)R_a \qquad \text{[Fig. 4.6(a)]} \qquad (4.30)$$

then

$$R_a' = (n-1)R_b' \qquad \text{[Fig. 4.6(b)]} \qquad (4.31)$$

where n is a positive number greater than one.

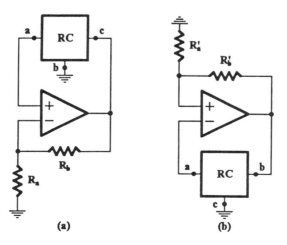

FIGURE 4.6
(a) The SAB EPF and (b) ENF configurations.

Under these conditions, the two configurations are said to be related by the complementary transformation (CT) [4]. The important characteristic of this situation is that the two circuits have identical poles and same sensitivities.

The process of applying the CT is the following: the terminals of the feedback network, which are connected to the output terminal of the opamp in the initial circuit, are connected to ground, and those connected to ground in the initial network are now connected to the opamp output terminal. At the same time, the opamp input terminals are interchanged. This is shown clearly in Fig. 4.7. It should be mentioned that with the application of the CT an ENF biquad is transformed to an EPF, and vice versa.

Application of the complementary transformation to a useful circuit gives rise to another that may also be useful. This idea has been successfully used [3, 4], and many SABs have been obtained that have interesting properties.

The SABs we shall present below will belong to one or other of the configurations in Fig. 4.6, and mostly they will be canonic. These are among the most useful ones in practice.

4.5.2 A Lowpass SAB

The most popular second-order lowpass circuit is that of Sallen-Key's [5] shown in Fig. 4.8, which employs an opamp in an arrangement of a VCVS with gain G. Clearly, it belongs to the EPF class of SABs. Its transfer voltage ratio, obtained by analyzing the circuit, is as follows:

$$H(s) = \frac{V_0}{V_i} = \frac{G/(C_1 C_2 R_1 R_2)}{s^2 + \left(\dfrac{1}{R_1 C_1} + \dfrac{1}{R_2 C_1} + \dfrac{1-G}{R_2 C_2}\right)s + \dfrac{1}{R_1 R_2 C_1 C_2}} \quad (4.32)$$

where

$$G = 1 + \frac{R_b}{R_a}$$

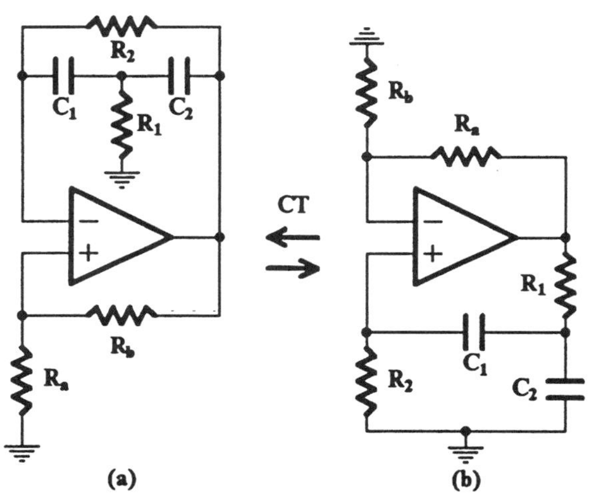

FIGURE 4.7
Application of the complementary transformation.

Realization of First- and Second-Order Functions Using Opamps

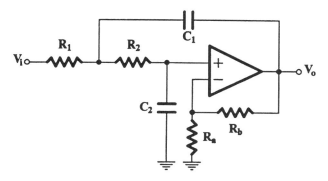

FIGURE 4.8
The Sallen-Key second-order lowpass circuit.

The realization of the second-order lowpass function,

$$F(s) = \frac{K}{s^2 + \beta s + \gamma} \tag{4.33}$$

by this circuit can be obtained by matching the coefficients of equal powers of s in Eqs. (4.32) and (4.33). Following this, one obtains the following:

$$\frac{1}{R_1 C_1} + \frac{1}{R_2 C_1} + \frac{1-G}{R_2 C_2} = \beta \tag{4.34a}$$

$$\frac{1}{R_1 R_2 C_1 C_2} = \gamma \tag{4.34b}$$

$$\frac{G}{R_1 R_2 C_1 C_2} = K \tag{4.34c}$$

Since the number of unknowns is larger than the number of the equations, some components will have to be selected *arbitrarily*. One popular choice is the following:

$$R_1 = R_2 = R \quad \text{and} \quad G = 1 (R_a = \infty) \tag{4.35}$$

With this selection, one takes advantage of the whole bandwidth of the opamp, i.e., up to frequency f_T. Substituting in Eqs. (4.34), we get

$$\frac{2}{RC_1} = \beta \tag{4.36a}$$

$$\frac{1}{R^2 C_1 C_2} = \gamma \tag{4.36b}$$

with the restriction that $K = \gamma$, which is not usually a problem in filter design.

From these equations, the following capacitance values are obtained:

$$C_1 = \frac{2}{R\beta} \tag{4.37a}$$

$$C_2 = \frac{\beta}{2\gamma R} \tag{4.37b}$$

As an example, consider the realization of the second-order Butterworth lowpass filter function

$$F(s) = \frac{1}{s^2 + \sqrt{2}s + 1} \tag{4.38}$$

Selecting $R = 1$ the (normalized) values of the passive components of the circuit in Fig. 4.7 will be as follows:

$$R_1 = R_2 = 1 \tag{4.39a}$$

$$C_1 = \sqrt{2} \tag{4.39b}$$

$$C_2 = \frac{1}{\sqrt{2}} \tag{4.39c}$$

The circuit can be denormalized to a cutoff frequency ω_c and an impedance level R_o. Two sets of passive component values can be the following:

Set 1	Set 2
$R_1 = R_2 = R_o$	$R_1 = R_2 = \dfrac{R_o}{\omega_c}$
$C_1 = \dfrac{\sqrt{2}}{\omega_c R_o}$	$C_1 = \dfrac{\sqrt{2}}{R_o}$
$C_2 = \dfrac{1}{\sqrt{2} R_o \omega_c}$	$C_2 = \dfrac{1}{\sqrt{2} R_o}$

In both cases, a proper value for R_b (in order to avoid dc offset voltage in the opamp output) will be

$$R_b = 2R_o$$

while $R_a = \infty$.

For example, if $R_o = 10$ kΩ, and the desired cutoff frequency is 1 kHz, one finds the following:

Realization of First- and Second-Order Functions Using Opamps

Set 1	Set 2
$R_1 = R_2 = 10\ k\Omega$	$R_1 = R_2 = \dfrac{10^4}{2\pi \times 10^3} = 1.59\ \Omega$
$C_1 = \dfrac{\sqrt{2}}{2\pi \times 10^3 \times 10^4} = 22.5\ nF$	$C_1 = \dfrac{\sqrt{2}}{10^4} = 141.4\ \mu F$
$C_2 = \dfrac{1}{2\sqrt{2}\pi \times 10^3 \times 10^4} = 11.25\ nF$	$C_2 = \dfrac{1}{\sqrt{2} \times 10^4} = 70.71\ \mu F$
$R_b = 20\ k\Omega$	$R_b = 3.18\ \Omega$

Clearly, the values in set 2 are not practical, but they can become so by raising the impedance level to $10\ \Omega$. Then we will have:

$$R_1 = R_2 = 15.9\ k\Omega$$

$$C_1 = 14.14\ nF$$

$$C_2 = 7.071\ nF$$

$$R_b = 31.8\ k\Omega$$

Using Eqs. (4.27) and (4.28) the ω_o and Q sensitivities with respect to variations in the passive component values (passive sensitivities) are found to be as follows ($\omega_o = \sqrt{\gamma}$, $Q = \sqrt{\gamma}/\beta$):

$$S^{\omega_o}_{R_1, R_2, C_1, C_2} = -\frac{1}{2} \qquad S^{\omega_o}_{R_a} = S^{\omega_o}_{R_b} = 0 \qquad S^{Q}_{R_1} = S^{Q}_{R_2} = 0$$

$$S^{Q}_{C_1} = -S^{Q}_{C_2} = \frac{1}{2} \qquad S^{Q}_{R_a} = S^{Q}_{R_b} = 0$$

This design is optimum with respect to passive sensitivities, but it is not optimum when variations of the open-loop gain of the opamp are taken into account.

The sensitivity of this circuit has also been studied for other design approaches. For example, selecting

$$R_1 = R_2 \quad \text{and} \quad C_1 = C_2$$

requires

$$G = 3 - \frac{1}{Q} = 3 - \frac{\beta}{\sqrt{\gamma}}$$

Although the ω_o sensitivities remain the same, the corresponding Q sensitivities increase considerably, becoming proportional to Q [6]. Thus, this design is not useful in practice.

Another design approach, followed by Saraga [7], starts with the requirement to minimize the sensitivity of Q with respect to variations in the open-loop gain of the opamp (active sensitivity). This approach leads to the following component values:

$$C_1 = \frac{\sqrt{3\gamma}}{\beta}, \qquad C_2 = 1, \qquad R_1 = \frac{\beta}{\gamma}, \qquad R_2 = \frac{1}{\sqrt{3\gamma}}, \qquad G = \frac{4}{3}$$

Although this design is optimum with respect to Q active sensitivity, it is not optimum with reference to Q passive sensitivities. We can find an overall optimum design with G between 1 and 4/3 taking into account the expected maximum variations both of passive and active components [6, 7].

4.5.3 A Highpass SAB

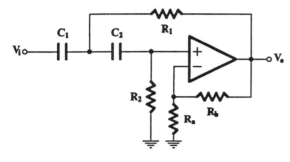

FIGURE 4.9
The Sallen-Key highpass SAB.

A useful circuit is the Sallen-Key [5] highpass SAB shown in Fig. 4.9, which belongs to the EPF class also.

Clearly, this circuit has the same topology as the Sallen-Key lowpass SAB, and it can be obtained from the latter by applying the RC:CR transformation to the RC section of the lowpass circuit. According to this transformation, each resistance R_L in the lowpass circuit is replaced by a capacitance C_H of value

$$C_H = \frac{1}{R_L}$$

and each capacitance C_L in the lowpass circuit is replaced by a resistance R_H of value

$$R_H = \frac{1}{C_L}$$

Normalized values are considered to apply in this transformation. Substituting R_{iH} for $1/sC_{iL}$ and sC_{iH} for $1/R_{iL}$, $i = 1,2$ in Eq. (4.32), we obtain the following transfer voltage ratio for the highpass circuit in Fig. 4.9:

$$H(s) \equiv \frac{V_0}{V_i} = \frac{Gs^2}{s^2 + \left(\dfrac{1}{R_2C_2} + \dfrac{1}{R_2C_1} + \dfrac{1-G}{R_1C_1}\right)s + \dfrac{1}{R_1R_2C_1C_2}} \qquad (4.40)$$

In Eq. (4.40) R_i, C_i, $i = 1,2$ are those in Fig. 4.9.

This result can be easily verified by a straightforward analysis of the circuit in Fig. 4.9. Clearly, the two networks, lowpass and highpass, have the same poles and therefore the same Q and ω_o sensitivities.

As an example, the normalized component values of the Sallen-Key highpass circuit realizing the second-order Butterworth highpass filter function

$$F(s) = \frac{s^2}{s^2 + \sqrt{2}s + 1} \tag{4.41}$$

will be the following:

$$C_{1H} = C_{2H} = 1$$

$$R_{1H} = \frac{1}{\sqrt{2}}$$

$$R_{2H} = \sqrt{2}$$

These values are obtained by applying the RC:CR transformation to the corresponding component values of the Sallen-Key lowpass filter, which are given by Eqs. (4.39).

Of course, the value of G remains unchanged, i.e., $G = 1$. The highpass circuit can be denormalized to an impedance level R_o and a cutoff frequency ω_c in exactly the same way as it was explained for the lowpass circuit.

4.5.4 A Bandpass SAB

An active RC circuit [8], useful in the realization of a second-order bandpass function is shown in Fig. 4.10. Clearly, this circuit belongs to the ENF class of SABs.

Assuming that the opamp is ideal, the transfer voltage ratio of the SAB is as following:

$$H(s) \equiv \frac{V_0}{V_i} = -\frac{hs}{s^2 + \left(\frac{C_1 + C_2}{R_2 C_1 C_2} - K \frac{1}{R_1 C_2}\right)s + \frac{1}{R_1 R_2 C_1 C_2}} \tag{4.42}$$

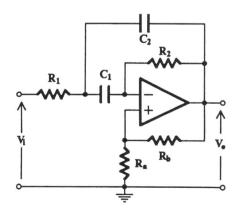

FIGURE 4.10
A bandpass SAB.

where

$$h = \frac{1+K}{R_1 C_2}, \quad K = \frac{R_a}{R_b} \tag{4.43}$$

The second-order bandpass transfer function

$$F(s) = \frac{\alpha s}{s^2 + \beta s + \gamma} \tag{4.44}$$

can be realized by this SAB to within a constant multiplier as the voltage ratio $-V_o/V_i$, if we rewrite $F(s)$ as follows:

$$F(s) = \frac{\alpha s}{s^2 + [(n+1)\beta - n\beta]s + \gamma} \tag{4.45}$$

where n is real and positive.

Let

$$r = \frac{R_1}{R_2} \quad q = \frac{C_1}{C_2}$$

Then, coefficient matching between Eqs. (4.42) and (4.45), after some simple mathematical manipulations, gives the following:

$$n = Q(q+1)\sqrt{\frac{r}{q}} - 1 \tag{4.46a}$$

$$K = \frac{n}{Q}\sqrt{\frac{r}{q}} \tag{4.46b}$$

$$R_2 C_1 = \sqrt{\frac{q}{\gamma r}} \tag{4.46c}$$

$$R_1 C_2 = \sqrt{\frac{r}{\gamma q}} \tag{4.46d}$$

$$R_2 C_2 = \frac{1}{\sqrt{\gamma r q}} \tag{4.46e}$$

Clearly, the Q-factor sensitivity of the SAB is affected by the choice of n, which has to be as low as possible for low Q-factor sensitivity.

In an attempt to minimize the Q-factor sensitivity, we may show, using Eq. (4.46a), that the value of n is minimal for any r, if

$$q = 1$$

i.e., when

$$C_1 = C_2$$

Then,

$$n = 2Q\sqrt{r} - 1 \qquad (4.47a)$$

which, depending on r, can be (within the practical limitations) as small as it is desired. With this choice of q, the rest of the design Eqs. (4.46) become as follows ($C_1 = C_2 = C$):

$$K = \frac{n}{Q}\sqrt{r} \qquad (4.47b)$$

$$R_1 C = \sqrt{\frac{r}{\gamma}} \qquad (4.47c)$$

$$R_2 C = \frac{1}{\sqrt{\gamma r}} \qquad (4.47d)$$

Since Eqs. (4.47) are not sufficient to give unique component values, the designer should select r and one of R_1, R_2, or C as well as R_a or R_b, taking into consideration the opamp specifications.

The ω_o and Q-factor sensitivities to variations in component values, assuming an ideal opamp, are given in Table 4.1.

TABLE 4.1
ω_o and Q-Factor Sensitivities ($C_1 = C_2$)

Component x_i	$S_{x_i}^{\omega_o}$	$S_{x_i}^{Q}$
R_1	$-\frac{1}{2}$	$\frac{1}{2} - 2Q\sqrt{r}$
R_2	$-\frac{1}{2}$	$-\frac{1}{2} + 2Q\sqrt{r}$
C_1	$-\frac{1}{2}$	$Q\sqrt{r} - \frac{1}{2}$
C_2	$-\frac{1}{2}$	$-\left(Q\sqrt{r} - \frac{1}{2}\right)$
K	0	$2Q\sqrt{r} - 1$

Clearly, the value of r has to be very small, but since n in Eq. (4.47a) should be positive for stability, the following condition must hold:

$$\sqrt{r} > \frac{1}{2Q} \qquad (4.48)$$

Fleischer [9] has shown that, when taking into consideration the effect of variation of the open-loop gain $A(s)$ of the opamp on ω_o and Q variations [assuming $A(s) \approx \omega_T/s$], the approximately optimum value of r is given by

$$r \approx 0.25 \frac{\omega_o}{\omega_T} \frac{\sigma_{\omega_T}}{\sigma_{R,C}} \qquad (4.49)$$

where $\sigma_{R,C}$ and σ_{ω_T} are the standard deviations of the passive elements (assumed equal) and of the gain bandwidth product ω_T of the opamp, respectively. In fact, the value of r need not be less than $1/60$, which results in practical values of the components R_1, R_2, R_a, and R_b. An analogous optimization approach followed by Daryanani [6] leads to a value of r very close to that given by Eq. (4.49).

Some important features of this SAB are the following:

a. Q factor, and hence bandwidth, can be varied independently of ω_o by varying K.
b. Successive stages can be cascaded without the need for isolating stages. This also holds for the SABs in Figs. 4.8 and 4.9.
c. All capacitors in all cascaded stages can be designed to be of the same value.
d. Any source resistance can be absorbed by R_1 to avoid errors in the frequency response. This is true for the SAB in Fig. 4.8 but not for the SAB in Fig. 4.9.
e. It is easy to reduce the output voltage at the center frequency to avoid bringing the opamp to the nonlinear region of its characteristics, i.e., to saturation, even for small input signals. Clearly, at the center frequency, the output voltage will be, from Eq. (4.42),

$$V_o\big|_{\omega_o} = \frac{hQ}{\omega_o} V_i\big|_{\omega_o}$$

where h is given by Eq. (4.43). This value can be quite high, even for moderate Q. The way to reduce this voltage is to split R_1 into two others, R'_1, R''_1, as shown in Fig. 4.11, such that

$$R_1 = \frac{R'_1 R''_1}{R'_1 + R''_1}$$

This will not affect the shape of the frequency response, i.e., ω_o and Q.

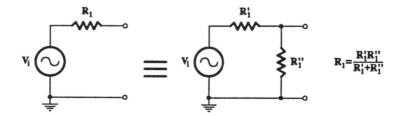

FIGURE 4.11
Some transformation to reduce the output voltage at the center frequency without altering the shape of the frequency response.

f. The ratio ω_T/ω_o in Eq. (4.49) gives the value of the amplifier gain at the center frequency of the filter. Since the sensitivities depend on the value of r, amplifiers with different ω_T will affect differently the performance of the circuit. Thus, a two-pole one-zero frequency compensated opamp will extend the useful range of the filter [9] further than that corresponding to the use of the 741-type opamps.

The form of the transfer function remains unchanged if R_1 and R_2 are interchanged with C_1 and C_2 respectively. However, in this case, feature d explained above is not applicable, if the signal source impedance is not zero. The same is true for feature e.

Example 4.1

Consider the use of this SAB to realize the function

$$F(s) = \frac{0.1s}{s^2 + 0.1s + 1}$$

with the center frequency at 10 krad/s and impedance level at 10 kΩ.

Clearly, the normalized ω_o and the Q-factor values are as following:

$$\omega_o = 1, \quad Q = 10$$

Selecting the practical resistance spread to be 1/100 results in $r = 10^{-2}$, which leads, through Eq. (4.47a) to

$$n = 2 \times 10 \times 10^{-1} - 1 = 1$$

Therefore,

$$K = \frac{R_a}{R_b} = \frac{1}{10} \times 10^{-1} = 10^{-2}$$

Selecting $C = 1$, we get from Eqs. (4.47)

$$R_2 = 10, \quad R_1 = 10^{-1}$$

and from Eq. (4.43),

$$h = \frac{1 + 10^{-2}}{10^{-1} \times 1} = 10.1$$

One set of denormalized component values can then be as follows:

$$C_1 = C_2 = 10 \text{ nF}$$

$$R_1 = 1 \text{ k}\Omega, R_2 = 100 \text{ k}\Omega, R_a = 1 \text{ k}\Omega, R_b = 100 \text{ k}\Omega$$

To achieve unity gain at the center frequency, we observe that since

$$\frac{hQ}{\omega_o} = 10.1 \times 10 = 101$$

we should arrange that

$$\frac{R''_1}{R'_1 + R''_1} = \frac{1}{101} \quad \text{and} \quad \frac{R'_1 R''_1}{R'_1 + R''_1} = R_1 = 1 \text{ k}\Omega$$

Solving then for R'_1 and R''_1, we get

$$R'_1 = 101 \text{ k}\Omega, \quad R''_1 = 1.01 \text{ k}\Omega$$

Let us now look for the optimum design taking into consideration the effect of the finite gain bandwidth product ω_T of the opamp. Assuming

$$\sigma_{\omega_T} = 0.25 \quad \sigma_{R,C} = 0.005$$

and $\omega_T = 2\pi \times 10^6$ rad/s, we obtain from (4.49)

$$r \approx 0.25 \frac{10^4 \times 0.25}{2\pi \times 10^6 \times 5 \times 10^{-3}} \approx \frac{1}{50}$$

To simplify the calculations a little, we may choose, with no problem,

$$r = \frac{1}{49}$$

Then, following the same procedure as before, we obtain the following set of denormalized values:

$$C_1 = C_2 = 10 \text{ nF}, \quad R_1 = 10/7 \text{ k}\Omega \ (R'_1 = 102.7 \text{ k}\Omega, \quad R''_1 = 1.449 \text{ k}\Omega),$$

$$R_2 = 70 \text{ k}\Omega, R_a = 2 \text{ k}\Omega, R_b = 75.4 \text{ k}\Omega$$

Note that R_a and R_b can be denormalized at a different impedance level from that for C_i, R_i $i = 1, 2$.

4.5.5 Lowpass- and Highpass-Notch Biquads

Friend [10] has introduced additional resistors to the SAB in Fig. 4.10, by means of which the input signal is added directly to the input terminals of the opamp as is shown in Fig. 4.12. Clearly, the circuit remains canonic, but it is now possible to realize complex zeros in addition to the complex poles. Depending on the values of these additional components all types of SABs except the lowpass are obtained, namely:

Bandpass when $\quad R_6, R_7, R_c = \infty$
Highpass when $\quad R_7 = \infty$

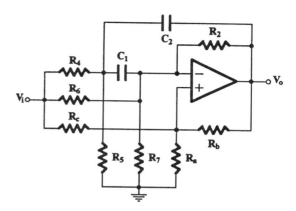

FIGURE 4.12
The generalized SAB developed by Friend.

Lowpass notch when	$R_6 = \infty$
Highpass notch when	$R_7 = \infty$
Allpass when	$R_6, R_7 = \infty$

Friend's generalization of the circuit in Fig. 4.10 has led to the design of the popular STAR building block [11]. It should also be mentioned that, if in addition to $R_6, R_7 = \infty$, also $R_b = \infty$ and $R_5 = \infty$, the circuit is reduced to that shown in Fig. 4.13 [12], which is studied separately below. Of the other four cases, the bandpass has already been studied above. The highpass is less economical than that of Sallen and Key, which we have examined already, and therefore we will not elaborate on it. This leaves us with the two notch cases, which are now examined a little further.

4.5.6 Lowpass Notch ($R_6 = \infty$)

For convenience in the analysis of the circuit we use conductances (G_i) instead of resistances (R_i). Assuming that

$$R_6 = \infty \tag{4.50a}$$

$$G_1 = G_4 + G_5 = 5 \tag{4.50b}$$

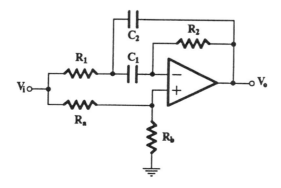

FIGURE 4.13
A simple allpass biquad.

$$K_c = \frac{G_c}{G_a + G_c} \tag{4.50c}$$

$$K_b = \frac{G_b}{G_a + G_c} \tag{4.50d}$$

$$C_1 = C_2 = C \tag{4.50e}$$

and an ideal opamp, straightforward analysis gives the following transfer function:

$$\frac{V_o}{V_i} = \frac{K_c s^2 + \dfrac{K_c G_1(G_7 + G_2)}{C^2}}{s^2 + \left[\dfrac{2G_2}{C} - \dfrac{K_b 2G_7 + G_1}{C}\right]s + \dfrac{G_1}{C^2}(G_2 - K_b G_7)} \tag{4.51}$$

under the conditions

$$G_c(2G_7 + 2G_2 + G_5) = G_4(G_a + G_b) \tag{4.52a}$$

$$\frac{G_7}{G_2} = \frac{G_a}{G_b + G_c} \tag{4.52b}$$

Clearly, this transfer function is of the form

$$F(s) = \frac{ds^2 + e}{s^2 + \beta s + \gamma} \tag{4.53}$$

which corresponds to a lowpass notch filter, provided that $d < e/\gamma$. This latter condition is in fact satisfied since, from the circuit transfer function, Eq. (4.51),

$$K_c < K_c \frac{G_2 + G_7}{G_2 - K_b G_7}$$

Equating coefficients of equal powers of s in Eqs. (4.51) and (4.53), we get the following four equations:

$$d = K_c \tag{4.54a}$$

$$e = \frac{K_c G_1(G_7 + G_2)}{C^2} \tag{4.54b}$$

$$\beta = \frac{2G_2}{C} - \frac{K_b(2G_7 + G_1)}{C} \tag{4.54c}$$

$$\gamma = \frac{G_1}{C^2}(G_2 - K_b G_7) \qquad (4.54d)$$

$$G_1 = G_4 + G_5$$

With the addition of Eqs. (4.52a) and (4.52b), we have only six equations—not enough to determine the values of all passive components of the circuit. We may then make the usual choices, as in the bandpass case.

$$C = 1 \qquad (4.55a)$$

$$r = \frac{G_2}{G_1} \approx 0.25 \frac{\omega_p}{\omega_T} \frac{\sigma_{\omega_T}}{\sigma_{R,C}} \qquad (4.55b)$$

where ω_p is the magnitude of the pole frequency.

Thus, the set of the necessary equations to determine all component values is completed.

4.5.7 Highpass Notch ($R_7 = \infty$)

A highpass notch SAB is obtained from the general circuit in Fig. 4.12 if $R_7 = \infty$. Working with conductances as previously, and assuming ideal opamp, the transfer voltage ratio of the circuit is found to be

$$\frac{V_o}{V_i} = \frac{K_c s^2 + \frac{G_1}{C^2}(K_c G_2 - G_6)}{s^2 + \left[\frac{2G_2}{C} - \frac{K_b}{C}(G_1 + 2G_6)\right]s + \frac{G_1}{C^2}(G_2 - K_b G_6)} \qquad (4.56)$$

under the condition

$$K_b \frac{G_c}{G_b} = \frac{G_4 + 2G_6}{2G_2 + G_5} \qquad (4.57)$$

where G_1, K_c, K_b, and C are given again by Eqs. (4.50b, c, d, e).

This transfer voltage ratio is again of the form of Eq. (4.53), but with $d > e/\gamma$, since

$$K_c > \frac{K_c G_2 - G_6}{G_2 - K_b G_6} = K_c \frac{G_2 - \frac{G_6}{K_c}}{G_2 - K_b G_6}$$

Therefore, the SAB can realize a highpass notch filter function. The design equations can be found as it was explained above, in the case of the lowpass notch function.

4.5.8 An Allpass SAB

Allpass biquads are useful in the realization of a high-order allpass function as the cascade connection of second-order sections, the transfer functions of which have the form

$$F(s) = \frac{s^2 - \beta s + \gamma}{s^2 + \beta s + \gamma} \tag{4.58}$$

Such a function is the (n,n) Padè approximation of the function e^{-s} [13].

A simple biquad [12] suitable for such a realization is shown in Fig. 4.13. Its transfer voltage ratio, assuming ideal opamp is as follows:

$$H(s) \equiv \frac{V_o}{V_i} = K \frac{s^2 - \left[\frac{R_a}{R_b}\frac{1}{R_1 C_2} - \frac{1}{R_2}\frac{C_1 + C_2}{C_1 C_2}\right]s + \frac{1}{R_1 R_2 C_1 C_2}}{s^2 + \frac{1}{R_2}\frac{C_1 + C_2}{C_1 C_2}s + \frac{1}{R_1 R_2 C_1 C_2}} \tag{4.59}$$

where

$$K = \frac{R_b}{R_a + R_b}$$

It can be shown that as the pole Q increases, so does the ratio $r = R_2/R_1$. For a given resistance ratio r, the maximum value of ω_i/σ_i, where $-\sigma_i \pm j\omega_i$ are the poles of $F(s)$ in Eq. (4.58), achieved when $C_1 = C_2$, is the following:

$$\left(\frac{\omega_i}{\sigma_i}\right)_{max} = \sqrt{r-1} \tag{4.60}$$

Thus, depending on the maximum acceptable range of resistance values, a limit is set on the position of the poles of $F(s)$ which are realized by this network.

Letting $C_1 = C_2 = C$ in Eq. (4.59), the transfer function becomes

$$H(s) = \frac{V_o}{V_i} = K \frac{s^2 - \left[\frac{R_a}{R_b}\frac{1}{R_1 C} - \frac{2}{R_2 C}\right]s + \frac{1}{R_1 R_2 C^2}}{s^2 + \frac{2}{R_2 C}s + \frac{1}{R_1 R_2 C^2}} \tag{4.61}$$

Equating coefficients of equal powers of s in Eqs. (4.58) and (4.61) we obtain the following component values:

$$R_1 = \frac{\beta}{2\gamma C}, \quad R_2 = \frac{2}{\beta C}, \quad \frac{R_a}{R_b} = \frac{\beta^2}{\gamma} = \frac{4R_1}{R_2} \tag{4.62}$$

with

$$K = \frac{R_b}{R_a + R_b}$$

Functions with higher ω_i/σ_i ratios can be realized by this network by connecting an additional resistor from the noninverting input of the opamp to the output, as was mentioned above in connection with Friend's work. Although this change makes the circuit more flexible, at the same time it increases its sensitivity to component values.

With or without this additional resistance, the network cannot realize a function whose poles and zeros are not equidistant from the origin in the s-plane. In such a case, other biquads [12, 14] should be used.

Another observation is the following: if the positions of resistors R_1 and R_2 are interchanged with those of capacitors C_1 and C_2, respectively, the resulting network is also an allpass biquad with a similar transfer function. As the reader can similarly show, the highest ω_i/σ_i ratio is also given by Eq. (4.60), achieved when $R_1 = R_2$ with $r = C_1/C_2$ this time.

A drawback of the circuit is its gain, which, given by K in Eq. (4.59), is lower than one. One way to increase the value of K is to reduce the amount of the output voltage V_o fed back to the input by means of R_2 and C_2. In practice, this may be achieved by means of a potentiometer as shown in Fig. 4.14. Thus, the value of K becomes

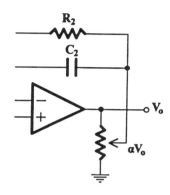

FIGURE 4.14
A practical way to enhance the gain of the SAB in Fig. 4.13.

$$K' = \frac{K}{\alpha}$$

where α is the potentiometer setting, this being between zero and one. Depending on the value of α, the circuit gain can be even greater than one. Naturally, the potentiometer resistance must be low with respect to R_2 and $1/\omega C_2$.

Finally, as can be easily seen from Eq. (4.59), if the coefficient of s in the numerator is made equal to zero, the circuit can be used as a bandstop *symmetrical notch* biquad realizing the function

$$F(s) = K\frac{s^2 + \omega_o^2}{s^2 + \beta s + \omega_o^2}$$

displaying the same behavior at zero and infinity.

As an example, let us use the circuit in Fig. 4.13 to obtain a delay of $\tau = 10$ ms by means of the realization of the second-order Padè approximation written for our purpose as follows:

$$F(s) = \frac{\tau^2 s^2 - 6\tau s + 12}{\tau^2 s^2 + 6\tau s + 12} = \frac{s^2 - 600s + 12\times 10^4}{s^2 + 600 + 12\times 10^4}$$

Selecting a convenient value for C, e.g., $C = 0.1$ µF, and using Eqs. (4.62), we get the following:

$$R_1 = \frac{600}{2 \times 12 \times 10^4 \times 10^{-7}} = 25 \text{ k}\Omega$$

$$R_2 = \frac{2}{600 \times 10^{-7}} = 33.3 \text{ k}\Omega$$

$$\frac{R_a}{R_b} = \frac{600^2}{12 \times 10^4} = 3$$

The actual values of R_a, R_b can be obtained either by assuming a suitable value of K or by optimizing some other circuit characteristic, e.g., the opamp offset. In the latter case, selecting $R_a//R_b = R_2$ we can obtain

$$R_a = 120 \text{ k}\Omega \qquad R_b = 40 \text{ k}\Omega$$

resulting in a $K = 0.25$. However, this value of K can become unity by means of a 1 kΩ potentiometer connected as shown in Fig. 4.14.

4.6 Realization of a Quadratic with a Positive Real Zero

In realizing certain lowpass non-minimum phase delay functions [14] as the cascade connection of second-order stages, one is involved with the realization of a quadratic with a positive real zero, i.e., with the function

$$F(s) = \frac{k(\alpha - s)}{s^2 + \beta s + \gamma} \tag{4.63}$$

An RC active network, which can realize $F(s)$ using one operational amplifier [15] and which has very low sensitivity to variations in the values of its components, is shown in Fig. 4.15. Its transfer voltage ratio is

$$\frac{V_o}{V_i} = \frac{\frac{g_1}{C_2}\left(\frac{g_a g_2}{g_b C_1} - s\right)}{s^2 + g_2 \frac{C_1 + C_2}{C_1 C_2}s + \frac{g_2}{C_1 C_2}\left[\left(1 + \frac{g_a}{g_b}\right)g_1 + g_a\right]} \tag{4.64}$$

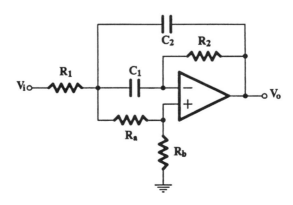

FIGURE 4.15
An active RC network for the realization of $F(s)$, Eq. (4.63).

The values of the components of the network realizing $F(s)$, Eq. (4.63), can be found by equating coefficients of equal powers of s in Eqs. (4.63) and (4.64). Since the number of the unknowns is larger than the number of the resulting equations, the values of any two components can be selected arbitrarily. Selecting, for example,

$$C_1 = C_2 = 1$$

coefficient matching in Eqs. (4.63) and (4.64) gives the following:

$$\alpha = \frac{g_a}{g_b} g_2 \qquad \beta = 2g_2 \qquad \gamma = g_2\left[\left(1 + \frac{g_a}{g_b}\right)g_1 + g_a\right] \qquad k = g_1$$

Solving for the unknown component values gives the following:

$$g_1 = k \tag{4.65a}$$

$$g_2 = \frac{\beta}{2} \tag{4.65b}$$

$$g_a = \frac{2\gamma - (\beta + 2\alpha)k}{\beta} \tag{4.65c}$$

$$g_b = \frac{2\gamma - (\beta + 2\alpha)k}{2\alpha} \tag{4.65d}$$

Clearly, for positive values of g_a and g_b, the following conditions should hold between the coefficients of the transfer function:

$$2\gamma = k(2\alpha + \beta)$$

On the other hand, if $C_1 \neq C_2$, one can show that, for the component values to be positive, k should be

$$k < \frac{\gamma}{\alpha + \beta \dfrac{C_2}{C_1 + C_2}} \tag{4.66}$$

It can also be shown that the Q-factor sensitivities are as follows:

$$S_{g_1}^Q, S_{g_a}^Q, (-S_{g_b}^Q) < \frac{1}{2}$$

$$S_{g_2}^Q = -\frac{1}{2}$$

$$S_{C_1}^Q = S_{C_2}^Q = 0 \text{ if } C_1 = C_2$$

It can be seen that the Q-factor sensitivities are extremely low and independent of the Q factor. Also, the Q factor is insensitive to variations in the values of the two capacitances, if $C_1 = C_2$. Therefore, this condition can be the starting point in the design of the stage.

As an example, consider the realization of the following function:

$$F(s) = \frac{0.5628(5.902 - s)}{s^2 + 4.117s + 6.9963} \tag{4.67}$$

which is part of a delay function. Selecting

$$C_1 = C_2 = 1$$

we have inequality (4.66) satisfied. The values of the other components are calculated by means of Eqs. (4.65). The circuit, denormalized to an impedance level of $10^6/3\,\Omega$ and a 0.1 s delay, is shown in Fig. 4.16.

4.7 Biquads Obtained Using the Twin-T RC Network

The twin-T (TT) RC network is shown in its simplest version in Fig. 4.17a. Its transfer voltage ratio V_2/V_1 is as follows:

$$\frac{V_2}{V_1} = \frac{s^2 + \frac{1}{T^2}}{s^2 + \frac{4}{T}s + \frac{1}{T^2}} \tag{4.68}$$

where $T = RC$.

Thus, the TT in Fig. 4.17(a) processes a pair of transmission zeros on the $j\omega$-axis or, in other words, it is a bandstop circuit. This pair of zeros can be moved out of the $j\omega$-axis, if a

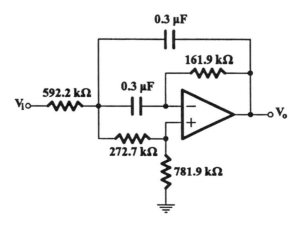

FIGURE 4.16
Realization of $F(s)$, Eq. (4.67), denormalized to $10^6/3\,\Omega$ and 0.1 s delay.

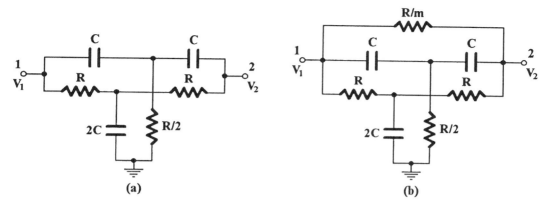

FIGURE 4.17
(a) The TT network and (b) the Bridged-TT network.

resistor is used to bridge the terminals 1 and 2, as shown in Fig. 4.17(b). In this case, the new transfer voltage ratio will be the following:

$$\frac{V_2}{V_1} = \frac{s^2 + \frac{2m}{T}s + \frac{1+2m}{T^2}}{s^2 + \frac{4+2m}{T}s + \frac{1+2m}{T^2}} \tag{4.69}$$

It can be seen from Eq. (4.69) that the transmission zeros have moved inside the LH of the s-plane. Either form of the TT shown in Fig. 4.17 (and, of course, more complicated ones [16]) can be used in conjunction with active elements to realize various biquadratic functions.

As an example, consider the circuit in Fig. 4.18. If

$$\frac{1}{R_3} = \frac{1}{R_1} + \frac{1}{R_2}$$

analysis of the circuit, assuming ideal opamp, gives the following voltage ratio:

$$\frac{V_o}{V_i} = \frac{G(s + \omega_o^2)}{s^2 + \frac{\omega_o}{Q}s + \omega_o^2} \tag{4.70}$$

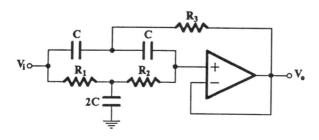

FIGURE 4.18
Example circuit.

where

$$G = 1, \quad \omega_o^2 = \frac{1}{R_1 R_2 C^2}, \quad \frac{\omega_o}{Q} = \frac{2}{R_2 C} \tag{4.71}$$

Thus, a useful bandstop circuit with zero output impedance has been obtained using the TT. A number of other biquads can be obtained using other forms of TT, and the interested reader is advised to refer to other sources [1, 16] for details.

4.8 Two-Opamp Biquads

A large number of two-opamp biquads can be obtained by a technique introduced below for the enhancement of the Q of certain SABs. Other useful two opamp biquads have also been suggested which possess interesting characteristics. In addition, very low sensitivity biquads may be obtained if the inductance of the LCR biquad is simulated by the two-opamp GIC [17, 18]. Although inductance simulation is explained in detail in Chapter 6, we include this possibility here because of its simplicity. In what follows, we examine first this possibility and then some other useful two-opamp biquads.

4.8.1 Biquads by Inductance Simulation

In Fig. 4.19, three possible biquads, obtained by combining an inductance, a capacitance, and a resistance, are shown. All of them have the same poles with $\omega_o = 1/\sqrt{LC}$ and $R\sqrt{C/L}$.

Substituting the component L in Fig. 4.19(a) by its GIC equivalent (see Sections 3.5.2 and 6.4), the corresponding active RC circuit in Fig. 4.20 is obtained without the buffer amplifier of gain k.

If the signal source is removed and placed at node b feeding R in series, while node a is earthed the bandpass biquad equivalent to that in Fig. 4.19(b) is obtained. Finally, to obtain the highpass equivalent, the signal source is placed between a broken node at c and earth, while nodes a and b are earthed.

It can be shown by referring to Fig. 4.20 that the value of the equivalent inductance is ideally as follows:

$$L_{eq} = C\frac{R_1 R_3 R_L}{R_2} \tag{4.72}$$

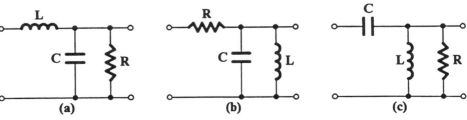

FIGURE 4.19
RLC biquads: (a) lowpass, (b) bandpass, and (c) highpass.

Realization of First- and Second-Order Functions Using Opamps

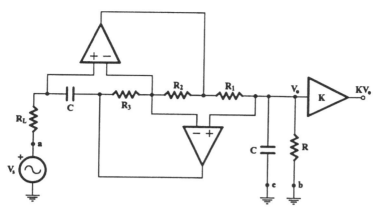

FIGURE 4.20
Lowpass biquad.

Substituting for L, we obtain the following ω_o and Q values:

$$\omega_o = \frac{1}{\sqrt{LC}} = \frac{1}{C(R_1 R_3 R_L/R_2)^{1/2}} \qquad (4.73)$$

$$Q = \omega_o CR = \frac{R}{(R_1 R_3 R_L/R_2)} \qquad (4.74)$$

In designing these filters, it is usual to select

$$R_1 = R_2 = R_3 = RL = r \qquad (4.75)$$

when, from Eqs. (4.73) and (4.74),

$$rC = \frac{1}{\omega_o} \qquad (4.76)$$

and

$$R = Qr \qquad (4.77)$$

The Q and ω_o sensitivities to variation in the passive component values are the following:

$$S_R^Q = -2S_{R_1}^Q = -2S_{R_3}^Q = 2S_{R_2}^Q = -2S_{R_L}^Q = 1$$

$$S_R^{\omega_o} = 0, \; S_C^{\omega_o} = 2S_{R_1}^{\omega_o} = -2S_{R_2}^{\omega_o} = 2S_{R_3}^{\omega_o} = 2S_{R_L}^{\omega_o} = -1$$

On the other hand, if we consider matched opamps with the one pole model describing their frequency response, we can determine that the error of ω_o is approximately [16] the following:

$$\frac{\Delta \omega_o}{\omega_o} \cong -2\frac{\omega_o}{\omega_T} \tag{4.78}$$

One important drawback of these biquads is that their output is not taken from a zero impedance node. Therefore in order to cascade such stages for realizing higher order filters isolation amplifiers should be inserted between successive stages, as shown in Fig. 4.20.

Inductance simulation can also be applied to obtain notch biquads, both lowpass (LPN) and highpass (HPN) notch. Suitable LCR biquads are those shown in Fig. 4.21. For the *allpass* notch, i.e., with $\omega_{oz} = \omega_{op}$, C_2 and L_2 in these networks should be deleted. The substitution of L by its GIC equivalent in Fig. 4.21(a) is straightforward, as it was achieved in the case of the lowpass biquad in Fig. 4.19(a). However, in the case of the HPN in Fig. 4.21(b), both L_1 and L_2 are simulated using the same GIC, as explained in Chapter 6. We do not include the active equivalents of the LPN and the HPN here, leaving them to the reader as an exercise.

Finally, the allpass biquad using inductance simulation is clearly a three-opamp biquad, and consequently it does not belong to the two opamp class of biquads. A simulation is given below.

4.8.2 Two-Opamp Allpass Biquads

The allpass biquadratic function

$$F(s) = \frac{s^2 - \beta s + \gamma}{s^2 + \beta s + \gamma} \tag{4.79}$$

is written in the following form:

$$F(s) = 1 - 2\frac{\beta s}{s^2 + \beta s + \gamma} = 1 - 2F_1(s) \tag{4.80}$$

where

$$F_1(s) = \frac{\beta s}{s^2 + \beta s + \gamma} \tag{4.81}$$

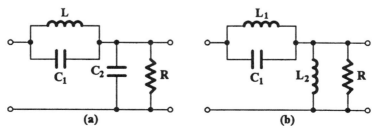

FIGURE 4.21
LCR notch biquads: (a) LPN and (b) HPN.

Then, $F_1(s)$ is realized by any bandpass SAB, and $F(s)$ is formed by using a second opamp to perform the summation in Eq. (4.80). Such an allpass active circuit is shown in Fig. 4.22, where the SAB in Fig. 4.10 has been used to realize $F_1(s)$.

The merits of the SAB in Fig. 4.10 can be used advantageously in the realization of the high-Q biquadratics in a high-order allpass function.

4.8.3 Selectivity Enhancement

There is a category of active RC networks with inherently low sensitivities but requiring excessive spread in component values, even when they realize relatively low (~10) Q values. These networks employ negative feedback exclusively around a finite- or infinite-gain amplifier.

To improve the selectivity of such circuits, one could employ positive feedback [8, 19], but this leads to sensitivity degradation of the circuits. We describe another method here, by means of an example, which does not lead to severe sensitivity degradation if it is applied carefully.

Consider the bandpass circuit in Fig. 4.23. It can be shown that the Q value of this circuit is the following:

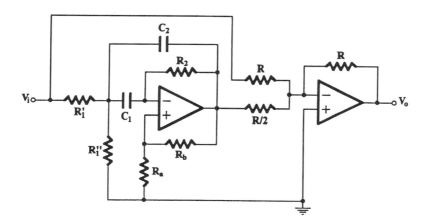

FIGURE 4.22
A two-opamp allpass biquad.

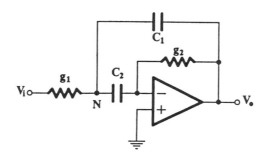

FIGURE 4.23
A low-Q bandpass circuit.

$$Q = \sqrt{\frac{g_1}{g_2} \frac{C_1 C_2}{(C_1 + C_2)^2}} \qquad (4.82)$$

We now introduce a VCVS of gain K at node N as shown in Fig. 4.24. Note that there are now available two zero impedance outputs.

By straightforward analysis, we can find that

$$\frac{V_o}{V_i} = -\frac{\left(\frac{g_1}{C_1}\right) s}{s^2 + \frac{g_2}{KC_2} s + \frac{g_1 g_2}{KC_1 C_2}} \qquad (4.83)$$

and

$$\frac{V_a}{V_i} = -\frac{g_1 g_2 / C_1 C_2}{s^2 + \frac{g_2}{KC_2} s + \frac{g_1 g_2}{KC_1 C_2}} \qquad (4.84)$$

i.e., the simultaneous realization of a bandpass and a lowpass function both having the same poles. From Eq. (4.83), the new Q factor will be

$$Q' = \sqrt{\frac{KC_2 g_1}{C_1 g_2}} \qquad (4.85)$$

It can be seen that for the same component values in the two circuits, there is an enhancement in Q since

$$\frac{Q'}{Q} = \left(1 + \frac{C_2}{C_1}\right)\sqrt{K} \qquad (4.86)$$

Since the introduction of the VCVS affects both the Q and ω_o values, their corresponding sensitivities will be increased. However, if the VCVS is realized by means of an opamp, and the value of K is very low (e.g., unity), its effect on the sensitivity will not be significant.

The Q enhancement technique suggested above can be applied to a large number of circuits [20–22]. In all cases, the availability of the second zero-impedance output is an advan-

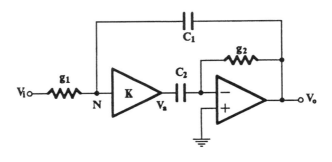

FIGURE 4.24
The enhanced-Q circuit.

tage. Note that, if the positions of resistors and capacitors in Fig. 4.24 are interchanged, the second transfer function of the network will be highpass instead of lowpass.

It is interesting to note that a circuit due to Bach [23] can be obtained from the Sallen and Key lowpass circuit using the Q enhancement technique. Thus, assuming that the gain of the voltage amplifier in the Sallen and Key filter is unity, Fig. 4.25(a), introducing an additional unity gain VCVS at node A, Bach's circuit in Fig. 4.25(b) is obtained.

Of course, this does not imply that the respective components in the two circuits have the same values. Applying the RC:CR transformation, the corresponding highpass circuits are obtained.

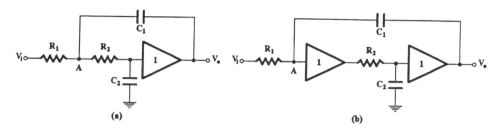

FIGURE 4.25
(a) Sallen and Key and (b) Bach's circuits.

4.9 Three-Opamp Biquads

The use of three opamps to realize second-order filter functions leads to multiple-output biquads with the additional advantage of versatility in that ω_o, Q, and the filter gain can be independently adjusted. Although they are not without problems, as we shall see later, technology has produced three opamp chips ready for use in realizing biquadratics.

The poles of the circuits we present here are obtained by means of two integrators in a feedback loop. That is why these are often referred to as the two *integrator-loop biquads*.

We may develop such a three-opamp biquad following the old analog computing technique for solving a differential equation. To show this, let us consider the second-order lowpass function realized as the ratio of two voltages V_o and V_i, i.e.,

$$\frac{V_o}{V_i} = \frac{K}{s^2 + \frac{\omega_o}{Q}s + \omega_o^2} \tag{4.87}$$

We can write this equation as follows:

$$\frac{s^2}{\omega_o^2}V_o = \frac{K}{\omega_o^2}V_i - \frac{1}{Q}\frac{s}{\omega_o}V_o - V_o \tag{4.88}$$

Clearly, V_o may be obtained from $(s^2/\omega_o^2)V_o$ by two successive integrations with proper time constants. On the other hand, the term $(s^2/\omega_o^2)V_o$ can be obtained as the sum of the three terms on the right-hand side of Eq. (4.88). We may thus realize Eq. (4.88) as it is shown in block diagram form in Fig. 4.26.

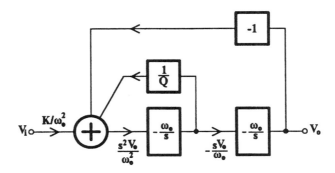

FIGURE 4.26
Block diagram for realizing V_o/V_i, Eq. (4.88).

Using opamps to perform the summing and integration operations, we obtain the circuit shown in Fig. 4.27. Straightforward analysis of this circuit shows that the function in Eq. (4.87) is realized with a minus sign. To avoid this, and also the use of a fourth amplifier, we make use of the circuit in Fig. 3.8, giving the difference of two voltages.

Following this, the circuit in Fig. 4.27 becomes as is shown in Fig. 4.28. Again, straightforward analysis of this circuit gives the following:

$$\frac{V_o}{V_i} = \frac{V_{LP}}{V_i} = \frac{K'/(RC)^2}{s^2 + \frac{1}{RC}\frac{1}{Q}s + \frac{1}{R^2C^2}} \tag{4.89}$$

$$\frac{V_{BP}}{V_i} = -\frac{(K'/RC)s}{s^2 + \frac{1}{RC}\frac{1}{Q}s + \frac{1}{R^2C^2}} \tag{4.90}$$

$$\frac{V_{HP}}{V_i} = \frac{K's^2}{s^2 + \frac{1}{RC}\frac{1}{Q}s + \frac{1}{R^2C^2}} \tag{4.91}$$

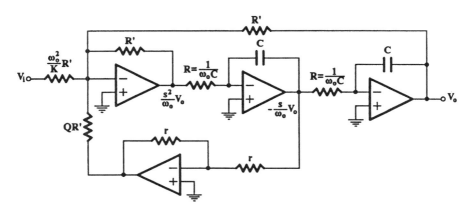

FIGURE 4.27
Implementation of the block diagram in Fig. 4.26.

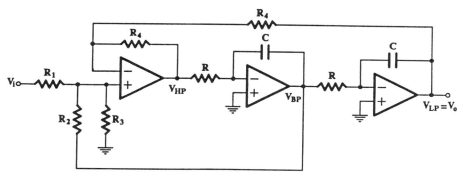

FIGURE 4.28
The KHN biquad.

where

$$K' = \frac{2R_2R_3}{R_1R_2 + R_1R_3 + R_2R_3} \quad (4.92a)$$

$$Q = \frac{2R_1R_3}{R_1R_2 + R_1R_3 + R_2R_3} \quad (4.92b)$$

and

$$\frac{K'}{R^2C^2} = K \quad (4.92c)$$

From these, we can obtain the following design equations:

$$RC = \frac{1}{\omega_o} \quad (4.93a)$$

$$R_2 = R_1 \frac{K}{\omega_o^2 Q} \quad (4.93b)$$

$$R_3 = \frac{R_1}{\dfrac{2\omega_o^2}{K} - 1 - \dfrac{Q\omega_o^2}{K}} \quad (4.93c)$$

with R_1 and R_4 taking suitably chosen values.

This circuit is referred to as the KHN (Kerwin, Huelsman, Newcomb) biquad [24], and it simultaneously displays lowpass, bandpass, and highpass behavior. All three filters have the same poles, of course.

A notch response can be obtained by adding the lowpass and highpass outputs using an extra opamp. The ω_o and Q passive sensitivities are very low. However, the excess phase shift introduced by each integrator and the summer, due to the finite GB product of the opamps, leads to Q enhancement with undesirable effects in the filter response. This is examined below, but first let us see another three-opamp biquad with some additional interesting characteristics.

4.9.1 The Tow-Thomas [25–27] Three-Opamp Biquad

A more versatile three-opamp biquad is shown in Fig. 4.29. Its transfer voltage ratio (V_o/V_i) is as follows:

$$\frac{V_o}{V_i} = -\frac{\frac{C_1}{C}s^2 + \frac{1}{CR}\left(\frac{R}{R_1} - \frac{r}{R_3}\right)s + \frac{1}{C^2RR_2}}{s^2 + \frac{1}{CR_4}s + \frac{1}{C^2R^2}} \tag{4.94}$$

Clearly, it is possible to obtain any kind of second-order filter function by a proper choice of the component values. Thus we may have the following:

LP: if $C_1 = 0$, $R_1 = R_3 = \infty$

BP: if $C_1 = 0$, $R_1 = R_2 = \infty$ (positive sign)

BP: if $C_1 = 0$, $R_2 = R_3 = \infty$ (negative sign)

HP: if $C_1 = C$, $R_1 = R_2 = R_3 = \infty$

Notch: if $C_1 = C$, $R_1 = R_3 = \infty$

Allpass: if $C_1 = C$, $R_1 = \infty$, $r = R_3/Q$

This circuit was initially proposed by Tow [25] and studied by Thomas [26, 27]. Its passive sensitivities are similar to those of the KHN three-opamp network. However, it is more versatile, realizing all kinds of second-order filter function without the use of an extra opamp, which is required in the case of the KHN circuit. It suffers, though, from the results of excess phase shift on Q enhancement, as explained below.

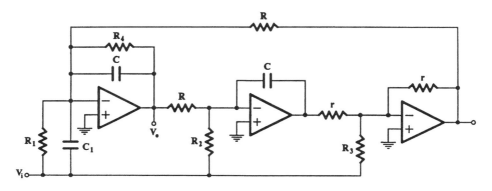

FIGURE 4.29
The Tow-Thomas three-opamp biquad.

4.9.2 Excess Phase and Its Compensation in Three-Opamp Biquads

Because of the finite gain-bandwidth product of the opamps, excess phase appears in the summer (beyond the 180°) and the integrators (beyond the 90°) in the three-opamp biquads, which finally leads to Q enhancement and to instabilities of the circuits. We may calculate this excess phase in the case of the simple sign-reversing amplifier in Fig. 4.30 as follows. Assuming that the opamp gain may be approximated with a model having a single pole at the origin (that is, at zero frequency), i.e.,

$$A = -\frac{\omega_T}{s}$$

simple analysis of the circuit in Fig. 4.30 gives the following transfer voltage ratio V_o/V_i:

$$\frac{V_o}{V_i} = -\frac{1}{1 + \frac{2s}{\omega_T}} \quad (4.95)$$

Thus, at frequency ω, there is an extra phase shift beyond the 180°,

$$\varphi_{ex} = -\tan^{-1}\frac{2\omega}{\omega_T} \cong \frac{2\omega}{\omega_T} \quad \text{for } \omega \ll \omega_T \quad (4.96)$$

This extra phase shift φ_{ex}, which in the case of the ideal opamp would be zero, is called the *excess phase*.

In a similar way, it can easily be shown that the excess phase for an integrator is approximately

$$\varphi_{ex} \cong -\frac{\omega}{\omega_T} \quad (4.97)$$

Since the two integrators and the summer are connected in cascade inside the loop, the overall excess phase becomes substantial. It can be shown that this leads to Q enhancement, which is undesirable. Thus, assuming that the opamps are identical (i.e., they have the same ω_T), the enhanced Q_{enh} of the biquad in Fig. 4.29 is approximately [28]

$$Q_{enh} \cong Q \frac{1}{1 - KQ\frac{\omega_o}{\omega_T}} \quad (4.98)$$

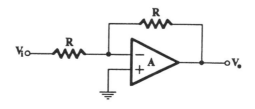

FIGURE 4.30
Sign-reversing amplifier.

where the value of Q is that for ideal opamps, ω_o is the pole frequency of the biquad, and K typically is 4. A similar Q enhancement is obtained in the case of the biquad in Fig. 4.28.

The excess phase, being a phase lag, can be cancelled by introducing an equal phase lead inside the loop. Passively, this can be easily achieved by connecting a capacitor of suitable value in parallel with one of resistors R in Fig. 4.28 or the resistor R in the integrator in Fig. 4.29. However, since the temperature coefficients of this capacitor and ω_T are not the same, this compensation cannot be perfect at different temperatures.

More successful is the active compensation as employed in the Åkerberg-Mossberg three-opamp biquad, which we examine next.

Clearly, to reverse the sign of an integrator, a second opamp should be employed as shown in Fig. 4.31(a). The excess phase for this non-inverting or *positive integrator* is approximately, according to Eqs. (4.96) and (4.97),

$$\varphi_{exc} \cong -\frac{3\omega}{\omega_T}$$

However, if the sign reversal of the integrator is achieved according to Fig. 4.31(b), then it can be easily shown that the excess phase of the integrator becomes

$$\varphi'_{exc} \cong \frac{\omega}{\omega_T}$$

i.e., phase lead instead of phase lag.

Thus, this phase lead will cancel out the phase lag of the other integrator inside the loop of the three-opamp biquads, and the overall excess phase will be zero. This scheme of excess phase compensation is called *active compensation*, and it is less temperature dependent than the passive compensation of excess phase that was mentioned above.

4.9.3 The Åkerberg-Mossberg Three-Opamp Biquad [29]

The Åkerberg-Mossberg three-opamp biquad is a modified version of the Tow-Thomas three-opamp biquad. To simplify matters we set the following values to some of the passive components in the circuit in Fig. 4.29:

$$C_1 = 0, \quad R_2 = R_3 = \infty$$

Then, the Tow-Thomas biquad becomes as shown in Fig. 4.32(a). By actively compensating the integrator as in Fig. 4.31, the Åkerberg-Mossberg biquad is obtained as shown in Fig. 4.32(b).

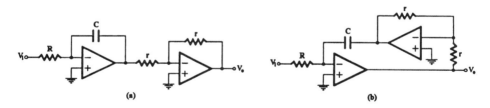

FIGURE 4.31
(a) A non-inverting integrator and (b) a non-inverting integrator with active compensation of excess phase.

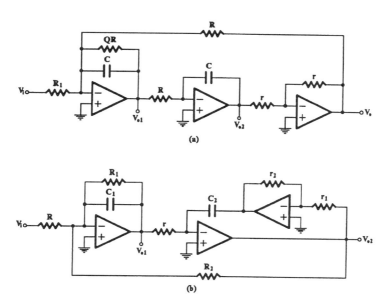

FIGURE 4.32
(a) The Tow-Thomas biquad actively compensated for excess phase, giving the Åkerberg-Mossberg biquad in (b).

Straightforward analysis of the biquad in Fig. 4.32(b) gives the following:

$$\frac{V_{o1}}{V_i} = -\frac{s/RC_1}{s^2 + s\frac{1}{R_1C_1} + \frac{r_1}{C_1C_2R_2rr_2}} \quad (4.99)$$

$$\frac{V_{o2}}{V_i} = -\frac{r_1/C_1C_2Rrr_2}{s^2 + s\frac{1}{R_1C_1} + \frac{r_1}{C_1C_2R_2rr_2}} \quad (4.100)$$

The Åkerberg-Mossberg biquad, in its simplified form, can thus simultaneously realize lowpass and bandpass functions with the same poles. In its more general version, it can realize any type of biquadratic function, if the input signal is fed properly weighted to the inputs of all amplifiers. In all cases, the poles are not affected.

It is clear from Eqs. (4.99) and (4.100) that both the Q factor and the pole frequency ω_o can be independently adjusted. Depending on the component values, voltages at internal nodes may differ substantially. This inevitably will lead to nonlinear operation of some amplifiers and, therefore, to reduced dynamic range even for small signals. For the minimization of such effects, r_1 should be equal to r_2, but this will make the active compensation ineffective, and thus Q enhancement will not be avoided.

4.10 Summary

Low-order (first and second) filter circuits were presented in this chapter, which have been chosen among an abundance that have appeared in the literature over the years. These are mostly canonic and have been proven useful in practice.

The biquads can be classified as SABs (single-amplifier biquads), two-opamp biquads, and three-opamp biquads. Also, the SABs can be classified as enhanced positive feedback (EPF) and enhanced negative feedback (ENF) biquads, depending on the feedback paths, positive or negative, in which the frequency dependent passive network is connected. By means of the application of the complementary transformation to one SAB, another SAB is obtained which has the same poles as the first one and, consequently, same pole sensitivities. This has led to the discovery of a number of useful biquads as well as to the rediscovery of others, previously suggested.

Among the important merits of a biquad, its sensitivity to variations in component values and economy are the most prominent. Sensitivity of Q, ω_o, and the gain-sensitivity product are the most useful measures that characterize the value of a biquad. It is because of this that high-Q biquadratic functions could better be realized by two- or three-opamp biquads instead of using SABs. This, of course, increases the cost and power consumption, and it is a matter of priority for the designer in solving a particular filtering problem to make the appropriate choice.

First- and second-order filter circuits are very useful in the realization of high-order filter functions. Some methods for such realizations are based on the use of lower-order sections, particularly biquads. Therefore, knowledge of the most suitable biquad in a particular case is of fundamental importance for achieving the "best" design. This will become apparent in the next chapter, where the design of high-order filters is considered.

References

[1] G. S. Moschytz. 1974. *Linear Integrated Networks: Design*, New York: Van Nostrand Reinhold.

[2] D. Hilberman. 1973. "An approach to the sensitivity and statistical variability of biquadratic filters," *IEEE Trans. on Circuit Theory*, CT-20, **4**: 382–390.

[3] A. S. Sedra. 1974. "Generation and classification of single amplifier filters," *Intl. J. Circuit Theory and Appl.* **2**(1), pp. 51–67.

[4] N. Fliege. 1973. "A new class of second-order RC-active filters with two operational amplifiers, *NTZ* **26**(4), pp. 279–282.

[5] R. P. Sallen and E. L. Key. 1955. "A practical method of designing RC active filters," *IRE Trans. on Circuit Theory* CT-2, pp. 74–85.

[6] G. Daryanani. 1976. *Principles of Active Network Synthesis and Design*, New York: John Wiley and Sons.

[7] W. Saraga, "Sensitivity of 2nd-order Sallen-Key-type active RC filters," *Electron. Lett.* **3**(10), pp. 442–444.

[8] T. Deliyannis. 1968. "High-Q factor circuit with reduced sensitivity," *Electron. Lett.* **4**, p. 577.

[9] P. E. Fleischer. 1976. "Sensitivity minimization in a single amplifier biquad circuit," *IEEE Trans. Circuits and Systems* CAS-23, **1**, pp. 45–55.

[10] J. J. Friend. 1970. "A single operational amplifier biquadratic filter section," *Proc. 1970 Int. Symposium on Circuit Theory*, Atlanta, Georgia, pp. 179–180.

[11] J. J. Friend, C. A. Harris, and D. Hilberman. 1975. "STAR: An active biquadratic filter section," *IEEE Trans. on Circuits and Systems* CAS-22, **2**, pp. 115–121.

[12] T. Deliyannis. 1969. "RC active allpass sections," *Electron. Lett.* **5**, p. 59.

[13] D. F. Tuttle. 1958. *Network Synthesis*, vol. I, New York: John Wiley and Sons.

[14] T. Deliyannis. 1970. "Six new delay functions and their realization using active RC networks," *The Radio and Electronic Engineer*, **39**(3), pp. 139–144.

[15] *ibid.* "Realization of a quadratic with a positive real zero," *ibid.*, pp. 271–272.

[16] A. S. Sedra and P. O. Brackett. 1978. *Filter Theory and Design: Active and Passive*, London: Pitman.

[17] A. Antoniou. 1969. "Realization of gyrators using operational amplifiers and their use in RC-active network synthesis," *Proc. IEE* (London) **116**, pp. 1838–1850.
[18] B. B. Bhattacharyya, W. S. Mikhael, and A. Antoniou. 1974. "Design of RC-active networks using generalized-immittance converters," *J. Frank. Inst.*, **297**(1), pp. 45–48.
[19] P.R. Geffe, "RC-amplifier resonators for active filters," *IEEE Trans. Circuit Theory* CT-15, pp. 415–419.
[20] T. Deliyannis. 1970. "A low-pass filter with extremely low sensitivity," *Proc. IEEE* **58**(9), pp. 1366–1367.
[21] T. Deliyannis and Y. Berdi. 1971. "Selectivity improvement in a useful second-order active RC section," *Int. J. Electronics* **31**(3), pp. 243–248.
[22] *ibid*. 1973. "Selectivity Enhancement of Certain Low-sensitivity RC Active Networks," **34**(4), pp. 513–526.
[23] R. E. Bach. 1960. "Selecting R-C values for active filters," *Electronics* **33**, pp. 82–83.
[24] W. Kerwin, L. P. Huelsman, and R. W. Newcomb. 1967. "State-variable synthesis for insensitive integrated circuit transfer functions," *IEEE J. Solid-State Circuits* SC-2, **3**, pp. 87–92.
[25] J. Tow. 1968. "Active RC filters—A state-space realization," *Proc. IEEE* **56**, pp. 1137–1139.
[26] L. C. Thomas. 1971. "The biquad: Part I—some practical design considerations," *IEEE Trans. on Circuit Theory* CT-18, pp. 350–357.
[27] *ibid*. "The biquad: Part II—a multipurpose active filtering system," pp. 358–361.
[28] A. S. Sedra. 1989. In *Miniaturized and Integrated Filters*, S. K. Mitra and C. F. Kurth (eds.), New York: John Wiley and Sons.
[29] D. Åkerberg and K. Mossberg. 1974. "A versatile active RC building block with inherent compensation for the finite bandwidth of the amplifier," *IEEE Trans. Circuits and Systems* CAS-21, **1**, pp. 75–78.

Chapter 5

Realization of High-Order Functions

5.1 Introduction

In most cases, the selectivity provided by a second-order filter is not adequate. Higher-order filter functions have to be realized in order to satisfy the stringent selectivity requirements in telecommunication systems, special instrumentation, and many other applications.

To realize such high-order filter functions, two main approaches have been found most useful in practice. The first is to cascade second-order stages without feedback (cascade filter) or through the application of negative feedback (multiple-loop feedback filters, MLFs). The second is to use combinations of active (e.g., opamps) and passive (resistors and capacitors) components in order to simulate either the inductances or the operation of a high-order LC ladder. Yet another approach, the use of just one opamp embedded in an RC network in order to realize the high-order function, although possible, has been dropped for reasons of high sensitivity.

The study of such high-order circuits requires some additional tools over those used in the previous chapter, which were suitable for second-order filters. For example, in an MLF or a simulated ladder filter, the value of each component does not affect only one pole or one zero of the filter, but more than one, making thus the filter tuning difficult. In this case, the Q sensitivity for, example, of a biquadratic section in the MLF circuit cannot be used as a criterion when comparing the MLF circuit to the simulated ladder. Therefore, more suitable sensitivity measures are required for such comparisons, the determination of which makes the use of a computer program unavoidable.

In this chapter, we first discuss briefly a number of criteria which characterize a useful, practical, active RC filter. Next, we introduce suitable sensitivity measures, which have been proven to be consistent with one another, as far as the information on sensitivity they give is concerned. Then, three high-order filters are discussed, all of them using the biquadratic circuit as a cell. These are the cascade, the multiple-loop feedback, and the cascade of biquartic stages filters, the latter being a mixture of the other two. The simulated LC ladder filter is examined in the next chapters.

5.2 Selection Criteria for High-Order Function Realizations

The realization of a high-order function can be achieved by a number of methods. Some of these have proved more advantageous in practice than others, and over the years they have prevailed. Before we see which methods are more acceptable in practice and therefore more useful in filter design, we should set a number of criteria that a design method must satisfy in order to be considered more suitable than others in solving a design problem. It must be emphasized, though, that there is not one method that is the best according to all criteria. So, we will consider the best method as the one that satisfies most of the criteria in a more satisfactory way than the rest.

The most important criteria that can be used in comparing the various methods of realization of a high-order function are the following:

- The possibility of realizing the required function using the available components.
- Sensitivity, i.e., stability of the filter characteristics. As we have already seen in the previous chapter, some biquads are more sensitive than others to variations in their component values. This is also true in high-order circuits.
- Economy. Some design methods lead to circuits that require fewer components than others and therefore are more advantageous from the economy point of view.
- Simplicity of design. The designer prefers to use an easy to understand and apply design method rather than a more complicated one.
- The possibility of producing the filter in integrated circuit form. Of course, this will depend on the number of filters to be manufactured; otherwise it will not be economical.
- Power dissipation. Lower power dissipation relaxes the power supply design and leads to lower heat produced by the filter.
- Tuning simplicity. Every circuit, after it has been built, requires tuning in order to satisfy the required specifications.
- Dynamic range. This determines the range between highest signal level that will pass undistorted through the filter and the lowest signal that can be distinguished from the noise. This is usually expressed in decibels and may be written as

$$\Delta R = \frac{\text{Maximum signal level}}{\text{Minimum signal level}} = \frac{\text{Distortion limit}}{\text{Noise floor}}$$

- Noise. Active elements produce their own noise, which is added to that of the passive components, thus decreasing the signal-to-noise ratio at the output of the filter.
- Other criteria, such as passband attenuation, etc., that the designer may set as applicable in the specific filter design problem.

Clearly, some of these criteria cannot be satisfied by the same circuit. For example, low sensitivity and small number of opamps (economy) used in the circuit cannot be satisfied simultaneously, as we have already seen in the realization of second-order functions. Also, a low-sensitivity circuit that employs a large number of opamps dissipates higher levels of dc power and produces more noise than other, more sensitive circuits that use a lower number of opamps. Thus, the task of the designer is to select a circuit design that satisfies most of the criteria that are considered more important for the filter design problem at hand.

5.3 Multiparameter Sensitivity

In the previous chapter (Section 4.4), we introduced various sensitivity measures useful in the case of the realization of second-order functions. These measures of sensitivity can, in some cases, be of some importance when studying the sensitivity of a high-order filter, too. However, they do not give a complete picture of the sensitivity of such a filter, due to the large number of its components.

More useful in sensitivity studies of high-order filters have been proved to be the so-called multiparameter sensitivity measures. Some of these are reviewed here below:

1. Worst-case sensitivity WS defined as follows:

$$WS = \sum_{i=1}^{n} \left| S_{xi}^{|H|} \right| \tag{5.1}$$

   where n is the number of elements, passive and active, and H the filter transfer function.

   Since H is a function of ω too, WS, is also a function of frequency. Worst-case sensitivity estimates the worst deviation from the nominal response when all components have the same percentage variation.

2. Schoeffler's sensitivity measure, in its simplified form, is defined as follows [1]:

$$\sigma_{\Delta|H|/|H|}^{2} = \sigma^{2} \sum_{i} \left| S_{xi}^{H} \right|^{2} \tag{5.2}$$

   where $\sigma_{\Delta|H|/|H|}$ is the standard deviation of $|H|$, and σ the standard deviation of resistors and capacitors, assumed to be the same for all these components, with the additional requirement that they are uncorrelated.

   Both these multiparameter sensitivity measures require for their determination the calculation of $\partial|H|/\partial x$ for all x_i, assuming they vary independently. However, since the variations of the components, in practice, are not infinitesimal, a more realistic picture, and therefore, more useful in engineering work, would be a sensitivity measure based on the real type of component variations, as is the following one.

3. Standard deviation of the amplitude response for a large number of measurements. Here, as a measurement, we consider the calculation of the amplitude response using one set of component values that have been obtained at random within the tolerance limits of the components. In doing this, we assume that the component values have a uniform or normal distribution around its nominal value. The limits of the distribution are set by the tolerance of the components.

   The standard deviation $\sigma_{\Delta|H|/|H|}$ is determined using the following formula [2]:

$$\sigma_{\Delta|H|/|H|}^{2} = \frac{1}{N}\sum_{i=1}^{N}|H_i|^2 - \left(\frac{1}{N}\sum_{i=1}^{N}|H_i|\right)^2 \tag{5.3}$$

or by the formula for a smaller number of measurements

$$\sigma^2_{\Delta|H|/|H|} = \frac{1}{N-1}\sum_{i=1}^{N}|H_i|^2 - \frac{1}{N(N-1)}\left[\sum_{i=1}^{N}|H_i|\right]^2 \quad (5.4)$$

where N, the number of measurements, is

$$100 < N < 10{,}000$$

Since $|H_i|$ is a function of ω, so is $\sigma_{\Delta|H|/|H|}$.

The component random values can be obtained as follows: the computer is instructed to give each time two random numbers, r_1 and r_2, both between 0 and 1. Assuming a uniform distribution of component values around the nominal value, if δ_x is the tolerance of the component with nominal value x, its random value x' will be either

$$x' = x(1 + \delta_x r_1) \quad \text{if} \quad r_2 \leq 0.5$$

$$x' = x(1 - \delta_x r_1) \quad \text{if} \quad r_2 > 0.5 \quad (5.5)$$

The standard deviation multiparameter sensitivity measure is used in comparing high-order circuits realizing the same filter function below. However, use of the other two measures leads to similar conclusions; therefore, they can also be applied in multiparameter sensitivity calculations.

5.4 High-Order Function Realization Methods

The most useful methods for the realization of high-order filter functions in practice fall into one of the following three general methods:

1. Cascade connection of second-order sections
2. Multiple-loop feedback circuits
3. Simulation of passive LC ladder networks

In Method 1, taking advantage of the useful biquadratic sections we examined in the previous chapter, we write the high-order function as the product of biquadratic factors that we realize accordingly. Next, we cascade these sections by connecting the output of each section to the input of the following one. This method has the advantages of simplicity in designing and aligning the filter, provided that the output of each section is of very low impedance—practically zero.

In Method 2, multiple feedback and, in some cases, multiple feed-forward is applied in a cascade connection of biquadratic sections. This coupling, as we shall see later, leads to a better sensitivity performance of the overall circuit compared to the corresponding circuit obtained by Method 1.

Simulation of passive resistively-terminated lossless ladder networks can be achieved by simulating either the inductances of the ladder, using GICs, PICs, gyrators, or functionally.

Functional simulation here implies that branch currents and node voltages in the ladder are modeled using analog computer simulation techniques. The ladder simulation method is attractive, because it leads to active circuits of lower sensitivities than the other two methods.

For simplicity reasons in design and filter alignment, a combination of Methods 1 and 2 may, in some cases, lead to useful circuits having the advantages of both methods. According to this method, the high-order function is written as the product of biquartics (fourth-order functions), which are realized as multiple-loop feedback sections and then cascaded.

In this chapter, Methods 1 and 2, as well as their combination, i.e., the cascade of biquartic sections, are explained to some detail. As previously mentioned, the simulation of resistively terminated ladder lossless filters is explained in following chapters.

5.5 Cascade Connection of Second-Order Sections

A high-order filter function $T(s)$ [we shall use $T(s)$ here for notational simplicity] can be realized as the ratio of the output voltage to the input voltage of a cascade connection of lower-order stages, each of which does not load the output of its proceeding one. For this to be true, the output impedance of each section must be much lower than the input impedance of the following section at all frequencies of interest.

The lower-order stages are preferably biquadratic. Their realization has been presented in the previous chapter. If the function under realization is of odd order, there will be a first-order term which, according to the type of $T(s)$ (lowpass, highpass, or allpass), can be realized by one of the first-order circuits suggested in Chapter 4.

Thus, the high-order filter function $T(s)$ will be written in the form

$$T(s) = t(s) \cdot \prod_{i=1}^{N} t_i(s) \tag{5.6}$$

where $t(s)$ is a first-order term, or simply unity, depending on the order n of the function, which is either odd ($n = 2N + 1$) or even $n = 2N$, respectively, with N being an integer and

$$t_i(s) = \frac{a_{i2}s^2 + a_{i1}s + a_{i0}}{s^2 + b_{i1}s + b_{i0}} \tag{5.7}$$

Depending on $T(s)$, one or two of the numerator coefficients in Eq.(5.7) may be zero, while in the case of an allpass function, the numerator coefficients will be equal to the corresponding coefficients of the denominator, with the additional constraint that $a_{i1} = -b_{i1}$, with $b_{i1} > 0$.

In forming each biquadratic term $t_i(s)$ and then cascading the biquad sections to obtain the overall circuit realizing $T(s)$, three degrees of freedom are at the designer's disposal. These are the following:

- Pole-zero pairing, i.e., which poles with which zeros of $T(s)$ will be paired to form each $t_i(s)$.
- Distribution of the overall gain in the various biquadratics.
- Physical position of each biquad in the cascade.

Clearly, the pole-zero pairing greatly affects the dynamic range of the corresponding biquad and consequently that of the whole filter. Also, the distribution of the filter gain among the various biquads influences their dynamic range, while the biquad sequence in the cascade has a significant effect on the total noise generation in the filter.

Consequently, the filter designer should take advantage of these degrees of freedom in order to optimize the design with regard to the following two main criteria:

- Maximization of dynamic range
- Maximization of the signal-to-noise ratio

In what follows, the optimization approach is explained to some detail with regard to the dynamic range of the filter, which has been shown to be the most relevant [3] in this case.

For reasons of clarity, we give here the universally acceptable definition of the dynamic range of a circuit. It is the ratio, expressed in decibels, of the maximum input signal (voltage) level V_{imax}, that passes undistorted through the circuit to the minimum input signal level V_{imin}, for which the signal at the output of the circuit is still above the output noise level. If the highest output voltage capability for undistorted operation is $V_{o,max}$ and K is the filter gain then the highest input voltage $V_{i,max}$ can be

$$V_{i,max} = \frac{V_{o,max}}{K} \quad (5.8)$$

Other less important points that may influence the designer's decisions could be the following:

- Minimization of the transmission sensitivity
- Minimization of the passband attenuation
- Simplification of tuning procedure

5.5.1 Pole-Zero Pairing

It will be noted that a complex pole near the $j\omega$-axis creates an elevation in the magnitude response of the corresponding biquadratic term, at frequencies around the imaginary part of the pole. On the other hand, a zero at a similar position creates a deep notch in the magnitude response at frequencies around the imaginary part of the zero. If such a pole and zero are very much apart in the s-plane, and they are paired to form a biquadratic function, the minimum value in the magnitude response inside the passband will be much lower than the maximum value, whether this is inside or outside the passband. In such a case, the input signal level cannot be very high in order to avoid nonlinear operation, and it can not be very low either, because then the signal at the output will be buried in noise at frequencies near the zero.

To avoid such a situation, the magnitude of the biquadratic response should be as flat as possible. To make this more clear, suppose that the magnitude response of the biquadratic term $t_i(j\omega)$ is as shown in Fig. 5.1. Let ω_L and ω_H be the filter passband edges, lower and upper, respectively. What we actually seek with the proper pole-zero pairing is to make the difference between the maximum value $|t_i|_{max}$ wherever in the response, and the minimum value $|t_i|_{min}$ inside the passband as small as possible. To achieve this, we should pair each complex pole with its nearest complex zero.

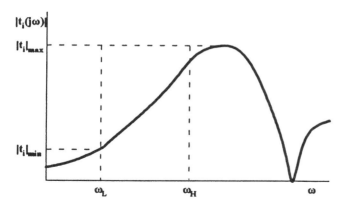

FIGURE 5.1
Biquadratic term.

This argument leads us to the following rule of thumb: to decompose a high-order filter function to the product of biquadratics for maximizing the dynamic range of each biquad (and consequently of the whole filter), we should pair each complex pole with its nearest zero, starting with the pole of highest Q factor.

As an example of the application of this rule, consider the pole-zero positions in Fig. 5.2. (The conjugate poles and zeros are supposed to be placed in the third quadrant). According to the rule of thumb, pole p_1 should be paired with zero z_1, pole p_2 with zero z_2, and pole p_3 with zero z_3. Based on the above argument, a certain algorithm has been suggested by Lueder [4] and discussed by Moschytz [3] for obtaining the optimum pole-zero pairing. The decomposition obtained using the rule of thumb in most cases is identical to the optimum decomposition. When it is not, the degradation in the dynamic range is not substantially different. For this reason, we do not explain the optimum decomposition algorithm here, but we advise the interested reader to consult the above-mentioned references as well as Reference 5.

There are cases, however, when the decomposition can be obtained on a different basis. Consider for example the following sixth-order bandpass function:

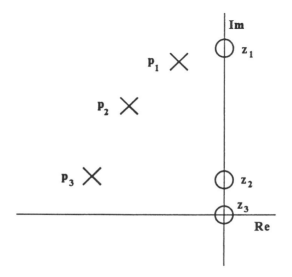

FIGURE 5.2
Illustration of pole-zero positions.

$$T(s) = K \frac{s^3}{(s^2 + b_{11}s + b_{01})(s^2 + b_{12}s + b_{02})(s^2 + b_{13}s + b_{03})} \quad (5.9)$$

Here, all zeros are at zero and infinity. Neglecting at present the distribution of the gain to the three stages, one may decompose $T(s)$ in the following way:

$$T(s) = K \frac{1}{s^2 + b_{11}s + b_{01}} \times \frac{s^2}{s^2 + b_{12}s + b_{02}} \times \frac{s}{s^2 + b_{13}s + b_{03}} \quad (5.10)$$

However, the following decomposition is also possible

$$T(s) = K \cdot t_1(s) \cdot t_2(s) \cdot t_3(s) \quad (5.11)$$

with

$$t_i(s) = \frac{s}{s^2 + b_{1i}s + b_{0i}} \quad (5.12)$$

Each of these two decompositions has practical advantages, and the designer may like to base the pole-zero pairing on these. For example, the lowpass (or the bandpass) section, if placed at the beginning in the cascade, will attenuate out-of-band high-frequency signals, which may otherwise lead to nonlinear operation of the opamps in the subsequent stages. On the other hand, the designer may choose the pole-zero pairing of Eq. (5.11), since a bandpass biquad is easier to tune than the lowpass and highpass configurations.

5.5.2 Cascade Sequence

The proper sequence of the biquads in the cascade is important for achieving maximum dynamic range in the cascade realization of a high-order filter. However, the determination of the best sequence may become a rather tedious procedure if the number of biquads to be cascaded is high. This is so because, for N biquads, there exist $N!$ different sequence possibilities, which will have to be examined.

An efficient algorithm has been described in the literature [6], but we will not explain it here. Fortunately, there exists a simple guide arising from experience that can help the designer to achieve a satisfactory result in practice much easier and quicker.

Thus, we start by determining the frequency of maximum magnitude in each biquad and then form the sequence in such a way that neighboring biquads have their frequencies of maxima as far apart as possible. If there is a lowpass or bandpass section, this is preferably placed in front, while if there exists a highpass, this is placed last. A bandpass section can also be placed last in the cascade, its action being similar to the highpass, namely to prevent any low-frequency noise generated by the leading stages inside the filter from appearing at the output.

A satisfactory solution can also be achieved if the biquads are placed in the cascade in increasing Q factor. It is interesting to note that these mostly intuitive suggestions may help the pole-zero pairing in cases like the example in the previous subsection. Thus, if we select to decompose the sixth-order bandpass function (with all zeros at the origin and infinity) in a lowpass, bandpass, and highpass biquadratics, we associate the lowest Q poles with

Realization of High-Order Functions

the lowpass function, the highest Q poles with the highpass, and the bandpass is left to be associated with the intermediate Q poles. Then, the proper sequence will be the lowpass biquad in front, followed by the bandpass, with the highpass last in the cascade.

Of course, if we had chosen to decompose this sixth-order bandpass function in three bandpass biquadratics, the proper sequence would be in order of increasing Q factor.

5.5.3 Gain Distribution [5]

Having optimized the pole-zero pairing and the biquad sequence in the cascade, we now turn to the distribution of the overall filter gain to the various stages to obtain as high a dynamic range as possible. We consider the filter transfer function of order $2N$ written as follows:

$$T(s) = \prod_{i=1}^{N} k_i t_i(s) \tag{5.13}$$

where $k_1 \cdot k_2 \ldots k_N = K$, with K being the overall gain of the filter and k_i the gain of the ith stage. Let also the biquad sequence be $k_1 t_1(s), k_2 t_2(s), \ldots k_N t_N(s)$.

We work here on the following idea: for the maximum input voltage that results in undistorted output voltage V_o of the filter, the output voltages of the intermediate stages should also be undistorted. To achieve this, we distribute the overall filter gain to the various stages in such a way that the maximum voltage at the output of each intermediate stage is also V_o, i.e.,

$$\max|V_{oi}(j\omega)| = \max|V_{oN}(j\omega)| = V_o \qquad i = 1, 2, \ldots, N-1$$

Let

$$T_i(s) = \prod_{\ell=1}^{i} k_\ell t_\ell(s) \qquad i = 1, 2, \ldots, N-1 \tag{5.14}$$

where $T_i(s)$ is the transfer function from the filter input to the output of the ith stage. Also let

$$\max\left|\prod_{i=1}^{N} t_i(j\omega)\right| = M_N \tag{5.15}$$

and

$$\max\left|\prod_{\ell=1}^{i} t_\ell(j\omega)\right| = M_i \qquad i = 1, 2, \ldots, N-1 \tag{5.16}$$

Then, the gain distribution should be such that

$$k_1 M_1 = K M_N$$
$$k_2 k_1 M_2 = K M_N$$
$$\dots\dots\dots\dots\dots\dots\dots\dots\dots\dots\dots$$
$$k_{N-1} k_{N-2} \dots k_1 M_{N-1} = K M_N$$

From these equations, we obtain the following values for all k_j:

$$k_1 = K \frac{M_N}{M_1} \tag{5.17}$$

$$k_2 = \frac{M_1}{M_2} \tag{5.18}$$

and in general,

$$k_j = \frac{M_{j-1}}{M_j} \qquad j = 2, 3, \dots, N \tag{5.19}$$

where

$$k_N = \frac{K}{\prod_{j=1}^{N-1} k_j} \tag{5.20}$$

As an example, let us design a bandpass filter having a center frequency at 1 krad/s and bandwidth 100rad/s, consistent with the Butterworth response of sixth-order.

Starting with the third-order Butterworth lowpass function,

$$t(s_n) = \frac{1}{s_n^3 + 2s_n^2 + 2s_n + 1} \tag{5.21}$$

we apply the lowpass-to-bandpass transformation

$$s_n = \frac{s^2 + 1}{0.1s}$$

to the lowpass function and obtain the following bandpass function:

$$T(s) = \frac{0.001 s^3}{s^6 + 0.2 s^5 + 3.02 s^4 + 0.401 s^3 + 3.02 s^2 + 0.2 s + 1} \tag{5.22}$$

If we choose to decompose $T(s)$ into three second-order bandpass functions, we will get, from (5.22)

$$T(s) = T_a(s) \cdot T_b(s) \cdot T_c(s)$$

where

$$T_a(s) = k_1 t_1(s) = \frac{k_1 s}{s^2 + 0.1s + 1}$$

$$T_b(s) = k_2 t_2(s) = \frac{k_2 s}{s^2 + 0.0478362s + 0.9170415}$$

$$T_c(s) = k_3 t_3(s) = \frac{k_3 s}{s^2 + 0.0521638s + 1.0904632}$$

with $K = 0.001$ and Q factors $Q_1 = 10$, $Q_2 = 20.02$, $Q_3 = 20.02$.

Having decided on the pole zero pairing, we now turn to the problem of the sequence in the cascade. We have the following possibilities:

$$T_a \cdot T_b \cdot T_c, \quad T_a \cdot T_c \cdot T_b, \quad T_b \cdot T_c \cdot T_a, \quad T_b \cdot T_a \cdot T_c, \quad T_c \cdot T_a \cdot T_b, \quad T_c \cdot T_b \cdot T_a$$

If we choose to follow the rule of thumb for forming the sequence in the order of increasing Q, then T_a should be in front, followed by either T_b or T_c, since T_b and T_c have equal Q factors. We choose the sequence T_a, T_b, T_c.

Next, we have to determine the gain distribution in such a way that the overall gain at the center frequency is unity, i.e., $K = 0.001$. Following the procedure outlined above, we find successively

$$k_1 = \frac{\max|T(j\omega)|}{\max|t_1(j\omega)|} = \frac{1}{10} = 0.1$$

$$k_2 = \frac{\max|t_1(j\omega)|}{\max|t_1(j\omega) \cdot t_2(j\omega)|} = \frac{10}{162.936} = 0.0613738$$

$$k_3 = \frac{K}{k_1 k_2} = \frac{0.001}{0.1 \times 0.0613738} = 0.162936$$

Thus, the final decomposition of the overall bandpass function will be as follows:

$$T_a = \frac{0.1s}{s^2 + 0.1s + 1} \quad (5.23)$$

$$T_b = \frac{0.0613738s}{s^2 + 0.0478362s + 0.9170415} \quad (5.24)$$

$$T_c = \frac{0.162936s}{s^2 + 0.0521638s + 1.0904632} \quad (5.25)$$

Each of these functions will be realized by the SAB shown in Fig. 5.3 and placed in the cascade in the order given in Fig. 5.4.

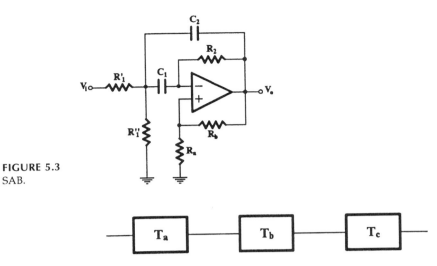

FIGURE 5.3
SAB.

FIGURE 5.4
Cascade sequence.

Following the procedure outlined in Section 4.5, the normalized and denormalized component values ($\omega_o = 1\text{krad/s}$, $R_o = 10\text{ k}\Omega$) are calculated. They are given on Table 5.1.

TABLE 5.1
Component Values

Component	Section T_1 Normalized	Section T_1 Denormalized kΩ, nF	Section T_2 Normalized	Section T_2 Denormalized kΩ, nF	Section T_3 Normalized	Section T_3 Denormalized kΩ, nF
R'_1	10.27	102.7	16.84	168.4	6.344	63.44
R''_1	0.145	1.45	0.1505	1.505	0.1398	1.40
R_2	7	70	7.3098	73.1	6.70336	67.03
C_1	1	100	1	100	1	100
C_2	1	100	1	100	1	100
R_a	0.265	2.65	0.3368	3.37	0.3368	3.37
R_b	10	10	10	100	10	100

5.6 Multiple-Loop Feedback Filters

We are concerned here with the application of negative feedback in a cascade connection of low-order sections. Two general topologies have been studied extensively:

- The leapfrog topology shown in Fig. 5.5
- The summed-feedback shown in Fig. 5.6

Realization of High-Order Functions

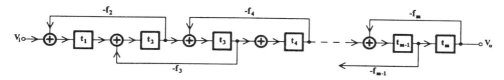

FIGURE 5.5
The leapfrog topology.

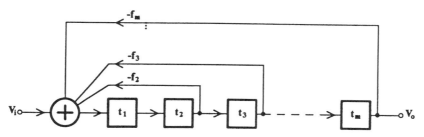

FIGURE 5.6
The summed-feedback topology.

The leapfrog topology is useful in the functional simulation of an LC ladder, and it is explained in the next chapter. The summed-feedback topology, as it appears in Fig. 5.6, is not suitable for realizing any finite transmission zeros. To overcome this problem, two useful techniques are the following:

a. The multiple- or distributed-input technique, shown in Fig. 5.7, in which the input signal is also feeding the input of all cascading sections, and
b. The summation of the input signal and the output signals from all cascaded sections, as shown in Fig. 5.8.

The topology in Fig. 5.6, and subsequently those in Figs. 5.7 and 5.8, are in fact generalizations (or adaptations) of similar analog computer methods for solving differential equations.

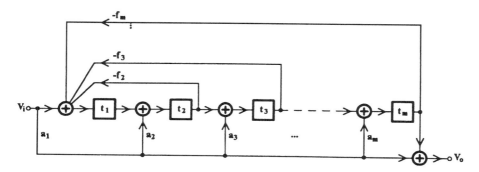

FIGURE 5.7
Summed-feedback distributed-input topology.

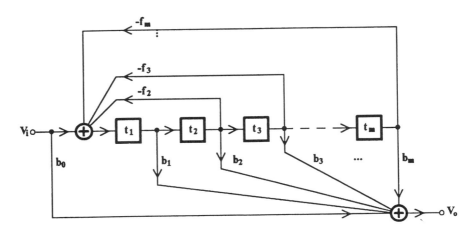

FIGURE 5.8
Summed-feedback summed-output topology.

To show this, consider for simplicity an nth-order lowpass function.

$$F(s) = \frac{K}{s^n + a_{n-1}s^{n-1} + \ldots + a_1 s + a_0} \quad (5.26)$$

This is to be realized as the voltage ratio V_o/V_i, in which case we will have, from (5.26)

$$V_o(s) = \frac{K}{s^n + a_{n-1}s^{n-1} + \ldots + a_1 s + a_0} V_i(s) \quad (5.27)$$

We can rewrite Eq. (5.27) in the following form:

$$s^n V_o = K V_i - (a_{n-1}s^{n-1} + a_{n-2}s^{n-2} + \ldots + a_1 s + a_0) V_o \quad (5.28)$$

Observe that V_o can be obtained from $s^n V_o$ by integrating $s^n V_o$ successively n times. If we then add KV_i and the output voltages from each integrator, weighted and signed according to Eq. (5.28), we will obtain $s^n V_o$. This is shown in Fig. 5.9 in block diagram form for n even. If the summation produces an extra sign reversal, the voltages should be summed with opposite signs. All voltages that take part in Eq. (5.28), with the opposite sign of that required, can have their sign reversed by summing them properly weighted, separately, using an opamp, the output of which is then connected to the input of the main summer in Fig. 5.9.

As an example, consider again the realization of the third-order Butterworth lowpass function

$$T(s) = \frac{1}{s^3 + 2s^2 + 2s + 1} = \frac{V_o(s)}{V_i(s)} \quad (5.29)$$

Writing this in the form of Eq. (5.28), we will have

$$s^3 V_o = V_i - (2s^2 + 2s + 1) V_o$$

Realization of High-Order Functions

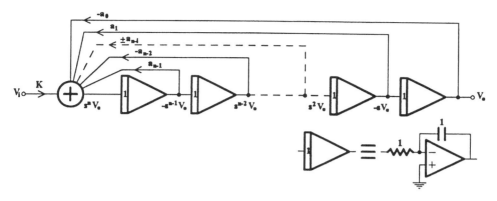

FIGURE 5.9
Realization of Eq. (5.28) using the analog computer technique. All time constants are normalized to unity.

Using single-input opamps, the complete circuit realizing the function will be as it is shown in Fig. 5.10. Notice that, from the output of the main summer, we additionally obtain the realization of the third-order Butterworth highpass function.

It can also be seen from Figs. 5.9 and 5.10 that any finite transmission zeros can be produced by summing voltages from the outputs of the various integrators properly signed and weighted. For example, the output of summer No. 2 in Fig. 5.10 gives the realization of the following function:

$$\frac{V'_o(s)}{V_i(s)} = -(2s^2 + 1)\frac{V_o}{V_i} = -\frac{2s^2 + 1}{s^3 + 2s^2 + 2s + 1}$$

Three other design methods based on the topology of Fig. 5.6 have been proposed and studied. These are the following:

- The primary-resonator block (PRB) [7, 8]
- The follow-the-leader feedback (FLF) [9, 10] and
- The shifted-companion form (SCF) [11].

Both the FLF and SCF networks are generalizations of the PRB network. In what follows, we review first the SCF method, in which we include the PRB, and then the FLF design.

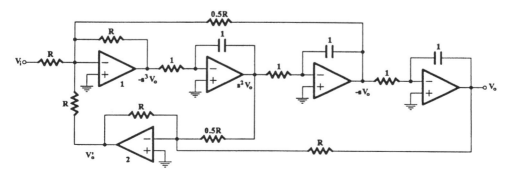

FIGURE 5.10
The analog computer approach to realizing the third-order Butterworth lowpass filter function.

5.6.1 The Shifted-Companion-Form (SCF) Design Method

For simplicity, we will explain this method by means of applying it to realize the third-order Butterworth lowpass function, Eq. (5.29). We proceed as follows.

First we select a parameter α, and we use it to shift the frequency variable s to a new frequency variable p such that

$$s = p - \alpha \tag{5.30}$$

We then introduce this into the expression for $T(s)$ Eq. (5.29) and obtain

$$T(p) = \frac{1}{(p-\alpha)^3 + 2(p-\alpha)^2 + 2((p-\alpha) + 1)}$$

or

$$T(p) = \frac{1}{p^3 + a_2 p^2 + a_1 p + a_o} \tag{5.31}$$

where

$$a_2 = 2 - 3\alpha$$
$$a_1 = 3\alpha^2 - 4\alpha + 2$$
$$a_o = 1 - \alpha^3 + 2\alpha^2 - 2\alpha \tag{5.32}$$

For an nth-order function, when the coefficient of p^n is 1, the usual selection of α is such that makes the coefficient of p^{n-1} equal to zero. This is, in fact, the ratio of the coefficient of s^{n-1} in the original denominator polynomial divided by n, the order of the function. In accordance with this, we get from the first of Eqs. (5.32) the following value of α:

$$\alpha = \frac{2}{3}$$

Using this in the rest of Eqs. (5.32), we get a_1 and a_o, i.e.,

$$a_1 = 2/3 \qquad a_o = 0.25926$$

Thus, $T(p)$ becomes

$$T(p) = \frac{1}{p^3 + \frac{2}{3}p + \frac{7}{27}} = \frac{1}{p^3 + 0.6667p + 0.25926} = \frac{V_o}{V_i}(p) \tag{5.33}$$

This transfer function can be realized by the block diagram (companion-form) of Fig. 5.11.

We now apply an opposite shift operation on the block diagram in Fig. 5.11 and obtain the block diagram in Fig. 5.12. Notice that, in order to obtain unity gain at dc, V_i is multiplied by $\alpha - 3$. This block diagram can be implemented in practice as shown in Fig. 5.13, assuming $C = 1$.

Realization of High-Order Functions 167

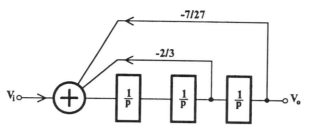

FIGURE 5.11
Realization of Eq. (5.33) in block diagram form.

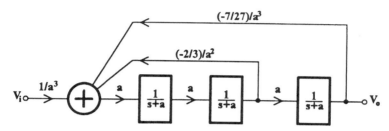

FIGURE 5.12
Realization of $T(s)$ in block diagram form.

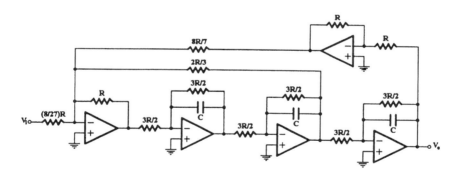

FIGURE 5.13
Final circuit realizing $T(s)$, Eq. (5.29).

A saving of one opamp (or two) in the circuit of Fig. 5.13 can be achieved if the summation of voltages, input, and feedback is performed using both inverting and noninverting inputs of the summer, or if the operation of summation is performed by the first lossy integrator with the opamp operating in differential mode.

The general SCF network has its first stage different from the others, because it performs the operation

$$\frac{1}{s + a_{n-1} + \alpha}$$

In this case, α is not selected as it was above, i.e., to make a_{n-1} equal to zero. However, when $a_{n-1} = 0$, because the value of α is selected for this purpose, the SCF network has all its stages identical, and the whole SCF circuit is identical to the PRB network.

By applying the usual lowpass-to-bandpass transformation

$$s_n = \frac{s^2 + \omega_o^2}{Bs}$$

to the block diagram in Fig. 5.12, the block diagram implementation of the geometrically, symmetric sixth-order Butterworth bandpass function will be obtained. In this case, each stage in the cascade becomes of order 2, and it requires a bandpass biquad for its realization. However, the feedback factors do not have to be changed.

Similarly, the highpass Butterworth filter function realization will be obtained if, in Fig. 5.12, the lowpass-to-highpass transformation is applied.

It should be mentioned that the PRB circuit cannot realize filters with finite transmission zeros, while the SCF can if α is not selected to make $a_{n-1} = 0$. In fact, by the general SCF network any transfer function, lowpass, highpass, bandpass, bandstop, and allpass can be realized. Of course, the summation or the feed-forward technique will be used for the realization of transmission zeros.

As a design example let us apply the transformation

$$s_n \to \frac{s^2 + 1}{0.1s}$$

as we did in the case of the CF. Then, each cascaded stage will become bandpass, as follows:

$$T(s) = \frac{\alpha}{\frac{s^2+1}{0.1s} + \alpha} = \frac{0.1\alpha s}{s^2 + 0.1\alpha s + 1}$$

With $\alpha = 2/3$, we finally get for $T(s)$

$$T(s) = \frac{0.0666667s}{s^2 + 0.0666667s + 1}$$

The Q factor of all stages is 15, thus the SAB in Fig. 5.3 can be used for the realization of each bandpass section, with all of these SABs being identical.

Since all three cascaded stages are tuned to the same center frequency and have equal Q factors and gains (unity), they will have the same maximum output voltage and there is no need to take any more steps to maximize the dynamic range of the filter.

Coming to the design of the SABs, we select $r = 1/49$ and, following the procedure outlined in Section 4.5, we obtain the component values, normalized and denormalized (ω_o = 1krad/s and R_o = 10 kΩ) given in Table 5.2. The overall circuit is given in Fig. 5.14(a) with each block representing the SAB in Fig. 5.14(b).

5.6.2 Follow-the-Leader Feedback Design (FLF)

The general FLF circuit is shown in block diagram form in Fig. 5.15. Clearly, the summation of the feedback voltages is responsible for the realization of the poles of the function, whereas the second summation is required for the realization of any finite transmission zeros. Here, $t_i(s)$ can be first-order lowpass or highpass functions or, alternatively, second-order functions.

Realization of High-Order Functions

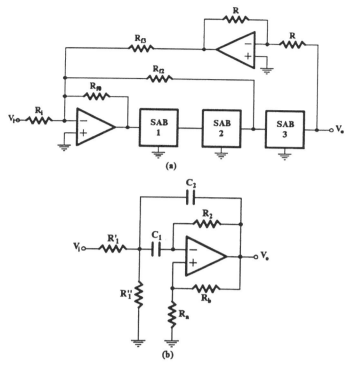

FIGURE 5.14
(a) The overall bandpass filter and (b) the circuit of each SAB.

TABLE 5.2

Component Values

Component	Values Normalized	Denormalized kΩ, nF
R'_1	15.47	154.7
R''_1	0.1442	1.442
R_2	7	70
C_1	1	100
C_2	1	100
R_a	0.3129	3.129
R_b	10	100
R_i	0.296	2.96
R_{f0}	1	10
R_{f2}	0.6667	6.667
R_{f3}	1.143	11.43
R	1	10

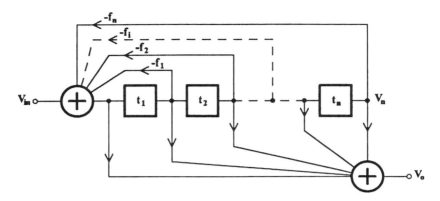

FIGURE 5.15
Block diagram of the general FLF circuit.

If we concentrate on all-pole functions the FLF block diagram can take the practical form shown in Fig. 5.16, where an opamp is used to perform the summation of the feedback voltages, assuming that there is no sign reversal in each t_i, $i = 1, 2,\ldots, n$, block.

This circuit is topologically similar to the SCF and PRB, except that there exists feedback from the output of stage t_1, which is missing in the case of the PRB circuit, although it can be considered part of the local feedback in the SCF circuit. It also differs from the PRB and the SCF circuits in that all t_i, $i = 1, 2,\ldots, n$ stages are not identical. However, it is possible to assume identical t_i and $R_1 = \infty$, in which case the FLF circuit becomes identical to the PRB circuit.

On the other hand the different t_i, $i = 1, 2,\ldots, n$ stages and the feedback from the t_1 stage can be used advantageously as additional degrees of freedom in order to improve the sensitivity of the FLF circuit. It has been shown [5], though, that this sensitivity improvement of the optimized FLF circuit is not so high as to force the designer to seek the optimized FLF circuit, if the PRB circuit can be used instead. For this reason, we do not include the optimization procedure here, but the reader can consult the relevant references [7–11] in order to satisfy any interest in the subject.

It was explained in Section 5.5 that the dynamic range is an important parameter in a high-order filter design using the cascade method. The same is true in the case of all multiple-loop feedback filters. Thus, the gain of the filter has to be properly distributed among the cascaded stages t_i so that the maximum voltage appearing at the output of each stage

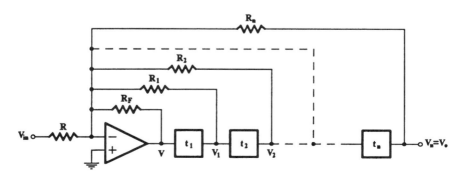

FIGURE 5.16
The practical FLF circuit with no finite transmission zeros.

Realization of High-Order Functions

is the same in all stages. The procedure to do this is the same as that followed in the cascade design, and it is not repeated here. However, we should note that in the PRB design, this gain distribution procedure is mathematically simpler than in the cascade or other FLF designs since, in the PRB design, all t_i stages are identical.

However, the optimized FLF design has been shown [12] to be the most practical multiple-loop feedback design based on sensitivity, dynamic range, and noise performance.

5.7 Cascade of Biquartics

As discussed in the previous sections, the CF filter is easy to design and tune, but its sensitivity in the passband is rather high compared to that of the MLF filters, when properly designed. MLFs, however, are difficult to adjust in practice. The cascade of biquartics filter, CBR, has been proposed [13] as an intermediate case, i.e., a filter with sensitivity in the passband lower than that of the CF filter, but which is easier to tune than the MLF filters. The design of CBR filters has been optimized [14–16] in the case of high-order geometrically symmetric bandpass filters with zeros at the origin and infinity. Therefore, here we will examine the design of this type of filters only. We refer to the stages of the CBR filter, which are of fourth order, as biquartic sections, or BR sections.

5.7.1 The BR Section

The block diagram of the BR stage is shown in Fig. 5.17. Each $t_i(s)$ stage, $i = 1, 2$, is a bandpass biquadratic function of the form

$$t_i(s) = \frac{h'_i s}{s^2 + \frac{\omega'_i}{Q'_i}s + \omega'^2_i} \tag{5.34}$$

Here, f is real and positive. If $f = 0$, the BR stage becomes the cascade of two biquadratics, which is of no interest to us here.

It should be mentioned that the topology of the BR section is the common topology of all MLF circuits, i.e., SCF, FLF, PRB, and LF, when these filters realize a fourth-order filter function. The transfer function of the BR stage is

$$T_{12}(s) = \frac{V_o}{V_i} = \frac{\alpha t_1(s) t_2(s)}{1 + f t_1(s) t_2(s)} = \alpha \frac{N(s)}{D(s)} \tag{5.35}$$

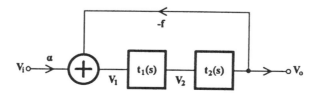

FIGURE 5.17
Block diagram of a BR section.

where $D(s)$ is a polynomial of fourth degree when $t_1(s)$ and $t_2(s)$ are given by Eq. (5.34) and

$$N(s) = h'_1 h'_2 s^2 \tag{5.36}$$

The gain coefficient is easily adjusted and can be useful to the designer in optimizing the dynamic range of the filter. We can set it equal to unity for reasons of simplicity.

Consider now the biquartic function

$$T_\alpha(s) = T_1(s) \cdot T_2(s) \tag{5.37}$$

with each $T_i(s)$, $i = 1, 2$ having the following form:

$$T_i(s) = \frac{h_i s}{s^2 + \frac{\omega_i}{Q_i} s + \omega_i^2} \tag{5.38}$$

If $T_\alpha(s)$ is to be realized by the biquartic section in Fig. 5.17, $D(s)$ in Eq. (5.35) has to be identified by the denominator of $T_\alpha(s)$ and similarly for $N(s)$. Thus, $D(s)$ will be

$$D(s) = \left(s^2 + \frac{\omega_1}{Q_1} s + \omega_1^2\right)\left(s^2 + \frac{\omega_2}{Q_2} s + \omega_2^2\right) \tag{5.39}$$

But from Eqs. (5.35) and (5.34), $D(s)$ is also given by

$$D(s) = \left(s^2 + \frac{\omega'_1}{Q'_1} s + \omega'^2_1\right)\left(s^2 + \frac{\omega'_2}{Q'_2} s + \omega'^2_2\right) + f h'_1 h'_2 s^2 \tag{5.40}$$

or

$$D(s) = D'(s) + \eta s^2 \tag{5.41}$$

where

$$D'(s) = \left(s^2 + \frac{\omega'_1}{Q'_1} s + \omega'^2_1\right)\left(s^2 + \frac{\omega'_2}{Q'_2} s + \omega'^2_2\right) \tag{5.42}$$

and

$$\eta = f h'_1 h'_2 \tag{5.43}$$

with η set in this form for convenience. Notice that $\eta > 0$.

Our task now is to determine $D'(s)$ and η (and consequently f) for the biquartic to realize $T_\alpha(s)$ in Eq. (5.37).

5.7.2 Effect of η on ω'_i and Q'_i

From Eqs. (5.41), (5.39), and (5.40), using simple algebra, the following equations can be obtained:

$$\frac{\omega'_1}{Q'_1} + \frac{\omega'_2}{Q'_2} = \frac{\omega_1}{Q_1} + \frac{\omega_2}{Q_2} \tag{5.44}$$

$$\omega'^2_1 + \omega'^2_2 + \frac{\omega'_1 \omega'_2}{Q'_1 Q'_2} = \omega_1^2 + \omega_2^2 + \frac{\omega_1 \omega_2}{Q_1 Q_2} - \eta \tag{5.45}$$

$$\frac{\omega'_1}{Q'_2} + \frac{\omega'_2}{Q'_1} = \frac{\omega_1}{Q_2} + \frac{\omega_2}{Q_1} \tag{5.46}$$

$$\omega'_1 \omega'_2 = \omega_1 \omega_2 \tag{5.47}$$

In the case of geometrically symmetric bandpass filters which have been obtained from the transformation of an all-pole lowpass function to bandpass, each pair of complex conjugate poles of the lowpass function transforms to two pairs of poles of the bandpass function which have identical Q factors.

Thus, depending on which two pairs of poles of the bandpass function the BR section is to realize, we distinguish two cases when referring to Eqs. (5.44) through (5.47).

a. $Q_1 = Q_2 = Q$ (symmetrical stage)
b. $Q_1 \neq Q_2$ (nonsymmetrical stage)

It has been shown [16] that the symmetrical stage is more advantageous in practice than the nonsymmetrical one. This will be explained later, when the realization of the overall function will be considered. Thus, the nonsymmetrical case will not be considered further here.

Referring to the symmetrical stage then, substituting Q for Q_1 and Q_2 in Eqs. (5.44) and (5.46) gives

$$\frac{\omega'_1}{Q'_1} + \frac{\omega'_2}{Q'_2} = \frac{\omega_1 + \omega_2}{Q} \tag{5.48}$$

and

$$\frac{\omega'_1}{Q'_2} + \frac{\omega'_2}{Q'_1} = \frac{\omega_1 + \omega_2}{Q} \tag{5.49}$$

Subtracting (5.49) from (5.48) or vice-versa and after some manipulation the following is obtained:

$$\left(\frac{1}{Q'_1} - \frac{1}{Q'_2}\right)(\omega'_1 - \omega'_2) = 0 \tag{5.50}$$

which gives the relationship between Q'_i and ω'_i, $i = 1, 2$.

Clearly, the values of Q'_i and ω'_i, $i = 1, 2$, will depend on the value of η as a consequence of Eq. (5.45). It can then be observed by means of Eq. (5.50), that there is a range of values of η for which

$$\omega'_1 = \omega'_2 \qquad Q'_1 \neq Q'_2$$

and another range of η values for which

$$\omega'_1 \neq \omega'_2 \qquad Q'_1 = Q'_2$$

and, finally, a value of $\eta = \eta_o$ for which

$$\omega'_1 = \omega'_2 \qquad Q'_1 = Q'_2$$

These can be clearly illustrated by means of an example. Consider pairing the equal Q factor biquadratics $T_b(s)$ and $T_c(s)$ of $T(s)$, Eq. (5.22), in order to form the biquartic function of interest here. In this case (however, see design example below with different indices)

$$Q_1 = Q_2 = Q = 20.0187$$

$$\omega_1 = 0.9576228$$

$$\omega_2 = 1.0442525$$

Using these values in Eqs. (5.45), (5.47), (5.48), and (5.49) the two diagrams in Fig. (5.18) can be obtained.

The value of $\eta = \eta_o$ is most interesting, because the two biquadratics $t_1(s)$ and $t_2(s)$ become identical and, consequently, the corresponding stages in the BR will be identical. This can be the starting point in any CBR filter design as it is considered here. We can determine this value η_o from Eq. (5.45) as follows.

Since, for identical $t_1(s)$ and $t_2(s)$,

$$\omega'_1 = \omega'_2 = \omega' \qquad Q'_1 = Q'_2 = Q'$$

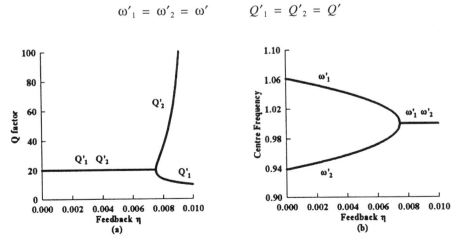

FIGURE 5.18
Effect of varying η on (a) the Q factors Q'_i and (b) the center frequencies ω'_i, $i = 1,2$ of biquads $t_1(s)$, $t_2(s)$, symmetrical stage.

Realization of High-Order Functions

substituting in Eq. (5.47) gives

$$\omega' = \sqrt{\omega_1 \omega_2} \tag{5.51}$$

Also from Eq. (5.48) using (5.51)

$$Q' = Q \frac{2\sqrt{\omega_1 \omega_2}}{\omega_1 + \omega_2} \tag{5.52}$$

Then, inserting these values in Eq. (5.45), the following value of η is obtained:

$$\eta = \eta_o = (\omega_1 - \omega_2)^2 \left(1 - \frac{1}{4Q^2}\right) \tag{5.53}$$

5.7.3 Cascading Biquartic Sections

Biquartic sections can be cascaded to realize high-order filter functions. There is no need for isolation stages between the BR sections, since their outputs are of low impedance, being the output of an opamp with negative voltage feedback. Thus, the cascade of biquartic sections filter or CBR filter is obtained.

As mentioned above, we consider here the realization of bandpass filter functions, which have been obtained from an all-pole lowpass via the usual lowpass-to-bandpass transformation. These functions will be of the following form:

$$T(s) = \prod_{i=1}^{N} T_i(s) \tag{5.54}$$

where each $T_i(s)$ will be given by Eq. (5.38).

Thus, the order of $T(s)$ will be $2N$, with N being even or odd.

If N is even, $T(s)$ can be written as the product of biquartic functions, each of them having in their numerator only a s^2 term multiplied by a constant. On the other hand, if N is odd, one bandpass biquadratic term can be separated from $T(s)$ and realized separately. Then, the rest of $T(s)$ will be of an order divisible by 4 and therefore will be treated as when N is even.

For an optimum CBR filter design, various degrees of freedom should be considered, namely, pairing the pole-pairs to obtain the biquartics, position of each BR section in the cascade, and distribution of the overall gain among the various stages. These degrees of freedom are to be considered in this and following sections.

Pairing pole pairs for obtaining symmetrical BR sections, apart from being practically more desirable than the nonsymmetrical sections, leads to further advantages [16] concerning sensitivity and noise. There is not much difference between the two cases, as far as dynamic range is concerned. Following this reasoning, pole-pairs are preferably paired in a way that symmetrical BR sections will be obtained.

5.7.4 Realization of Biquartic Sections

There is a flexibility in the realization of each biquartic section (Fig. 5.17), depending on the Q factor of the biquadratic blocks, t_i [Eq. (5.34)]. Thus, for Q factors ≤ 30 and in the audio frequencies regime, SABs [17] can be used [18] in the realization. Then, the overall BR section will be as shown in Fig. 5.19(a). However, since the possibility exists with this SAB to use both inputs of the operational amplifier to obtain summation of voltages, the summer can be eliminated, and the BR circuit will be as shown in Fig. 5.19(b) using two operational amplifiers.

In BR sections, in which the Q factors of the biquadratic blocks are greater than 30, two-opamp or even three-opamp biquads should be used. A suitable two-opamp biquad is the GIC-type biquad [19] shown in Fig. 5.20, which will take the place of each SAB in Fig. 5.19(a).

The two biquads, SAB and GIC-type, have been studied [20–22] and optimized from the sensitivity and noise points of view and, for this reason, they are used in the realization of filter functions here.

5.7.4.1 Design Example

For reasons of comparison, let us consider the use of this technique to design the sixth-order Butterworth bandpass filter that we have also designed as a CF and a PRB (SCF) filter.

Since it is a sixth-order function, we choose to pair the two equal-Q pole biquadratics in order to form the BR function, while the remaining lower-Q second-order bandpass function will be realized by the SAB in Fig. 5.3. Thus, we will have, from Eq. (5.22)

$$T_2(s)T_3(s) = \frac{k_2\, 0.047832s}{s^2 + 0.0478362s + 0.9170415} \times \frac{k_3\, 0.0521638s}{s^2 + 0.521638s + 1.0904632}$$

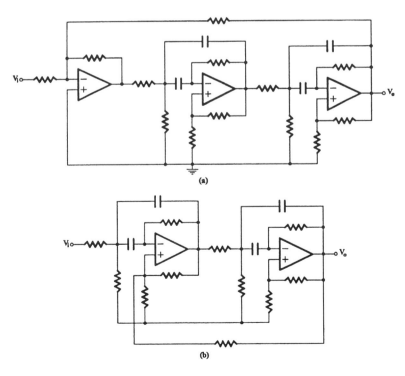

FIGURE 5.19
Realization of a BR section using (a) three and (b) two operational amplifiers.

Realization of High-Order Functions

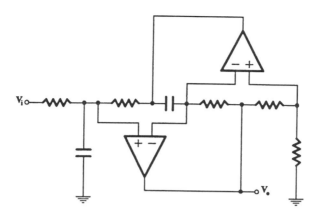

FIGURE 5.20
GIC-type biquad.

with

$$\omega_2 = 0.9576228 \qquad Q_2, Q_3 = 20.0187 = Q$$

$$\omega_3 = 1.0442525$$

and

$$D(s) = s^4 + 0.1s^3 + 2s^2 + 0.11s + 1$$

Then,

$$\omega'_2 = \omega'_3 = \sqrt{\omega_2 \omega_3} = 1$$

$$Q'_2 = Q'_3 = \frac{2\sqrt{\omega_2 \omega_3}}{\omega_2 + \omega_3} Q = \frac{2 \times 20.0187}{2.0018753} = 20.00$$

and

$$\eta_o = (\omega_2 - \omega_3)^2 \left(1 - \frac{1}{4Q^2}\right) = 0.0075$$

It can be seen that Q'_2, Q'_3 are slightly less than Q, while the two biquads in the BR stage will be tuned to the center frequency of the filter. Then,

$$t_2(s) = t_3(s) = \frac{0.04999689s}{s^2 + 0.04999689s + 1}$$

In practice it will be impossible to realize the coefficient of s using components even of 0.1 percent tolerance. So, to a very good approximation, we can write this function practically as follows:

$$t_2(s) = t_3(s) = \frac{0.05s}{s^2 + 0.05s + 1}$$

Two identical SABs will be used to realize $t_2(s)$ and $t_3(s)$.

To complete the design, we must calculate the feedback ratio f using the value of η_o. From Eq. (5.43), we will have

$$f = \frac{\eta_o}{h'_2 h'_3} = \frac{0.0075}{0.05 \times 0.05} = 3$$

If we choose to have the biquad (SAB) leading in the cascade, the overall filter will be as shown in Fig. 5.21 in block diagram form. The summation is performed using an opamp as is shown in Fig. 5.22.

Component values for the overall filter are given in Table 5.3, both normalized, and denormalized to $R_o = 10$ kΩ and $\omega_o = 1$ krad/s.

5.7.5 Sensitivity of CBR Filters

Having dealt with the problem of realization of the CBR filters, we can now proceed to examine their sensitivity and, if possible, optimize their design from this point of view. It is clear that the filter sensitivity will not depend on the sequence of the BR sections in the cascade. It will, however, depend on the feedback ratios η, on the types and the design of biquads used, and on the particular pairing of pole pairs to form the biquartic functions.

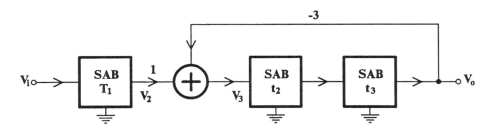

FIGURE 5.21
Realization of $F(s)$ using a SAB and a BR section connected in cascade. SABs t_2 and t_3 are identical.

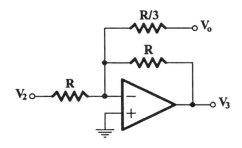

FIGURE 5.22
Opamp for performing summation.

TABLE 5.3

Component Values

	SAB T_1		SABs t_2, t_3	
Component	Normalized	Denormalized kΩ, nF	Normalized	Denormalized kΩ, nF
R'_1	10.27	102.7	20.67	206.7
R''_1	0.145	1.45	0.1439	1.44
R_2	7	70	7	70
C_1	1	100	1	100
C_2	1	100	1	100
R_a	0.265	2.65	0.33678	3.37
R_b	10	100	10	100
R	–	–	1	10

The sensitivity of the CBR circuit has been examined [16] realizing an eighth-order, 0.5 dB ripple Chebyshev bandpass filter function, which has been obtained from the corresponding all-pole lowpass via the transformation

$$S_n = \frac{s^2 + 1}{0.1s}$$

The standard deviation of the magnitude response of the filter was considered as the sensitivity measure. This was calculated according to a Monte Carlo method for 1000 tries. All passive components were assumed to have values uniformly distributed within their tolerance limits ±1 percent, and operational amplifiers were assumed to correspond to one-pole model with 10 percent tolerance in their gain-bandwidth product ($f_T = 1$ MHz). Results are as follows.

The sensitivity is reduced as the feedback ratio η increases up to the η_o value. Further increase in $\eta > \eta_o$ does not lead to substantial further reduction in sensitivity. It appears that sensitivity reduction of the BR section as η increases follows the reduction in the difference $\omega'_1 - \omega'_2$ and becomes nearly constant when $\omega'_1 = \omega'_2$ for $\eta \geq \eta_o$. In all cases ($0 < \eta < \eta_{max}$), the sensitivity of the BR section is lower than the sensitivity of this section with $f = 0$ (cascade filter). Here, by η_{max}, the value of η is denoted for which one of the Q'_i factors becomes infinite. This value can be obtained from Eq. (5.45) for the symmetrical case to be

$$\eta_{max} = (\omega_1 - \omega_2)^2 + \frac{\omega_1 \omega_2}{Q^2} \tag{5.55}$$

SABs can be used instead of GIC-type biquads, if the Q factors of the biquadratics in a BR stage are low (≤ 30), thus saving in operational amplifiers. It has been shown in this case [16] that the two CBR circuits are equivalent from the sensitivity point of view, but in the circuits with SABs, there is a saving of at least two operational amplifiers. Also, noise performance should be superior since, as has been shown [15], CBR circuits with SABs are less noisy than CBR circuits with GIC-type biquads.

Referring to the position of each BR section in the cascade, it has been shown [15, 16] that, as a rule of thumb, the BR sections should be placed in ascending Q factor order starting with the lowest Q section. This result is in agreement with the optimum ordering in the CF case.

5.8 Summary

The realization of a high-order filter function is necessary when the designer's filter problem is to satisfy the stringent selectivity requirements in telecommunication systems, special instrumentation and many other applications. Direct methods of realization of such filter functions using only one opamp are not practical, because they result in highly sensitive active circuits.

In this chapter, three practical methods of realizing high-order filter functions have been reviewed. A fourth method is explained in the next chapter. The three methods, namely, the cascade connection of second-order stages, CF; the multiple-loop feedback, MLF; and the cascade connection of fourth-order stages, CBR, have advantages and disadvantages. Thus, the CF can realize any type of stable filter function, is easy to design and tune, and requires fewer opamps than the other filters. Its disadvantage is the higher sensitivity in the passband compared to the other filters.

Three MLF circuits were reviewed. The primary-resonator block, the follow-the-leader feedback, and the shifted-companion form. Their common characteristic is the application of negative feedback in a cascade connection of low-order stages, first- or second-order, depending on the type of filter function, whether it is lowpass (highpass) or bandpass (bandstop), respectively. The MLF circuits have low sensitivity in the passband—much lower than the CF—but their design and tuning are more involved than in the case of the corresponding CF. The SCF is the most general of the three, while the FLF can be optimized to have lowest sensitivity and noise. The PRB is practically suitable for the realization of geometrically symmetric bandpass filters when the design of the SCF and the nonoptimized FLF result in the same PRB circuit.

The cascade connection of biquartic stages, CBR, is an alternative and useful approach for the design of geometrically symmetric bandpass filters. It combines the advantages of the CF (easy to design and tune) with the low sensitivity characteristics of the MLF. It has been proven also to display noise performance similar to that of the FLF circuit, which is the best among all the MLF circuits. Lowpass active filters of special form [23] as well as other filter functions [13, 24] can also be realized as CBR circuits, but the optimization procedure outlined above was derived [15] for the case of the geometrically symmetric bandpass filters only.

From the previous discussion, one can conclude that, when the selectivity demand is relatively low, the CF can be the preferable solution. However, when this demand is more stringent, the PRB or the CBR filters should be the choice provided, of course, that they can realize the pertinent function. Otherwise, the designer should look for an optimized FLF circuit or for a LC ladder simulated circuit provided, of course, that a suitable LC ladder realizing the required transfer function exists.

The method of LC ladder simulation leads to active RC filters of very low sensitivity,—lower than that of the MLF circuits—and is examined in the following two chapters.

References

[1] J. D. Schoeffler. 1976. "The synthesis of minimum sensitivity networks," *IEEE Trans. Circuit Theory* CT-11, 272–276.

[2] I. M. Sobel. 1975. *The Monte-Carlo Method,* Moscow: Mir Publishers, Moscow.
[3] G. S. Moschytz. 1975. *Linear Integrated Networks: Design,* New York: Van Nostrand-Reinhold.
[4] E. Lueder. 1970. "A decomposition of a transfer function minimization distortion and in band losses," *Bell Syst. Tech. J.* **49**, pp. 455–569.
[5] A. S. Sedra and P. O. Bracket. 1978. *Filter Theory and Design: Active and Passive,* London: Pitman.
[6] S. Halfin. 1970. "An optimization method for cascaded filters," *Bell Syst. Tech. J.* **44**, pp. 185–190.
[7] G. Hurtig, III. 1972. "The primary resonator block techniques of filter synthesis," *Proc. Int. Filter Symposium,* p. 84.
[8] G. Hurtig, III. 1973. Filter network having negative feedback loops, U.S. patent B, 720,881.
[9] K. R. Laker and M. S. Gausi. 1974. "Synthesis of a low sensitivity multiloop feedback active RC filter," *IEEE Trans. Circuit and Systems* CAS-21, pp. 252–259.
[10] ibid. 1974. "A comparison of active multiple-loop feedback techniques for realising high order bandpass filters," pp. 774–783.
[11] J. Tow. 1975. "Design and evaluation of shifted companion form active filters," *Bell Syst. Tech. J.* **54**, pp. 545–568.
[12] C. F. Chiou and R. Schaumann. 1980. "Comparison of dynamic range properties of high-order active bandpass filters," *Proc. IEEE* **127**, Pt, G, pp. 101–108.
[13] J. Tow. 1978. "Some results on two-section generalized FLF active filters," *IEEE Trans. Circuits and Systems,* 181–184.
[14] T. Deliyannis and S. Fotopoulos. 1981. "Noise in the cascade of biquartic section filter," *Proc. IEE* **128**, Pt G, 192–194.
[15] T. Deliyannis and S. Fotopoulos. 1982. "Sensitivity and noise considerations in the cascade of biquartic sections filters," *Proc. 1982 Intl. Symposium on Circuits and Systems,* ISCAS 82, Rome, 1102–1105.
[16] S. Fotopoulos and T. Deliyannis. 1984. "Active RC realisation of high order bandpass filter functions by cascading biquartic sections," *Int. J. Circuit theory and Applications* **12**, pp. 223–238.
[17] T. Deliyannis. 1968. "High Q factor circuit with reduced sensitivity," *Electron. Lett.* **4**, 577–579.
[18] P. E. Fleischer. 1976. "Sensitivity minimization in a single amplifier biquad circuit," *IEEE Trans. Circuits and Systems* CAS-23, 45–55.
[19] B. B. Bhattacharyya, W. S. Mikhael, and A. Antoniou. 1974. "Design of RC active networks using generalised-immitance converters," *J. Frank. Inst.* **297**(1), pp. 45–48.
[20] A. S. Sedra and J. L. Espinoza. 1975. "Sensitivity and frequency limitations of biquadratic active filters," *IEEE Trans. Circuits and Systems* CAS-22, 122–130.
[21] H. J. Bachler and W. Guggenbuhl. 1979. "Noise and sensitivity optimization of a single amplifier biquad," *IEEE Trans. Circuits and Systems* CAS-26, 30–36.
[22] C. F. Chiou and R. Schaumann. 1981. "Performance of GIC-derived active RC biquads with variable gain," *Proc. IEE* **128**, Pt G, pp. 46–52.
[23] M. Biey. 1983. "Design of lowpass two-section generalised FLF active filters," *Electron. Lett.* **19**, 639–640.
[24] A. N. Gonuleren. 1982. "Multiloop feedback unsymmetrical active filters using quads," *Intl. J. Cir. Theor. Appl.* **10**, 1–18.

Further Reading

R. Schaumann, M.S. Ghausi, and K.R. Laker. 1990. *Design of Analog Filters: Passive RC, and Switched Capacitor,* Englewood Cliffs, NJ: Prentice-Hall.

Chapter 6

Simulation of LC Ladder Filters Using Opamps

6.1 Introduction

In Chapter 5, we examined two general methods of high-order filter design, namely the cascade of low-order sections and the multiple-loop feedback method. We will now explain ways for the simulation of passive LC ladder filters as an alternative but, at the same time, popular and very useful approach to active RC filter design.

For reasons that we will explain in Section 6.2, the passive LC filter in the form of a ladder has to be resistively terminated as shown in Fig. 1.19, repeated here as Fig. 6.1 for convenience. The transfer voltage ratio of this filter obtained from Table 1.2 is as follows:

$$H(s) = \frac{V_o}{V_2} = \frac{Z_{21}R_L}{(R_s + Z_{11})(R_L + Z_{22}) - Z_{21}Z_{22}} \tag{6.1}$$

where Z_{ij}, $i, j = 1, 2$ are the z-parameters of the LC two-port.

Clearly, a filter function can be realized by this circuit, provided that it is written in the form of Eq. (6.1), so that the LC two-port parameters can be suitably identified and subsequently synthesized. The terminating resistances are taken into consideration during the design, and their ratio influences the amount of signal power transferred from source to load via the LC two-port.

In this chapter, we do not intend to design the LC two-port, as this has been done long ago for all the filter functions that have been obtained as the solution to the approximation problem. The values of the inductors and capacitors for various orders and R_L/R_s ratios of these functions have been tabulated and appear in many text and reference books [1–4].

The purpose of this chapter is to explain some ways for simulating either the impedance of the inductors or the operation of the passive LC ladder by means of active RC circuits. Thus, in Section 6.2, we give the motivation for using the simulation of the LC ladder filter in order to obtain useful high-order active RC filters. In Sections 6.3 and 6.4, we use gyra-

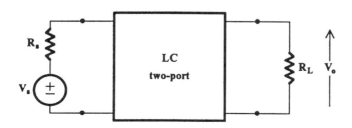

FIGURE 6.1
Resistively terminated LC two-port.

tors and generalized-immittance converters, first introduced in Chapter 3, to simulate the impedance of the inductors and the supercapacitors, the latter appearing in the passive ladder through a suitable impedance transformation of the prototype filter. Next, in Section 6.5, the method of simulating the actual operation of the passive ladder filter is presented.

6.2 Resistively-Terminated Lossless LC Ladder Filters

We are interested here in LC filters that are in the form of an LC ladder resistively terminated at both ends, as is shown in Fig. 6.2. In this circuit, X_i, $i = 1, 2, \ldots, n$ are LC impedances or admittances. X_1 and/or X_n may be missing, in some cases.

The important characteristic of the ladder is that there is a single path of signal transmission from source to load. Orchard has shown [5] that when the ladder is properly designed, at the frequencies of maximum power transfer, the first-order sensitivity of the magnitude of the transfer function to each inductor and capacitor is zero, while it remains low in the intermediate frequencies throughout the passband. This can be intuitively explained as follows: since the LC ladder is lossless at the frequencies of maximum power transfer, a change in an L or a C value can only increase the loss of the filter. Thus, the derivative of the frequency response (magnitude) with respect to each L and each C will be zero at these frequencies and so will be the corresponding sensitivity.

Furthermore, it has been shown [6, 7] that, for this type of filter, the sensitivity to the inductors and capacitors can be near zero throughout the passband rather than at only a few frequencies. It is therefore logical to expect that this low sensitivity of the lossless ladder filter will be retained in an active RC network which simulates the "operation" of the LC ladder or which contains active RC subcircuits that simulate the impedance of the inductances. This idea has resulted in the popularity of this method in active RC filter design.

It should be mentioned that the sensitivity of the ladder LC filter is not low in the transition and stopbands. However by proper design [6, 7] this can be close to a lower bound. On the other hand, as long as the loss in these bands remains higher than the filter requirements dictate, this sensitivity will not be of any practical importance.

6.3 Methods of LC Ladder Simulation

The simulation of a resistively terminated LC ladder can be achieved by the following four methods:

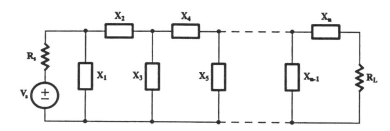

FIGURE 6.2
A resistively terminated ladder network.

1. Inductance substitution by a gyrator-C combination.
2. Impedance transformation of part or the whole of the LC ladder. In this case generalized-immitance converters are employed.
3. Simulation of currents and voltages in the ladder. The leapfrog (LF), coupled-biquad (CB) as well as the signal-flow-graph methods are names that have been used in the past to express essentially the same method.
4. The linear transformation (LT), which includes the wave active filter (WAF) method, approaches the simulation of the LC ladder the same way as c, but it uses transformed variables instead of simulating the voltages and currents of the LC ladder.

Methods 1 and 2 are usually considered to constitute the topological approach to LC ladder simulation, whereas the last two (3 and 4) constitute the functional or operational approach. In this chapter, we will explain the first three of these methods, while the fourth will be treated in the next chapter.

6.4 The Gyrator

The gyrator was introduced in Section 3.3. It is, in effect, a positive impedance inverter defined by its transmission matrix as follows:

$$[A] = \begin{bmatrix} 0 & \pm 1/g_2 \\ \pm g_1 & 0 \end{bmatrix}$$

Its symbol is shown again in Fig. 6.3(a). If it is terminated at port 2 in an impedance Z, Fig. 6.3(b), the impedance seen at port 1 will be

$$Z_1 = \frac{1}{g_1 g_2} \cdot \frac{1}{Z} \tag{6.2}$$

Clearly g_1, g_2 have the dimensions of a conductance and are called, most appropriately, gyration conductances. For inductance simulation, Z will be the impedance of a capacitor C, when Z_1 becomes, from (6.2)

$$Z_1 = s \frac{C}{g_1 g_2} = s L_{eq} \tag{6.3}$$

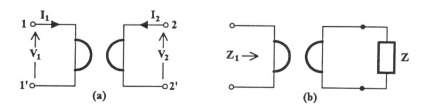

FIGURE 6.3
(a) The gyrator symbol and (b) the gyrator terminated in an impedance Z at port 2.

where

$$L_{eq} = \frac{C}{g_1 g_2} \tag{6.4}$$

The gyrator, strictly speaking, is an active two-port. However, if

$$g_1 = g_2 = g$$

it behaves as a lossless passive one, since then the power flowing in, $V_1 I_1$, is equal to the power flowing out, $V_2(-I_2)$, from the gyrator. However, the active gyrator, depending on the value of g_1/g_2, will act as an amplifier for the signal in one direction and as an attenuator in the opposite direction.

In the ideal case, we consider g_1 and g_2 independent of frequency which, in practice, is true only for a limited frequency range.

It can be seen from Fig. 6.3(b) that this simulated inductance seen at port 1 is grounded. However, in lowpass, bandpass, and bandstop LC ladders, floating inductors, i.e., inductors not connected to ground, are present. To simulate such an inductor, two gyrators are required to be connected as shown in Fig. 6.4. The two gyrators have to be matched; otherwise, this arrangement will not simulate a pure floating inductance.

The quality of the simulated inductor depends greatly on the quality of the capacitor C. Thus, if g_c is the leakage conductance of the capacitor, then the associated loss resistance RL of the simulated inductor can be calculated from Eq. (6.2) as follows:

$$Z_1 = \frac{1}{g_1 g_2} \cdot \frac{1}{Z_c} = \frac{1}{g_1 g_2}(sC + g_c)$$

Thus,

$$L_{eq} = \frac{C}{g_1 g_2}, \qquad R_L = \frac{g_c}{g_1 g_2} \tag{6.5}$$

Clearly, g_c should be as small as possible.

6.4.1 Gyrator Imperfections

Let us assume that the gyrator in Fig. 6.5 is not ideal and that its admittance matrix is as follows:

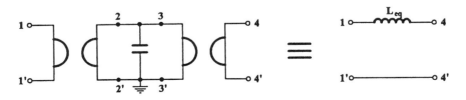

FIGURE 6.4
Use of two gyrators to simulate a floating inductor.

Simulation of LC Ladder Filters Using Opamps

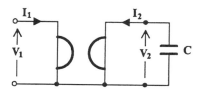

FIGURE 6.5
Nonideal gyrator terminated in a capacitor.

$$[Y] = \begin{bmatrix} g_a & g_1 \\ -g_2 & g_b \end{bmatrix} \tag{6.6}$$

where g_a, g_b are non-zero, pure conductances. Since

$$I_2 = -sCV_2$$

straightforward analysis gives that the input impedance Z_{in} will be the following:

$$Z_{in} \equiv \frac{V_1}{I_1} = \frac{sC + g_b}{g_a(sC + g_b) + g_1 g_2} \tag{6.7}$$

Substituting $j\omega$ for s, we get

$$Z_{in}(j\omega) = \frac{j\omega C g_1 g_2 + g_b(g_1 g_2 + g_a g_b) + \omega^2 C^2 g_a}{(g_1 g_2 + g_a g_b)^2 + \omega^2 C^2 g_a^2} \tag{6.8}$$

Thus, the input impedance represents the series connection of an inductance L_{eq} and an unwanted resistance R_u, where

$$L_{eq} = \frac{C g_1 g_2}{(g_1 g_2 + g_a g_b)^2 + \omega^2 C^2 g_a^2} \tag{6.9}$$

$$R_u = \frac{g_b(g_1 g_2 + g_a g_b) + \omega^2 C^2 g_a}{(g_1 g_2 + g_a g_b)^2 + \omega^2 C^2 g_a^2} \tag{6.10}$$

Therefore, the quality factor Q of the simulated inductance is finite, while both L_{eq} and its associated resistance R_u are functions of ω^2. Using Eqs. (6.9) and (6.10) this quality factor can be determined to be

$$Q = \frac{L_{eq}\omega}{R_u} = \frac{C g_1 g_2 \omega}{g_b(g_1 g_2 + g_a g_b) + \omega^2 C^2 g_a} \tag{6.11}$$

which has the following maximum value Q_{max}:

$$Q_{max} = \frac{g_1 g_2}{g_1 g_2 + g_a g_b} \sqrt{1 + \frac{g_1 g_2}{g_a g_b}} \qquad (6.12)$$

The value of Q_{max} is independent of frequency and the capacitance C (provided the capacitor is ideal) and occurs at the following frequency:

$$\omega_o = \frac{1}{C}\sqrt{(g_1 g_2 + g_a g_b)\frac{g_b}{g_a}} \qquad (6.13)$$

Thus, for high-quality simulated inductances, the gyrator parasitic conductances g_a, g_b should be as small as possible compared to g_1, g_2.

6.4.2 Use of Gyrators in Filter Synthesis

It is implied from the above discussion that the gyrator-capacitor combination can take the place of an inductor in the LC ladder. As an example, consider the third-order highpass filter shown in Fig. 6.6(a). Let $g_1 = g_2 = 10^{-3}$S.

The capacitance required for the simulation of the 30 mH inductance will be determined using Eq. 6.4. Thus, solving for C, we get

$$C = g_1 g_2 L_{eq} = 10^{-6} \times 3 \times 10^{-2} F = 30 \text{ nF}$$

The circuit using the simulated inductance is shown in Fig. 6.6(b).

LC ladders most suitable for inductance simulation using gyrators are those with no floating inductors in their structure. Such are highpass filters and bandpass filters with no transmission zeros in the upper stopband. The structures of these types of filters are shown in Fig. 6.7(a) and (b), respectively.

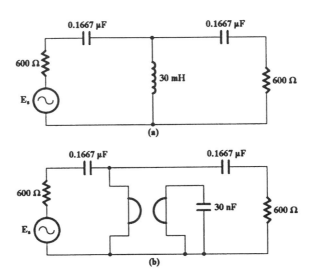

FIGURE 6.6
(a) Simulation of the inductance in an LC filter using (b) a gyrator-capacitor combination.

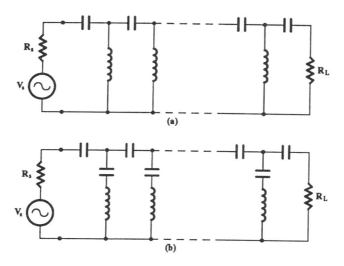

FIGURE 6.7
Optimum ladder structures for inductance simulation using gyrators: (a) highpass and (b) bandpass with no transmission zeros in the upper stopband.

It should be mentioned that the impedance-inverting property of the gyrator can be used to transform an RC impedance to an RL impedance. For example, consider the gyrator terminated at port 2 by the RC impedance

$$Z_{RC} = R = \frac{1}{sC} \tag{6.14}$$

The impedance presented at its input will be

$$Z_{in} = \frac{1}{g_1 g_2} \cdot \frac{1}{Z_{RC}}$$

or

$$Z_{in} = \frac{1}{g_1 g_2} \cdot \frac{1}{R + \frac{1}{sC}} = \frac{1}{g_1 g_2} \cdot \frac{sC}{sCR + 1} \tag{6.15}$$

Clearly, Z_{in} represents the equivalent impedance of a resistance $(1/Rg_1g_2)$ in parallel with an inductance (C/g_1g_2), as the reader can easily see.

This suggests that a filter function can be decomposed into an RC impedance (or admittance) function and an RL impedance (or admittance) function, the latter being realized as the input impedance (or admittance) at one port of a gyrator terminated by the appropriate RC impedance (or admittance) at the other port. Then, the two impedances (or admittances) are combined to give the overall circuit using a gyrator, resistors and capacitors. This method of using the gyrator in the synthesis of active RC networks was very popular in the 1960s, when saving in active elements was considered a figure of merit in active RC filter design. This is not so true today, though, when the low price of the opamps allows for the relaxation of this condition in favor of resorting to simpler methods of active RC synthesis such as the inductance simulation method.

6.5 Generalized Impedance Converter, GIC

The concept of the GIC was introduced in Section 3.3. As a reminder, it is a two-port defined by its transmission matrix

$$[A] = \begin{bmatrix} k & 0 \\ 0 & k/f(s) \end{bmatrix} \qquad (6.16)$$

where $f(s)$ is the impedance conversion function and k a positive constant.

The GIC is very useful in LC ladder filter simulation by active RC networks, when it is used either as a positive-impedance converter (PIC) or to produce a frequency-dependent negative resistance of type-D (D-FDNR). In the first case, $f(s) = s$, and the GIC is terminated at port 2 by a resistance R. Then, the input impedance at port 1 is the following:

$$Z_{i1} = f(s)Z_L = sR \qquad (6.17)$$

This is recognized as the impedance of an inductance R, in henries.

On the other hand, if $f(s) = 1/s$ and a capacitor C is connected across port 2, the input impedance at port 1 will be

$$Z_{i1} = \frac{1}{s} \cdot \frac{1}{sC} = \frac{1}{s^2 C} \qquad (6.18)$$

This is recognized as an FDNR of type-D, which is a negative resistance in effect, since, if $j\omega$ is substituted for s in Eq. (6.18), Z_{i1} becomes

$$Z_{i1} = -\frac{1}{\omega^2 C} \qquad (6.19)$$

which is resistive, negative, and frequency dependent.

It is usual to give the GIC the symbol shown in Fig. 6.8 with the dot always indicating the side of port 1 and the conversion function $f(s)$ written inside the box.

6.5.1 Use of GICs in Filter Synthesis

According to the previous discussion, a GIC with conversion function $f(s) = s$ and a resistor connected across its port 2 presents in its port 1 the impedance of an inductor (Fig. 6.9), thus operating as a PIC.

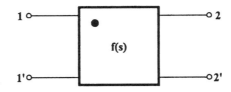

FIGURE 6.8
Usual symbol of the GIC.

Simulation of LC Ladder Filters Using Opamps

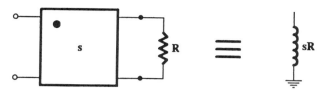

FIGURE 6.9
The GIC operating as PIC.

Two matched GICs of this type are required in order to simulate a floating inductance as shown in Fig. 6.10. This is in agreement with the corresponding case for gyrators.

Following this approach, we can say that optimum ladders for inductance simulation are, as in the case of gyrators, highpass filters, and bandpass filters with no zeros of transmission in the upper stopband. These filter structures are shown in Fig. 6.7(a) and (b), respectively. This type of inductance simulation technique was first introduced by Gorski-Popiel [8].

It has been shown [9] that the optimum GIC circuit for this application is that shown in Fig. 6.11 terminated at port 2 by the resistance R_5. The conversion function of this GIC can be easily shown (Section 3.5.2) to be the following:

$$\frac{V_1}{I_1} = sCRR_5 = ks \tag{6.20}$$

with
$$k = CRR_5 \tag{6.21}$$

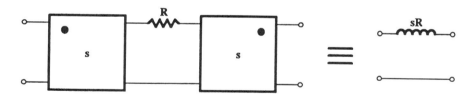

FIGURE 6.10
Simulation of a floating inductor.

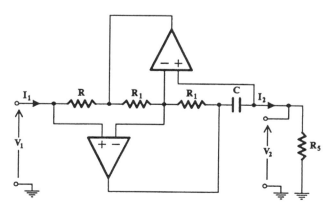

FIGURE 6.11
Most practical GIC circuit for inductance simulation used as a PIC.

Extending this, it can be shown [8, 10], that any $(n + 1)$-terminal network consisting of inductors can be simulated only using n GICs and a resistive network of the same topology. The importance of this statement is that it can lead to savings in the numbers of GICs required in the simulation of LC ladder filters, provided that such subnetworks can be separated from the corresponding ladders.

Consider, for example, the ladder filter in Fig. 6.12(a). The inductor subnetwork is as shown in Fig. 6.12(b) and can be simulated using three GICs [Fig. 6.12(c)] instead of five that would have been required otherwise (two for each floating inductor and one for L_3, the position of which can be exchanged with that of C to avoid its being floating).

As a second example, consider realizing a bandpass active filter from the Butterworth third-order lowpass filter

$$F(s) = \frac{1}{s^3 + 2s^2 + 2s + 1}$$

From tables [1, 2] or otherwise, we get the passive realization in LC ladder form shown in Fig. 6.13(a). Since there is a floating inductance in the circuit, two gyrators or two GICs are required for its realization arranged as is shown in Fig. 6.4 and Fig. 6.10, respectively.

From this, we can obtain the bandpass filter by applying the usual lowpass-to-bandpass transformation to the reactive components (Section 2.6.3)

$$s_n \Rightarrow \frac{\omega_o}{B}\left(\frac{s^2 + \omega_o^2}{\omega_o s}\right)$$

For $\omega_o = 1$ krad/s, $B = 100$ rad/s and an impedance level of 600 Ω, the sixth-order bandpass LC ladder filter will be as shown in Fig. 6.13(b). Clearly, it is preferable, for reasons of economy, to simulate this filter using PICs as is depicted in Fig. 6.13(c), since using gyrators

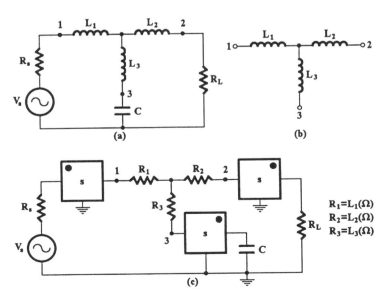

FIGURE 6.12
The LC ladder in (a), with the LC subnetwork in (b) simulated in (c) using GICs as PICs.

Simulation of LC Ladder Filters Using Opamps

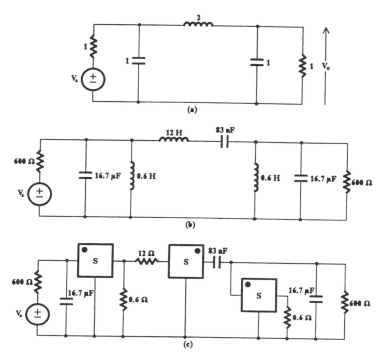

FIGURE 6.13
The third-order Butterworth lowpass filter in (a) is transformed into the bandpass in (b), which is simulated using PICS as in (c).

would require a larger number of opamps to be employed for the simulation. Notice that this simulated circuit employs three resistances of very low values, which make the circuit impractical. If the values of the terminating resistances can be raised further without any effect on the use of the circuit, the impedance level of the whole filter can be raised to make all the component values practical. On the other hand, if the terminating resistance values have to remain 600 Ω, the conversion constant k of the PICs will be smaller than 1, i.e., 10^{-3}, which will lead to increased resistance values of the PICs terminated resistors, namely 600 Ω and 12 kΩ instead of 0.6 Ω and 12 Ω, respectively.

One should note here that the PICs used in the simulation of the inductance subnetwork must be matched. Two different inductance subnetworks may employ for their simulation two different sets of PICs, but all the PICs in each set should be matched.

It can be seen that, by changing the inductance-subnetwork to the topologically similar resistive subnetwork using PICs, we have in effect performed a complex impedance scaling on part of the LC ladder. Complex impedance scaling of the whole ladder is explained immediately below.

6.6 FDNRs: Complex Impedance Scaling

This technique of inductance simulation has been treated extensively by Bruton [11]. It is most suitable for lowpass filters of the minimum capacitor realization. The reason for this will become apparent below.

Bruton's technique amounts to complex impedance scaling of the entire filter and not to part of it as we saw in the case of using PICs. It is based on the fact that the filter transfer function will remain unchanged if the impedance of each element is divided by the same quantity, in this case by s. We can demonstrate this simulation approach by means of an example.

The filter in Fig. 6.12(a) is scaled by dividing each impedance by s. Then, the resistance is transformed to the impedance of a capacitance, while the impedance of the capacitor is transformed to the impedance of a *supercapacitor* (Section 3.3.1), as is shown in Fig. 6.14(a). The simulated circuit is shown in Fig. 6.14(b), where a FDNR D-element is used to simulate the supercapacitor.

A GIC circuit useful in realizing the FDNR D-element is shown in Fig. 6.15, terminated in a capacitor C_5. It can be easily shown that its input admittance is

$$\frac{I_1}{V_1} = s^2 C_1 C_5 R_4 \qquad (6.22)$$

In cases where the capacitive terminations of the ladder are undesirable in Fig. 6.14(a), we can use two extra PICs terminated by two resistances R_s and R_L, as shown in Fig. 6.16.

Comparing the circuit in Fig. 6.14(b) to that in Fig. 6.12(c), it can be seen that the former requires fewer opamps than the latter, but its terminations are capacitive. However, if this is unacceptable, because of the existence of a source impedance or a resistive load, buffer

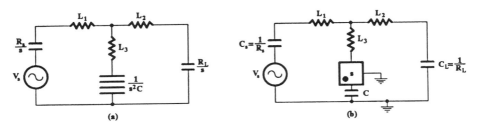

FIGURE 6.14
(a) Transformation of the ladder in Fig. 6.12(a) to another employing a D-element, which is realized by a GIC with $f(s) = s$ terminated in a capacitor (b).

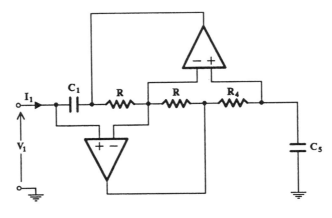

FIGURE 6.15
A useful FDNR D-element.

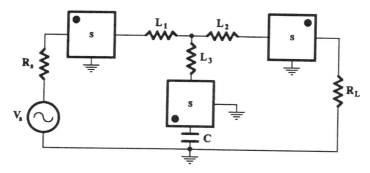

FIGURE 6.16
Realization of the ladder of Fig. 6.12(a) using a D-element and resistive terminations.

amplifiers can be used in the circuit, one between the signal source and the $1/R_s$ capacitor and the other between capacitor $1/R_L$ and the load resistance. This approach avoids using the circuit in Fig. 6.16, which employs three GICs also.

Lowpass LC ladder filters can be designed using either a minimum number of inductors or a minimum number of capacitors [1]. Of the two, the latter, when simulated using Bruton's transformation, leads to a filter with minimum number of D-elements, thus saving opamps. Unfortunately, this is not true in the case of bandpass filters, where there is no saving in opamps, even as compared to the PIC design.

One serious practical problem with this ladder simulation technique is that there is no dc return path for the noninverting inputs of the two opamps in the FDNR circuit embedded in the circuit in Fig. 6.14(b). To avoid this, one solution [9] is to connect two large resistors, as large as practically possible, in parallel with the two terminating capacitors C_s and C_L. Their values should be chosen such that the required value of the transfer function at dc will not change. To demonstrate this, let R_a and R_b be the two resistances connected across the capacitors C_s and C_L, respectively, in Fig. 6.14(b). The value of V_o/V_s at dc is equal to 0.5 for all equally terminated lowpass LC ladders of interest. Therefore, the voltage drop across R_b should be $V_s/2$, which means that at dc

$$\frac{V_o}{V_s} = \frac{R_b}{R_a + L_1 + L_2 + R_b} = 0.5 \tag{6.23}$$

from which we get

$$R_b = R_a + L_1 + L_2 \tag{6.24}$$

6.7 Functional Simulation

In this approach to ladder simulation [12, 13], we seek to simulate the "operation" of the LC ladder, i.e., the equations that describe the topology of the LC ladder, rather than simulate the impedance of the inductances. In other words, instead of using an active circuit that simulates the impedance sL of inductance L, we try to simulate the voltage and the current that exist in sL, which are related by

$$I = \frac{V}{sL} \tag{6.25}$$

In so doing, we use voltages that are analogous to each inductor current and voltage. For example, we can simulate the above I, V relationship, by means of an integrator with time constant dependent on L, the output voltage of which is analogous to I.

Similarly, we treat the operation of a capacitance C. Here, an integrator with time constant dependent on C will be used to integrate a voltage, which is analogous to current I, in order to produce another voltage V according to the relationship

$$V = \frac{I}{sC} \tag{6.26}$$

It will be recalled that we use integrators, and not differentiators, for reasons related to the excessive noise behavior of the latter.

The functional simulation method that we are to explain here is known as the leapfrog (LF) method, because it leads to a circuit structure resembling that of the so-named children's game (see Fig. 5.5).

We will explain the LF method by means of an example. Consider the fifth-order lowpass filter shown in Fig. 6.17(a), where all node voltages as well as the currents in the inductors have been suitably named. We write the relationships connecting these currents and the node voltages using Kirchhoff's current rule as well as Ohm's law. Referring to node No. 1, we have

$$\frac{V_s - V_1}{R_s} - sC_1 V_1 - I_2 = 0$$

which can be also written as follows:

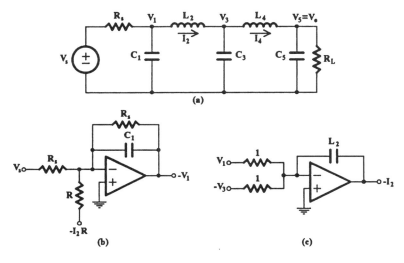

FIGURE 6.17
Fifth-order lowpass filter.

$$-V_1 = \frac{1}{sC_1}\left(\frac{V_s - V_1}{R_s} - I_2\right) \tag{6.27}$$

Thus, $-V_1$ can be obtained at the output of an inverting lossy integrator as shown in Fig. 6.17(b). In a similar way, we may write for I_2

$$-I_2 = -\frac{1}{sL_2}(V_1 - V_3) \tag{6.28}$$

Then, a voltage analogous to $-I_2$ can be obtained at the output of a lossless inverting integrator, in which the capacitance in farads is arithmetically equal to the value of L_2 in henrys. This is shown in Fig. 6.17(c).

Working on an analogous basis, we will obtain the rest of the required equations, which are as follows:

$$V_3 = -\frac{1}{sC_3}(I_4 - I_2) \tag{6.29}$$

$$I_4 = -\frac{1}{sL_4}(V_5 - V_3) \tag{6.30}$$

$$-V_5 = -\frac{1}{sC_5}\left(I_4 - \frac{V_5}{R_L}\right) \tag{6.31}$$

Clearly, V_3 and I_4 will be obtained at the output of lossless integrators, while $V_5(=V_o)$ at the output of a lossy integrator.

In building up the overall active RC ladder, it is usually helpful to produce it first in block diagram form and then insert the integrators and the other components in the places of the corresponding blocks. Thus, from Eqs. (6.27) through (6.31), we obtain the block diagram shown in Fig. 6.18.

It can be seen that integrators, sign changers, and summers are required for the implementation. However, summation of voltages can either be performed by the integrators or by the sign changers, preferably the former. Following this, we can now proceed to produce the actual active RC ladder. This is shown in Fig. 6.19.

It can be seen that the "horizontal" branches of the active ladder consist of inverting integrators alternating with inverting integrators and sign inverters in cascade, the latter being in effect noninverting integrators. The top and bottom integrators are lossy, because of the passive ladder terminating resistors, while the rest are lossless integrators.

In the case of a ladder of even order, the last capacitor will be missing [C_5 in Fig. 6.17(a)]. Then, the current in the last L will also pass through the load resistor R_L, which will give rise to a simple V, I relationship. Thus, in the case of Fig. 6.17(a) with C_5 missing the load end of the ladder would be as shown in Fig. 6.20(a). We will then have

$$-I_4 = \frac{1}{R_L}(-V_5) \tag{6.32}$$

and the active RC ladder would terminate as it is shown in Fig. 6.20(b).

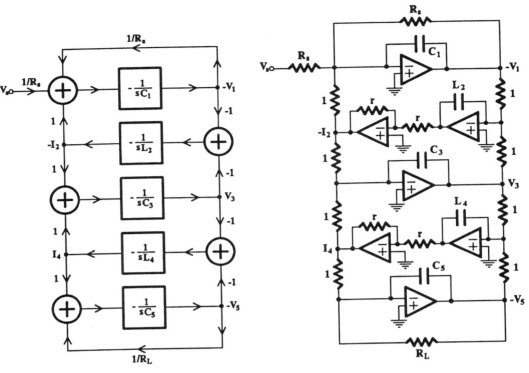

FIGURE 6.18
Block diagram of the active ladder simulating the operation of the filter in Fig. 6.17(a).

FIGURE 6.19
The overall LF or active RC ladder simulating the operation of the LC ladder filter in Fig. 6.17(a).

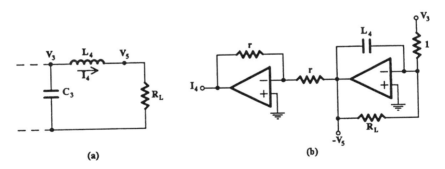

FIGURE 6.20
When capacitor C_5 in Fig. 6.17(a) is missing (a), the last branch of the active ladder will be as in (b).

6.7.1 Example

As an example, let us consider the realization of the third-order Butterworth lowpass function in the form of an LF structure. The lowpass prototype LC filter terminated by equal resistances in a normalized form is as shown in Fig. 6.21. Following the procedure outlined above, we find first the equations for voltages V_1, V_2 and current I_2. These are as follows:

$$-V_1 = -\frac{1}{sC_1}\left(\frac{V_s - V_1}{R_s} - I_2\right)$$

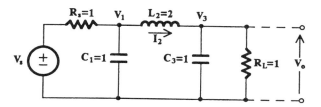

FIGURE 6.21
The lowpass LC filter of third-order.

$$-I_2 = -\frac{1}{sL_2}(V_1 - V_3)$$

$$V_3 = -\frac{1}{sC_3}\left(\frac{V_3}{R_L} - I_2\right)$$

Then, the LF structure in block diagram form, as well as the practical circuit, will be as shown in Fig. 6.22(a) and (b), respectively. The circuit in Fig. 6.22(b) can be denormalized to any convenient impedance level and any required cutoff frequency.

The use of so many inverting and noninverting integrators in the active ladder inevitably creates problems due to the excess phase associated with each of them. These are more pronounced at higher frequencies and require attention. We can use the methods explained in Section 4.9 and compensate the integrators, particularly the noninverting ones, since the excess phase created by them is higher than that created by the inverting integrators.

6.7.2 Bandpass Filters

Applying the usual lowpass-to-bandpass transformation

$$s_n = \frac{s^2 + \omega_o^2}{sB}$$

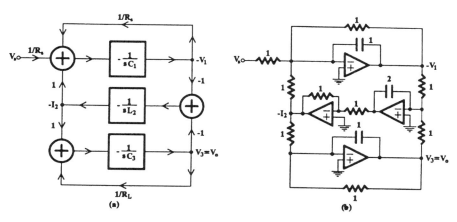

FIGURE 6.22
LF simulation of the third-order Butterworth lowpass LC filter shown in Fig. 6.21.

to the block diagram in Fig. 6.18, each integrator block will be transformed to a biquad of infinite Q factor, apart from the two lossy integrators at the beginning and the end of the ladder, which will have finite Q factors. We can redraw the block diagram of the general form shown in Fig. 5.5 repeated here in Fig. 6.23.

The bandpass filters obtained this way are geometrically symmetric or all-pole, as they are often called. Szentirmai [14] has generalized the method so that it can realize almost any filter function, i.e., bandpass and band-reject filters, which are not symmetrical, having arbitrary stopband requirements. If the bandpass filter is symmetrical, then all biquads can be chosen to resonate at the same frequency, which gives the circuit the characteristic of modularity.

The LF in its general form with each $t_i(s)$ block being a biquad is usually called the coupled-biquad structure (CB). The infinite-Q biquadratics will be in practice realized by biquads that can be adjusted to have as high a Q factor as possible. In many cases, we can use SABs, but in some cases, three-opamp biquads (see Chapter 4) may be more suitable. However, in the latter case, the number of opamps becomes excessive while, if these are not compensated for their excess phase, there will be serious distortion in their frequency response, when they operate at higher frequencies (say, above 100 kHz).

The LF active filters retain the low-sensitivity characteristic of the passive ladder in the passband, while in the transition or stopbands they are no better than the corresponding cascade filter.

As an example of obtaining a geometrically symmetric bandpass filter from a lowpass active ladder, let us apply the transformation

$$s_n \Rightarrow 10\left(\frac{s^2+1}{s}\right)$$

to the integrators of the lowpass filter in Fig. 6.22(a).

The two lossy integrators have the same transfer function

$$t_1(s), t_3(s) = -\frac{1}{s+1} \qquad (6.33)$$

while the lossless one in the middle, being noninverting, has the transfer function

$$t_2(s) = \frac{1}{2s} \qquad (6.34)$$

Applying the above lowpass-to-bandpass transformation to Eqs. (6.33) and (6.34) gives the following biquadratic functions, respectively:

$$T_1(s), T_3(s) = -\frac{0.1s}{s^2 + 0.1s + 1} \qquad (6.35)$$

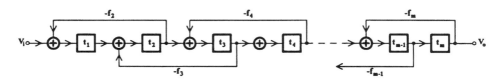

FIGURE 6.23
The leapfrog structure, repeated.

Simulation of LC Ladder Filters Using Opamps

$$T_2(s) = \frac{1}{20} \frac{s}{s^2 + 1} \tag{6.36}$$

Each one of T_1 and T_3 may be realized by the sign-inverting bandpass SAB in Fig. 4.10 or by the three-opamp Åkerberg-Mossberg biquad. However for the realization of T_2 a non-inverting bandpass biquad is required that is capable of having an infinite Q factor. Such a biquad can be either the Åkerberg-Mossberg suitably modified [9] or the Sallen and Key [15] bandpass biquad. Also, the SAB in Fig. 4.10 could be used followed by or, better, following a sign inverter.

If we choose to use the SAB in Fig. 4.10, T_1 and T_3, being identical, will be realized by the SAB in Fig. 6.24(a) and T2 by the circuit in Fig. 6.24(b). These biquads are then coupled according to the scheme in Fig. 6.25 to obtain the overall bandpass filter. Each biquad in this figure can be denormalized to the required center frequency ω_o and to a suitable impedance level. Note that the gain of the filter is 0.5 as in the LC prototype.

6.7.3 Dynamic Range of LF Filters

We have pointed out in Chapter 5 that the dynamic range of high-order active filters is an important characteristic of the various structures and should always be maximized during

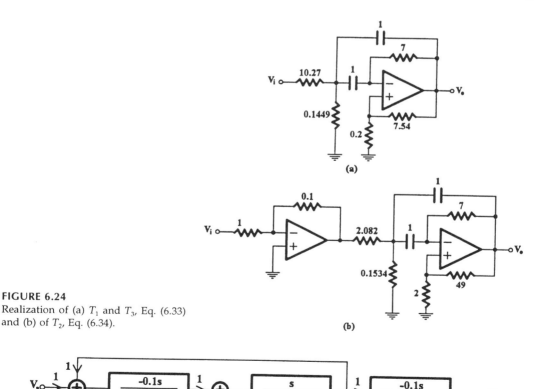

FIGURE 6.24
Realization of (a) T_1 and T_3, Eq. (6.33) and (b) of T_2, Eq. (6.34).

FIGURE 6.25
The overall LF or CB filter realizing the sixth-order Butterworth bandpass function.

the design stage. Luckily, the LF ladder can be adjusted for maximum dynamic range. For this purpose, one tries to adjust the gain of each t_i block to obtain the same maximum value at the output of every opamp in the circuit. Then the value of each feedback ratio f_i is changed to the value that keeps the product $f_i t_{i-1} t_i$ unchanged.

This is an advantage of the LF over the GIC-type simulation, where optimization of the dynamic range can be achieved only by properly intervening during the phase of the original passive synthesis [9].

6.8 Summary

The simulation of a resistively terminated ladder LC filter is desirable, because it leads to active RC filters with very low sensitivity in the passband. This simulation can be achieved either by simulating the inductances by means of opamps, resistors and capacitors, or by simulating the operation of the ladder. The opamps in the first case are used to realize gyrators, GICs (PICs), or FDNRs. In the second case, the opamps are used as integrators, lossless and lossy, summers, and sign-reversing elements (LF) or as parts of active biquads (CB).

The low passband sensitivity of the filters designed using the latter simulation approach is their most important characteristic. This allows for looser tolerances of the components when trimming these filters.

Another very important advantage of this kind of active RC filters is the availability of the prototype ladder LC designs. The well known lowpass Butterworth, Chebyshev, Bessel, and Cauer filter functions have been designed as resistively terminated ladder LC filters and tabulated. From these tables, using the element transformation table of Section 2.7.1 other filter types (namely highpass, bandpass, and bandstop) can be obtained, depending on the requirements. Of course, this does not prevent anyone from designing a custom ladder LC filter, something that one must certainly do when the requirements call for the design of an equalizer.

The main important disadvantages of the ladder simulation method are the following:

1. The large number of opamps required which, apart from the economical problem associated with it, leads to high dc power consumption, thus creating heat in the circuit. In some cases though, the number of opamps required may be reduced if the use of one-opamp grounded gyrators and FDNRs [16, 17] can satisfy the filter requirements.
2. The limited dynamic range of the filters obtained by the topological method of simulation. This is not important in the case of the LF type of simulation since, as we have mentioned, maximization of the dynamic range of the corresponding circuits is possible. In the case of the topological method of simulation, maximizing the dynamic range is rather involved and should be applied during the design of the passive ladder circuit.

References

[1] L. Weinberg. 1962. *Networks Analysis and Synthesis*, New York: McGraw-Hill.
[2] A. I. Sverev. 1967. *Handbook of Filter Synthesis*, New York: John Wiley and Sons.

[3] E. Christian and E. Eisenman. 1966. *Filter Design Tables and Graphs,* New York: John Wiley & Sons.
[4] J. K. Skwirzynski. 1965. *Design Theory and Data for Electrical Filters,* Princeton, NJ: D. Van Nostrand.
[5] H. J. Orchard. 1966. "Inductorless filters," *Electron. Lett.* **2**, pp. 224–225.
[6] M. L. Blostein. 1963. "Some bounds on sensitivity in RLC networks," *Proc. 1st Allerton Conf. on Circuits and System Theory,* Urbana, Illinois, pp. 488–501.
[7] G. C. Temes and H. J. Orchard. 1973. "First order sensitivity and worst-case analysis of doubly terminated reactance two-ports," *IEEE Trans. on Circuit Theory* **20**(5), pp. 567–571.
[8] J. Gorski-Popiel. 1967. "RC-active synthesis using positive-immittance converters," *Electron. Lett.* **3**, pp. 381–382.
[9] A. S. Sedra and P. O. Brackett. 1979. *Filter Theory and Design: Active and Passive,* London: Pitman.
[10] A. Antoniou. 1970. "Novel RC-active network synthesis using generalized-immittance converters," *IEEE Trans. on Circuit Theory* CT-17, pp. 212–217.
[11] L. T. Bruton. 1980. *RC-Active Circuits: Theory and Design,* Englewood Cliffs, NJ: Prentice-Hall.
[12] F. E. I. Girling and E. F. Good. 1970. "Active filters 12. The leap-frog or active ladder synthesis," *Wireless World* **76**, pp. 341–345.
[13] F. E. I. Girling and E. F. Good. 1970. "Active filters 13. Applications of the active ladder synthesis," *Wireless World* **76**, pp. 445–450.
[14] G. Szentirmai. 1973. "Synthesis of multiple-feedback active filters," *Bell Syst. Tech. J.* **52**, pp. 527–555.
[15] R. P. Sallen and E. L. Key. 1955. "A practical method of designing RC active filters," *IRE Trans. Circuit Theory* CT-2, pp. 74–85.
[16] H. J. Orchard and A. N. Wilson, Jr. 1974. "New active gyrator circuit," *Electron. Lett.* **10**, pp. 261–262.
[17] C. E. Schmidt and M. S. Lee. 1975. "Multipurpose simulation network with a single amplifier," *Electron. Lett.* **11**, pp. 9–10.

Chapter 7
Wave Active Filters

7.1 Introduction

Wave active filter (WAF) design is an alternative approach to the simulation of resistively terminated LC ladder filters in the effort to obtain active RC filters of low sensitivity [1]. Their development [2–4] followed the introduction of the wave digital filters initially by Fettweis [5] and later by Constantinides [6]. Some difficulties observed in the earlier work on WAF were overcome by the introduction of a more general approach [7] based on the derivation of a general wave two-port for a floating impedance and another for a shunt admittance. Thus, it was possible to indicate the one-one correspondence between the passive ladder elements and these two-ports, which can be easily implemented using resistors, capacitors, and opamps. In further developments [8–10], the somehow excessive number of opamps employed in the initially proposed WAFs was greatly reduced by the introduction of certain modified techniques, whereas their sensitivity and high-frequency performance was studied [11].

The material presented in this chapter is based mainly on the content of References 7 and 11. Also, the design of linear transformation active (LTA) filters [12] is included. Their development, which followed that of the WAFs, led to a general design approach that may be interpreted to include the designs of WAFs, leapfrog, and signal-flow-graph filters as special cases.

7.2 Wave Active Filters

These filters simulate the resistively terminated LC ladder filters by means of active equivalent subnetworks of each series-arm impedance and each parallel arm admittance of the passive ladder. Each element of the passive ladder is treated as an elementary two-port, and its active RC equivalent is determined after its voltage and current port variables have been linearly transformed to another sets of variables, which will be subsequently referred to as the *wave variables*.

To develop this simulation method, consider the resistively terminated LC ladder shown in Fig. 7.1. Without loss of generality, let each Z_i and each Y_{i+1} be a simple reactive element L or C. Of course, in some cases, Z_1 and/or Y_n may not be present, but this does not matter as far as the subsequent development of the method is concerned.

Consider now such an elementary two-port N shown in Fig. 7.2(a) and defined by its V and I port variables.

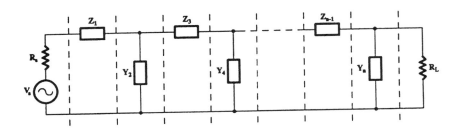

FIGURE 7.1
Terminated LC ladder.

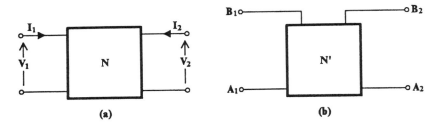

FIGURE 7.2
(a) An elementary two-port N and (b) its wave equivalent N'.

For reasons to be justified later, we consider that these V, I port variables are related by means of a modified transmission matrix A defined by the following convention:

$$\begin{bmatrix} V_1 \\ I_1 \end{bmatrix} = \begin{bmatrix} a_{11} & a_{12} \\ a_{21} & a_{22} \end{bmatrix} \begin{bmatrix} V_2 \\ I_2 \end{bmatrix} = [A] \begin{bmatrix} V_2 \\ I_2 \end{bmatrix} \quad (7.1)$$

where

$$[A] = \begin{bmatrix} a_{11} & a_{12} \\ a_{21} & a_{22} \end{bmatrix} \quad (7.2)$$

Clearly, in Eq. (7.1), parameters a_{12} and a_{22} have the opposite signs to those they would normally have.

Let us now introduce the following linear transformation to port variables:

$$A_i = V_i + I_i R_i$$

$$B_i = V_i - I_i R_i \quad i = 1, 2 \quad (7.3)$$

Variables A_i, $i = 1, 2$ are considered to be the incident waves, while B_i the reflected waves. R_i, $i = 1, 2$ are the port normalization resistances used to translate the currents I_i, $i = 1, 2$ to voltages. This transformation can be represented by the wave equivalent two-port in Fig. 7.2(b). Combining Eqs. (7.3) and (7.1) gives the following wave equations:

$$\begin{bmatrix} B_1 \\ B_2 \end{bmatrix} = \begin{bmatrix} s_{11} & s_{12} \\ s_{21} & s_{22} \end{bmatrix} \begin{bmatrix} A_1 \\ A_2 \end{bmatrix} \tag{7.4}$$

Here s_{ij}, $i, j = 1, 2$ are the scattering parameters, which take the following values:

$$s_{11} = (a_{11} - a_{21}R_1 - a_{12}G_2 + a_{22}R_1G_2)/\Delta$$

$$s_{12} = R_1G_2/\Delta$$

$$s_{21} = 1/\Delta$$

$$s_{22} = (a_{11} + a_{21}R_1 + a_{12}G_2 + a_{22}R_1G_2)/\Delta$$

$$\Delta = a_{11} + a_{21}R_1 - a_{12}G_2 - a_{22}R_1G_2$$

$$G_2 = 1/R_2 \tag{7.5}$$

For a series-arm impedance Z and a parallel-arm admittance Y, we have, in terms of the modified transmission matrix description, the following:

$$\begin{bmatrix} a_{11} & a_{12} \\ a_{21} & a_{22} \end{bmatrix}_Z = \begin{bmatrix} 1 & -Z \\ 0 & -1 \end{bmatrix} \tag{7.6}$$

$$\begin{bmatrix} a_{11} & a_{12} \\ a_{21} & a_{22} \end{bmatrix}_Y = \begin{bmatrix} 1 & 0 \\ Y & -1 \end{bmatrix} \tag{7.7}$$

Substituting in Eqs. (7.5) and subsequently in Eqs. (7.4) we get the following equations relating the wave variables for the two cases:

$$\begin{bmatrix} B_1 \\ B_2 \end{bmatrix} = \begin{bmatrix} \dfrac{R_2 - R_1 + Z}{R_1 + R_2 + Z} & \dfrac{2R_1}{R_1 + R_2 + Z} \\ \dfrac{2R_2}{R_1 + R_2 + Z} & \dfrac{R_1 - R_2 + Z}{R_1 + R_2 + Z} \end{bmatrix} \begin{bmatrix} A_1 \\ A_2 \end{bmatrix} \tag{7.8}$$

and

$$\begin{bmatrix} B_1 \\ B_2 \end{bmatrix} = \begin{bmatrix} \dfrac{G_1 - G_2 - Y}{G_1 + G_2 + Y} & \dfrac{2G_2}{G_1 + G_2 + Y} \\ \dfrac{2G_1}{G_1 + G_2 + Y} & \dfrac{G_2 - G_1 - Y}{G_1 + G_2 + Y} \end{bmatrix} \begin{bmatrix} A_1 \\ A_2 \end{bmatrix} \tag{7.9}$$

with $G_i = 1/R_i$.

If we consider that the incident waves A_1, A_2 and the reflected waves B_1, B_2 are voltages, Eqs. (7.8) and (7.9) can be represented by the circuits in Figs. 7.3(a) and (b), respectively. Equations (7.8) and (7.9) do not change if both the numerator and the denominator of each scattering parameter is divided by Z/R_o and Y/G_o, respectively. It is then possible to use this approach to simulate LC ladder filters using active RC equivalent circuits for each series-arm impedance and parallel-arm admittance as we explain below.

To this end, we have to develop active RC equivalents for the various elements in the ladder, namely, for each L, C, the signal source, and the terminating resistances. Then, we should solve the problem of interconnecting them in order to obtain the overall ladder.

7.3 Wave Active Equivalents (WAEs)

In this section, we develop WAE circuits for each elementary two-port in the ladder, whether it be an L, or a C, or a tuned circuit.

7.3.1 Wave Active Equivalent of a Series-Arm Impedance

If, in Fig. 7.3(a), Z is the impedance of an inductance L, we may divide R_1, R_2, and Z by Z/R_o, i.e. by sL/R_o and obtain the WAE in Fig. 7.4(a). This impedance scaling has transformed the inductor into a resistor, and the two resistors into capacitors.

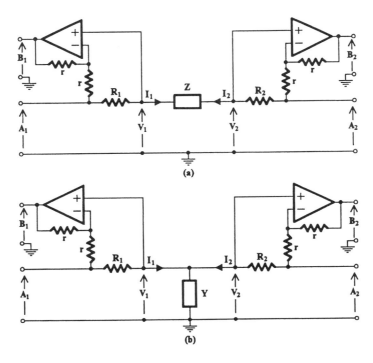

FIGURE 7.3
Active representation of Eqs. (7.8) and (7.9) in (a) and (b), respectively.

Wave Active Filters

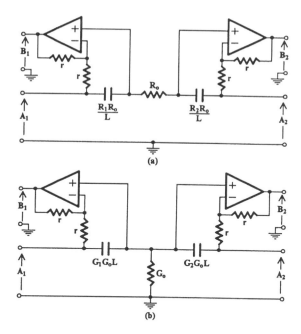

FIGURE 7.4
WA equivalents of an inductor (a) in a series arm, and (b) in a shunt arm.

On the other hand, if Z is the impedance of a capacitance C, no impedance scaling is required, since the wave equivalent in Fig. 7.3(a) will already be an active RC circuit.

7.3.2 Wave Active Equivalent of a Shunt-Arm Admittance

If the shunt element is a capacitor, its wave active equivalent will be that in Fig. 7.3(b), and no impedance scaling will be necessary for active RC realization. However, if the shunt element is an inductor of admittance $1/sL$, we can apply impedance scaling by dividing G_1, G_2 and Y by Y/G_o to obtain the WAE of Fig. 7.4(b), which is active RC realizable.

7.3.3 WAEs for Equal Port Normalization Resistances

The WAEs in Figs. 7.3(a) and (b) also apply when the port-normalization resistances R_1 and R_2 are equal. However, if we start with Eqs. (7.8) and (7.9) we can obtain additional WAEs, which can be more useful in some cases.

If we let $R_1 = R_2 = R$ in Eqs. (7.8) we obtain the following in the case of the series-arm impedance Z:

$$B_1 = \frac{Z}{2R+Z}A_1 + \frac{2R}{2R+Z}A_2$$

$$B_2 = \frac{2R}{2R+Z}A_1 + \frac{Z}{2R+Z}A_2$$

(7.10)

If we consider voltages A_1 and A_2 as the excitations, then B_1 and B_2 will be equal to the node voltages V_a and V_b in Figs. 7.5(a) and (b), respectively. In the case of Fig. 7.5(a), it can be seen that B_2 may be obtained as follows:

$$B_2 = A_1 + A_2 - V_a \tag{7.11}$$

Similarly, in the case of Fig. 7.5(b), B_1 may be obtained as

$$B_1 = A_1 + A_2 - V_b \tag{7.12}$$

Thus, depending on the relative position of 2R and Z with respect to the excitations A_1 and A_2, we can obtain the alternative WAE circuits shown in Fig. 7.6. The two WAEs in Fig. 7.6 can be combined in one circuit as shown in Fig. 7.7.

Applying then the impedance scaling, i.e., dividing Z and 2R by Z_o/R_o in Figs. 7.6(a) and (b) and in Fig. 7.7, when Z is the impedance of an inductance, active RC realizable wave equivalent circuits can be obtained. This is possible because, by dividing both numerators and denominators in Eqs. (7.8) by Z/R_o, the values of B_1 and B_2 do not change. No impedance scaling is required if the series element is a capacitor.

7.3.4 Wave Active Equivalent of the Signal Source

Consider the situation in Fig. 7.8. The voltage V, expressed in terms of V_S and I, is as follows:

$$V = V_S + I R_S \tag{7.13}$$

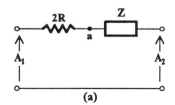
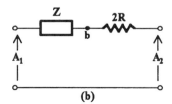

FIGURE 7.5
Circuits with A_1 and A_2 as excitations.

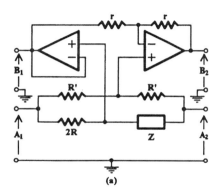
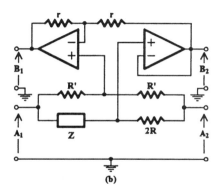

FIGURE 7.6
Alternative WAEs for a series-arm impedance.

Wave Active Filters

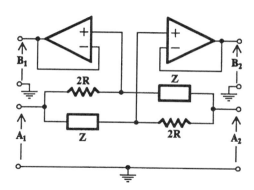

FIGURE 7.7
Alternative WAE obtained by combining Figs. 7.6(a) and (b) in one.

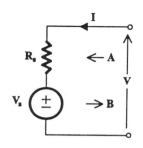

FIGURE 7.8

We also have that

$$A = V + IR$$

$$B = V - IR$$

where R is the port normalization resistance. Substituting for V from Eq. (7.13), we get

$$A = V_S + I(R_S + R) \qquad (7.14a)$$

$$B = V_S - I(R - R_S) \qquad (7.14b)$$

It can be seen from Eq. (7.14b) that if $R = R_S$

$$B = V_S \qquad (7.15)$$

This means that the reflected voltage from the signal source end of the ladder is equal to the open-circuit output voltage of the source.

7.3.5 Wave Active Equivalent of the Terminating Resistance

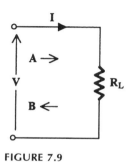

FIGURE 7.9

The situation is shown in Fig. 7.9. We may write

$$V = I R_L$$

while

$$A = V + IR = I(R_L + R)$$

$$B = V - IR = I(R_L - R)$$

Again, for $R_L = R$

$$B = 0 \qquad (7.16)$$

i.e., there is no reflection wave, and the incident power of the signal is completely absorbed by R.

7.3.6 WAEs of Shunt-Arm Admittances

The above procedure concerning the series-arm impedance can be repeated for the case of the shunt-arm admittance when $G_1 = G_2 = G$. Thus, starting with Eqs. (7.9) we can find alternative WAEs to that in Fig. 7.3(b). However, the circuit of Fig. 7.3(b) has been found to be directly usable and gives practically acceptable results. Thus, we do not intend to pursue this matter further, and leave it to the reader as an exercise.

7.3.7 Interconnection Rules

Having derived wave equivalents for the various elements in the passive ladder, the next step is to determine proper interconnection rules for the adjacent WAEs in order to avoid any errors that may arise from loading effects.

Consider the two adjacent passive two-ports N_a and N_b shown in Fig. 7.10(a). In the passive ladder, port 2 of N_a is directly connected to port 1 of N_b so that V_{1b} is equal to V_{2a} and $I_{1b} = -I_{2a}$. This can also be expressed mathematically as follows:

$$\begin{bmatrix} V_{1b} \\ I_{1b} \end{bmatrix} = \begin{bmatrix} 1 & 0 \\ 0 & -1 \end{bmatrix} \begin{bmatrix} V_{2a} \\ I_{2a} \end{bmatrix} \quad (7.17)$$

It is evident that the corresponding WAEs N'_a and N'_b cannot have their adjacent ports directly connected. Let us assume that a matching two-port N_c is going to be used in order to achieve the correct interconnection of the adjacent ports of N'_a and N'_b as shown in Fig. 7.10(b). Mathematically, where $[P]$ is a 2×2 nonsingular matrix, this can be described as follows:

$$\begin{bmatrix} A_{1b} \\ B_{1b} \end{bmatrix} = [P] \begin{bmatrix} A_{2a} \\ B_{2a} \end{bmatrix} \quad (7.18)$$

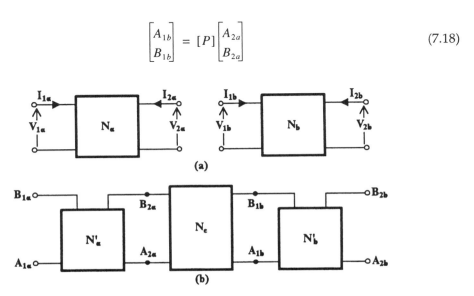

FIGURE 7.10
(a) Two adjacent two-ports and (b) their WAEs interconnected.

Assuming equal port normalization resistances at ports 2a and 1b, and using Eqs. (7.3), we can write for each one of the adjacent ports the following:

$$\begin{bmatrix} A_{1b} \\ B_{1b} \end{bmatrix} = \begin{bmatrix} 1 & R \\ 1 & -R \end{bmatrix} \begin{bmatrix} V_{1b} \\ I_{1b} \end{bmatrix} \quad (7.19)$$

$$\begin{bmatrix} A_{2a} \\ B_{2a} \end{bmatrix} = \begin{bmatrix} 1 & R \\ 1 & -R \end{bmatrix} \begin{bmatrix} V_{2a} \\ I_{2a} \end{bmatrix} \quad (7.20)$$

Substituting in (7.19) for $[V_{1b}\ I_{1b}]^T$ from Eq. (7.17) we get

$$\begin{bmatrix} A_{1b} \\ B_{1b} \end{bmatrix} = \begin{bmatrix} 1 & R \\ 1 & -R \end{bmatrix} \begin{bmatrix} 1 & 0 \\ 0 & -1 \end{bmatrix} \begin{bmatrix} V_{2a} \\ I_{2a} \end{bmatrix}$$

or

$$\begin{bmatrix} A_{1b} \\ B_{1b} \end{bmatrix} = \begin{bmatrix} 1 & -R \\ 1 & R \end{bmatrix} \begin{bmatrix} V_{2a} \\ I_{2a} \end{bmatrix} \quad (7.21)$$

Finally, solving Eq. (7.20) for $[V_{2a}\ I_{2a}]^T$ and inserting its value in Eq. (7.21) gives

$$\begin{bmatrix} A_{1b} \\ B_{1b} \end{bmatrix} = \begin{bmatrix} 1 & -R \\ 1 & R \end{bmatrix} \begin{bmatrix} 1 & R \\ 1 & -R \end{bmatrix}^{-1} \begin{bmatrix} A_{2a} \\ B_{2a} \end{bmatrix}$$

or

$$\begin{bmatrix} A_{1b} \\ B_{1b} \end{bmatrix} = \begin{bmatrix} 0 & 1 \\ 1 & 0 \end{bmatrix} \begin{bmatrix} A_{2a} \\ B_{2a} \end{bmatrix} \quad (7.22)$$

On comparing Eq. (7.22) with Eq. (7.18), we obtain

$$[P] = \begin{bmatrix} 0 & 1 \\ 1 & 0 \end{bmatrix} \quad (7.23)$$

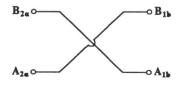

FIGURE 7.11
The cross-cascade connection of adjacent WAE ports with equal normalization resistances.

Therefore, the operation of the matching two-port N_c is to connect terminal A_{1b} to B_{2a} and the terminal B_{1b} to A_{2a}. Physically, this means that the reflected wave at port 2a is the incident wave at port 1b, and the incident wave at port 2a is the reflected wave at port 1b. This constitutes the required interconnection rule of two adjacent WAE ports and will be subsequently referred to as the *cross-cascade connection*. Schematically, this is shown in Fig. 7.11.

However, if the adjacent port normalization resistances are not equal, say R_1 and R_2, the two ports cannot be directly cross-cascaded. Since, in the passive LC ladder, the two adjacent terminals of the elements are connected by a short, Z = 0, we can replace Z in the WAE of Fig. 7.3(a) by a short-circuit and thus obtain the required matching network in Fig. 7.12.

The same matching network can be obtained from Fig. 7.3(b), if we set Y = 0, i.e., an open circuit. However, such a network may not be required, since the general circuits in Fig. 7.3 provide for different port normalization resistances, which can be always chosen at will, thus saving in active and passive components.

7.3.8 WAEs of Tuned Circuits

The tuned circuit in the LC ladder can be either series or parallel and connected as a series-arm impedance or a parallel-arm admittance.

There is no problem in the cases of a series-tuned in series-arm or a parallel tuned in a shunt-arm. In both cases, the two elements L and C can be treated as different two-ports and their WAEs determined as was explained above. However, the previously derived WAEs cannot be applied when the parallel tuned circuit is in a series-arm or the series-tuned circuit is in a shunt-arm of the LC ladder. In these cases, we work as follows.

Consider the case of the parallel-tuned circuit in a series-arm [Fig. 7.13(a)]. We may write Z in the following form:

$$Z = \frac{1}{sC_P + \dfrac{1}{sL_P}} \tag{7.24}$$

Substituting in Eqs. (7.8), we get (assuming $R_1 = R_2 = R$)

$$B_{1P} = \frac{1}{1 + 2R(sC_P + 1/sL_P)}A_1 + \frac{2R(sC_P + 1/sL_p)}{1 + 2R(sC_p + 1/sL_P)}A_2$$

$$B_{2P} = \frac{2R(sC_P + 1/sL_P)}{1 + 2R(sC_P + 1/sL_P)}A_1 + \frac{1}{1 + 2R(sC_p + 1/sL_P)}A_2 \tag{7.25}$$

On the other hand, for a series-tuned circuit in a series-arm, we have

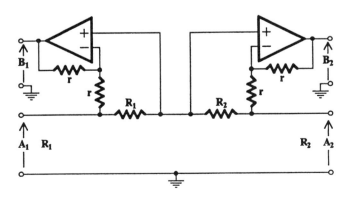

FIGURE 7.12
Matching network for nonequal port normalization resistances.

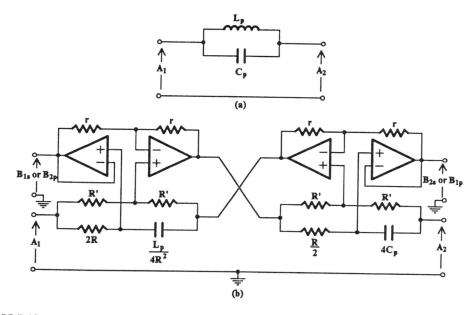

FIGURE 7.13
WAE of the parallel tuned circuit in a series-arm. Values of resistors r and R' may not be different and can be chosen conveniently.

$$Z = sL_s + \frac{1}{sC_s} \tag{7.26}$$

and substituting in Eqs. (7.9) we obtain the following:

$$B_{1S} = \frac{sL_s + 1/sC_s}{2R + (sL_s + 1/sC_s)}A_1 + \frac{2R}{2R + (sL_s + 1/sC_s)}A_2$$

$$B_{2P} = \frac{2R}{2R + (sL_s + 1/sC_s)}A_1 + \frac{sL_s + 1/sC_s}{2R + (sL_s + 1/sC_s)}A_2 \tag{7.27}$$

On comparing Eqs. (7.25) with Eqs. (7.27), it is observed that B_{1P} and B_{2S} are of the same form. They can be equal provided that $L_S = 4R^2C_P$ and $C_S = L_P/4R^2$. The same is true for B_{2P} and B_{1S}. We may then conclude that the WAE of a parallel-tuned circuit in a series-arm is the same with the WAE of a series-tuned circuit in a series-arm, under the condition that the values of L and C in the series-tuned circuit are given by the above mentioned relationships in terms of R, L_P, and C_P.

Following this and the interconnection rules explained previously, we obtain the WAE of the parallel tuned circuit in a series-arm to be as is shown in Fig. 7.13(b). The values $R/2$ and $4C_P$ in the second half of this WAE have been obtained by means of impedance scaling using the factor Z/R, where $Z = s2L_S = 8R^2C_Ps$.

The case of a series-tuned circuit in a shunt-arm is found to be simpler. One can simulate the inductance by a grounded gyrator, and this in series with the capacitor will take the place of admittance Y in Fig. 7.3(b). Since this case is rather simple, we will not consider it any further.

7.3.9 WA Simulation Example

As an example of a LC ladder simulation according to the wave active method consider the lowpass prototype in Fig. 7.14(a) with $R_S = R_L$. First, we split the ladder in simple two-port elements as the broken lines indicate, and for each one of these elementary two-ports we find the corresponding WAE circuit. These have been determined above and appear in Figs. 7.3(b) and 7.13. Choosing equal port normalization resistances in all WAEs and normalized to $R_S = 1$, we can easily obtain the WA filter shown in Fig. 7.14(b).

7.3.10 Comments on the Wave Active Filter Approach

It would be useful and constructive at this point if we made some comments on the wave active filter method of LC ladder simulation before proceeding to any further developments on this theory.

1. One interesting point about this method is that there is no restriction on the type of LC ladder filter that can be simulated. Once we have determined WAE circuits for any type of element, and their combination in shunt or in series as tuned circuits, these can be accordingly cross-cascaded to form the overall ladder as an active RC circuit. And this can be achieved for equal as well as nonequal resistive terminations.

2. It is also interesting and useful to note, looking at the WA filter in Fig. 7.14(b), that there are two outputs available, namely V_{o1} and V_{o2}. Still more interesting is the fact that the two transfer functions V_{o1}/V_s and V_{o2}/V_s are power complementary, when $R_S = R_L$, i.e.,

$$\left|\frac{V_{o1}}{V_s}\right|^2 + \left|\frac{V_{o2}}{V_s}\right|^2 = 1 \tag{7.28}$$

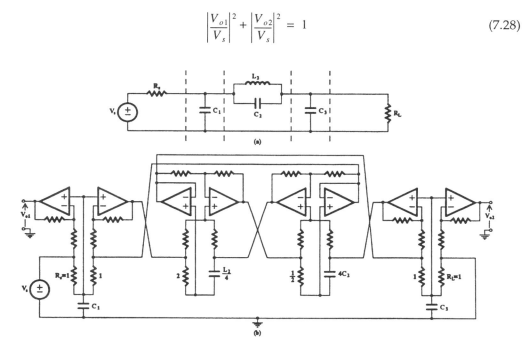

FIGURE 7.14
(a) The LC ladder prototype and (b) its WA realization. All resistor values not indicated are equal and can assume any convenient value in practice.

The meaning of this is that, if V_{o2}/V_s is a lowpass function, V_{o1}/V_s will be its power complementary highpass, and vice-versa. We will leave the proof of this statement until later when we will examine it in more detail.

3. It can be seen from Fig. 7.14(b) and deduced from the theory presented so far that the ladder simulation by this method requires a large number of opamps, larger than any other simulation method and indeed any other method of filter synthesis we have examined up to now. This is an important drawback of this elegant approach, which makes it uneconomical both from the component count and the power consumption points of view. It is this major problem of WAF that we will try to solve in the following sections.

7.4 Economical Wave Active Filters

The main disadvantage of the wave active filters is the large number of opamps they employ, namely, $2N$ opamps for a LC ladder with N storage elements. Among the various methods for reducing the number of opamps used in the design of WAFs, which have been suggested in References 8 through 11, we choose to describe the last one because of its simplicity and its similarity to the method of LC ladder simulation using GICs (PICs). In certain structures, the reduction in the number of the required opamps can be substantial.

Consider a passive two-port network N described by its y-parameters. As in the case of single elements, we introduce two new pairs of variables A_i, B_i $i = 1, 2$ such that

$$\begin{bmatrix} A_i \\ B_i \end{bmatrix} = \begin{bmatrix} 1 & R_i \\ 1 & -R_i \end{bmatrix} \begin{bmatrix} V_i \\ I_i \end{bmatrix} \quad i = 1, 2 \tag{7.29}$$

where R_i, $i = 1, 2$ are the port normalization resistances. Using in these equations, the relationships between I_i and V_i by means of the y-parameters, Y_{ij}, $i, j = 1, 2$, of the two-port we obtain

$$\begin{bmatrix} A_1 \\ A_2 \end{bmatrix} = \begin{bmatrix} 1 + R_1 Y_{11} & R_1 Y_{12} \\ R_2 Y_{21} & 1 + R_2 Y_{22} \end{bmatrix} \begin{bmatrix} V_1 \\ V_2 \end{bmatrix} = [Q_1] \begin{bmatrix} V_1 \\ V_2 \end{bmatrix} \tag{7.30a}$$

$$\begin{bmatrix} B_1 \\ B_2 \end{bmatrix} = \begin{bmatrix} 1 - R_1 Y_{11} & -R_1 Y_{12} \\ -R_2 Y_{21} & 1 - R_2 Y_{22} \end{bmatrix} \begin{bmatrix} V_1 \\ V_2 \end{bmatrix} = [Q_2] \begin{bmatrix} V_1 \\ V_2 \end{bmatrix} \tag{7.30b}$$

where

$$Q_1 = \begin{bmatrix} 1 + R_1 Y_{11} & R_1 Y_{12} \\ R_2 Y_{21} & 1 + R_2 Y_{22} \end{bmatrix} \tag{7.31a}$$

$$Q_2 = \begin{bmatrix} 1 - R_1 Y_{11} & -R_1 Y_{12} \\ -R_2 Y_{21} & 1 - R_2 Y_{22} \end{bmatrix} \tag{7.31b}$$

Then, solving Eq. (7.30a) for V_1, V_2 and substituting in Eq. (7.30b) gives the following relationship between B_i and A_i, $i = 1, 2$:

$$\begin{bmatrix} B_1 \\ B_2 \end{bmatrix} = [S] \begin{bmatrix} A_1 \\ A_2 \end{bmatrix} \tag{7.32}$$

with

$$[S] = [Q_1]^{-1}[Q_2] \tag{7.33}$$

Matrix Q_1 should be nonsingular.

If the new variables A_i, B_i, $i = 1, 2$ are voltages, each pair of them can be obtained through the use of an operational amplifier, as shown in Fig. 7.15. This active network is the wave active equivalent (WAE) of the passive two-port network N.

The above theory can be extended to the case of an n-port. For simplicity, consider that the n-ports have a common terminal. With the n-port being described through its y-parameters, the n pairs of variables A_i, B_i, $i = 1, 2..., n$ are given as follows:

$$[A] = [Q_1] \cdot [V]$$

$$[B] = [Q_2] \cdot [V] \tag{7.34}$$

where

$$[Q_1] = \begin{bmatrix} 1 + R_1 Y_{11} & R_1 Y_{12} & \cdots & R_1 Y_{1n} \\ R_2 Y_{21} & 1 + R_2 Y_{22} & \cdots & R_2 Y_{2n} \\ \cdots & \cdots & \cdots & \cdots \\ R_n Y_{n1} & R_n Y_{n2} & \cdots & 1 + R_n Y_{nn} \end{bmatrix} \tag{7.35}$$

$$[Q_2] = \begin{bmatrix} 1 - R_1 Y_{11} & -R_1 Y_{12} & \cdots & -R_1 Y_{1n} \\ -R_2 Y_{21} & 1 - R_2 Y_{22} & \cdots & -R_2 Y_{2n} \\ \cdots & \cdots & \cdots & \cdots \\ -R_n Y_{n1} & -R_n Y_{n2} & \cdots & 1 - R_n Y_{nn} \end{bmatrix} \tag{7.36}$$

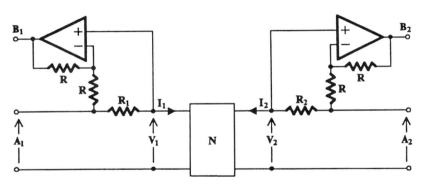

FIGURE 7.15
Wave active equivalent of passive two-port N.

If $[Q_1]$ is nonsingular, Eqs. (7.34) give

$$[B] = [Q_2] \cdot [Q_1]^{-1} \cdot [A]$$

or

$$[B] = [S] \cdot [A] \qquad (7.37)$$

where

$$[S] = [Q_2] \cdot [Q_1]^{-1} \qquad (7.38)$$

The WAE of the n-port can be obtained in a manner similar to that of the two-port.

This theory can be used to simulate a passive LC ladder filter as follows: the ladder is split into one kind of element (L or C) n-port subnetworks. For each subnetwork, a WAE is determined, and this is transformed into an active RC circuit, if it is not in that form already, and then all these RC WAEs are connected in the usual WAF manner according to the position of the corresponding passive subnetworks in the initial ladder. The terminating resistors can be associated with the first and last subnetworks and therefore can be included in the corresponding WAEs.

As an example, consider the ladder in Fig. 7.16(a), which is split into subnetworks, as shown by the broken lines. The four-port L-subnetwork consisting of inductors L_1, L_2, and L_3 has the active RC subnetwork in Fig. 7.16(b) as its WAE. The overall WAF is shown in Fig. 7.16(c). It should be mentioned that there is a saving in operational amplifiers of 50 per-

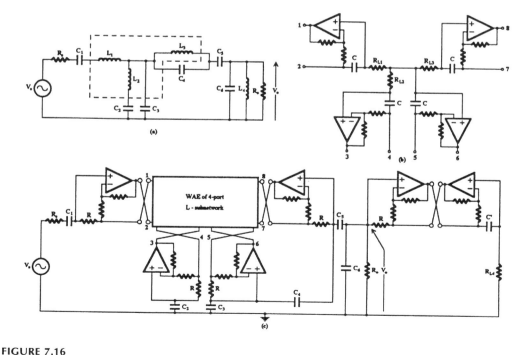

FIGURE 7.16
(a) Eighth-order LC ladder bandpass filter, (b) WAE of subnetwork consisting of inductors L_1, L_2, and L_3, and (c) overall WAE of ladder.

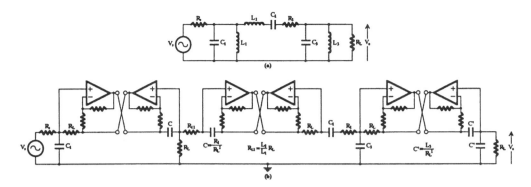

FIGURE 7.17
(a) Sixth-order RLC bandpass ladder and (b) its wave active equivalent.

cent when this is compared to the corresponding WAF of the initially proposed design. On the other hand, this WAF employs two operational amplifiers fewer than that of Brackett [8], who approaches the design differently. More specifically, Brackett (a) simulates the parallel-tuned circuit L_3C_4 using the WAE of a series-tuned circuit, in which he interchanges the outputs B_1 and B_2, and (b) uses gyrators to simulate grounded inductors L_2 and L_4.

This extended WAF method of simulation is general and independent of the type of the passive ladder, namely, whether this is lossless or lossy. Thus, in the case of the lossy one, each resistor in the ladder can be associated with one neighboring L- or C- subnetwork. It is preferable that all resistors be associated with C-subnetworks, since then they remain resistors in the corresponding RC WAEs, whereas, if they are associated with L-subnetworks, they become capacitors when the complex impedance scaling is applied to make the WAE RC realizable.

Thus, the RC WAE of the RLC doubly-terminated ladder in Fig. 7.17(a) is as shown in Fig. 7.17(b). Resistor R_2 has been associated with the two-port subnetwork consisting of capacitors C_2 and C_3.

7.5 Sensitivity of WAFs

In this section, the sensitivity of WAFs is examined and compared to the sensitivity of LF filters.

Without loss of generality, consider a passive ladder that, for the purpose of being simulated by a WAF, is split into two-port subnetworks. The corresponding WAF will be in block diagram form as shown in Fig. 7.18. The blocks labeled MAT (for *matching active transformer*) represent the network in Fig. 7.19, with its passive components suitably

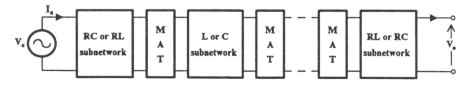

FIGURE 7.18
Block diagram of general WAF.

labelled for better identification. Note that, normally, $R_{ai} = R_{bi}$, $i = 1, 2$. For active RC realization, the application of proper impedance scaling will turn each L- or RL-subnetwork into a R- or CR-subnetwork, respectively. Then, depending on whether the L-subnetwork precedes or follows the MAT, resistor R_2 or R'_1 will become a capacitor. However, for the sake of argument, we keep the block diagram in the form shown in Fig. 7.18.

It is clear from Fig. 7.18 that if all MATs behave ideally, because of the one-to-one correspondence between the subnetworks in the passive ladder and the WAF, the sensitivities of the two filters to the elements of these subnetworks will be the same. Since these sensitivities for doubly terminated LC ladders are known to be low, the expected higher sensitivity of the WAF over that of the passive ladder will be due to the nonideal behavior of each MAT and the mismatch that this is creating in the WAF.

The operation of each MAT is affected mostly by the finite gain-bandwidth (GB) product of the operational amplifiers, which becomes most important when the filter is designed to operate at high frequencies. Ignoring for the moment the effect of the MAT sections on the WAF sensitivity at low frequencies, it has been shown [11] through worst sensitivity (WS) studies that

a. The WS of the earlier introduced and the economical WAF are of the same order but, as it may be expected, higher than that of the passive ladder.
b. The WS of the WAFs are of the same order but smaller than that of the corresponding LF.

In this study, both passive and active components were included in the calculation of the WS measure with the opamps considered to be of the one-pole model and its GB = 1 MHz.

Thus the sensitivity of the WAF at low frequencies, where the opamps behave nearly as ideal devices, is similar to the sensitivity of corresponding active RC filters resulting by any other ladder simulation techniques. It remains, of course, the problem of high frequency operation which we examine next.

7.6 Operation of WAFs at Higher Frequencies

We are concerned here with the problems arising in the operation of the WAFs at higher frequencies due to the finite GB product of the opamps and ignoring any slew-rate effects.

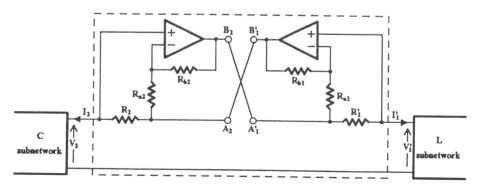

FIGURE 7.19
Matching active transformer (MAT).

Since MATs are the only active components of the WAF, we can study their behavior at higher frequencies, since at these frequencies resistors and capacitors are considered to behave ideally.

Because the WAF has been drawn in Fig. 7.18 as the cascade connection of MATs and passive subnetworks, it is more appropriate to study the operation of each MAT through its transmission matrix. Ideally this matrix (nonmodified) should be as follows:

$$[A] = \begin{bmatrix} 1 & 0 \\ 0 & 1 \end{bmatrix} \tag{7.39}$$

Owing to the finite GB product of the opamps the matrix elements change, but detailed study [11] has shown that only the a_{12} element is substantially affected, taking the following value:

$$a_{12} = 1 - \frac{\gamma_1 \gamma_2}{\kappa_i} R'_1 \tag{7.40}$$

where

$$\kappa_i = \frac{\mu_i}{1 + \beta_i \mu_i}, \qquad \gamma_i = (1 - \beta_i)\kappa_i \qquad i = 1, 2 \tag{7.41}$$

with μ_i being the open-loop gain of the opamp and γ_i the feedback ratio $R_{ai}/(R_{ai} + R_{bi})$, the nominal value of which is 0.5. As an indication of this effect let us calculate the value of a_{12} for the following typical values:

$$\mu_{io} = 10^5, \qquad f_c = 10 \text{ Hz} \qquad f_T = 1 \text{ MHz}$$

$$R_{ai} = R_{bi} = 1.2 k\Omega, \qquad R'_1 = 10 \text{ k}\Omega$$

At the frequency of 50 kHz, Eq. (7.40) gives the following value of a_{12}:

$$a_{12} \approx 946$$

This is an extremely high value compared to the values of the other elements of the transmission matrix, calling for frequency compensation of the MAT. To make a_{12} as small as possible the product $\gamma_1 \gamma_2$ in Eq. (7.40) should be kept as close to unity as possible. In practice, this has been found [11] to be achieved by connecting a capacitor of about 260 pF across each resistor R_{ai} in the circuit of the MAT, when working with opamps of the 741 type. Fortunately, this compensation does not affect the element a_{21}, while it improves slightly the values of elements a_{11}, a_{22} to be even closer to unity at higher frequencies. With this type of compensation, the operating range of WAFs can be extended up to at least 100 kHz.

7.7 Complementary Transfer Functions [7]

It is interesting and useful to note that in a WAF, there are two outputs available leading to two different transfer functions being realized at the same time. This can be easily seen in Fig. 7.15, where output voltages B_1 and B_2 are available from opamp outputs. Viewing the WAF as a whole wave equivalent, the signal excitation is the incident wave voltage A_1, which is equal to the voltage of the signal source, while B_1 and B_2 are the reflected waves, with $A_2 = 0$ when $R_2 = R_L$.

Let us consider that the two-port N is entirely reactive. If we remove the two amplifiers, the remaining will represent a resistively terminated LC ladder. Let a_{ij}, $i, j = 1, 2$ be the parameters of its modified transmission matrix as we have considered before. We can see that

$$\left.\frac{B_1}{A_1}\right|_{A_1 = V_s} = S_{11}$$

$$\left.\frac{B_2}{A_1}\right|_{A_1 = V_s} = S_{21} \tag{7.42}$$

which by means of Eqs. (7.5) can be written as follows:

$$S_{11} = \frac{(a_{11} - a_{12}G) - (a_{21} - a_{22}G)R}{(a_{11} - a_{12}G) + (a_{21} - a_{22}G)R}$$

$$S_{21} = \frac{2}{(a_{11} - a_{12}G) + (a_{21} - a_{22}G)R} \tag{7.43}$$

Since N is reactive, a_{11} and a_{22} are even functions, while a_{12} and a_{21} are odd functions. Therefore,

$$S_{11}S_{11}^* = \frac{a_{11}^2 + a_{22}^2 - (a_{12}^2 G^2 + a_{21}^2) - 2}{a_{11}^2 + a_{22}^2 - (a_{12}^2 G^2 + a_{21}^2 R^2) + 2}$$

$$S_{21}S_{21}^* = \frac{4}{a_{11}^2 + a_{22}^2 - (a_{12}^2 G^2 + a_{21}^2 R^2) + 2} \tag{7.44}$$

To obtain Eqs. (7.44), use has been made of the fact that the LC ladder two-port is reciprocal. Adding Eqs. (7.44) gives

$$S_{11}S_{11}^* + S_{21}S_{21}^* = 1 \tag{7.45}$$

or

$$|S_{11}|^2 + |S_{21}|^2 = 1 \tag{7.46}$$

The meaning of Eq. (7.46) is that the two functions B_2/A_1 and B_1/A_1 for $R_S = R_L = R$ are power complementary. This means that if, for example, B_2/A_1 is a lowpass function, its power complementary function B_1/A_1 is highpass obtained simultaneously with the former.

7.8 Wave Simulation of Inductance

The reader may have observed that there exists some degree of similarity between the simulation of a passive ladder through the wave active theory and the method using GICs. To clarify this point, let us consider the action of the MAT circuit in Fig. 7.19, when the L-subnetwork is a grounded inductor L. With the application of impedance scaling (division of impedances by s) in order to obtain active RC realization, the inductor L will become a resistor R_L, while resistor R'_1 will become a capacitor C'_1. Then the input impedance at the left port of the MAT will be

$$\frac{V_2}{-I_2} = sC'_1 R_2 R_L \qquad (7.47)$$

Clearly, this is the impedance of an inductance of value $C'_1 R_2 R_L$.

On the other hand, if the C subnetwork is a grounded capacitor C, the input impedance at the right port of the MAT can be found to be

$$\frac{V'_1}{-I'_1} = R_2 \frac{C'_1}{C} \qquad (7.48)$$

which is purely resistive.

Thus, the circuit of the MAT (within the broken lines in Fig. 7.19) with a capacitor C'_1 in place of resistor R'_1 behaves as a PIC. A PIC can also be obtained if R_2 is replaced by a capacitor instead of R'_1. Compared to the Antoniou PIC [12] it uses two extra resistors, but all resistor values can be equal thus reducing to zero the component spread. Also, as in the case of the Antoniou GIC, if R_2 is replaced by a capacitor instead of R'_1, while the right-hand-port is terminated by a capacitor, the circuit will behave as an FDNR type-D.

This resemblance of the economical WAFs to the GIC filters though should not lead to the conclusion that the former do not have to offer anything new in practice. In fact, it has been shown [11] that in some cases the economical WAFs may employ fewer opamps than their GIC counterparts.

7.9 Linear Transformation Active Filters (LTA Filters)

These filters [13–15], in a way similar to that of the WAF, simulate the resistively terminated LC ladder filters by means of active equivalent subnetworks of each series- and parallel-arm passive element of the ladder. Each element of the ladder is again treated as a two-port, and its active RC equivalent is determined through linear transformations of its port voltage and current variables, hence their name.

Wave Active Filters

To develop this simulation method of resistively terminated LC ladder filters, let us again consider the ladder in Fig. 7.1, repeated here as Fig. 7.20 for convenience. Without loss of generality, let each Z_i and $Y_i + 1$ be a simple reactive element, L or C. The ladder is split in elementary two-ports as implied by the broken vertical lines in the figure.

Consider now one of these elementary two-ports, shown in Fig. 7.21(a), defined by its port V and I variables, which are related by means of the modified transmission matrix A', as in the WAF case. Thus,

$$\begin{bmatrix} V_1 \\ I_1 \end{bmatrix} = [A'] \begin{bmatrix} V_2 \\ I_2 \end{bmatrix} = \begin{bmatrix} a_{11} & a_{12} \\ a_{21} & a_{22} \end{bmatrix} \begin{bmatrix} V_2 \\ I_2 \end{bmatrix} \qquad (7.49)$$

We introduce now another set of variables $x_i, y_i, i = 1, 2$ obtained from the corresponding $V_i, I_i, i = 1, 2$ variables by means of the following linear transformation:

$$\begin{bmatrix} x_i \\ y_i \end{bmatrix} = [Q_i] \begin{bmatrix} V_i \\ I_i \end{bmatrix} \qquad (7.50)$$

where all

$$[Q_i] = \begin{bmatrix} \alpha_i & \beta_i \\ \gamma_i & \delta_i \end{bmatrix} \qquad (7.51)$$

are nonsingular matrices. These are the so-called *transformation matrices*.

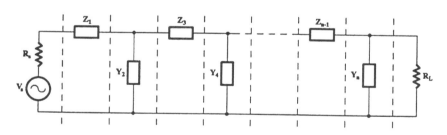

FIGURE 7.20
Terminated LC ladder.

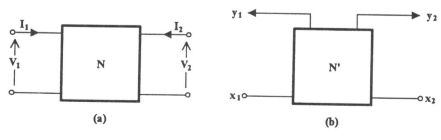

FIGURE 7.21
(a) An element of the ladder treated as a two-port, and (b) the symbol of its LTA equivalent.

From Eqs. (7.49) and (7.50), we can obtain the following relationship among the new sets of variables:

$$\begin{bmatrix} x_1 \\ y_1 \end{bmatrix} = [Q_1][A'][Q_2]^{-1} \begin{bmatrix} x_2 \\ y_2 \end{bmatrix} \tag{7.52}$$

Clearly, for active RC simulation of the passive ladder, the transformation matrices should be suitably selected.

We may consider the x_i, $i = 1, 2$ variables as inputs and the y_i, $i = 1, 2$ variables as outputs. Then, from Eq. (7.52) we can obtain the following relationship:

$$\begin{bmatrix} y_1 \\ y_2 \end{bmatrix} = \begin{bmatrix} K & L \\ M & N \end{bmatrix} \begin{bmatrix} x_1 \\ x_2 \end{bmatrix} \tag{7.53}$$

In the general case, K, L, M, and N are functions of the complex variable s. The usual LTA equivalent of the elementary two-port in Fig. 7.21(a) is as shown in Fig. 7.21(b). Notice that a little circle indicates the input terminals, while a little arrow the output terminals.

As an example, consider the determination of the LTA equivalent of the series inductor shown in Fig. 7.22(a). Its modified transmission matrix is as follows:

$$[A'] = \begin{bmatrix} 1 & -sL \\ 0 & -1 \end{bmatrix} \tag{7.54}$$

A useful choice for $[Q_1]$ and $[Q_2]$ is the following [14]:

$$[Q_1] = \begin{bmatrix} \gamma & 0 \\ 0 & R \end{bmatrix} \quad [Q_2] = \begin{bmatrix} 0 & -R \\ \gamma & 0 \end{bmatrix} \tag{7.55}$$

where γ and R are freely selectable parameters.

Substituting in Eq. (7.52) from Eqs. (7.54) and (7.55), we obtain the following:

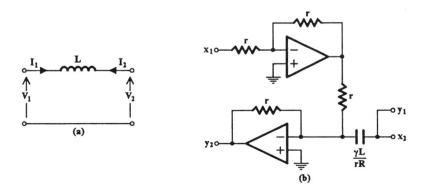

FIGURE 7.22
(a) A series inductor and (b) its possible LTA equivalent.

Wave Active Filters

$$[Q_1][A'][Q_2]^{-1} = \begin{bmatrix} \gamma & 0 \\ 0 & R \end{bmatrix} \begin{bmatrix} 1 & -sL \\ 0 & -1 \end{bmatrix} \begin{bmatrix} 0 & -R \\ \gamma & 0 \end{bmatrix} \quad (7.56)$$

or

$$\begin{bmatrix} x_1 \\ y_1 \end{bmatrix} = \begin{bmatrix} s\dfrac{\gamma L}{R} & 1 \\ 1 & 0 \end{bmatrix} \begin{bmatrix} x_2 \\ y_2 \end{bmatrix} \quad (7.57)$$

Assuming that $x_i, y_i, i = 1, 2$ are voltages, a possible implementation of Eq. (7.57) by an active RC circuit is shown in Fig. 7.22(b).

Following this procedure, LTA equivalent circuits of an inductor in a parallel-arm, a series- and a parallel-arm capacitor have also been found [13, 14]. It is usual to describe an LTA equivalent for the combination of the first reactive element in the ladder and the signal source V_S with its output resistance R_S. This reactive element, of course, can be a series or parallel inductor or capacitor. Similarly, a LTA equivalent is determined for the load resistor R_L combined with the last reactive element of the ladder, which can also be a series or parallel inductor or capacitor depending on the passive ladder and its order.

7.9.1 Interconnection Rule

Having derived LTA equivalents for each element of the ladder, the next step is to determine a suitable way for interconnecting the equivalents of adjacent elements in order to build up the overall active ladder. For this purpose, we consider the two adjacent two-ports in the V-I domain shown in Fig. 7.23(a) and proceed as in the case of WAFs. In the passive ladder prototype in the V-I domain, the two adjacent ports of N_a and N_b are directly connected, so that

$$\begin{bmatrix} V_{1b} \\ I_{0b} \end{bmatrix} = \begin{bmatrix} 1 & 0 \\ 0 & -1 \end{bmatrix} \begin{bmatrix} V_{2a} \\ I_{2a} \end{bmatrix} \quad (7.58)$$

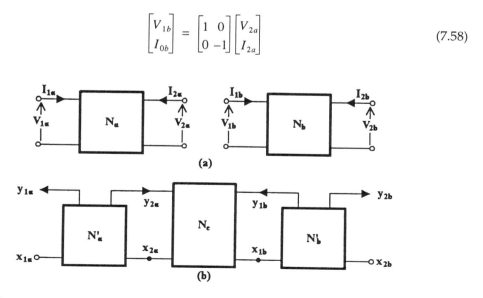

FIGURE 7.23
(a) Adjacent ladder elements and (b) the interconnection of their LTA equivalents.

In order to interconnect the LTA equivalents N'_a and N'_b of Fig. 7.23(b), a matching interconnecting two-port N_c should be determined that will keep the original parameters of N'_a and N'_b unaltered, thus avoiding any loading effects.

But,

$$\begin{bmatrix} x_{1b} \\ y_{1b} \end{bmatrix} = [Q_{1b}] \begin{bmatrix} V_{1b} \\ I_{1b} \end{bmatrix} \tag{7.59}$$

Substituting for $[V_{1b}\ I_{1b}]^T$ from Eq. (7.58) gives

$$\begin{bmatrix} x_{1b} \\ y_{1b} \end{bmatrix} = [Q_{1b}] \begin{bmatrix} 1 & 0 \\ 0 & -1 \end{bmatrix} \begin{bmatrix} V_{2a} \\ I_{2a} \end{bmatrix} \tag{7.60}$$

which, by virtue of Eq. (7.50), written for N_a, finally gives

$$\begin{bmatrix} x_{1b} \\ y_{1b} \end{bmatrix} = [Q_{1b}] \begin{bmatrix} 1 & 0 \\ 0 & -1 \end{bmatrix} [Q_{2a}]^{-1} \begin{bmatrix} x_{2a} \\ y_{2a} \end{bmatrix} \tag{7.61}$$

Therefore, the matching network N_c should be described by the following matrix P:

$$[P] = [Q_{1b}] \begin{bmatrix} 1 & 0 \\ 0 & -1 \end{bmatrix} [Q_{2a}]^{-1} \tag{7.62}$$

When $[Q_{1b}]$ and $[Q_{2a}]$ are chosen as in Eqs. (7.55), substitution in (7.62) gives matrix P to be the following:

$$[P] = \begin{bmatrix} 0 & 1 \\ 1 & 0 \end{bmatrix} \tag{7.63}$$

Substituting in Eq. (7.61) gives

$$x_{1b} = y_{2a}$$

$$y_{1b} = x_{2a} \tag{7.64}$$

Thus, with this specific selection of $[Q_{1b}]$ and $[Q_{2a}]$, the interconnection rule of the adjacent ports of N'_a and N'_b should be the cross-cascade.

Clearly, the interconnection rule depends on $[Q_{1b}]$ and $[Q_{2a}]$. In order to make this as simple as possible—specifically, the cross-cascade connection—we can proceed as follows: we select $[Q_{1b}]$ to be as simple as possible and then determine $[Q_{2a}]$ through Eq. (7.62) written as

$$[Q_{2a}] \begin{bmatrix} 0 & 1 \\ 1 & 0 \end{bmatrix} = [Q_{1b}] \begin{bmatrix} 1 & 0 \\ 0 & -1 \end{bmatrix} \tag{7.65}$$

This relationship has been called the *compatibility relationship* [15]. Once $[Q_{2a}]$ has been found, then $[Q_{1a}]$ can be determined or obtained from the existing list [15] in order to make Eq. (7.53) easily active RC implementable. To avoid complications, it is advisable to follow the above procedure starting from the load end of the ladder and moving toward the source end.

7.9.2 General Remarks on the Method

It is evident that the choice of the transformation matrices Q_{1i} and Q_{2i} for the ith series- or shunt-arm of the ladder will strongly determine the complexity of the overall LTA structure. Initially proposed LTA equivalents required a large number of opamps but, most important, they employed some of them in the form of differentiators, which made LTA filters quite "noisy."

However, later LTA developments [15] resulted in greatly improved LTA filters on both these aspects. Thus, by a more "suitable" choice of the transformation matrices, which includes making some of their elements frequency dependent, reduction in the complexity of the LTA filter can be achieved, while use of differentiators is avoided using integrators instead. These most suitable transformation matrices have been given in tabulated form both for series- and shunt-arm elements of the ladder [15].

The LTA method is a general approach to simulation of resistively terminated LC ladder filters. Due to its generality, other methods of LC-ladder simulation, such as wave active filters, leapfrog, and signal-flow-graph (SFG), may be interpreted as special cases of the LTA method. It must be remembered, though, that the development of LTA filters eventually followed that of the WAFs.

7.10 Summary

The WAF design method is an alternative approach to the simulation of resistively terminated LC ladder filters. The initially proposed WAFs employed a large number of operational amplifiers, but later developments resulted in more economical structures employing one opamp per storage element. These canonic structures are very similar to the structures of GIC (PIC) filters but, in some cases, they may use two opamps fewer than the latter. The sensitivity of WAFs is low, as low as the sensitivity of the other ladder simulation methods. Also, WAFs can be easily compensated in order to improve their useful frequency range. Another important feature of these active filters is that they provide two signal outputs that are power complementary.

The development of WAFs was followed by the introduction of the LTA filters which work on the same principles. However, their design approach is more general than that of the WAF, and it may be interpreted that the WAF design as well as the design of other active filters, such as the LF and the SFG, are special cases of that of the LTA filters.

References

[1] H. J. Orchard. 1966. "Inductorless filters," *Electron. Lett.* **2**(6), pp. 224–225.

[2] A. G. Constantinides and I. Haritantis. 1975. "Wave active filters," *Electron. Lett.* **11**, pp. 254–256.

[3] H. Wupper and K. Meerkötter. 1975. "New active filter synthesis based on scattering parameters," *Proc. IEEE Symposium on Circuits and Systems*, pp. 254–257.

[4] H. Wupper and K. Meerkötter. 1975. "New active filter synthesis based on scattering parameters," *IEEE Trans. Circuits Syst.* CAS-22, pp. 594–602.

[5] A. Fettweis. 1971. "Digital filter structures related to classical filter networks," *Arch. Elektr. Übertrag* **25**, pp. 29–89.

[6] A. G. Constantinides. 1974. "Alternative approach to design of wave digital filters," *Electron. Lett.* **10**, pp. 59–60.

[7] I. Haritantis, A. G. Constantinides, and T. Deliyannis. 1976. "Wave active filters," *Proc. IEE* **123**(7), pp. 676–682.

[8] P. O. Brackett. 1976. "Circuits and limitations of wave filters," *Proc. IEEE ISCAS*, pp. 69–72.

[9] I. Haritantis, T. Deliyannis, and G. Alexiou. 1978. "Wave active filters, some recent results," *Proc. 6th Microcoll.*, Budapest, I, II-5/28.8.

[10] I. Haritantis and T. Deliyannis. 1978. "Wave active filters with reduced number of operational amplifiers," *Proc. Intl. Symposium MECO 78*, Athens, 1, pp. 311–315.

[11] G. Alexiou, T. Deliyannis, and I. Haritantis. 1981. "Sensitivity and high-frequency performance of new wave active filters," *IEE Proc.* **128**, Pt. G, No. 5, pp. 251–256.

[12] A. Antoniou. 1970. "Novel RC-active-network synthesis using generalized-immittance converters," *IEEE Trans. Circuit Theory* CT-17, pp. 212–217.

[13] A. G. Constantinides and H. Dimopoulos. 1976. "Active RC filters derivable from LC ladder filters via linear transformations," *IEE J. Electronic Circuits and Syst.* **I**(1), pp. 17–21.

[14] A. G. Constantinides and I. Haritantis. 1977. "Realisation of LT (linear transformation) active filters that simulate LC filters," in *IEE Colloquium on Electronic Filters Digest 1977/37*, London.

[15] H. Dimopoulos and A. G. Constantinides. 1978. "Linear transformation active filters," *IEEE Trans. Circuits Syst.* CAS-25, pp. 845–852.

Chapter 8

Single Operational Transconductance Amplifier (OTA) Filters

8.1 Introduction

In the previous chapters active RC filters using the operational amplifier (opamp) have been discussed extensively. These filters have been widely used in various low frequency applications in telecommunication networks, signal processing circuits, communication systems, control, and instrumentation systems for a long time. However, active RC filters cannot work at higher frequencies (over 200kHz) due to opamp frequency limitations and are not suitable for full integration. They are also not electronically tunable and usually have complex structures. Many attempts have been made to overcome these drawbacks [1]–[8]. The most successful approach is to use the operational transconductance amplifier (OTA) to replace the conventional opamp in active RC filters [9]–[45], as predicted in [9]. In recent years OTA-based high frequency integrated circuits, filters and systems have been widely investigated.

As seen in Chapter 3, an ideal operational transconductance amplifier is a voltage-controlled current source, with infinite input and output impedances and constant transconductance. The OTA has two attractive features: its tranconductance can be controlled by changing the external dc bias current or voltage, and it can work at high frequencies. The OTA has been implemented widely in CMOS and bipolar and also in BiCMOS and GaAs technologies. The typical values of transconductances are in the range of tens to hundreds of μS in CMOS and up to mS in bipolar technology. The CMOS OTA, for example, can work typically in the frequency range of 50 MHz to several 100 MHz. Linearization techniques make the OTA able to handle input signals of the order of volts with nonlinearities of a fraction of one percent. We will not discuss the OTA design in this book, although it is very important. The reader can look at References [2]–[5] on this topic.

Programmable high-frequency active filters can therefore be achieved by incorporating the OTA. These OTA filters also have simple structures and low sensitivity. In Chapter 3 the OTA and some simple OTA-based building blocks were introduced. In this chapter we will discuss how to construct filters using a single OTA, because single OTA active filters have advantages such as low power consumption, noise, parasitic effects, and cost. Commercially widely available OTAs are very easy to access for one to build filters with resistors and capacitors.

However, single OTA filters may not be suitable for full integration as they contain resistors which demand large chip area. These filter structures may also not be fully programmable, as only one OTA is utilized. It should be emphasized that on-chip tuning is the most effective way to overcome fabrication tolerances, component nonidealities, aging, and changing operating conditions such as temperature. Therefore, in monolithic design we should also further avoid using resistors. In recent years, active filters which use only OTAs and capacitors have been widely studied [12]–[23], [26]–[43]. These filters are intuitively called OTA-C filters, which will also be the subject of the remaining chapters. Fortunately, the single OTA filter structures can be readily converted into fully integrated OTA-C counterparts by using OTAs to simulate the resistors. This will be shown in the chapter.

It should be noted that practical OTAs will have finite input and output impedances. For the CMOS OTA, for example, the input resistance is usually very large, being neglectable, but the output resistance is in the range of $50k\Omega$ to $1M\Omega$, and the input and output capacitances are typically of the order of $0.05pF$ [7]. Also, at very high frequencies, the OTA transconductance will be frequency dependent due to its limited bandwidth. These nonideal impedance and transconductance characteristics will influence the stability and frequency performances of OTA filters. Practical OTAs will also exhibit nonlinearity for large signals and have noise, which will affect the dynamic range of OTA filters.

In this chapter a large number of first-order and second-order single OTA filter structures are generated systematically. Design methods and equations are derived. Sensitivity analysis is conducted, and OTA nonideality effects are investigated. Performances of the generated OTA filter architectures are also compared. Knowledge of the OTA in Chapter 3 and single opamp active RC filters in Chapter 4 should be of help in understanding this chapter.

8.2 Single OTA Filters Derived from Three-Admittance Model

Consider the general circuit model in Fig. 8.1. It contains one OTA and three admittances. With the indicated input and output voltages it can be simply shown that

$$H_1(s) = \frac{V_{o1}}{V_i} = \frac{g_m Y_2}{Y_1 Y_2 + Y_1 Y_3 + Y_2 Y_3 + g_m Y_2} \tag{8.1}$$

$$H_2(s) = \frac{V_{o2}}{V_i} = \frac{g_m (Y_1 + Y_2)}{Y_1 Y_2 + Y_1 Y_3 + Y_2 Y_3 + g_m Y_2} \tag{8.2}$$

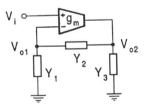

FIGURE 8.1
General model with three admittances.

Using these expressions we can readily derive different first-order and second-order filter structures from the general three-admittance model in Fig. 8.1 by assigning different components to Y_i and checking the corresponding transfer functions in Eqs. (8.1) and (8.2). For example, Y_i can be a resistor ($Y_i = g_i$), a capacitor ($Y_i = sC_i$), an open circuit ($Y_i = 0$), or a short circuit ($Y_i = \infty$). It can also be a parallel combination of two components ($Y_i = g_i + sC_i$).

8.2.1 First-Order Filter Structures

In this section we use the general model to generate first-order filters.

First-Order Filters with One or Two Passive Components

Selecting $Y_1 = sC_1$, $Y_2 = \infty$ and $Y_3 = 0$ gives rise to the simplest structure as shown in Fig. 8.2(a), which has a lowpass filter function given by

$$H_1(s) = \frac{g_m}{sC_1 + g_m} \tag{8.3}$$

with the dc gain equal to unity and the cutoff frequency equal to g_m/C_1.

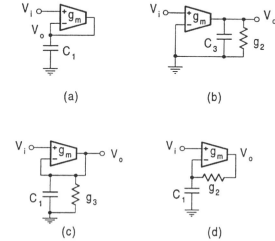

FIGURE 8.2
Simple first-order lowpass (a, b, c) and general (d) filters.

Figure 8.2(b) shows another simple lowpass filter corresponding to $Y_1 = \infty$, $Y_2 = g_2$, and $Y_3 = sC_3$. The transfer function is derived as

$$H_2(s) = \frac{g_m}{sC_3 + g_2} \tag{8.4}$$

with the dc gain equal to g_m/g_2 and the cutoff frequency being g_2/C_3.

The circuit in Fig. 8.2(c), corresponding to $Y_1 = sC_1$, $Y_2 = \infty$ and $Y_3 = g_3$, has the lowpass characteristic as

$$H_1(s) = \frac{g_m}{sC_1 + (g_3 + g_m)} \tag{8.5}$$

When $Y_1 = sC_1$, $Y_2 = g_2$, and $Y_3 = 0$, the output from V_{o2} is a general type, given by

$$H_2(s) = \frac{sg_m C_1 + g_m g_2}{sg_2 C_1 + g_m g_2} \tag{8.6}$$

which has the standard form of

$$H(s) = K \frac{s + \omega_z}{s + \omega_p} \tag{8.7}$$

The circuit is shown in Fig. 8.2(d). The circuits in Fig. 8.2 were also discussed, for example, in Ref. [15], here we show that they can be derived from the model in Fig. 8.1.

First-Order Filters with Three Passive Components

Observe that all the circuits in Fig. 8.2 contain less than three passive elements. In Fig. 8.3 we present a set of first-order filters with three passive components, which are derived from Fig. 8.1.

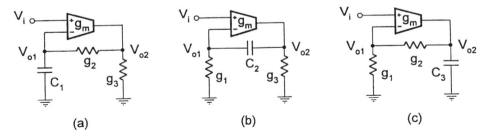

FIGURE 8.3
First-order filter configurations with three passive components.

It is first verified that when choosing $Y_1 = sC_1$, $Y_2 = g_2$ and $Y_3 = g_3$, the general model produces a lowpass filter from V_{o1}, that is

$$H_1(s) = \frac{g_m g_2}{s(g_2 + g_3)C_1 + g_2(g_3 + g_m)} \tag{8.8}$$

and a general transfer function from V_{o2}, given by

$$H_2(s) = \frac{sg_m C_1 + g_m g_2}{s(g_2 + g_3)C_1 + g_2(g_3 + g_m)} \tag{8.9}$$

The circuit is shown in Fig. 8.3(a).

Then consider the circuit in Fig. 8.3(b), which is obtained by setting $Y_1 = g_1$, $Y_2 = sC_2$ and $Y_3 = g_3$. It is found that a highpass filter is derived whose transfer function is given by

$$H_1(s) = \frac{sg_m C_2}{s(g_1 + g_3 + g_m)C_2 + g_1 g_3} \tag{8.10}$$

with the gain at the infinite frequency being $g_m/(g_1 + g_3 + g_m)$ and the cutoff frequency equal to $g_1 g_3/[(g_1 + g_3 + g_m)C_2]$.

This circuit also offers a general first-order characteristic, as can be seen from its transfer function

$$H_2(s) = \frac{sg_m C_2 + g_m g_1}{s(g_1 + g_3 + g_m)C_2 + g_1 g_3} \tag{8.11}$$

Finally, if Y_1 and Y_2 are resistors and Y_3 a capacitor, then both $H_1(s)$ and $H_2(s)$ are of lowpass characteristic. The circuit is presented in Fig. 8.3(c) and the transfer functions are given below.

$$H_1(s) = \frac{g_m g_2}{s(g_1 + g_2)C_3 + g_2(g_1 + g_m)} \tag{8.12}$$

$$H_2(s) = \frac{g_m(g_1 + g_2)}{s(g_1 + g_2)C_3 + g_2(g_1 + g_m)} \tag{8.13}$$

It is interesting to note from Eqs. (8.8) and (8.12) that the filters in Figs. 8.3(a) and (c) have similar characteristics from output V_{o1} or $H_1(s)$. The circuits in Figs. 8.2(a–c) and 8.3(c) will also be used as lossy integrators to construct integrator-based OTA-C filters in Chapter 9.

8.2.2 Lowpass Second-Order Filter with Three Passive Components

It should be pointed out that the model in Fig. 8.1 can also support many second-order filters. In this section however we only derive and discuss the simplest lowpass filter in order for the reader to appreciate some advantages of OTA filters before a comprehensive investigation of structure generation, design, and performance analysis of various second-order filters using a single OTA. Choosing in Fig. 8.1 $Y_1 = sC_1$, $Y_2 = g_2$, $Y_3 = sC_3$, the transfer function in Eq. (8.1) becomes

$$H_1(s) = \frac{g_m g_2}{s^2 C_1 C_3 + s g_2 (C_1 + C_3) + g_m g_2} \tag{8.14}$$

which is a lowpass filter characteristic. The corresponding circuit is shown in Fig. 8.4, which has only one resistor and two capacitors.

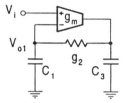

FIGURE 8.4
Simplest second-order lowpass filter derived from Fig. 8.1.

It will be recalled from Chapter 4 that the standard form of the lowpass characteristic is normally written as

$$H_d(s) = \frac{K \omega_o^2}{s^2 + \frac{\omega_o}{Q} s + \omega_o^2} \tag{8.15}$$

where K is the dc gain, ω_o is the undamped natural frequency, and Q is the quality factor, representing the selectivity, that is, the initial steepness of the transition band.

Comparison of Eqs. (8.14) and (8.15) indicates that the dc gain of the filter, K, is unity and

$$\omega_o = \sqrt{\frac{g_m g_2}{C_1 C_3}}, \quad Q = \sqrt{\frac{g_m}{g_2}} \frac{\sqrt{C_1 C_3}}{C_1 + C_3} \tag{8.16}$$

For convenience of design and also from the viewpoint of cost we set $C_1 = C_3$. This permits the development of simple design formulas for the component values, given by

$$C_1 = C_3 = C, \quad g_2 = \frac{\omega_o C}{2Q}, \quad g_m = 2Q \omega_o C \tag{8.17}$$

where C can be arbitrarily assigned.

As an example, we design the filter for the specifications of

$$f_o = 4 MHz, \quad Q = 1/\sqrt{2}, \quad K = 1$$

This is a Butterworth filter. Choosing $C_1 = C_3 = C = 5pF$, using Eq. (8.17) we can compute $g_2 = 88.86\mu S$ and $g_m = 177.72\mu S$.

Now we consider the filter sensitivity performance. Using the relative sensitivity definition introduced in Chapter 4, namely,

$$S_x^Q = \frac{x}{Q}\frac{\partial Q}{\partial x}, \quad S_x^{\omega_o} = \frac{x}{\omega_o}\frac{\partial \omega_o}{\partial x} \tag{8.18}$$

for the lowpass filter in Fig. 8.4 it is found that:

$$S_{g_m}^{\omega_o} = S_{g_2}^{\omega_o} = -S_{C_1}^{\omega_o} = -S_{C_3}^{\omega_o} = \tfrac{1}{2} \tag{8.19}$$

$$S_{g_m}^Q = -S_{g_2}^Q = \tfrac{1}{2}, \quad -S_{C_1}^Q = S_{C_3}^Q = \tfrac{1}{2}\tfrac{C_1-C_3}{C_1+C_3} = 0 \tag{8.20}$$

and these results indicate superior sensitivity performance. Note that setting $C_1 = C_3$ leads not only to practical convenience, but also to a decrease in the sensitivity of the filter to deviations in the capacitor design values, as can be seen from Eq. (8.20).

It is therefore clear from the above discussion that the OTA lowpass filter has a very simple structure, minimum component count, very simple design formulas, and extremely low sensitivity. As will be seen, this is generally true for other OTA filters.

8.2.3 Lowpass Second-Order Filters with Four Passive Components

It is quite straightforward to treat each admittance in the general model as a single passive component, either a resistor or capacitor as seen above. If more components are used for a single admittance, then more filter architectures can be obtained. In the following we generate useful lowpass second-order filters with four passive components, using again the model in Fig. 8.1.

The lowpass filter with $Y_1 = sC_1$, $Y_2 = g_2$, $Y_3 = g_3 + sC_3$ is depicted in Fig. 8.5(a). Its transfer function is derived as

$$H_1(s) = \frac{g_m g_2}{s^2 C_1 C_3 + s[(g_2+g_3)C_1 + g_2 C_3] + g_2(g_m+g_3)} \tag{8.21}$$

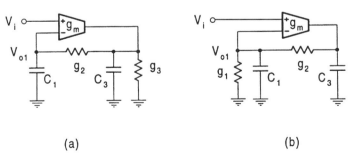

(a) (b)

FIGURE 8.5
Lowpass filters with four passive components.

Comparing the transfer function in Eq. (8.21) with the desired function in Eq. (8.15) yields the following equations

8.2. SINGLE OTA FILTERS FROM THREE-ADMITTANCE MODEL

$$\omega_o = \sqrt{\frac{g_2(g_3+g_m)}{C_1 C_3}}, \quad Q = \frac{\sqrt{g_2(g_3+g_m)C_1 C_3}}{(g_2+g_3)C_1+g_2 C_3},$$

$$K = \frac{g_m}{g_3+g_m} \tag{8.22}$$

A convenient design is to select $C_1 = C_3 = C$ and $g_2 = g_3 = g$. That is, all capacitances are equal and all conductances are identical, which makes the design easy and economical. With this selection, only three component values need to be decided. Generally, we can determine the component values for given ω_o, Q and K. We can also assign a value to any of C, g, or g_m and determine the other two in terms of ω_o and Q for a not specified K.

For the equal capacitances and conductances Eqs. (8.22) accordingly become

$$\omega_o C = \sqrt{g(g+g_m)}, \quad Q = \frac{\sqrt{g(g+g_m)}}{3g}, \quad K = \frac{g_m}{g+g_m} \tag{8.23}$$

From Eqs. (8.23) it can be determined that

$$g = \frac{\omega_o C}{3Q}, \quad g_m = 3Q\omega_o C\left(1 - \frac{1}{9Q^2}\right), \quad K = 1 - \frac{1}{9Q^2} \tag{8.24}$$

It is very interesting to note, see Eqs. (8.24), that

$$Q = \frac{1}{3}, \quad K = 0, \quad g_m = 0 \tag{8.25}$$

$$Q > \frac{1}{3}, \quad K > 0, \quad g_m > 0 \tag{8.26}$$

$$Q < \frac{1}{3}, \quad K < 0, \quad g_m < 0 \tag{8.27}$$

Equation (8.26) indicates that the circuit can realize large Q and positive gain, while Eq. (8.27) implies that with the interchange of the OTA input terminals the resulting circuit will complementarily implement small Q and negative gain. Equation (8.25) means that the design method cannot implement $Q = 1/3$. However, this does not represent a problem, since Q of $1/2$ or lower can be realized straightforwardly with a passive RC circuit. We should stress that throughout the chapter, for $g_m > 0$, the OTA is connected just as it appears in figures, while $g_m < 0$ simply means the interchange of the OTA input terminals.

Using the sensitivity definition in Eq. (8.18) it can be derived from Eqs. (8.22) that the general sensitivity expressions are given by

$$S_{C_1}^{\omega_o} = S_{C_3}^{\omega_o} = -S_{g_2}^{\omega_o} = -\frac{1}{2}, \quad S_{g_3}^{\omega_o} = \frac{1}{2}\frac{g_3}{g_3+g_m},$$

$$S_{g_m}^{\omega_o} = \frac{1}{2}\frac{g_m}{g_3+g_m} \tag{8.28}$$

$$S_{C_1}^{Q} = \frac{1}{2} - \frac{(g_2+g_3)C_1}{(g_2+g_3)C_1+g_2 C_3},$$

$$S^Q_{C_3} = \frac{1}{2} - \frac{g_2 C_3}{g_2 C_3 + (g_2 + g_3) C_1},$$

$$S^Q_{g_2} = \frac{1}{2} - \frac{g_2 (C_1 + C_3)}{g_2 (C_1 + C_3) + g_3 C_1},$$

$$S^Q_{g_3} = \frac{g_3}{2(g_3 + g_m)} - \frac{g_3 C_1}{g_3 C_1 + g_2 (C_1 + C_3)},$$

$$S^Q_{g_m} = S^{\omega_o}_{g_m} \tag{8.29}$$

$$S^K_{C_1} = S^K_{C_3} = S^K_{g_2} = 0, \quad -S^K_{g_3} = S^K_{g_m} = \frac{g_3}{g_3 + g_m} \tag{8.30}$$

For the design with $C_1 = C_3 = C$ and $g_2 = g_3 = g$, substituting the design formulas in Eqs. (8.24) we have further

$$S^{\omega_o}_{C_1} = S^{\omega_o}_{C_3} = -S^{\omega_o}_{g_2} = -\frac{1}{2}, \quad S^{\omega_o}_{g_3} = \frac{1}{18Q^2},$$

$$S^{\omega_o}_{g_m} = \frac{1}{2}\left(1 - \frac{1}{9Q^2}\right) \tag{8.31}$$

$$S^Q_{C_1} = -S^Q_{C_3} = S^Q_{g_2} = -\frac{1}{6}, \quad S^Q_{g_3} = -\frac{1}{3} + \frac{1}{18Q^2},$$

$$S^Q_{g_m} = S^{\omega_o}_{g_m} \tag{8.32}$$

$$S^K_{C_1} = S^K_{C_3} = S^K_{g_2} = 0, \quad -S^K_{g_3} = S^K_{g_m} = \frac{1}{9Q^2} \tag{8.33}$$

It can be seen from these results that the structure in Fig. 8.5(a) has very low sensitivity.

Another lowpass filter can be obtained, which corresponds to $Y_1 = g_1 + sC_1$, $Y_2 = g_2$, $Y_3 = sC_3$, as shown in Fig. 8.5(b). It has the transfer function

$$H_1(s) = \frac{g_m g_2}{s^2 C_1 C_3 + s [g_2 C_1 + (g_1 + g_2) C_3] + g_2 (g_m + g_1)} \tag{8.34}$$

This lowpass filter is similar to the one discussed above, as can be seen from Eqs. (8.21) and (8.34). The same design technique can be used, and the sensitivity performance is also similar.

8.2.4 Bandpass Second-Order Filters with Four Passive Components

The bandpass filter with $Y_1 = g_1$, $Y_2 = sC_2$, $Y_3 = g_3 + sC_3$ is shown in Fig. 8.6(a). The circuit transfer function is derived as

$$H_1(s) = \frac{s g_m C_2}{s^2 C_2 C_3 + s [(g_1 + g_3 + g_m) C_2 + g_1 C_3] + g_1 g_3} \tag{8.35}$$

8.2. SINGLE OTA FILTERS FROM THREE-ADMITTANCE MODEL

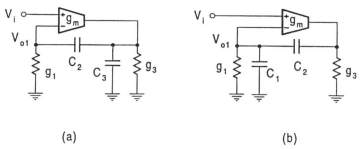

(a)   (b)

FIGURE 8.6
Bandpass filters with four passive components.

The ideal bandpass characteristic is typically written as

$$H_d(s) = \frac{K \frac{\omega_o}{Q} s}{s^2 + \frac{\omega_o}{Q} s + \omega_o^2} \tag{8.36}$$

where ω_o is the geometric center frequency of the passband, ω_o/Q is the $3dB$ bandwidth, which can also be denoted by B, and Q is again the quality factor.

Comparing Eq. (8.35) with Eq. (8.36) leads to the following design equations:

$$\omega_o = \sqrt{\frac{g_1 g_3}{C_2 C_3}}, \quad Q = \frac{\sqrt{g_1 g_3 C_2 C_3}}{(g_1 + g_3 + g_m) C_2 + g_1 C_3},$$

$$K = \frac{g_m C_2}{(g_1 + g_3 + g_m) C_2 + g_1 C_3} \tag{8.37}$$

We set $C_2 = C_3 = C$ and $g_1 = g_3 = g$ and obtain from Eqs. (8.37)

$$g = \omega_o C, \quad g_m = \frac{\omega_o C}{Q}(1 - 3Q), \quad K = 1 - 3Q \tag{8.38}$$

It can be seen from Eq. (8.38) that for practical Q values, $g_m < 0$ and $K < 0$ which mean that the OTA input terminals need to be interchanged and negative gain will be achieved.

The sensitivities of the filter are found to be

$$S_{C_2}^{\omega_o} = S_{C_3}^{\omega_o} = -S_{g_1}^{\omega_o} = -S_{g_3}^{\omega_o} = -\frac{1}{2}, \quad S_{g_m}^{\omega_o} = 0 \tag{8.39}$$

$$-S_{C_2}^{Q} = S_{C_3}^{Q} = \frac{1}{2} - \frac{g_1 C_3}{g_1 C_3 + (g_1 + g_3 + g_m) C_2},$$

$$S_{g_1}^{Q} = \frac{1}{2} - \frac{g_1 (C_2 + C_3)}{g_1 (C_2 + C_3) + (g_3 + g_m) C_2},$$

$$S_{g_3}^{Q} = \frac{1}{2} - \frac{g_3 C_2}{g_3 C_2 + g_1 (C_2 + C_3) + g_m C_2},$$

$$S_{g_m}^Q = -\frac{g_m C_2}{g_1 C_3 + (g_1 + g_3 + g_m) C_2} \quad (8.40)$$

$$S_{C_2}^K = -S_{C_3}^K = \frac{g_1 C_3}{g_1 C_3 + (g_1 + g_3 + g_m) C_2},$$

$$S_{g_1}^K = -\frac{g_1 (C_2 + C_3)}{g_1 C_3 + (g_1 + g_3 + g_m) C_2},$$

$$S_{g_3}^K = -\frac{g_3 C_2}{g_1 C_3 + (g_1 + g_3 + g_m) C_2},$$

$$S_{g_m}^K = 1 - \frac{g_m C_2}{g_1 C_3 + (g_1 + g_3 + g_m) C_2} \quad (8.41)$$

When $C_1 = C_3 = C$ and $g_2 = g_3 = g$, we have the following simple expressions:

$$S_{C_2}^{\omega_o} = S_{C_3}^{\omega_o} = -S_{g_1}^{\omega_o} = -S_{g_3}^{\omega_o} = -\frac{1}{2}, \quad S_{g_m}^{\omega_o} = 0 \quad (8.42)$$

$$-S_{C_2}^Q = S_{C_3}^Q = S_{g_3}^Q = \frac{1}{2} - Q,$$

$$S_{g_1}^Q = \frac{1}{2} - 2Q, \quad S_{g_m}^Q = -1 + 3Q \quad (8.43)$$

$$S_{C_2}^K = -S_{C_3}^K = -S_{g_3}^K = Q, \quad S_{g_1}^K = -2Q, \quad S_{g_m}^K = 3Q \quad (8.44)$$

From the sensitivity results, it can be observed that the design using the circuit in Fig. 8.6(a) with the OTA input terminals interchanged has very low ω_o sensitivity. However, the Q and K sensitivities display a modest Q dependence, although this is no problem for low Q design. The realization of large Q may cause an increase in the sensitivity. But considering that the ω_o sensitivity contributes more to response deviation than the Q sensitivity [47], the design is still useful for not very large Q, since the ω_o sensitivities are extremely low. Also, note that for filter design, the gain sensitivity is of less concern than the ω_o and Q sensitivities. Therefore when commenting the filter sensitivity performance, we mainly consider the ω_o and Q sensitivities.

It is also worthwhile mentioning that in bandpass filter design the design formulas can also be expressed in terms of ω_o and B only and the bandwidth sensitivities can be calculated by using $S_x^B = S_x^{\omega_o} - S_x^Q$. This can be practiced readily for the bandpass filter in Fig. 8.6(a) using the above results.

Another bandpass filter is associated with $Y_1 = g_1 + sC_1$, $Y_2 = sC_2$, $Y_3 = g_3$, shown in Fig. 8.6(b). Its transfer function is given by

$$H_1(s) = \frac{s g_m C_2}{s^2 C_1 C_2 + s[g_3 C_1 + (g_1 + g_3 + g_m) C_2] + g_1 g_3} \quad (8.45)$$

This filter function is similar to that of the above bandpass filter in Eq. (8.35). Thus similar performances are expected.

8.3 Second-Order Filters Derived from Four-Admittance Model

In this section we consider another two general single-OTA models and filter structures derived from them. We first consider the model in Fig. 8.7, which consists of an OTA and four admittances. This model may be looked upon as a result of grounding the non-inverting terminal of the OTA and applying a voltage input through an admittance to the inverting terminal of the OTA in Fig. 8.1. It can be shown that the transfer function of the new model in Fig. 8.7 is given by

$$H(s) = \frac{Y_1(Y_3 - g_m)}{Y_1Y_3 + Y_1Y_4 + Y_2Y_3 + Y_2Y_4 + Y_3Y_4 + g_m Y_3} \tag{8.46}$$

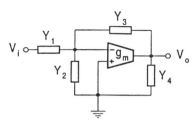

FIGURE 8.7
General model with four admittances.

Similarly, filter structures can be generated by selecting proper components in the model and the corresponding transfer functions can be obtained from Eq. (8.46).

8.3.1 Filter Structures and Design

The filter structures derived from the general model will be presented in this section. We will show how to design the filters to meet given specifications and analyze the corresponding sensitivity performance.

Lowpass Filter

When choosing $Y_1 = g_1$, $Y_2 = sC_2$, $Y_3 = g_3$, $Y_4 = sC_4$, we have a lowpass filter as shown in Fig. 8.8, which has the transfer function

$$H(s) = \frac{g_1(g_3 - g_m)}{s^2 C_2 C_4 + s[g_3 C_2 + (g_1 + g_3) C_4] + (g_1 + g_m) g_3} \tag{8.47}$$

Comparing its transfer function in Eq. (8.47) with the desired function in Eq. (8.15) yields the following equations:

$$\omega_o = \sqrt{\frac{(g_1 + g_m)g_3}{C_2 C_4}}, \quad Q = \frac{\sqrt{(g_1 + g_m)g_3 C_2 C_4}}{g_3 C_2 + (g_1 + g_3) C_4},$$

$$K = \frac{g_1 g_3 - g_1 g_m}{g_1 g_3 + g_3 g_m} \tag{8.48}$$

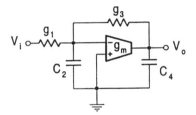

FIGURE 8.8
Lowpass filter derived from Fig. 8.7.

Based on these expressions we can design and analyze the filter. But we want first to draw the reader's attention to the similarity and difference of Eq. (8.22) and Eq. (8.48). The two filters have the same ω_o and Q expressions in form, the difference being only in the subscripts of g_j and C_j, although the gain expressions are different. The same design method can be used and the same design formulas and sensitivity performance of ω_o and Q will be achieved. To show this, we select $C_2 = C_4 = C$ and $g_1 = g_3 = g$. Using Eq. (8.48) we can obtain the design formulas as

$$g = \frac{\omega_o C}{3Q}, \quad g_m = 3Q\omega_o C \left(1 - \frac{1}{9Q^2}\right), \quad K = -\left(1 - \frac{2}{9Q^2}\right) \tag{8.49}$$

and the sensitivity expressions of the filter as

$$S_{C_2}^{\omega_o} = S_{C_4}^{\omega_o} = -S_{g_3}^{\omega_o} = -\frac{1}{2}, \quad S_{g_1}^{\omega_o} = \frac{1}{18Q^2},$$

$$S_{g_m}^{\omega_o} = \frac{1}{2}\left(1 - \frac{1}{9Q^2}\right) \tag{8.50}$$

$$-S_{C_2}^{Q} = S_{C_4}^{Q} = S_{g_3}^{Q} = -\frac{1}{6}, \quad S_{g_1}^{Q} = -\frac{1}{3} + \frac{1}{18Q^2},$$

$$S_{g_m}^{Q} = S_{g_m}^{\omega_o} \tag{8.51}$$

$$S_{C_2}^{K} = S_{C_4}^{K} = 0, \quad S_{g_1}^{K} = 1 - \frac{1}{9Q^2},$$

$$S_{g_3}^{K} = -\frac{1 - \frac{1}{9Q^2}}{1 - \frac{2}{9Q^2}}, \quad S_{g_m}^{K} = \frac{2}{9Q^2} \frac{1 - \frac{1}{9Q^2}}{1 - \frac{2}{9Q^2}} \tag{8.52}$$

Just as we expected, the designed lowpass filter has very low sensitivity and simple design formulas like the filter in Fig. 8.5(a).

8.3. SECOND-ORDER FILTERS FROM FOUR-ADMITTANCE MODEL

Bandpass Filter

A bandpass filter will result for $Y_1 = sC_1$, $Y_2 = g_2$, $Y_3 = g_3$, $Y_4 = sC_4$ as shown in Fig. 8.9(a). The corresponding transfer function is given by

$$H(s) = \frac{s(g_3 - g_m)C_1}{s^2 C_1 C_4 + s[g_3 C_1 + (g_2 + g_3)C_4] + (g_2 + g_m)g_3} \quad (8.53)$$

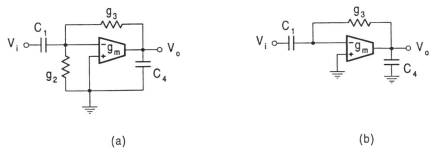

(a) (b)

FIGURE 8.9
Bandpass filters derived from Fig. 8.7.

Comparing Eq. (8.53) with Eq. (8.36) leads to

$$\omega_o = \sqrt{\frac{(g_2+g_m)g_3}{C_1 C_4}}, \quad Q = \frac{\sqrt{(g_2+g_m)g_3 C_1 C_4}}{g_3 C_1 + (g_2+g_3)C_4},$$

$$K = \frac{(g_3 - g_m)C_1}{g_3 C_1 + (g_2 + g_3)C_4} \quad (8.54)$$

Setting $C_1 = C_4 = C$ and $g_2 = g_3 = g$, for example, we can obtain g and g_m, being the same as those in Eq. (8.49) of the lowpass filter, but $K = -(9Q^2 - 2)/3$.

As a numerical example, for the bandpass filter of $f_o = 1MHz$ and $Q = 5$ choosing $C = 10pF$ we can determine $g = 4.2\mu S$ and $g_m = 938.3\mu S$. The filter gain is equal to 74.3.

As is obvious from their ω_o and Q expressions, the bandpass filter in Fig. 8.9(a) has the same ω_o and Q sensitivities as those of the lowpass filter in Fig. 8.8. As demonstrated above, these sensitivities are very low, less than or equal to 1/2. The gain sensitivities of the bandpass filter are given below:

$$S_{C_1}^K = -S_{C_4}^K = \tfrac{2}{3}, \quad S_{g_2}^K = -\tfrac{1}{3},$$

$$S_{g_3}^K = -\tfrac{2}{3} + \tfrac{1}{2-9Q^2}, \quad S_{g_m}^K = \tfrac{1-9Q^2}{2-9Q^2} \quad (8.55)$$

The gain sensitivities are also as low as those of the lowpass filter in Fig. 8.8.

We must emphasize the attractive low sensitivity feature of the bandpass filter. Especially the sensitivities will become smaller as Q increases, which makes it particularly suitable for large Q applications. Recalling that the bandpass filters generated in Section 8.2.4 are not suitable for large Q applications, because the Q sensitivities are proportional to Q.

Notice that for $g_2 = 0$, the transfer function in Eq. (8.53) becomes

$$H(s) = \frac{s(g_3 - g_m)C_1}{s^2 C_1 C_4 + sg_3(C_1 + C_4) + g_m g_3} \quad (8.56)$$

This reveals that eliminating the g_2 resistor in Fig. 8.9(a), the circuit can still support the bandpass function. This simplified circuit is given in Fig. 8.9(b).

For the simplified bandpass filter without the g_2 resistor in Fig. 8.9(b), we have

$$\omega_o = \sqrt{\frac{g_m g_3}{C_1 C_4}}, \quad Q = \sqrt{\frac{g_m}{g_3}} \frac{\sqrt{C_1 C_4}}{C_1 + C_4}, \quad K = \frac{(g_3 - g_m) C_1}{g_3 (C_1 + C_4)} \tag{8.57}$$

Selecting $C_1 = C_4 = C$, we can obtain

$$g_3 = \frac{\omega_o C}{2Q}, \quad g_m = 2Q\omega_o C, \quad K = \frac{1}{2}\left(1 - 4Q^2\right) \tag{8.58}$$

which are similar to the formulas in Eq. (8.17) for the lowpass filter in Section 8.2.2.

It can also be observed that the bandpass filter with $g_2 = 0$ in Fig. 8.9(b) has the same ω_o and Q sensitivities as those of the lowpass filter in Section 8.2.2. The gain sensitivities are shown as

$$S_{g_m}^K = -S_{g_3}^K = -\frac{4Q^2}{1 - 4Q^2}, \quad S_{C_1}^K = -S_{C_4}^K = \frac{1}{2} \tag{8.59}$$

which are also low.

Other Considerations on Structure Generation

Throughout this chapter, we are mainly concerned with canonic second-order structures containing only two capacitors. Of course, if more capacitors are used, then more structures may be obtained. For example, if $Y_1 = sC_1$, $Y_2 = sC_2$, $Y_3 = g_3$, $Y_4 = sC_4$, then the bandpass filter in Fig. 8.10(a) will arise, which has the transfer function

$$H(s) = \frac{s(g_3 - g_m) C_1}{s^2 (C_1 + C_2) C_4 + s g_3 (C_1 + C_2 + C_4) + g_m g_3} \tag{8.60}$$

Comparison of Eq. (8.60) with Eq. (8.36) yields ω_o, Q and K expressions, from which design can be carried out. Two design methods are given below. One method is to set $C_1 = C_2 = C_4 = C$. The following formulas are then obtained.

$$g_m = 3Q\omega_o C, \quad g_3 = \frac{2\omega_o C}{3Q}, \quad K = \frac{1}{3} - \frac{3}{2}Q^2 \tag{8.61}$$

The other method is to set $C_1 + C_2 = C_4 = C$ and specify K. This yields

$$g_m = 2Q\omega_o C, \quad g_3 = \frac{\omega_o C}{2Q}, \quad C_1 = \frac{2KC}{1 - 4Q^2}, \quad C_2 = C - C_1 \tag{8.62}$$

From the C_1 formula we can see that for practical Q values ($Q > 1/2$), only negative gain K can be achieved.

It is also possible to obtain other filter configurations by using a combination of more elements for an admittance. For example, if $Y_1 = sC_1$, $Y_2 = g_2$, $Y_3 = g_3 + sC_3$, $Y_4 = g_4$ (two components

8.3. SECOND-ORDER FILTERS FROM FOUR-ADMITTANCE MODEL

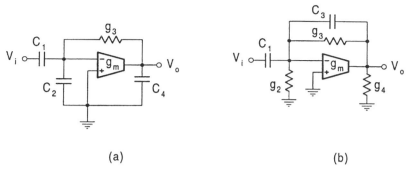

(a)   (b)

FIGURE 8.10
Bandpass with three capacitors and highpass filter using component matching.

are used for Y_3) as shown in Fig. 8.10(b), we have the transfer function as

$$H(s) = \frac{s^2 C_1 C_3 + s C_1 (g_3 - g_m)}{s^2 C_1 C_3 + s\left[(g_3 + g_4) C_1 + (g_2 + g_4 + g_m) C_3\right] + (g_2 g_3 + g_2 g_4 + g_3 g_4 + g_3 g_m)} \quad (8.63)$$

When $g_3 = g_m$, a highpass filter will result. This realization is however not particularly attractive, due to the use of difference matching. This problem for the highpass filter realization can be overcome by using the models in Section 8.3.2 and Section 8.7.

8.3.2 Second-Order Filters with the OTA Transposed

The second model with four admittances is displayed in Fig. 8.11. This model may be considered as a modification of Fig. 8.1 by grounding the non-inverting terminal of the OTA and applying a voltage input through an admittance to the output node of the OTA. It can also be reckoned as a consequence of transposing the OTA, that is, interchanging the input and output of the OTA in Fig 8.7. The general transfer function of the model can be demonstrated as

$$H(s) = \frac{Y_1 Y_3}{Y_1 Y_3 + Y_1 Y_4 + Y_2 Y_3 + Y_2 Y_4 + Y_3 Y_4 + g_m Y_3} \quad (8.64)$$

Note that the transfer function misses the term of $-g_m$ in the numerator, but has the same denominator compared with the function in Eq. (8.46). As will be seen, the former leads to some advantages such as more filter functions and better programmability while retaining low sensitivity. Also, similar design methods can be used. For example, the capacitances can take the same value and the resistances may be set to be identical. A number of filter configurations can be produced from the model.

Highpass Filter

A highpass characteristic is achieved by setting $Y_1 = sC_1$, $Y_2 = g_2$, $Y_3 = sC_3$, $Y_4 = g_4$. The circuit is shown in Fig. 8.12, with the transfer function given by

$$H(s) = \frac{s^2 C_1 C_3}{s^2 C_1 C_3 + s\left[g_4 C_1 + (g_2 + g_4 + g_m) C_3\right] + g_2 g_4} \quad (8.65)$$

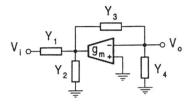

FIGURE 8.11
General four-admittance model with the OTA transposed.

Note that there are no difference nulling conditions involved in this highpass realization which also saves one resistor, compared with the one in Fig. 8.10(b).

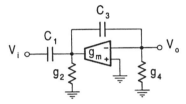

FIGURE 8.12
Highpass filter with transposed OTA.

Design can be carried out by comparing Eq. (8.65) with the standard highpass characteristic

$$H_d(s) = \frac{Ks^2}{s^2 + \frac{\omega_o}{Q}s + \omega_o^2} \qquad (8.66)$$

where K is the gain at the infinite frequency, ω_o is the undamped natural frequency, and the quality factor Q relates to the transition sharpness. Design equations are as follows ($K = 1$):

$$\omega_o = \sqrt{\frac{g_2 g_4}{C_1 C_3}}, \quad Q = \frac{\sqrt{g_2 g_4 C_1 C_3}}{g_4 C_1 + (g_2 + g_4 + g_m) C_3} \qquad (8.67)$$

Choosing $C_1 = C_3 = C$ and $g_2 = g_4 = g$ we can determine that

$$g = \omega_o C, \quad g_m = \frac{\omega_o C}{Q}(1 - 3Q) \qquad (8.68)$$

The ω_o and Q sensitivities are similar to those in Section 8.2.4 as can be inspected from the similarity between the two denominators of Eqs. (8.35) and (8.65). From the sensitivity results in Eqs. (8.42) and (8.43). It can be seen that for this design, the highpass circuit has very low ω_o sensitivities, but Q sensitivities will increase with Q. The filter thus may not suit very high Q applications. The design also requires interchanging the OTA input terminals. A highpass filter which has very low Q sensitivity will be presented in Section 8.7.

Lowpass Filter

A lowpass filter is attained by choosing $Y_1 = g_1$, $Y_2 = sC_2$, $Y_3 = g_3$, $Y_4 = sC_4$. The corresponding circuit is exhibited in Fig. 8.13 and its transfer function is given by

$$H(s) = \frac{g_1 g_3}{s^2 C_2 C_4 + s[g_3 C_2 + (g_1 + g_3) C_4] + (g_1 + g_m) g_3} \qquad (8.69)$$

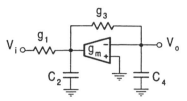

FIGURE 8.13
Lowpass filter with transposed OTA.

The denominator of the transfer function in Eq. (8.69) is the same as that in Eq. (8.47). The design formulas for $C_2 = C_4 = C$ and $g_1 = g_3 = g$ are hence the same as those in Eq. (8.49), with the only difference being $K = 1/9Q^2$. The ω_o and Q sensitivities are also the same as those in Eqs. (8.50) and (8.51), which are very low.

Bandpass Filters

A bandpass filter can be obtained by selecting $Y_1 = sC_1$, $Y_2 = g_2$, $Y_3 = g_3$, $Y_4 = sC_4$ which is shown in Fig. 8.14(a) and has a transfer function as

$$H(s) = \frac{s g_3 C_1}{s^2 C_1 C_4 + s[g_3 C_1 + (g_2 + g_3) C_4] + (g_2 + g_m) g_3} \qquad (8.70)$$

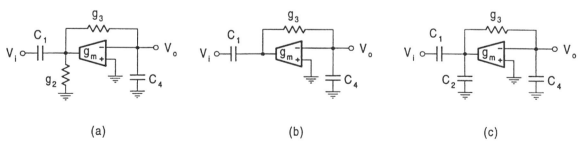

FIGURE 8.14
Bandpass filters with transposed OTA.

One design method is to set $C_1 = C_4 = C$ and $g_2 = g_3 = g$, which gives the formulas the same as those for the bandpass filter in Fig. 8.9(a), as Eqs. (8.70) and (8.53) have exactly the same denominator, but $K = 1/3$. Another method for the bandpass filter design is to set $C_1 = C_4 = C$ only. The filter gain K can then be used as a design parameter. The design formulas are derived as

$$g_3 = K \frac{\omega_o C}{Q}, \qquad g_2 = (1 - 2K) \frac{\omega_o C}{Q},$$

$$g_m = \frac{Q\omega_o C}{K}\left[1 - \frac{K(1-2K)}{Q^2}\right] \quad (8.71)$$

The condition is $K < 1/2$ to ensure a positive g_2. When $K = 1/2$, we have

$$g_3 = \frac{\omega_o C}{2Q}, \quad g_2 = 0, \quad g_m = 2Q\omega_o C \quad (8.72)$$

Similar to the discussion in Section 8.3.1, this reveals that the g_2 resistor can be removed. Generally, a simpler bandpass filter can be obtained by removing the g_2 resistor from Fig. 8.14(a), as shown in Fig. 8.14(b). This simple filter has a transfer function

$$H(s) = \frac{sg_3 C_1}{s^2 C_1 C_4 + sg_3 (C_1 + C_4) + g_m g_3} \quad (8.73)$$

another circuit which is as simple as the lowpass filter in Fig. 8.4.

A bandpass filter with three capacitors is also obtained by assigning $Y_1 = sC_1$, $Y_2 = sC_2$, $Y_3 = g_3$, $Y_4 = sC_4$ as shown in Fig. 8.14(c). The transfer function is derived as

$$H(s) = \frac{sg_3 C_1}{s^2 (C_1 + C_2) C_4 + sg_3 (C_1 + C_2 + C_4) + g_m g_3} \quad (8.74)$$

With $C_2 = 0$ this circuit will also reduce to Fig. 8.14(b). It should be noted that the bandpass filters in Fig. 8.14 all have very low sensitivities as their counterparts in Section 8.3.1.

The model in Fig. 8.11 can also support another bandpass filter which corresponds to $Y_1 = g_1$, $Y_2 = sC_2$, $Y_3 = sC_3$, $Y_4 = g_4$ as shown in Fig. 8.15. This bandpass filter has a transfer function

$$H(s) = \frac{sg_1 C_3}{s^2 C_2 C_3 + s[g_4 C_2 + (g_1 + g_4 + g_m) C_3] + g_1 g_4} \quad (8.75)$$

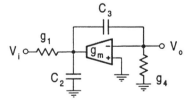

FIGURE 8.15
Another bandpass filter from Fig. 8.11.

Assuming $C_2 = C_3 = C$ we determine g_1, g_4 and g_m in terms of ω_o, Q and K, given by

$$g_1 = K\frac{\omega_o}{Q}, \quad g_4 = \frac{\omega_o Q}{K}, \quad g_m = \frac{\omega_o Q}{K}\left[-2 + \frac{(1-K)K}{Q^2}\right] \quad (8.76)$$

We can also further assign $g_1 = g_4 = g$, which will result in the same g and g_m as those for the highpass filter in Eq. (8.68), but K is fixed to be Q.

8.4 Tunability of Active Filters Using Single OTA

It is well known that the transconductance of an OTA is controllable by the bias dc current or voltage. For instance, the relationship between the transconductance and bias current of the bipolar OTA, CA3080, is given by [9]

$$g_m = \frac{1}{2V_T} I_B \tag{8.77}$$

where V_T is the thermal voltage and has a value of $26mV$ at room temperature. I_B is the bias current. If voltage is preferred to be the controlling variable, then a bias circuit can be used to convert the voltage to the current.

It is obvious that when design has determined g_m, the bias current needed can also be decided by Eq. (8.77), given by

$$I_B = 2V_T g_m \tag{8.78}$$

For example, if $g_m = 19.2mS$, then $I_B = 1mA$.

Programmability is one of the most attractive features of the OTA, since this makes it possible to tune filters electronically, which is especially important for on-chip tuning of fully integrated filters [5, 6, 37, 38, 39, 41, 42]. From the transfer functions of the OTA filters developed, it can be demonstrated that some structures are indeed tunable. For example, the center frequency ω_o of the bandpass filters in Figs. 8.9, 8.10, and 8.14 can be tuned independently of their bandwidth B, while the bandpass filters in Figs. 8.6 and 8.15 have the bandwidth B separately tunable from the center frequency ω_o. The quality factor Q can be controlled independently from the cutoff frequency ω_o for the highpass filter in Fig. 8.12.

8.5 OTA Nonideality Effects

Having considered filter structure generation, design, and sensitivity analysis we can now discuss some of the more practical problems in OTA filter design. In particular we will deal with the effects of OTA nonidealities on filter performance. The methods for the evaluation and reduction of the effects will be proposed.

8.5.1 Direct Analysis Using Practical OTA Macro-Model

It will be recalled from Chapter 3 that an OTA macro-model with finite input and output impedances and transconductance frequency dependence is shown in Fig. 8.16. We use G_i and C_i to represent the differential input conductance and capacitance and drop subscript d (for differential) for simplicity. G_o and C_o are those at the output. The common-mode input conductance G_{ic} and capacitance C_{ic} are ignored because they are usually very small in practice compared with differential counterparts and can be absorbed as most filter structures have a grounded capacitor or a grounded OTA resistor from OTA input terminals to ground. This will be assumed throughout all remaining chapters, unless otherwise stated. The input and output admittances can be written as $Y_i = G_i + sC_i$ and $Y_o = G_o + sC_o$. The transconductance frequency dependence can be described using a single pole model, as mentioned in Chapter 3 and repeated below:

$$g_m(s) = \frac{g_{m0}}{1 + \frac{s}{\omega_b}} \tag{8.79}$$

where ω_b is the finite bandwidth of the OTA and g_{m0} is the dc transconductance. The phase shift model is also often used, which is given, in the frequency domain, by [12]

$$g_m(j\omega) = g_{m0} e^{-j\phi} \tag{8.80}$$

where ϕ is the phase delay. Both models can be approximated as

$$g_m(s) \approx g_{m0}(1 - s\tau) \tag{8.81}$$

where $\tau = 1/\omega_b$ is the time delay and $\phi = \omega\tau$, when $\omega \ll \omega_b$. In the following the related terminologies may be used alternatively.

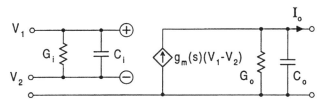

FIGURE 8.16
Practical OTA macro-model.

To give the reader some numerical order of OTA parameter values, a CMOS OTA, for example, may have the following data:

$$g_{m0} = 56\mu S, \quad f_b = 100 MHz \ (\tau = 1.59 ns), \quad G_i = 0,$$

$$G_o = 1\mu S \ (R_o = 1M\Omega), \quad C_i = 0.05 pF, \quad C_o = 0.1 pF$$

Now we consider the effects of OTA nonidealities on filters in detail. For the circuit in Fig. 8.1, incorporating the OTA macro-model we can derive the following modified transfer function

$$H_1'(s) = \frac{Y_2 g_m(s) + (Y_2 + Y_3 + Y_o) Y_i}{(Y_1 + Y_i)(Y_2 + Y_3 + Y_o) + Y_2(Y_3 + Y_o) + Y_2 g_m(s)} \tag{8.82}$$

Noting that if only the OTA frequency dependence is of concern, the associated transfer function can be simply obtained by substituting $g_m(s)$ for g_m in the ideal expression in Eq. (8.1).

Using the general equation, the impact of the OTA nonidealities on any derived filter structures can be evaluated. Take the lowpass filter in Fig. 8.4 as an example. With finite OTA impedances and bandwidth taken into account, the transfer function of the filter becomes

$$H_1'(s) = K \frac{s^2 + \frac{\omega_z}{Q_z} s + \omega_z^2}{s^2 + \frac{\omega_o'}{Q'} s + \omega_o'^2} \tag{8.83}$$

where

$$\omega_o' = \omega_o \sqrt{\frac{1 + \frac{G_i}{g_{m0}} + \frac{G_o}{g_{m0}}}{1 + \frac{C_i}{C_1} + \frac{C_o}{C_3}}} \tag{8.84}$$

8.5. OTA NONIDEALITY EFFECTS

$$Q' = Q \frac{\sqrt{\left(1 + \frac{G_i}{g_{m0}} + \frac{G_o}{g_{m0}}\right)\left(1 + \frac{C_i}{C_1} + \frac{C_o}{C_3}\right)}}{1 + \frac{C_3}{C_1+C_3}\frac{G_i}{g_2} + \frac{C_1}{C_1+C_3}\frac{G_o}{g_2} + \frac{C_i}{C_1+C_3} + \frac{C_o}{C_1+C_3} - \frac{g_{m0}\tau}{C_1+C_3}} \tag{8.85}$$

$$K = \frac{C_3 C_i}{C_1 C_3 + C_3 C_i + C_1 C_o} \tag{8.86}$$

$$\omega_z = \sqrt{\frac{g_2(g_{m0} + G_i)}{C_3 C_i}} \tag{8.87}$$

$$Q_z = \frac{\sqrt{g_2(g_{m0} + G_i) C_3 C_i}}{g_2 C_i + C_3 G_i - g_{m0} g_2 \tau} \tag{8.88}$$

Note that K is the gain at the infinity frequency, that is, $H'_1(\infty) = K$. The dc gain can be derived as

$$H'_1(0) = H_1(0) \frac{1 + \frac{G_i}{g_{m0}}}{1 + \frac{G_i}{g_{m0}} + \frac{G_o}{g_{m0}}} \tag{8.89}$$

In the above equations, ω_o and Q are as shown in Eq. (8.16). $H_1(0)$ represents the ideal dc gain, which is unity.

During the formulation of Eq. (8.83), for simplicity and without loss of insight into the problem, we use a first-order approximation. The first glance at the equation indicates that the ideal all-pole lowpass function in Eq. (8.14), now becomes a general biquadratic function with finite transmission zeros and all coefficients are changed.

Of all the parasitics contributing to the change of the transfer function, the input and output conductances (especially the latter) seem to have greater influence on the low frequency response than others and introduce losses causing reduction of the pole and zero quality factors and the low-frequency gain. For example, the dc gain in Eq. (8.89) is totally dependent on the finite conductances, being less than unity. The finite input and output capacitances affect more the high-frequency response. At the extreme infinite frequency the magnitude, as shown in Eq. (8.86), is no longer zero, but a finite value determined completely by the nonideal capacitances, especially the input capacitance. Note in particular that the input conductance and capacitance provide extra signal paths, as can be seen from the numerator parameters. Therefore the differential input application of the OTA may not be favorable in some cases.

Two major effects of $g_m(s)$ should be emphasized. From the pole quality factor Q' expression in Eq. (8.85), we can see that transconductance frequency dependence can enhance the Q, which is known as the Q enhancement effect. The other is the stability problem, that is, the finite ω_b may cause the circuit to oscillate by shifting the poles to the right plane.

To appreciate the change more clearly, we further write the parameters in the relative change form (a first-order approximation is adopted during the whole simplification). Using Eq. (8.84) and denoting $\Delta \omega_o = \omega'_o - \omega_o$ we can obtain

$$\frac{\Delta \omega_o}{\omega_o} = \frac{1}{2}\left(\frac{G_i}{g_{m0}} + \frac{G_o}{g_{m0}} - \frac{C_i}{C_1} - \frac{C_o}{C_3}\right) \tag{8.90}$$

In a similar way, from Eq. (8.85) and with $\Delta Q = Q' - Q$ we have

$$\frac{\Delta Q}{Q} = \left(\frac{1}{2g_{m0}} - \frac{1}{g_2}\frac{C_3}{C_1+C_3}\right)G_i + \left(\frac{1}{2g_{m0}} - \frac{1}{g_2}\frac{C_1}{C_1+C_3}\right)G_o$$

$$+ \frac{g_{m0}}{C_1+C_3}\tau + \frac{C_3-C_1}{2C_1(C_1+C_3)}C_i + \frac{C_1-C_3}{2C_3(C_1+C_3)}C_o \tag{8.91}$$

Finally, from Eq. (8.89) and with $\Delta H_1(0) = H_1'(0) - H_1(0)$ we can derive

$$\frac{\Delta H_1(0)}{H_1(0)} = -\frac{G_o}{g_{m0}} \tag{8.92}$$

Equation (8.90) clearly shows that G_i and G_o increase ω_o, while C_i and C_o decrease ω_o. The excess phase has no effect on ω_o (for the first-order approximation). Equation (8.92) reveals that G_o has a reduction impact on the dc gain. The effects on Q depend on how the circuit is designed. For the design in Section 8.2.2, with normalized $C_1 = C_3 = C = 1F$, Eq. (8.91) reduces to

$$\frac{\Delta Q}{Q} = \frac{g_2 - g_{m0}}{2g_{m0}g_2}(G_i + G_o) + \frac{g_{m0}}{2}\tau \tag{8.93}$$

Further substituting the design formulas in Eq. (8.17) with $C = 1F$ gives

$$\frac{\Delta Q}{Q} = \frac{1-4Q^2}{4\omega_o Q}(G_i + G_o) + Q\omega_o\tau \tag{8.94}$$

Therefore, τ has a Q enhancement effect. C_i and C_o have no impact on Q for the first-order approximation and $C_1 = C_3$. G_i and G_o will cause Q reduction. We should stress that the contribution of excess phase ($\phi = \omega_o \tau$) to the Q enhancement is multiplied by Q^2, that is ΔQ(due to ϕ)$= Q^2\phi$, as can be seen from Eq (8.94). Therefore, for large Q applications, even a very small phase shift can cause a very big increase in Q and thus instability. From this example we also see that a good design can reduce nonideality effects. In particular, using equal design capacitances also reduces the influence of finite OTA input and output capacitances on the pole quality factor, besides the benefits mentioned in Section 8.2.2 such as the zero sensitivities of Q to the capacitances.

It should also be noted that OTAs using different IC technologies may have different performances. For instance, MOS and CMOS OTAs have a very large input resistance, which may thus be assumed infinite in most cases. However, the input resistance of bipolar OTAs is quite low. The above analysis is general, which could be simplified for the CMOS OTA by dropping off G_i, for example.

Similarly, taking the OTA nonidealities into consideration, the general transfer functions of Figs. 8.7 and 8.11 become, respectively,

$$H'(s) = \frac{Y_1(Y_3 - g_m(s))}{\begin{array}{c}Y_1Y_3+Y_1(Y_4+Y_o)+(Y_2+Y_i)Y_3+(Y_2+Y_i)(Y_4+Y_o)\\+Y_3(Y_4+Y_o)+Y_3g_m(s)\end{array}} \tag{8.95}$$

and

$$H'(s) = \frac{Y_1Y_3}{\begin{array}{c}Y_1Y_3+Y_1(Y_4+Y_i)+(Y_2+Y_o)Y_3+(Y_2+Y_o)(Y_4+Y_i)\\+Y_3(Y_4+Y_i)+Y_3g_m(s)\end{array}} \tag{8.96}$$

8.5. OTA NONIDEALITY EFFECTS

Using the respective equation we can analyze the influence of OTA nonidealities on the filters derived from the general models in Figs. 8.7 and 8.11. The difference of the two expressions in terms of Y_i and Y_o is due to the different connection of the OTA in the models.

8.5.2 Simple Formula Method

A simple method for evaluation of the effects of finite bandwidth has been proposed in Ref. [46]. This method uses the sensitivity to the amplifier gain to assess the effects of phase shift, which simplifies the analysis. Using this method we can, for example, assess the influence of the OTA finite bandwidth on the filter. The associated formulas are given below:

$$\frac{\Delta\omega_o}{\omega_o} = \frac{\omega_o}{2Q\omega_b}\left(S_{g_m}^{\omega_o} - S_{g_m}^Q\right) \tag{8.97}$$

$$\frac{\Delta Q}{Q} = \frac{\omega_o}{2Q\omega_b}\left[(4Q^2-1)S_{g_m}^{\omega_o} + S_{g_m}^Q\right] \tag{8.98}$$

For the simple lowpass structure in Fig. 8.4, $S_{g_m}^{\omega_o} = S_{g_m}^Q = \frac{1}{2}$, as given in Eqs. (8.19) and (8.20). It can be shown that the effect of the finite bandwidth ω_b of the OTA is to cause fractional deviations in Q and ω_o, given approximately by

$$\frac{\Delta\omega_o}{\omega_o} = 0, \quad \frac{\Delta Q}{Q} = Q\frac{\omega_o}{\omega_b} \tag{8.99}$$

Recognizing that it is deviations in ω_o which frequently cause the greatest deviation in the amplitude response of the filter (see Section 4.4), another attractive feature of this filter is observed from the result. Equation (8.99) can also be derived from Eqs. (8.90) and (8.94), as expected.

Similarly, for the lowpass filter in Fig. 8.8, using the results of $S_{g_m}^{\omega_o}$ and $S_{g_m}^Q$ in Eqs. (8.50) and (8.51), we have

$$\frac{\Delta\omega_o}{\omega_o} = 0, \quad \frac{\Delta Q}{Q} = \frac{Q\omega_o}{\omega_b}\left(1 - \frac{1}{9Q^2}\right) \tag{8.100}$$

8.5.3 Reduction and Elimination of Parasitic Effects

It is possible to reduce the effects of OTA input and output impedances by absorption and those of transconductance frequency dependence by phase lead compensation. To show the former we consider the second-order filter model in Fig. 8.7. The latter will be handled in Chapter 9.

From Eq. (8.95) we can see that if Y_2 and Y_4 are a parallel of a resistor and a capacitor, that is, $Y_2 = g_2 + sC_2$, and $Y_4 = g_4 + sC_4$, then the effects of Y_i and Y_o can be completely eliminated by absorption design, that is, G_i and C_i are absorbed by g_2 and C_2, respectively, and G_o and C_o by g_4 and C_4. Figure 8.17 shows the lowpass circuit which can absorb the OTA input and output impedances and all node parasitic capacitances. The circuit has the following ideal transfer function:

$$H(s) = \frac{g_1(g_3 - g_m)}{s^2 C_2 C_4 + s[(g_3+g_4)C_2+(g_1+g_2+g_3)C_4] + (g_1g_3+g_1g_4+g_2g_3+g_2g_4+g_3g_4+g_mg_3)} \tag{8.101}$$

For the circuit in Fig. 8.17, the OTA finite conductances and capacitances cause a change in design capacitances and conductances as

$$\Delta C_2 = C_i, \quad \Delta C_4 = C_o, \quad \Delta g_2 = G_i, \quad \Delta g_4 = G_o$$

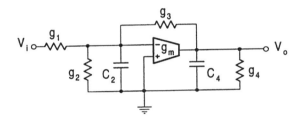

FIGURE 8.17
Lowpass filter that can absorb all parasitic resistances and capacitances.

The absorption approach determines the real component values by subtracting the nominal values with the increments due to nonideal OTA parameters, that is

$$C_{\text{real}} = C_{\text{nominal}} - \Delta C, \quad g_{\text{real}} = g_{\text{nominal}} - \Delta g \qquad (8.102)$$

This requires that

$$C_{\text{nominal}} > \Delta C, \quad g_{\text{nominal}} > \Delta g$$

For example, the nominal values for relevant capacitances and conductances must be much bigger than the respective parasitic values. It should be noted that at very high frequencies this may not be always met.

Similar methods for the elimination of the effects of finite OTA input and output impedances can also be discussed based on Eq. (8.96) for the filters derived from Fig. 8.11.

In most cases in this chapter each admittance is treated as a single component, resistor or capacitor. Only in the cases in which we want to achieve additional functions or performances do we consider them as a combination of two components. This will also be the case for the remaining sections of the chapter.

8.6 OTA-C Filters Derived from Single OTA Filters

In the above, many interesting filters using a single OTA have been developed. These single OTA filter structures may not be fully integratable and fully programmable due to the fact that they contain resistors and use only one OTA. But they are still useful for monolithic implementation, because by replacing the discrete resistor with the simulated OTA resistor, they can be very easily converted into the counterparts using OTAs and capacitors only. The derived OTA-C filters should be suitable for full integration. In the following we first discuss how to simulate resistors using OTAs only and then selectively illustrate some OTA-C filters thus derived from the single OTA counterparts.

8.6.1 Simulated OTA Resistors and OTA-C Filters

Resistors can be simulated using OTAs. Figure 8.18(a) shows a simple single OTA connection. This circuit is equivalent to a grounded resistor with resistance equal to the inverse of the OTA transconductance, that is, $R = 1/g_m$ [12]. Floating resistor simulation may require more OTAs. Figure 8.18(b) shows a circuit with two identical OTAs [15]. It can be shown that it is equivalent to a floating resistor of resistance equal to $R = 1/g_m$. Finally, for the ideal voltage input, the first OTA in the input terminated floating resistor simulation is redundant and can thus be eliminated, as

8.6. OTA-C FILTERS DERIVED FROM SINGLE OTA FILTERS

shown in Fig. 8.18(c). This simulation not only saves one OTA but also has a high input impedance, a feature useful for cascade design.

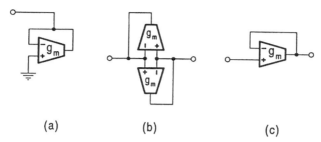

FIGURE 8.18
OTA simulation of resistors.

For simplicity from now on throughout the remaining chapters we will drop subscript m off transconductance g_m in almost all cases except when for some special cases in which the subscript m must be used. The reader should keep in mind that in OTA-C filters, g really means g_m since only OTA and capacitors are used. The function of resistors is simulated by OTAs as discussed above.

We now derive OTA-C filters from some single OTA filters using the resistor substitution method. To stress that they are based on single OTA filter prototypes we keep the g_m symbol for this OTA. A lowpass OTA-C filter is obtained from Fig. 8.4, by simply replacing the floating resistor by the OTA equivalent in Fig. 8.18(b), which is depicted in Fig. 8.19(a). Figure 8.19(b) shows the OTA-C bandpass filter derived from Fig. 8.6(a) using the OTA grounded resistor in Fig. 8.18(a). We give the OTA-C equivalents of the lowpass filter in Fig. 8.8 and the bandpass filter in Fig. 8.9(a), as shown in Figs 8.19(c) and 8.19(d), respectively. The lowpass OTA-C filter in Fig. 8.19(c) uses an input terminated OTA resistor in Fig. 8.18(c) and the grounded OTA resistor in Fig. 8.18(a). The bandpass OTA-C filter in Fig. 8.19(d) consists of an OTA grounded resistor and an OTA floating resistor. The single OTA bandpass filter in Fig. 8.15 and the highpass filter in Fig. 8.12 are also converted into the OTA-C counterparts, which are shown in Figs. 8.19(e) [36] and 8.19(f), respectively.

8.6.2 Design Considerations of OTA-C Structures

The transfer functions of the OTA-C filters are the same as those of the single OTA counterparts. The difference is only that in OTA-C filters, the gs are all OTA transconductances. The resistor substitution method also retains the sensitivity property of the original single OTA filter. Therefore the structures that have minimum sensitivity should be first considered in OTA-C realization. It is evident that the number of OTAs in the derived OTA-C filters will depend on how many resistors are in the original circuits. The architectures with fewer resistors may be attractive in the sense of reducing the number of OTAs. Also, note that the grounded resistor needs fewer OTAs to simulate than the floating resistor, and thus the single OTA filter structures using grounded resistors may be preferable in terms of reduction in the number of OTAs in the derived OTA-C filters. As will be discussed immediately, the grounded resistor will also introduce fewer parasitic elements into the filter circuit than the floating resistor when the nonidealities of the OTA(s) simulating them are taken into consideration. It should also be noted that structures using grounded capacitors are advantageous with respect to reducing parasitic effects and the chip area, as the floating capacitor has bigger parasitic capacitances and requires larger chip area.

For the OTA-RC filters we have discussed the effects of nonidealities of the OTA g_m. When dealing with the OTA-C equivalent we must also consider the nonidealities of the OTAs simulating resistors. For the grounded OTA resistor in Fig. 8.18(a), the equivalent grounded admittance due to the OTA nonidealities can be demonstrated as (to be general, we include the OTA common-mode

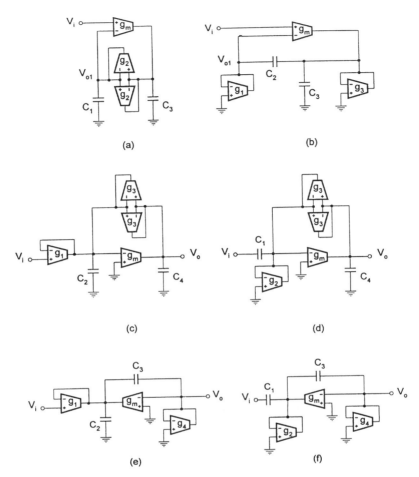

FIGURE 8.19
Examples of OTA-C filters derived from single OTA counterparts.

impedance)

$$Y_{GR} = Y_{id} + Y_{ic} + Y_o + g_m(s) = (G_{id} + G_{ic} + G_o + g_{m0})$$
$$+ s(C_{id} + C_{ic} + C_o - g_{m0}/\omega_b) \qquad (8.103)$$

which is a complex admittance, no longer a pure conductance.

For the floating resistor simulation in Fig. 8.18(b), the nonidealities of the two identical OTAs will have more complex effects. We can draw the equivalent circuit taking the OTA nonidealities into account and use the current source shift theorem (in a loop) to simplify the equivalent circuit. The resulting circuit can be further proved to be equivalent to a π type admittance network with the series arm admittance given by

$$Y_{FR\pi s} = 2Y_{id} + g_m(s) = (2G_{id} + g_{m0}) + s(2C_{id} - g_{m0}/\omega_b) \qquad (8.104)$$

8.6. OTA-C FILTERS DERIVED FROM SINGLE OTA FILTERS

and the two equal parallel arm admittances, given by

$$Y_{FR\pi p} = 2Y_{ic} + Y_o = (2G_{ic} + G_o) + s(2C_{ic} + C_o) \qquad (8.105)$$

Unlike the grounded resistor, in this case it is impossible to write an equivalent floating admittance.

Now we can consider the effects on OTA-C filters of the nonidealities from the resistor simulation OTAs. For example, in the lowpass OTA-C filter in Fig. 8.19(a) the two identical OTAs simulating the floating resistor of conductance g_2 will have a π equivalent circuit due to their nonidealities as shown above. As can be seen from the circuit structure and the expressions of the series and parallel arm admittances of the π network in Eqs. (8.104) and (8.105), respectively, the finite differential input conductances ($2G_{id2}$) can be absorbed by transconductance g_{20} and the common-mode capacitances and output capacitance ($2C_{ic2} + C_{o2}$) can also be absorbed by C_1 and C_3. But the effects of the finite differential input capacitances and the finite bandwidth will produce a parasitic floating capacitance equal to $2C_{id2} - g_{20}/\omega_{b2}$, and the effect of the common-mode input conductances and the output conductance will generate two parasitic grounded resistors of equal conductances of $2G_{ic2} + G_{o2}$ in parallel with C_1 and C_3. Such parasitic elements will affect the filter poles and zeros. Further analysis can be easily carried out by substituting

$$Y_1' = (2G_{ic2} + G_{o2}) + s[C_1 + (2C_{ic2} + C_{o2})]$$

$$Y_3' = (2G_{ic2} + G_{o2}) + s[C_3 + (2C_{ic2} + C_{o2})]$$

$$Y_2' = (2G_{id2} + g_{20}) + s(2C_{id2} - g_{20}/\omega_{b2})$$

for Y_1, Y_3 and Y_2 in Eq. (8.1). The reader may formulate the corresponding practical expression of the transfer function and compare it with the ideal one in Eq. (8.14) to study the effects in details.

As a second example, the bandpass OTA-C filter with two grounded OTA resistors in Fig. 8.19(b) is considered. Taking the nonidealities of the g_1 and g_3 OTAs into account and using Eq. (8.103) we have the changed grounded admittances as

$$Y_1' = [(G_{id1} + G_{ic1} + G_{o1}) + g_{10}] + s(C_{id1} + C_{ic1} + C_{o1} - g_{10}/\omega_{b1})$$

$$Y_3' = [(G_{id3} + G_{ic3} + G_{o3}) + g_{30}] + s[C_3 + (C_{id3} + C_{ic3} + C_{o3} - g_{30}/\omega_{b3})]$$

It can be seen that the finite conductances can be absorbed by the respective transconductances of the g_1 and g_3 OTAs. Also, the finite capacitances and bandwidth of the g_3 OTA can be absorbed by C_3. But a parasitic capacitor from the output node to ground will be produced by the finite capacitances and bandwidth of the g_1 OTA, which cannot be absorbed. Again a detailed evaluation can be conducted by substituting Y_1' and Y_3' for Y_1 and Y_3 in Eq. (8.1) and comparing the resulting equation with the ideal transfer function in Eq. (8.35). For example, if only the finite capacitances and bandwidth of the g_1 OTA are considered, we can readily demonstrate that their effect is to produce extra terms in the denominator of the transfer function in Eq. (8.35), which are

$$\left[s^2(C_2 + C_3) + sg_3 \right] (C_{id1} + C_{ic1} + C_{o1} - g_{10}/\omega_{b1})$$

The nonideality effects of the input termination OTA can also be similarly evaluated. The reader may, for example, consider the g_1 OTA in the lowpass circuit in Fig. 8.19(c).

Tuning may need reconsideration. As we have already found in Section 8.4, it is not possible to tune the frequency and quality factor independently in some single OTA filters. By replacing fixed resistors by tunable OTAs the programmability can be enhanced. For instance, the single OTA bandpass filter in Fig. 8.9(a) has only ω_o tunable, while the OTA-C simulation in Fig. 8.19(d) has also tunable B. The tuning process simply involves the tuning of B by the g_2 or g_3 OTA, followed by the adjusting of ω_o by the g_m OTA. It is noted that in the original single OTA bandpass circuit in Fig. 8.15, only the bandwidth or the quality factor is tunable, but now the OTA-C derivative in Fig. 8.19(e) has also the tunable center frequency, as can be seen from Eq. (8.75). We can first tune ω_o by the g_1 or g_4 OTA and then B or Q by the g_m OTA. The final example is the highpass OTA-C filter in Fig. 8.19(f), whose ω_o can be tuned by the g_2 or g_4 OTA and Q then by the g_m OTA, compared with the single OTA prototype in Fig. 8.12 which has only Q electronically adjustable.

8.7 Second-Order Filters Derived from Five-Admittance Model

In this section a more complex one OTA and five-admittance model is considered. The general model with complete feedback is shown in Fig. 8.20. This will be seen to be a development for Fig. 8.1 with two additional admittances. Because more admittances are used, more filter structures and design flexibility can be achieved.

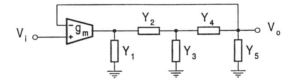

FIGURE 8.20
Five-admittance model with complete output feedback.

The circuit transfer function can be shown as

$$H(s) = \frac{g_m Y_2 Y_4}{\substack{Y_1 Y_2 Y_4 + Y_1 Y_2 Y_5 + Y_1 Y_3 Y_4 + Y_1 Y_3 Y_5 + Y_1 Y_4 Y_5 \\ + Y_2 Y_3 Y_4 + Y_2 Y_3 Y_5 + Y_2 Y_4 Y_5 + g_m Y_2 Y_4}} \quad (8.106)$$

Different filter characteristics can be realized using the general model. This can be done by trying different combinations of passive components in Eq. (8.106). Suppose that each admittance is realized with one element. Exhaustive search shows that a total of 13 different structures can be derived: one highpass, four bandpass and three lowpass filters with five passive components; two bandpass and two lowpass filters with four passive components; as well as one lowpass filter with three passive components. The combinations of components for the 13 structures are presented in Table 8.1. The corresponding configurations and transfer functions can be derived from the general model in Fig. 8.20 and the general expression in Eq. (8.106), which will be presented in the following. These filter structures are suitable for cascade design due to their high input impedance. Note that the four passive element lowpass and bandpass filters derived are actually the same as the counterparts in Figs. 8.5 and 8.6. The three passive component lowpass filter is the same as that in Fig. 8.4. This is no surprise, as the general three-admittance model with output V_{o1} in Fig. 8.1 can be derived from the five-admittance model in the above. We therefore will not repeat them here, although the reader

8.7. SECOND-ORDER FILTERS FROM FIVE-ADMITTANCE MODEL

is encouraged to check this. In the following we will concentrate on the filters with five passive components. These filters can realize the lowpass, highpass, and bandpass functions.

Table 8.1 Generation of Filter Structures Based on Model in Fig. 8.20

Type	Components					Circuit Figure	Function Equation
	Y_1	Y_2	Y_3	Y_4	Y_5		
General	Y_1	Y_2	Y_3	Y_4	Y_5	8.20	8.106
HP	g_1	sC_2	g_3	sC_4	g_5	8.21	8.108
BP1	g_1	sC_2	sC_3	g_4	g_5	8.22(a)	8.113
BP2	g_1	g_2	sC_3	sC_4	g_5	8.22(b)	8.117
BP3	g_1	sC_2	g_3	g_4	sC_5	8.22(c)	8.118
BP4	sC_1	g_2	g_3	sC_4	g_5	8.22(d)	8.119
BP5*	g_1	∞	sC_3	sC_4	g_5	8.6(a)	8.35
BP6*	g_1	sC_2	sC_3	∞	g_5	8.6(b)	8.45
LP1	sC_1	g_2	sC_3	g_4	g_5	8.23(a)	8.120
LP2	g_1	g_2	sC_3	g_4	sC_5	8.23(b)	8.122
LP3	sC_1	g_2	g_3	g_4	sC_5	8.23(c)	8.123
LP4*	g_1	∞	sC_3	g_4	sC_5	8.5(a)	8.21
LP5*	sC_1	g_2	sC_3	∞	g_5	8.5(b)	8.34
LP6*	sC_1	g_2	sC_3	∞	0	8.4	8.14

* Note that the symbol subscriptions used here are different from those in Section 8.2.

8.7.1 Highpass Filter

A highpass filter can be obtained by selecting $Y_1 = g_1$, $Y_2 = sC_2$, $Y_3 = g_3$, $Y_4 = sC_4$, $Y_5 = g_5$ as shown in Fig. 8.21.

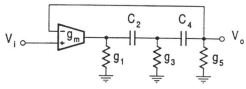

FIGURE 8.21
Highpass filter derived from Fig. 8.20.

We first manipulate Eq. (8.106) according to Y_2 and Y_4 into

$$H(s) = \frac{g_m Y_2 Y_4}{(Y_1 + Y_3 + Y_5 + g_m) Y_2 Y_4 + (Y_1 + Y_3) Y_5 Y_2 + Y_1 (Y_3 + Y_5) Y_4 + Y_1 Y_3 Y_5} \quad (8.107)$$

The transfer function is then easily derived as

$$H(s) = \frac{s^2 g_m C_2 C_4}{s^2 (g_1 + g_3 + g_5 + g_m) C_2 C_4 + s[(g_1 + g_3) g_5 C_2 + g_1 (g_3 + g_5) C_4] + g_1 g_3 g_5} \quad (8.108)$$

Comparison of Eqs. (8.108) and (8.66) will give rise to design equations of ω_o, Q, and K in terms of gs and Cs. Using these equations we can determine component values and analyze sensitivity performance. For the setting up of $C_2 = C_4 = C$ and $g_1 = g_3 = g_5 = g$ we can obtain the component values as

$$g = 4Q\omega_o C, \quad g_m = 64Q^3\omega_o C\left(1 - \frac{3}{16Q^2}\right), \quad K = 1 - \frac{3}{16Q^2} \quad (8.109)$$

and the sensitivities of the design as

$$S_{g_1}^{\omega_o} = S_{g_3}^{\omega_o} = S_{g_5}^{\omega_o} = \frac{1}{2}\left(1 - \frac{1}{16Q^2}\right),$$

$$S_{g_m}^{\omega_o} = -\frac{1}{2}\left(1 - \frac{3}{16Q^2}\right), \quad S_{C_2}^{\omega_o} = S_{C_4}^{\omega_o} = -\frac{1}{2} \quad (8.110)$$

$$S_{g_1}^{Q} = S_{g_5}^{Q} = -\frac{1}{4}\left(1 - \frac{1}{8Q^2}\right), \quad S_{g_3}^{Q} = \frac{1}{32Q^2},$$

$$S_{g_m}^{Q} = \frac{1}{2}\left(1 - \frac{3}{16Q^2}\right), \quad S_{C_2}^{Q} = S_{C_4}^{Q} = 0 \quad (8.111)$$

$$S_{C_2}^{K} = S_{C_4}^{K} = 0, \quad S_{g_m}^{K} = \frac{3}{16Q^2},$$

$$S_{g_1}^{K} = S_{g_3}^{K} = S_{g_5}^{K} = -\frac{1}{16Q^2} \quad (8.112)$$

The sensitivities of the filter are extremely low, the maximum value being $1/2$. Recalling the highpass filter in Section 8.3.2 which has large Q sensitivities for high Q design, the above highpass filter has the advantage of being suitable for any practical Q values in term of sensitivity.

The highpass filter in Fig. 8.21 contains two floating capacitors and three grounded resistors which will determine its performance to the OTA nonidealities and circuit parasitics, which will be discussed with comparison with other filter structures in Section 8.7.4.

A 100 kHz highpass filter is now designed which has a normalized characteristic of

$$H_d(s) = \frac{s^2}{s^2 + 0.5s + 1}$$

which reveals that $Q = 2$. Let $C_2 = C_4 = 10pF$. We can obtain $g_1 = g_3 = g_5 = 50.265\mu S$ and $g_m = 3.066mS$. The designed filter will have a gain of $K = 0.953$.

8.7.2 Bandpass Filter

Four bandpass filter structures are presented in this section. The first bandpass filter is derived from Fig. 8.20 by setting $Y_1 = g_1$, $Y_2 = sC_2$, $Y_3 = sC_3$, $Y_4 = g_4$, $Y_5 = g_5$ as shown in Fig. 8.22(a).

8.7. SECOND-ORDER FILTERS FROM FIVE-ADMITTANCE MODEL

The transfer function can be found, by sorting out Eq. (8.106) according to Y_2 and Y_3, as

$$H(s) = \frac{sg_m g_4 C_2}{s^2 (g_4 + g_5) C_2 C_3 + s\{[g_1(g_4 + g_5) + g_4(g_5 + g_m)]C_2 + g_1(g_4 + g_5)C_3\} + g_1 g_4 g_5}$$

(8.113)

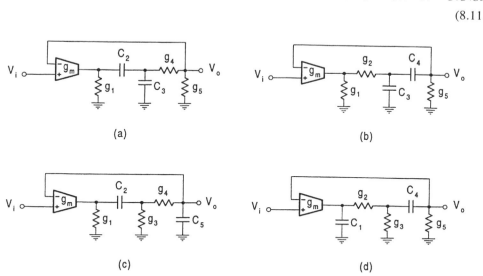

FIGURE 8.22
Four bandpass filters derived from Fig. 8.20.

Similarly we can also derive the design formulas and sensitivity results for this circuit. When $C_2 = C_3 = C$ and $g_1 = g_4 = g_5 = g$, the design formulas are found to be

$$g = \sqrt{2}\omega_o C, \quad g_m = \sqrt{2}\omega_o C\left(-5 + \frac{\sqrt{2}}{Q}\right),$$

$$K = \frac{Q}{\sqrt{2}}\left(-5 + \frac{\sqrt{2}}{Q}\right) \quad (8.114)$$

The OTA input terminals should be interchanged for practical Q values. The sensitivities are derived as

$$S^{\omega_o}_{g_m} = 0, \quad S^{\omega_o}_{g_4} = S^{\omega_o}_{g_5} = \frac{1}{4}, \quad S^{\omega_o}_{g_1} = -S^{\omega_o}_{C_2} = -S^{\omega_o}_{C_3} = \frac{1}{2} \quad (8.115)$$

$$S^Q_{g_m} = -\frac{Q}{\sqrt{2}}\left(-5 + \frac{\sqrt{2}}{Q}\right), \quad S^Q_{g_4} = -\frac{1}{4} + \sqrt{2}Q, \quad S^Q_{g_5} = \frac{3}{4} - \frac{3Q}{\sqrt{2}},$$

$$S^Q_{g_1} = \frac{1}{2} - 2\sqrt{2}Q, \quad -S^Q_{C_2} = S^Q_{C_3} = \frac{1}{2} - \sqrt{2}Q \quad (8.116)$$

The second bandpass filter structure is shown in Fig. 8.22(b), which corresponds to $Y_1 = g_1, Y_2 =$

g_2, $Y_3 = sC_3$, $Y_4 = sC_4$, $Y_5 = g_5$. The transfer function is given by

$$H(s) = \frac{sg_m g_2 C_4}{s^2 (g_1 + g_2) C_3 C_4 + s\{(g_1 + g_2) g_5 C_3 + [(g_1 + g_2) g_5 + (g_1 + g_m) g_2] C_4\} + g_1 g_2 g_5} \quad (8.117)$$

Comparing Eq. (8.117) with Eq. (8.113) we can see that the bandpass filters in Figs. 8.22(a) and (b) have similar transfer functions and therefore similar design procedures and sensitivity performance.

The third bandpass filter with $Y_1 = g_1$, $Y_2 = sC_2$, $Y_3 = g_3$, $Y_4 = g_4$, $Y_5 = sC_5$ is revealed in Fig. 8.22(c). The fourth bandpass filter corresponding to the choice of $Y_1 = sC_1$, $Y_2 = g_2$, $Y_3 = g_3$, $Y_4 = sC_4$, $Y_5 = g_5$ is drawn in Fig. 8.22(d). These two bandpass filters have similar transfer functions. The transfer function of Fig. 8.22(c) is formulated as

$$H(s) = \frac{sg_m g_4 C_2}{s^2 (g_1 + g_3 + g_4) C_2 C_5 + s[(g_1 + g_3 + g_m) g_4 C_2 + g_1 (g_3 + g_4) C_5] + g_1 g_3 g_4} \quad (8.118)$$

and the transfer function of Fig. 8.22(d) is given by

$$H(s) = \frac{sg_m g_2 C_4}{s^2 (g_2 + g_3 + g_5) C_1 C_4 + s[(g_2 + g_3) g_5 C_1 + g_2 (g_3 + g_5 + g_m) C_4] + g_2 g_3 g_5} \quad (8.119)$$

It can be shown that all the bandpass filters in Fig. 8.22 have similar sensitivity performance. Also they contain one grounded and one floating capacitor and one floating and two grounded resistors. The OTA-C equivalents will have similar performances to the nonidealities of the g_m OTA and the OTAs simulating the resistors. Section 8.7.4 will further discuss these issues.

8.7.3 Lowpass Filter

Three lowpass filter configurations are now generated. The first lowpass filter is obtained by selecting $Y_1 = sC_1$, $Y_2 = g_2$, $Y_3 = sC_3$, $Y_4 = g_4$, $Y_5 = g_5$. Substitution into Eq. (8.106) leads to

$$H(s) = \frac{g_m g_2 g_4}{s^2 (g_4 + g_5) C_1 C_3 + s[(g_2 g_4 + g_2 g_5 + g_4 g_5) C_1 + (g_2 g_4 + g_2 g_5) C_3] + g_2 g_4 (g_m + g_5)} \quad (8.120)$$

which compares to the standard lowpass filter characteristic. The corresponding lowpass filter circuit is shown in Fig. 8.23(a).

If $C_1 = C_3 = C$ and $g_2 = g_4 = g_5 = g$, then it can be derived

$$g = \frac{2\omega_o C}{5Q}, \quad g_m = 5Q\omega_o C \left(1 - \frac{2}{25Q^2}\right), \quad K = 1 - \frac{2}{25Q^2} \quad (8.121)$$

The second interesting structure shown in Fig. 8.23(b) comes from the setting $Y_1 = g_1$, $Y_2 = g_2$, $Y_3 = sC_3$, $Y_4 = g_4$, $Y_5 = sC_5$. The transfer function is given by

$$H(s) = \frac{g_m g_2 g_4}{s^2 (g_1 + g_2) C_3 C_5 + s[(g_1 + g_2) g_4 C_3 + (g_1 g_2 + g_1 g_4 + g_2 g_4) C_5] + (g_1 + g_m) g_2 g_4} \quad (8.122)$$

This transfer function is very similar to the one in Eq. (8.120). So the lowpass filter in Fig. 8.23(b) will have similar performances as the one in Fig. 8.23(a).

8.7. SECOND-ORDER FILTERS FROM FIVE-ADMITTANCE MODEL

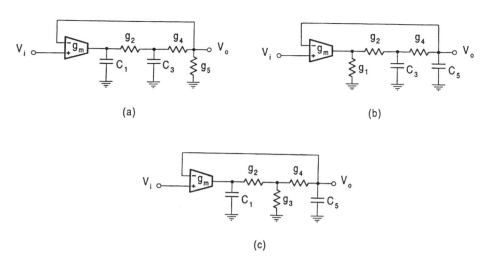

FIGURE 8.23
Three lowpass filters derived from Fig. 8.20.

The third lowpass filter is given in Fig. 8.23(c), which corresponds to $Y_1 = sC_1$, $Y_2 = g_2$, $Y_3 = g_3$, $Y_4 = g_4$, $Y_5 = sC_5$. The transfer function is derived as

$$H(s) = \frac{g_m g_2 g_4}{s^2 (g_2 + g_3 + g_4) C_1 C_5 + s [(g_2 + g_3) g_4 C_1 + g_2 (g_3 + g_4) C_5] + (g_m + g_3) g_2 g_4}$$

(8.123)

It can be shown that the lowpass structures in Fig. 8.23 all have low sensitivities. They contain two grounded capacitors and one grounded and two floating resistors and have similar performances for the nonidealities of the g_m OTA and the OTAs simulating the resistors, as will be seen in the next section.

8.7.4 Comments and Comparison

As discussed in Section 8.5, in integrated filter design, grounded capacitors are usually preferred because they have smaller parasitic capacitances and need less chip area than floating ones. The highpass filter contains two floating capacitors, bandpass filters use one grounded and one floating capacitors, and lowpass filters contain only grounded capacitors. Thus, the lowpass filters are better than the bandpass filters, which are better than the highpass filter in terms of the use of grounded capacitors.

In filter design the number of OTAs should be small, as more OTAs means larger chip area, larger power consumption, more noise, and more parasitic effects. As developed in Section 8.6, a grounded resistor needs one OTA to simulate, but a floating resistor requires two OTAs to simulate. Note also that the floating resistor when simulated using OTAs will introduce more equivalent parasitic elements (a π network, not an admittance). The highpass filter has three grounded resistors, bandpass filters have two grounded and one floating resistor, and lowpass filters embrace one grounded and two floating resistors. The numbers of OTAs needed for simulation of resistors in the highpass, bandpass, and lowpass filters are three, four, and five, respectively. Therefore, in terms of the number of OTAs the derived highpass structure is better than the bandpass filters, which are better than the lowpass filters.

The number of grounded capacitors and the number of grounded resistors are in conflict; if one is big, then the other must be small, as the total number is three. In real design some compromise may have to be made in order to achieve the global optimum.

The OTA is used as a differential input OTA in all highpass, bandpass, and lowpass structures. The nonideality effects of the g_m OTA will be similar for all the structures due to the similarity among the structures, although for example, in some structures such as Figs. 8.21, 8.22(a–c) and 8.23(b), the finite output conductance may be absorbed and in others such as Figs. 8.22(d), 8.23(a) and 8.23(c), the finite output capacitance may be absorbed. (A similar observation for the finite OTA input conductance and capacitance can also be discussed.)

The effects of nonidealities of the resistor simulation OTAs will be quite different. After absorption (see Section 8.6), the highpass filter in Fig. 8.21 will have three grounded parasitic capacitors in parallel with respective grounded resistors; the bandpass filters in Fig. 8.22 will have one floating and two grounded parasitic capacitors in parallel with the corresponding floating and grounded design resistors and one grounded parasitic resistor in parallel with the grounded capacitor; the lowpass filters in Fig. 8.23 will have two floating and one grounded parasitic capacitors in parallel with the related design resistors and two grounded parasitic resistors in parallel with respective design capacitors. It is thus clear that with respect to the effects of nonidealities of the OTAs simulating resistors, the highpass filter is the best, followed by the bandpass filters and then the lowpass filters. Again detailed analysis can be conducted by using the changed admittances due to the nonidealities of the resistor simulation OTAs to replace the ideal ones in Eq. (8.106) for any filter architectures.

8.8 Summary

In this chapter, we have used the operational transconductance amplifier to construct active filters. We have in particular presented systematic methods for generating second-order filters using a single OTA with a reasonable number of resistors and capacitors. The transfer functions, design formulas, and sensitivity results have been formulated. These OTA filters are insensitive to tolerance and parasitics, of high frequency capability, electronically tunable, and simple in structure. They are suitable for discrete implementation using commercially available OTAs and also useful for IC fabrication, when resistors are replaced by OTA equivalents, resulting in OTA-C filters. We have investigated OTA-C filters derived from the single OTA filters by resistor substitution. The effects of OTA nonidealities such as finite input and output impedances and transconductance frequency dependence have also been considered for both discrete and IC filters. It has been proved that these nonidealities influence filter performance. Some techniques have been suggested to reduce the effects from the structural standpoint.

It is noted that there are some other OTA filter structures. References [23] and [24] gave some single OTA structures with current input and voltage output. Filter architectures based on an OTA and an opamp were studied in Ref. [25]. The opamp may limit the working frequency, but in most cases it can be eliminated. OTA-C filters can also be obtained from single opamp active RC filters (as well as multiple opamp architectures) either by direct replacement of the opamp and resistors by OTAs or by some transformation [17]. Many more useful OTA-C filters will be introduced in the following chapters. In the next chapter we will investigate two integrator loop OTA-C filters. The current-mode equivalents of the single OTA filters developed in this chapter will be studied in Chapter 12.

References

[1] Mitra, S.K. and Hurth, C.F., Eds., *Miniatured and Integrated Filters*, John Wiley & Sons, New York, 1989.

REFERENCES

[2] Laker, K.R. and Sansen, W., *Design of Analog Integrated Circuits and Systems*, McGraw-Hill, New York, 1994.

[3] Johns, D.A. and Martin, K., *Analog Integrated Circuit Design*, John Wiley & Sons, New York, 1997.

[4] Toumazou, C., Lidgey, F.J., and Haigh, D.G., Eds., *Analogue IC Design: the Current-Mode Approach*, Peter Peregrinus, London, 1990.

[5] Tsividis, Y.P. and Voorman, J.O., Eds., *Integrated Continuous-Time Filters: Principles, Design and Applications*, IEEE Press, 1993.

[6] Schaumann, R., Ghausi, M.S., and Laker, K.R., *Design of Analog Filters: Passive, Active RC and Switched Capacitor*, Prentice-Hall, NJ, 1990.

[7] Schaumann, R., Continuous-Time Integrated Filters, in *The Circuits and Filters Handbook*, Chen, W.K., Ed., CRC Press, Boca Raton, FL, 1995.

[8] Huelsman, L., *Active and Passive Analog Filter Design*, McGraw-Hill, New York, 1993.

[9] Wheatley, C.F. and Wittlinger, H.A., OTA obsoletes op. amp., *Proc. Nat. Electron. Conf.*, 152–157, 1969.

[10] Franco, S., Use transconductance amplifiers to make programmable active filters, *Electronic Design*, 98–101, Sep. 1976.

[11] Malvar, H.S., Electronically tunable active filters with operational transconductance amplifiers, *IEEE Trans. Circuits Syst.*, 29(5), 333–336, 1982.

[12] Bialko, M. and Newcomb, R.W., Generation of all finite linear circuits using the integrated DVCCS, *IEEE Trans. Circuit Theory*, 18, 733–736, 1971.

[13] Urbaś, A. and Osiowski, J., High-frequency realization of C-OTA second order active filters, *Proc. IEEE Intl. Symp. Circuits Syst.*, 1106–1109, 1982.

[14] Malvar, H.S., Electronically controlled active-C filters and equalizers with operational transconductance amplifiers, *IEEE Trans. Circuits Syst.*, 31(7), 645–649, 1984.

[15] Geiger, R.L. and Sánchez-Sinencio, E., Active filter design using operational transconductance amplifiers: a tutorial, *IEEE Circuits and Devices Magazine*, 20–32, Mar. 1985.

[16] Sánchez-Sinencio, E., Geiger, R.L., and Nevarez-Lozano, H., Generation of continuous-time two integrator loop OTA filter structures, *IEEE Trans. Circuits Syst.*, 35(8), 936–946, 1988.

[17] Ananda Mohan, P.V., Generation of OTA-C filter structures from active RC filter structures, *IEEE Trans. Circuits Syst.*, 37, 656–660, 1990.

[18] Acar, C., Anday, F., and Kuntman, H., On the realization of OTA-C filters, *Intl. J. Circuit Theory Applications*, 21(4), 331–341, 1993.

[19] Sun, Y. and Fidler, J.K., Novel OTA-C realizations of biquadratic transfer functions, *Intl. J. Electronics*, 75, 333–348, 1993.

[20] Sun, Y. and Fidler, J.K., Resonator-based universal OTA-grounded capacitor filters, *Intl. J. Circuit Theory Applications*, 23, 261–265, 1995.

[21] Tan, M.A. and Schaumann, R., Design of a general biquadratic filter section with only transconductance and grounded capacitors, *IEEE Trans. Circuits Syst.*, 35(4), 478–480, 1988.

[22] Nawrocki, R. and Klein, U., New OTA-capacitor realization of a universal biquad, *Electron. Lett.*, 22(1), 50–51, 1986.

[23] Al-Hashimi, B.M. and Fidler, J.K., Novel high-frequency continuous-time low-pass OTA based filters, *Proc. IEEE Intl. Symp. Circuits Syst.*, 1171–1172, 1990.

[24] Al-Hashimi, B.M., Fidler, J.K., and Garner, P., High frequency active filters using OTAs, *Proc. IEE Colloquium on Electronic Filters*, 3/1–3/5, London, 1989.

[25] Deliyannis, T., Active RC filters using an operational transconductance amplifier and an operational amplifier, *Intl. J. Circuit Theory Applications*, 8, 39–54, 1980.

[26] Sun, Y., Jefferies, B., and Teng, J., Universal third-order OTA-C filters, *Intl. J. Electronics*, 85(5), 597–609, 1998.

[27] Nawrocki, R., Building set for tunable component simulation filters with operational transconductance amplifiers, *Proc. Midwest Symp. Circuits and Systems*, 227–230, 1987.

[28] Tan, M.A. and Schaumann, R., Simulating general-parameter filters for monolithic realization with only transconductance elements and grounded capacitors, *IEEE Trans. Circuits Syst.*, 36(2), 299–307, 1989.

[29] de Queiroz, A.C.M., Caloba, L.P., and Sánchez-Sinencio, E., Signal flow graph OTA-C integrated filters, *Proc. IEEE Intl. Symp. Circuits Syst.*, 2165–2168, 1988.

[30] Nawrocki, R., Electronically tunable all-pole low-pass leapfrog ladder filter with operational transconductance amplifier, *Intl. J. Electronics*, 62(5), 667–672, 1987.

[31] Sun, Y. and Fidler, J.K., Synthesis and performance analysis of a universal minimum component integrator-based IFLF OTA-grounded capacitor filter, *IEE Proceedings: Circuits, Devices and Systems*, 143, 107–114, 1996.

[32] Sun, Y. and Fidler, J.K., Structure generation and design of multiple loop feedback OTA-grounded capacitor filters, *IEEE Trans. on Circuits and Systems, Part-I: Fundamental Theory and Applications*, 44(1), 1–11, 1997.

[33] Nawrocki, R., Electronically controlled OTA-C filter with follow-the-leader-feedback structures, *Intl. J. Circuit Theory Applications*, 16, 93–96, 1988.

[34] Nevarez-Lozano, H., Hill, J.A., and Sánchez-Sinencio, E., Frequency limitations of continuous-time OTA-C filters, *Proc. IEEE Intl. Symp. Circuits Syst.*, 2169–2172, 1988.

[35] Sun, Y. and Fidler, J.K., Performance analysis of multiple loop feedback OTA-C filters, *Proc. IEE 14th Saraga Colloquium on Digital and Analogue Filters and Filtering Systems*, 9/1–9/7, London, 1994.

[36] Ramírez-Angulo, J. and Sánchez-Sinencio, E., Comparison of biquadratic OTA-C filters from the tuning point of view, *Proc. IEEE Midwest Symp. Circuits Syst.*, 510–514, 1988.

[37] Park, C.S. and Schaumann, R., Design of a 4-MHz analog integrated CMOS transconductance-C bandpass filter, *IEEE J. Solid-State Circuits*, 23(4), 987–996, 1988.

REFERENCES

[38] Silva-Martinez, J., Steyaert, M.S.J., and Sansen, W., A 10.7-MHz 68-dB SNR CMOS continuous-time filter with on-chip automatic tuning, *IEEE J. Solid-State Circuits*, 27, 1843–1853, 1992.

[39] Loh, K.H., Hiser, D.L., Adams, W.J., and Geiger, R.L., A versatile digitally controlled continuous-time filter structure with wide range and fine resolution capability, *IEEE Trans. Circuits Syst.*, 39, 265–276, 1992.

[40] Nedungadi, A.P. and Geiger, R.L., High-frequency voltage-controlled continuous-time low-pass filter using linearized CMOS integrators, *Electron. Lett.*, 22, 729–731, 1986.

[41] Gopinathan, V., Tsividis, Y.P., Tan, K.S., and Hester, R.K., Design considerations for high-frequency continuous-time filters and implementation of an antialiasing filter for digital video, *IEEE J. Solid-State Circuits*, 25(6), 1368–1378, 1990.

[42] Wang, Y.T. and Abidi, A.A., CMOS active filter design at very high frequencies, *IEEE J. Solid-State Circuits*, 25(6), 1562–1574, 1990.

[43] Sun, Y. and Fidler, J.K., Structure generation of current-mode two integrator loop dual output-OTA grounded capacitor filters, *IEEE Trans. on Circuits and Systems, Part II: Analog and Digital Signal Proc.*, 43(4), 1996.

[44] Fidler, J.K., Mack, R.J., and Noakes, P.D., Active filters incorporating the voltage-to-current transactor, *Microelectronics J.*, 8, 19–22, 1977.

[45] Al-Hashimi, B.M. and Fidler, J.K., A novel VCT-based active filter configuration, *Proc. 6th Intl. Symp. Networks, Systems and Signal Proc.*, Yugoslavia, Jun. 1989.

[46] Fidler, J.K., Sensitivity assessment of parasitic effects in second-order active-filter configurations, *Electronic Circuits and Systems*, 2(6) 181–185, 1978.

[47] Weyton, L., A useful sensitivity measure for second order RC active filter configuration, *IEEE Trans. Circuits and Syst.*, 23, 506–508, 1976.

[48] Sallen, R.P. and Key, E.L., A practical method of designing RC active filters, *IRE Trans. Circuit Theory*, 2, 74–85, 1955.

Chapter 9

Two Integrator Loop OTA-C Filters

9.1 Introduction

As discussed in Chapter 8, using the operational transconductance amplifier (OTA) to replace the conventional opamp in active RC filters results in several benefits. OTA-C filters offer improvements in design simplicity, parameter programmability, circuit integrability, and high-frequency capability when compared to opamp-based filters, as well as reduced component count. They are also insensitive to tolerance. Hence OTA-C filter structures have received great attention from both academia and industry and have become the most important technique for high-frequency continuous-time integrated filter design [1]–[49]. Note that OTA-C filters are also widely known as g_m-C filters in the literature, especially in solid-state circuit implementation. The term transconductance-C filters is also often used. In this book we adopt the term OTA-C filters.

This chapter deals with second-order OTA-C filters of two integrator loop configuration. Two integrator loop filters are a very popular category of filters, their opamp realization being discussed in the context of biquads in Chapter 4. They have very low sensitivity, and they can be used alone and can also be used as a section for cascade high-order filter design. The early papers on two integrator loop OTA-C filters include [6]–[11]. It was first proved in Ref. [6] that OTAs and capacitors can be used to construct all building blocks for active filter design, laying the foundation for OTA-C filters. Active filters using only OTAs and capacitors (without any opamps and resistors) were then investigated practically in [9]. The authors of [9] not only proposed a very interesting two integrator loop structure which has been simplified or generalized to develop more filter structures later [10], but also proposed methods for some practical high frequency (MHz range) design problems such as compensation and tuning which have been used in other integrated OTA-C filter designs [12, 46]. The term OTA-C filters was also first used in [9] in terms of then C-OTA filters. Another important publication on (two integrator loop) OTA-C filters is reference [11], where versatile filter functions were achieved by switches. This paper has led to many further similar publications such as [15, 19]. Since 1985 when a tutorial paper on OTA-C filters was published [12], more papers have been published [13]–[40], [42]–[49]. In particular, Reference [16] has systematically summarized and extended the work on two integrator loop OTA-C filters published previously [6]–[14] using the block diagram method. OTA-C filters based on passive LC ladder simulation were also investigated [26]–[28] at the early stages of the development of OTA-C filters, which will be discussed in details in the next chapter.

In this chapter we will classify and study two integrator loop OTA-C filters in a systematic and comprehensive way. Some further or new results will be given. Throughout the chapter the diversity of structures, functions, methods, and performances is emphasized and many design choices are given for a particular application.

9.2 OTA-C Building Blocks and First-Order OTA-C Filters [6, 12]

The basic building blocks with which we shall work comprise the ideal integrator, amplifier, and summer. The OTA realizations of these blocks are depicted in Fig. 9.1, which were also discussed in Chapter 3.

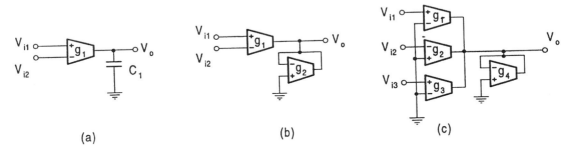

FIGURE 9.1
(a) Ideal integrator, (b) amplifier, and (c) summer.

With $\tau_1 = C_1/g_1$ the ideal integrator in Fig. 9.1(a) has the characteristic of

$$H(s) = \frac{V_o}{V_{i1} - V_{i2}} = \frac{1}{s\tau_1} \tag{9.1}$$

The amplifier in Fig. 9.1(b) has gain

$$k = \frac{V_o}{V_{i1} - V_{i2}} = \frac{g_1}{g_2} \tag{9.2}$$

and the summer in Fig. 9.1(c) has the relation of

$$V_o = \beta_1 V_{i1} - \beta_2 V_{i2} + \beta_3 V_{i3} \tag{9.3}$$

where $\beta_j = g_j/g_4$.

Some lossy integrators and first-order filters are shown in Fig. 9.2. These circuits can be constructed using the basic building blocks in Fig. 9.1. They were also derived from the single OTA and three admittance model in Chapter 8. In particular, the structure in Fig. 9.2(a) relates to the circuit in Fig. 8.3(c). In practice they are used directly as building blocks in the design of second-order and higher-order filters.

For the feedback circuit in Fig. 9.2(a), with $\tau_1 = C_1/g_1$ and $k_1 = g_2/g_3$ we have the transfer functions as

$$V_{o1} = \frac{V_{i1} + k_1 V_{i2} - V_{i3}}{\tau_1 s + k_1} \tag{9.4}$$

$$V_{o2} = \frac{k_1 V_{i1} - k_1 \tau_1 s V_{i2} + \tau_1 s V_{i3}}{\tau_1 s + k_1} \tag{9.5}$$

9.2. OTA-C BUILDING BLOCKS

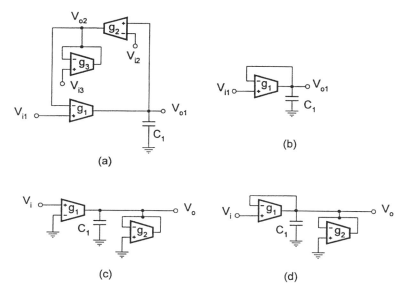

FIGURE 9.2
Lossy integrators and first-order filters.

The two relations in Eqs. (9.4) and (9.5) show that the filter in Fig. 9.2(a) can fulfill lowpass and highpass specifications, depending on the voltages applied to V_{i1}, V_{i2} and V_{i3}. We can also obtain a canonical lowpass structure with feedback being achieved by direct connection, that is, $k_1 = 1$, which is shown in Fig. 9.2(b). The corresponding relation of this circuit becomes

$$V_{o1} = \frac{1}{\tau_1 s + 1} V_{i1} \qquad (9.6)$$

Note that the circuit of Fig. 9.2(b) is called a lossy integrator, as is that of Fig. 9.2(a) with input V_{i1} and output V_{o1}, V_{i2} and V_{i3} being left disconnected.

Another different lossy integrator is given in Fig. 9.2(c), which has the lowpass function

$$V_o = \frac{g_1}{sC_1 + g_2} V_i = \frac{1}{\tau_1 s + k_1} V_i \qquad (9.7)$$

where $\tau_1 = C_1/g_1$ and $k_1 = g_2/g_1$.

The last lossy integrator is shown in Fig. 9.2(d), which has the lowpass function given by

$$V_o = \frac{1}{\tau_1 s + k_1} V_i \qquad (9.8)$$

where $\tau_1 = C_1/g_1$ and $k_1 = 1 + g_2/g_1$. This one differs from Fig. 9.2(b) in that it has an extra OTA with g_2 and from Fig. 9.2(c) in that it has a feedback loop.

9.3 Two Integrator Loop Configurations and Performance

In this section we will generally introduce two integrator loop structures and their performance. The OTA-C realizations will be discussed in the following sections.

9.3.1 Configurations

Figure 9.3 shows two configurations of two integrator loop systems. They both have two feedback loops consisting of two ideal integrators and two amplifiers. The difference between Figs. 9.3(a) and (b) is in the feedback structure; the former is of the summed-feedback (SF) type and the latter of the distributed-feedback (DF). This is because in Fig. 9.3(a) the outputs of all the integrators are fedback to the first integrator input; while in Fig. 9.3(b) the last (second in this case) integrator output is fedback to the inputs of all the integrators. It is also apparent that Fig. 9.3(b) with τ_1 and τ_2 being interchanged and k_{22} being represented by k_{11} is the transpose of Fig. 9.3(a).

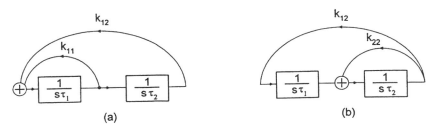

FIGURE 9.3
Two integrator loop configurations.

Using the basic building blocks in Fig. 9.1 and the first-order circuits in Fig. 9.2 we can readily realize the two integrator loop configurations in Fig. 9.3. Before doing this, we describe some common features briefly.

9.3.2 Pole Equations

The loops in Fig. 9.3 determine the pole characteristics of the systems. The structure in Fig. 9.3(a) has the system pole polynomial $D_{SF}(s)$ (denominator of the transfer function), the pole angular frequency ω_o and the pole quality factor Q as shown below, respectively.

$$D_{SF}(s) = \tau_1 \tau_2 s^2 + k_{11} \tau_2 s + k_{12} = \tau_1 \tau_2 \left(s^2 + \frac{\omega_o}{Q} s + \omega_o^2 \right) \tag{9.9}$$

$$\omega_o = \sqrt{\frac{k_{12}}{\tau_1 \tau_2}}, \quad \frac{\omega_o}{Q} = \frac{k_{11}}{\tau_1}, \quad Q = \frac{1}{k_{11}} \sqrt{k_{12} \frac{\tau_1}{\tau_2}} \tag{9.10}$$

Similarly Fig. 9.3(b) has the pole equation and parameters, given by Eqs. (9.11) and (9.12).

$$D_{DF}(s) = \tau_1 \tau_2 s^2 + k_{22} \tau_1 s + k_{12} = \tau_1 \tau_2 \left(s^2 + \frac{\omega_o}{Q} s + \omega_o^2 \right) \tag{9.11}$$

9.3. TWO INTEGRATOR LOOP CONFIGURATIONS

$$\omega_o = \sqrt{\frac{k_{12}}{\tau_1 \tau_2}}, \quad \frac{\omega_o}{Q} = \frac{k_{22}}{\tau_2}, \quad Q = \frac{1}{k_{22}}\sqrt{k_{12}\frac{\tau_2}{\tau_1}} \quad (9.12)$$

9.3.3 Design

The polynomial equation determines the system poles, thus the performance. The coefficients k_{ij} and time constants τ_j are determined by individual circuit components. For the given ω_o and Q of Fig. 9.3(a), the design will require deciding on k_{ij} and τ_j. This can be carried out using Eq. (9.10) for the summed-feedback configuration. For example, setting $\tau_1 = \tau_2 = \tau$ yields the design formulas, given by

$$k_{11} = \frac{\omega_o \tau}{Q}, \quad k_{12} = (\omega_o \tau)^2 \quad (9.13)$$

We can also select $k_{11} = k_{12} = k$ and obtain

$$\tau_1 = k\frac{Q}{\omega_o}, \quad \tau_2 = \frac{1}{\omega_o Q} \quad (9.14)$$

We can further determine component values for the particular realization. The design method is also suitable for the distributed-feedback configuration.

9.3.4 Sensitivity

It can be readily shown that the two integrator loop systems have very low sensitivity. For instance, using Eq. (9.10) the ω_o, $\omega_o/Q (= B)$ and Q sensitivities to τs and ks are calculated as

$$S_{\tau_1}^{\omega_o} = S_{\tau_2}^{\omega_o} = -S_{k_{12}}^{\omega_o} = -1/2, \quad S_{k_{11}}^{\omega_o} = 0 \quad (9.15)$$

$$S_{\tau_1}^{\omega_o/Q} = -S_{k_{11}}^{\omega_o/Q} = -1, \quad S_{\tau_2}^{\omega_o/Q} = S_{k_{12}}^{\omega_o/Q} = 0 \quad (9.16)$$

$$S_{\tau_1}^{Q} = -S_{\tau_2}^{Q} = S_{k_{12}}^{Q} = 1/2, \quad S_{k_{11}}^{Q} = -1 \quad (9.17)$$

The calculated sensitivities are all very low. Most of them are either 0 or $\pm 1/2$ and only a few are ± 1. (This is compared with the Q sensitivities of some bandpass filters which are Q-dependent in Chapter 8.) The parameter sensitivities to circuit components can be further easily computed for the particular realization.

9.3.5 Tuning

Tuning is a major problem in continuous-time filter design. In practice the constants τ_1, τ_2, k_{11}, k_{22}, k_{12} can be properly implemented using OTAs and are thus electronically controllable since the associated transconductances g_j are related to bias voltages or currents. At this stage it is clear from Eqs. (9.10) and (9.12) that Q can be adjusted by k_{11} (or k_{22}) independently from ω_o, a very attractive property. We should stress that k_{11} (or k_{22}) plays a key role in independent tuning. We shall see soon that most structures have this independent tuning capability.

9.3.6 Biquadratic Specifications

In filter design, different types of filter may be required. The most common are the lowpass (LP), bandpass (BP), highpass (HP), bandstop (BS, also called symmetrical notch), lowpass notch (LPN), highpass notch (HPN), and allpass (AP) characteristics. The standard expressions for these functions are given in Table 9.1 for the convenience of the discussion to come. The general biquadratic transfer function may also be required in some cases. Taking the development of Section 4.3 of Chapter 4 further, it can be written in the form

$$H(s) = \frac{K_2 s^2 + K_1 \frac{\omega_o}{Q} s + K_0 \omega_o^2}{s^2 + \frac{\omega_o}{Q} s + \omega_o^2} \quad (9.18)$$

The typical specifications can be derived from this general expression using the conditions in the table.

Table 9.1 Some Popular Biquadratic Filter Specifications

Type	Numerators	Derivative conditions
LP	$K_{LP}\omega_o^2$	$K_2 = K_1 = 0, K_0 = K_{LP}$
BP	$K_{BP}\frac{\omega_o}{Q}s$	$K_2 = K_0 = 0, K_1 = K_{BP}$
HP	$K_{HP}s^2$	$K_1 = K_0 = 0, K_2 = K_{HP}$
BS	$K_{BS}(s^2 + \omega_n^2), \omega_n = \omega_o$	$K_1 = 0, K_2 = K_0 = K_{BS}$
LPN	$K_{LPN}(\frac{\omega_o}{\omega_n})^2(s^2 + \omega_n^2), \omega_n > \omega_o$	$K_1 = 0, K_2 = K_{LPN}(\frac{\omega_o}{\omega_n})^2$
		$K_0 = K_{LPN}$
HPN	$K_{HPN}(s^2 + \omega_n^2), \omega_n < \omega_o$	$K_1 = 0, K_2 = K_{HPN},$
		$K_0 = K_{HPN}(\frac{\omega_n}{\omega_o})^2$
AP	$K_{AP}(s^2 - \frac{\omega_o}{Q}s + \omega_o^2)$	$K_2 = -K_1 = K_0 = K_{AP}$

Note: The denominators for all the functions have the same form as $s^2 + \frac{\omega_o}{Q}s + \omega_o^2$.

In the following we will show how filter structures can be generated, how many functions can be realized, how the filters can be best designed and how the performances can be evaluated and enhanced. As a large number of different OTA-C filter structures can be obtained, we will only selectively introduce some of the more interesting architectures.

9.4 OTA-C Realizations of Distributed-Feedback (DF) Configuration

In this section we realize the DF configuration using OTAs and capacitors. The general realization with arbitrary feedback coefficients is first discussed and then the structures with special feedback coefficients are presented.

9.4.1 DF OTA-C Circuit and Equations

The DF configuration in Fig. 9.3(b) has two ideal integrators which can be realized using the ideal OTA-C integrator in Fig. 9.1(a) and two feedback coefficients which may be realized using the OTA voltage amplifier in Fig. 9.1(b). Therefore, the general realization will require two capacitors and six OTAs, as shown in Fig. 9.4 [16]. We wish to use this biquad to show how to obtain different

9.4. OTA-C REALIZATIONS OF DF CONFIGURATION

functions by using different inputs and outputs. Now we formulate the equations of this general structure.

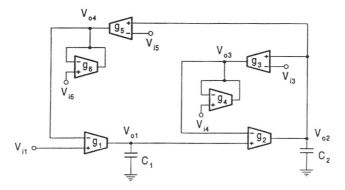

FIGURE 9.4
DF OTA-C realization with arbitrary k_{12} and k_{22}.

With $\tau_1 = C_1/g_1$, $\tau_2 = C_2/g_2$, $k_{12} = g_5/g_6$, and $k_{22} = g_3/g_4$, nodal equations of the circuit can be written as

$$V_{o1} = (V_{i1} - V_{o4})/\tau_1 s \tag{9.19}$$

$$V_{o2} = (V_{o1} - V_{o3})/\tau_2 s \tag{9.20}$$

$$V_{o3} = (V_{o2} - V_{i3})k_{22} + V_{i4} \tag{9.21}$$

$$V_{o4} = (V_{o2} - V_{i5})k_{12} + V_{i6} \tag{9.22}$$

From these equations we can derive the general input and output voltage relations as

$$D_{DF1}(s)V_{o1} = (\tau_2 s + k_{22})V_{i1} - k_{12}k_{22}V_{i3} + k_{12}V_{i4} + (\tau_2 s + k_{22})k_{12}V_{i5}$$
$$- (\tau_2 s + k_{22})V_{i6} \tag{9.23}$$

$$D_{DF1}(s)V_{o2} = V_{i1} + k_{22}\tau_1 s V_{i3} - \tau_1 s V_{i4} + k_{12}V_{i5} - V_{i6} \tag{9.24}$$

$$D_{DF1}(s)V_{o3} = k_{22}V_{i1} - \left(\tau_1\tau_2 s^2 + k_{12}\right)k_{22}V_{i3} + \left(\tau_1\tau_2 s^2 + k_{12}\right)V_{i4}$$
$$+ k_{12}k_{22}V_{i5} - k_{22}V_{i6} \tag{9.25}$$

$$D_{DF1}(s)V_{o4} = k_{12}V_{i1} + k_{12}k_{22}\tau_1 s V_{i3} - k_{12}\tau_1 s V_{i4} - \left(\tau_1\tau_2 s^2 + k_{22}\tau_1 s\right)k_{12}V_{i5}$$
$$+ \left(\tau_1\tau_2 s^2 + k_{22}\tau_1 s\right)V_{i6} \tag{9.26}$$

where

$$D_{DF1}(s) = \tau_1\tau_2 s^2 + k_{22}\tau_1 s + k_{12} \qquad (9.27)$$

9.4.2 Filter Functions

The equations formulated above indicate that this general circuit can offer a variety of filter specification with different inputs and outputs. Table 9.2 presents all possible functions. To illustrate how to use the table, we look at row V_{o3} and column V_{i4} and the corresponding intersection BS. If we apply the input voltage to V_{i4} only, grounding all the other input voltages, and take the output voltage from V_{o3} only, then the circuit will give a BS transfer function from V_{i4} to V_{o3}. The lowpass and bandpass filters can be easily seen from Eqs. (9.23–9.27).

Table 9.2 Filter Functions from DF OTA-C Structure in Fig. 9.4

	V_{i1}	V_{i3}	V_{i4}	V_{i5}	V_{i6}	V_{i35}	V_{i14}	V_{i46}	V_{i146}
V_{o1}		LP	LP			BP			LP
V_{o2}	LP	BP	BP	LP	LP				BP
V_{o3}	LP	BS	BS	LP	LP	HP	LPN	HPN	BS
V_{o4}	LP	BP	BP			HP			AP
$V_{o1}-V_{o3}$	BP	HP	HP	BP	BP				HP

It is a very attractive feature that the circuit can directly give the bandstop function without need of extra components or difference component matching as in most cases. The BS filter is given by output V_{o3} for input V_{i3} or V_{i4}, the transfer functions being given by

$$-\frac{V_{o3}}{V_{i3}} = k_{22}\frac{V_{o3}}{V_{i4}} = \frac{k_{22}(\tau_1\tau_2 s^2 + k_{12})}{\tau_1\tau_2 s^2 + k_{22}\tau_1 s + k_{12}} \qquad (9.28)$$

More interesting, we can also obtain a lowpass notch filter (see Table 9.1 for its definition) from V_{o3} when $V_{i1} = V_{i4} = V_{i14}$ meaning that V_{i1} and V_{i4} are connected together and provided by the same source voltage V_{i14}, with the transfer function given by

$$\frac{V_{o3}}{V_{i14}} = \frac{\tau_1\tau_2 s^2 + (k_{12} + k_{22})}{\tau_1\tau_2 s^2 + k_{22}\tau_1 s + k_{12}} \qquad (9.29)$$

and a highpass notch filter from V_{o3} when $V_{i4} = V_{i6} = V_{i46}$, given by

$$\frac{V_{o3}}{V_{i46}} = \frac{\tau_1\tau_2 s^2 + (k_{12} - k_{22})}{\tau_1\tau_2 s^2 + k_{22}\tau_1 s + k_{12}} \qquad (9.30)$$

For the realization of highpass characteristics, from the V_{o1} and V_{o3} expressions we can see that

$$D_{DF1}(s)(V_{o1} - V_{o3}) = \tau_2 s V_{i1} + k_{22}\tau_1\tau_2 s^2 V_{i3} - \tau_1\tau_2 s^2 V_{i4}$$
$$+ k_{12}\tau_2 s V_{i5} - \tau_2 s V_{i6} \qquad (9.31)$$

This reveals that the voltage output across nodes 1 and 3 ($V_{o1} - V_{o3}$) can support the HP function for inputs V_{i3} or V_{i4}. However, this method may not be favorable in some cases, because of the floating

9.4. OTA-C REALIZATIONS OF DF CONFIGURATION

output. But as will be seen in later sections it can be readily converted to the output to the ground by using extra OTAs.

A further observation of Eqs. (9.23–9.27) indicates that when $V_{i3} = V_{i5} = V_{i35}$ the voltages of nodes 3 and 4 offer the HP function. The transfer functions are rewritten as

$$\frac{V_{o3}}{V_{i35}} = -\frac{k_{22}\tau_1\tau_2 s^2}{\tau_1\tau_2 s^2 + k_{22}\tau_1 s + k_{12}} \tag{9.32}$$

$$\frac{V_{o4}}{V_{i35}} = -\frac{k_{12}\tau_1\tau_2 s^2}{\tau_1\tau_2 s^2 + k_{22}\tau_1 s + k_{12}} \tag{9.33}$$

In this case V_{o1} supports the BP characteristic with the numerator of $k_{12}\tau_2 s$.

More complex functions can also be realizable. When $V_{i1} = V_{i4} = V_{i6} = V_{i146}$, a general biquadratic function can be attained from V_{o4}, given by

$$\frac{V_{o4}}{V_{i146}} = \frac{\tau_1\tau_2 s^2 + (k_{22} - k_{12})\tau_1 s + k_{12}}{\tau_1\tau_2 s^2 + k_{22}\tau_1 s + k_{12}} \tag{9.34}$$

For example, the BS and AP characteristics will result from $k_{12} = k_{22}$ and $k_{12} = 2k_{22}$, respectively. The difference equality or matching may not be easy to maintain in discrete implementation, but it can be quite readily achieved in integrated circuit implementation with on-chip tuning.

From the above discussion it can be seen that the circuit with all capacitors grounded, all outputs with respect to the ground, and no need of difference component matching, can offer all the LP, BP, HP, BS, LPN, and HPN characteristics. With use of difference component matching, the AP and other complex functions can also be achieved.

Filter functions can also be produced by capacitor injection. Voltage injection through the ungrounded capacitor technique is however not suitable for integrated circuit implementation and cascade design of high-order filters. The former is because the resulting floating capacitor will increase the chip area and parasitic effects, and the latter is due to undesirable capacitive coupling between cascaded stages caused by the injection capacitors and the need for ideal buffers. Even for individual use the injection requires an ideal voltage source with zero source resistance. Practical considerations therefore lead one to avoid using capacitor voltage injection and ground all capacitors. In this chapter we therefore consider the realizations with only grounded capacitors. This also implies that the capacitor injection technique will not be used.

9.4.3 Design Examples

We have presented various design examples of lowpass, bandpass, and highpass filters in Chapter 8. For the DF OTA-C structure in Fig. 9.4, the LP, BP, and HP filters may be conveniently realized using the transfer functions obtained in the above. Here we just show two more complex examples, that is, the design of the BS and AP filters. First we want to see how the standard BS characteristic

$$H_{BS}(s) = K_{BS}\frac{s^2 + \omega_o^2}{s^2 + \frac{\omega_o}{Q}s + \omega_o^2} \tag{9.35}$$

can be implemented using the transfer function $-V_{o3}/V_{i3}$.

Comparing Eqs. (9.28) and (9.35) and setting $\tau_1 = \tau_2 = \tau$ we have

$$\tau = K_{BS}\frac{Q}{\omega_o}, \quad k_{22} = K_{BS}, \quad k_{12} = (K_{BS}Q)^2 \tag{9.36}$$

Another method is to let $k_{22} = k_{12} = k$. The corresponding design formulas are derived as

$$k = K_{BS}, \quad \tau_1 = \frac{1}{\omega_o Q}, \quad \tau_2 = K_{BS}\frac{Q}{\omega_o} \tag{9.37}$$

It is very flexible to further determine gs and Cs according to their relations with ks and τs. Consider the following specifications for the bandstop filter:

$$f_o = 500kHz, \quad Q = 8, \quad K_{BS} = 2$$

Based on the design formulas in Eq. (9.37), for example, with the selection of

$$g_1 = g_2 = g_3 = g_5 = g = 40\mu S$$

we can calculate the other component values as

$$g_4 = g_6 = g/K_{BS} = 20\mu S, \quad C_1 = g/\omega_o Q = 1.59 pF,$$

$$C_2 = gK_{BS}Q/\omega_o = 203.7 pF$$

In this design only two different transconductance values are used, a useful feature for integrated circuit implementation.

To design an AP filter having the transfer function

$$H_{AP}(s) = K_{AP}\frac{s^2 - \frac{\omega_o}{Q}s + \omega_o^2}{s^2 + \frac{\omega_o}{Q}s + \omega_o^2} \tag{9.38}$$

we use the transfer function V_{o4}/V_{i146} in Eq. (9.34). For the arbitrary k_{22} we determine the design formulas as

$$K_{AP} = 1, \quad \tau_1 = \frac{2}{\omega_o Q}, \quad \tau_2 = k_{22}\frac{Q}{\omega_o}, \quad k_{12} = 2k_{22} \tag{9.39}$$

The reader is encouraged to consider other interesting realizations, for example the LPN and HPN filters.

9.4.4 DF OTA-C Realizations with Special Feedback Coefficients

We further consider the most important three special cases of the realization of the DF configuration in Fig. 9.3(b). The first structure is the one in Fig. 9.5, which is the realization corresponding to $k_{12} = 1$ and this unity feedback is realized by pure connection, saving two OTAs. This circuit was first published in 1983 [10] and has also been investigated later on.

The equations of the circuit can be easily derived as (there remain $\tau_1 = C_1/g_1$, $\tau_2 = C_2/g_2$ and $k_{22} = g_3/g_4$)

$$D_{DF2}(s)V_{o1} = (\tau_2 s + k_{22})V_{i1} - k_{22}V_{i3} + V_{i4} \tag{9.40}$$

$$D_{DF2}(s)V_{o2} = V_{i1} + k_{22}\tau_1 s V_{i3} - \tau_1 s V_{i4} \tag{9.41}$$

9.4. OTA-C REALIZATIONS OF DF CONFIGURATION

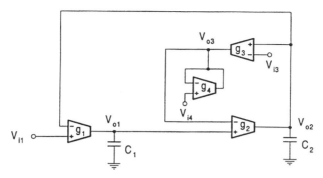

FIGURE 9.5
DF OTA-C realization with $k_{12} = 1$ and arbitrary k_{22}.

$$D_{DF2}(s)V_{o3} = k_{22}V_{i1} - \left(\tau_1\tau_2 s^2 + 1\right)k_{22}V_{i3}$$

$$+ \left(\tau_1\tau_2 s^2 + 1\right)V_{i4} \tag{9.42}$$

$$D(s)_{DF2}(V_{o1} - V_{o3}) = \tau_2 s V_{i1} + k_{22}\tau_1\tau_2 s^2 V_{i3} - \tau_1\tau_2 s^2 V_{i4} \tag{9.43}$$

where

$$D_{DF2}(s) = \tau_1\tau_2 s^2 + k_{22}\tau_1 s + 1 \tag{9.44}$$

As can be seen from Eqs. (9.40 through 9.44) this structure has the LP, BP, HP, and BS characteristics. One interesting realization of the HP function is correspondent to $V_{i1} = V_{i3} = V_{i13}$ for V_{o3}, giving

$$\frac{V_{o3}}{V_{i13}} = -\frac{k_{22}\tau_1\tau_2 s^2}{\tau_1\tau_2 s^2 + k_{22}\tau_1 s + 1} \tag{9.45}$$

The other outputs have for V_{i13}: $D_{DF2}(s)V_{o1} = \tau_2 s V_{i13}$ and $D_{DF2}(s)V_{o2} = (k_{22}\tau_1 s + 1)V_{i13}$.

The second important case is that $k_{12} = k_{22} = k$ and k is realized using a single voltage amplifier, saving two OTAs, which is shown in Fig. 9.6. This structure was derived in [20] based on the modification of the biquad proposed in [13]. The latter biquad is however not included in this chapter because it is in nature based on current integrators. The current-mode OTA-C filters will be handled in Chapter 12.

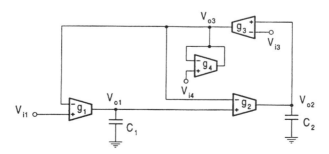

FIGURE 9.6
DF OTA-C realization with $k_{12} = k_{22} = k$.

This structure offers the LP, BP, and HP functions, which can be seen from the following equations with $\tau_1 = C_1/g_1$, $\tau_2 = C_2/g_2$ and $k = g_3/g_4$:

$$D_{DF3}(s)V_{o1} = (\tau_2 s + k) V_{i1} + k\tau_2 s V_{i3} - \tau_2 s V_{i4} \qquad (9.46)$$

$$D_{DF3}(s)V_{o2} = V_{i1} + k(\tau_1 s + 1) V_{i3} - (\tau_1 s + 1) V_{i4} \qquad (9.47)$$

$$D_{DF3}(s)V_{o3} = k V_{i1} - k\tau_1\tau_2 s^2 V_{i3} + \tau_1\tau_2 s^2 V_{i4} \qquad (9.48)$$

where

$$D_{DF3}(s) = \tau_1\tau_2 s^2 + k\tau_1 s + k \qquad (9.49)$$

A further look at Eq. (9.48) reveals that this filter structure also supports the BS function from V_{o3} when $V_{i1} = V_{i4} = V_{i14}$. The function is given by

$$\frac{V_{o3}}{V_{i14}} = \frac{\tau_1\tau_2 s^2 + k}{\tau_1\tau_2 s^2 + k\tau_1 s + k} \qquad (9.50)$$

V_{o1} and V_{o2} will give the LP (numerator = k) and BP (numerator = $-\tau_1 s$) for V_{i14}, respectively.

The simplest or canonical second-order filter is correspondent to $k_{22} = k_{12} = 1$ that is realized with pure connection as shown in Fig. 9.7 [9, 10]. The transfer functions are given by

$$V_{o1} = \frac{(\tau_2 s + 1) V_{i1}}{\tau_1\tau_2 s^2 + \tau_1 s + 1} \qquad (9.51)$$

$$V_{o2} = \frac{V_{i1}}{\tau_1\tau_2 s^2 + \tau_1 s + 1} \qquad (9.52)$$

A LP function can be obtained from V_{o2}, whilst $V_{o1} - V_{o2}$ has a BP characteristic.

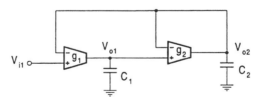

FIGURE 9.7
DF OTA-C realization with $k_{12} = k_{22} = 1$.

9.5 OTA-C Filters Based on Summed-Feedback (SF) Configuration

This section is now concerned with the OTA-C synthesis of the SF two integrator loop configuration in Fig. 9.3(a). Both arbitrary and special feedback coefficients are considered.

9.5.1 SF OTA-C Realization with Arbitrary k_{12} and k_{11}

Both general and specific OTA-C realizations of the SF configuration in Fig. 9.3(a) can be achieved. A very important two integrator two loop OTA-C structure is the one similar to the KHN active RC biquad (see Chapter 4 and [50]), as shown in Fig. 9.8 [8, 16]. Note that the two feedback coefficients are not separately realized and they share the same OTA resistor, saving one OTA compared with use of separate amplifiers. We now analyze the circuit.

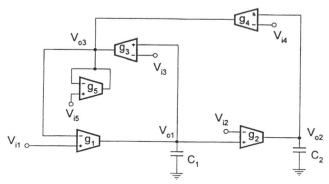

FIGURE 9.8
SF OTA-C realization with arbitrary k_{12} and k_{11}.

With $\tau_1 = C_1/g_1$, $\tau_2 = C_2/g_2$, $k_{11} = g_3/g_5$ and $k_{12} = g_4/g_5$, we can derive from the circuit

$$s\tau_1 V_{o1} = V_{i1} - V_{o3} \tag{9.53}$$

$$s\tau_2 V_{o2} = V_{o1} - V_{i2} \tag{9.54}$$

$$V_{o3} = k_{11}(V_{o1} - V_{i3}) + k_{12}(V_{o2} - V_{i4}) + V_{i5} \tag{9.55}$$

Solution of the equations gives

$$D_{SF1}(s)V_{o1} = \tau_2 s V_{i1} + k_{12} V_{i2} + k_{11}\tau_2 s V_{i3}$$
$$+ k_{12}\tau_2 s V_{i4} - \tau_2 s V_{i5} \tag{9.56}$$

$$D_{SF1}(s)V_{o2} = V_{i1} - (\tau_1 s + k_{11}) V_{i2} + k_{11} V_{i3}$$
$$+ k_{12} V_{i4} - V_{i5} \tag{9.57}$$

$$D_{SF1}(s)V_{o3} = (k_{11}\tau_2 s + k_{12}) V_{i1} - k_{12}\tau_1 s V_{i2} - k_{11}\tau_1\tau_2 s^2 V_{i3}$$
$$- k_{12}\tau_1\tau_2 s^2 V_{i4} + \tau_1\tau_2 s^2 V_{i5} \tag{9.58}$$

$$D_{SF1}(s)(V_{i1} - V_{o3}) = \tau_1\tau_2 s^2 V_{i1} + k_{12}\tau_1 s V_{i2} + k_{11}\tau_1\tau_2 s^2 V_{i3}$$
$$+ k_{12}\tau_1\tau_2 s^2 V_{i4} - \tau_1\tau_2 s^2 V_{i5} \tag{9.59}$$

where
$$D_{SF1}(s) = \tau_1\tau_2 s^2 + k_{11}\tau_2 s + k_{12} \quad (9.60)$$

It can be seen that the circuit can realize the LP, BP, and HP functions. The output as a difference of $V_{i1} - V_{o3}$ under the excitation of V_{i1} also has an HP characteristic. A very important feature of the circuit is that it can simultaneously output the LP, BP, and HP functions, which is similar to the KHN active RC biquad (see Chapter 4 or [50]) and is often called the KHN OTA-C biquad.

This circuit can also give a general function from V_{o3} for $V_{i1} = V_{i2} = V_{i5} = V_{i125}$, that is

$$\frac{V_{o3}}{V_{i125}} = \frac{\tau_1\tau_2 s^2 + (k_{11}\tau_2 - k_{12}\tau_1)s + k_{12}}{\tau_1\tau_2 s^2 + k_{11}\tau_2 s + k_{12}} \quad (9.61)$$

From this expression it can be seen that when $k_{12}\tau_1 = k_{11}\tau_2$ the circuit has a BS function, while if $k_{12}\tau_1 = 2k_{11}\tau_2$ an AP will arise.

Design Example of KHN OTA-C Biquad

We now design the KHN biquad to realize the three standard LP, BP, and HP specifications in Table 9.1. We choose V_{i5} as the input (all the other input voltages are grounded) and use V_{o2}, V_{o1}, and V_{o3} as respective outputs. To start the design, we write the relevant transfer functions from Eqs. (9.56 through 9.58) as

$$\frac{V_{o2}}{V_{i5}} = \frac{-1}{\tau_1\tau_2 s^2 + k_{11}\tau_2 s + k_{12}}, \quad \frac{V_{o1}}{V_{i5}} = \frac{-\tau_2 s}{\tau_1\tau_2 s^2 + k_{11}\tau_2 s + k_{12}},$$

$$\frac{V_{o3}}{V_{i5}} = \frac{\tau_1\tau_2 s^2}{\tau_1\tau_2 s^2 + k_{11}\tau_2 s + k_{12}} \quad (9.62)$$

The minus sign in Eq. (9.62) will be left out in the following discussion for convenience. Using Eq. (9.62) we can determine the design formulas as

$$k_{12} = \frac{1}{K_{LP}}, \quad k_{11} = \frac{1}{K_{BP}}, \quad \tau_1 = \frac{1}{K_{BP}}\frac{Q}{\omega_o}, \quad \tau_2 = \frac{K_{BP}}{K_{LP}}\frac{1}{\omega_o Q} \quad (9.63)$$

Suppose that $f_o = 200kHz$, $Q = 1$, and $K_{LP} = K_{BP} = K_{HP} = 1$ are required. Selecting $g_1 = g_2 = g_5 = 62.83\mu S$ we can calculate $g_3 = g_4 = 62.83\mu S$ and $C_1 = C_2 = 50.00pF$. For the given requirements this is an excellent design, since all transconductances are identical, making the biquad easy to tune; all capacitances are equal, being economical; and also the three transfer functions have the same maximum amplitude value equal to 1, leading to the largest dynamic range (this will be further investigated in Section 9.11.6).

9.5.2 SF OTA-C Realization with $k_{12} = k_{11} = k$

Another interesting structure is obtained when $k_{12} = k_{11} = k$ and a single voltage amplifier is utilized to realize the identical feedback coefficient, saving two OTAs, as shown in Fig. 9.9 [32]. Note that there is a change in the polarity of the g_2 OTA due to the use of the differential input OTA of g_3, compared with the general SF structure in Fig. 9.8.

Nodal analysis with $k = g_3/g_4$, $\tau_1 = C_1/g_1$ and $\tau_2 = C_2/g_2$ yields

$$D_{SF2}(s)V_{o1} = \tau_2 s V_{i1} + kV_{i2} - \tau_2 s V_{i3} \quad (9.64)$$

9.6. BIQUADRATIC OTA-C FILTERS USING LOSSY INTEGRATORS

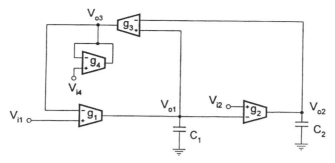

FIGURE 9.9
SF OTA-C realization with $k_{12} = k_{11} = k$.

$$D_{SF2}(s)V_{o2} = -V_{i1} + (\tau_1 s + k) V_{i2} + V_{i3} \quad (9.65)$$

$$D_{SF2}(s)V_{o3} = k(\tau_2 s + 1) V_{i1} - k\tau_1 s V_{i2} + \tau_1\tau_2 s^2 V_{i3} \quad (9.66)$$

$$D_{SF2}(s)(V_{i1} - V_{o3}) = \tau_1\tau_2 s^2 V_{i1} + k\tau_1 s V_{i2} - \tau_1\tau_2 s^2 V_{i3} \quad (9.67)$$

where

$$D_{SF2}(s) = \tau_1\tau_2 s^2 + k\tau_2 s + k \quad (9.68)$$

Obviously this circuit supports the LP, BP, and HP characteristics and the simultaneous output of all the three functions is available for V_{i3}. The general function can also be obtained for $V_{i1} = V_{i2} = V_{i3} = V_{i123}$, given by

$$\frac{V_{o3}}{V_{i123}} = \frac{\tau_1\tau_2 s^2 + k(\tau_2 - \tau_1)s + k}{\tau_1\tau_2 s^2 + k\tau_2 s + k} \quad (9.69)$$

This expression reveals that $\tau_1 = \tau_2$ can lead to a BS characteristic, while $\tau_1 = 2\tau_2$ gives rise to an AP function.

It is interesting to note the difference of the three general expressions in Eqs. (9.34), (9.61), and (9.69) in the term of s in the respective numerators. From Eqs. (9.34), (9.69), and (9.61), different functions can be obtained by setting k_{12} and k_{22}, τ_1 and τ_2, or the mixture of k_{ij} and τ_j, respectively.

9.6 Biquadratic OTA-C Filters Using Lossy Integrators

In the above, we have discussed the two ideal integrator two loop structures. We now consider the configurations given in Fig. 9.10. These configurations contain one overall loop with one ideal integrator and one lossy integrator. For Fig. 9.10(a) if the lossy integrator is realized using the single loop OTA-C integrators in Figs. 9.2(a) and (b), then the derived OTA-C filters will be the same as those derived from the two ideal integrator two loop DF configurations. In this section we want to emphasize another implementation, that is, using the lossy integrators in Figs. 9.2(c) and (d) to realize the configurations in Fig. 9.10.

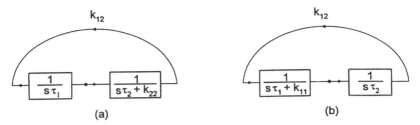

FIGURE 9.10
Configurations with lossy integrator.

9.6.1 Tow-Thomas OTA-C Structure

The realization of Fig. 9.10(a) with $k_{12} = 1$ using the lossy integrator in Fig. 9.2(c) is shown in Fig. 9.11 [7, 10, 11]. The voltage transfer functions with $\tau_1 = C_1/g_1$, $\tau_2 = C_2/g_2$ and $k_{22} = g_3/g_2$ are derived as

$$V_{o1} = \frac{(\tau_2 s + k_{22}) V_{i1} + V_{i2} - k_{22} V_{i3}}{\tau_1 \tau_2 s^2 + k_{22} \tau_1 s + 1} \quad (9.70)$$

$$V_{o2} = \frac{V_{i1} - \tau_1 s V_{i2} + k_{22} \tau_1 s V_{i3}}{\tau_1 \tau_2 s^2 + k_{22} \tau_1 s + 1} \quad (9.71)$$

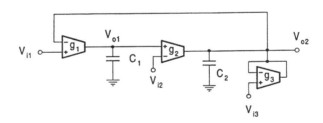

FIGURE 9.11
Tow-Thomas OTA-C filter.

This OTA-C architecture consists of two integrators in a single loop: one ideal integrator and the other lossy integrator of the type in Fig. 9.2(c). This biquad is the single most popular biquad in practice. It can be considered to be the OTA-C equivalent of the Tow-Thomas (TT) active RC biquad (see Chapter 4 or [51]). It can also be generated by OTA-C simulation of resistors and inductors of a passive RLC resonator, thus being often called the active OTA-C resonator. This will be discussed in detail in the next chapter. The circuit has very low sensitivity and low parasitic effects, and is simple in structure. The functions of this OTA-C biquad have also been thoroughly investigated with the aid of switches in [11, 15, 19]. The design and practical performance analysis of the filter will be discussed in detail in Section 9.11.

9.6.2 Feedback Lossy Integrator Biquad

Another realization of Fig. 9.10(a) with $k_{12} = 1$ is the one using the lossy integrator in Fig. 9.3(d), as shown in Fig. 9.12 [12]. This circuit differs from the TT filter in that it has two feedback loops and differs from the canonical DF filter in Section 9.4.4 in that it uses a lossy integrator. The circuit is a combination of two categories, although we discuss it in this section. The transfer functions of the circuit with the inputs and outputs indicated are derived as

$$V_{o1} = \frac{(\tau_2 s + k_{22})V_{i1} - (k_{22} - 1)V_{i3}}{\tau_1 \tau_2 s^2 + k_{22}\tau_1 s + 1} \tag{9.72}$$

$$V_{o2} = \frac{V_{i1} + (k_{22} - 1)\tau_1 s V_{i3}}{\tau_1 \tau_2 s^2 + k_{22}\tau_1 s + 1} \tag{9.73}$$

where $k_{22} = 1 + \frac{g_3}{g_2}$, $\tau_1 = C_1/g_1$ and $\tau_2 = C_2/g_2$.

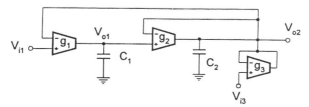

FIGURE 9.12
Feedback lossy integrator OTA-C filter.

Similar discussion can be conducted based on the configuration in Fig. 9.10(b), but only the lossy integrator in Fig. 9.2(c) can be used. The resulting OTA-C circuit is similar to that discussed above in function and performance and is therefore not dealt with here.

9.7 Comparison of Basic OTA-C Filter Structures

In the above we have generated a number of OTA-C filter structures based on summed-feedback and distributed-feedback two integrator loop configurations. This section compares performances of these different architectures.

9.7.1 Multifunctionality and Number of OTA

From the formulated transfer expressions for different architectures it is seen that the proposed filters all have a multifunction feature, supporting at least two functions at different input or output positions. All the architectures support the LP and BP functions. The HP and BS filters are obtainable from Figs. 9.4, 9.5, 9.6, 9.8, and 9.9, noting that using Figs. 9.8 and 9.9 to realize the BS function involves difference matching although an on-chip tuning scheme may be used to achieve accurate matching. Figures 9.4, 9.8, and 9.9 also offer the AP function. Finally, Fig. 9.4 can further supplies the LPN and HPN characteristics.

The role of circuit node 3 makes a very special contribution to the multifunctionality, as can be seen from the expressions related to output V_{o3} or the inputs relevant to node 3 such as V_{i3} and V_{i4} in Figs. 9.4 through 9.6, 9.8, and 9.9 and V_{i5} in Fig. 9.8.

Inspecting the different structures we can also see that Fig. 9.7 contains two OTAs; Figs. 9.11 and 9.12 both require three; Figs. 9.5, 9.6, and 9.9 all embrace four; Fig. 9.8 five; and Fig. 9.4 six. All the architectures use two grounded capacitors.

In Table 9.3 we present the filter functions and the number of OTAs of all the structures. Two extremes are the simplest two OTA structure with the LP and BP functions and the most complex six OTA circuit containing all filter functions.

Table 9.3 Comparison of Different OTA-C Filter Architectures

Structure of filter	Number of OTA	Functions with direct realization	Functions with difference matching	Sensitivity of ω_o and Q	Tuning of ω_o and Q
Fig. 9.4	6	LP, BP, HP, BS, LPN, HPN	AP	Low	Independent
Fig. 9.5	4	LP, BP, HP, BS		Low	Independent
Fig. 9.6	4	LP, BP, HP, BS		Low	Not
Fig. 9.7	2	LP, BP		Low	Not
Fig. 9.8	5	LP, BP, HP	BS, AP	Low	Independent
Fig. 9.9	4	LP, BP, HP	BS, AP	Low	Not
Fig. 9.11	3	LP, BP		Low	Independent
Fig. 9.12	3	LP, BP		Low	Independent

9.7.2 Sensitivity

We have generally discussed the sensitivity. To appreciate this for individual architectures, we summarize the parameter relations of ω_o and Q in Table 9.4. From this table we can very easily calculate the sensitivities of ω_o and Q. These are found to be very low for all the structures generated, with the values being 0, 0.5, 1, or between 0 and 1.

Table 9.4 Parameter Relations of OTA-C Realizations of SF and DF Configurations

Structure	τ_1	τ_2	k_{11}	k_{22}	k_{12}	ω_o	$\frac{\omega_o}{Q}$	Q
Fig. 9.4	$\frac{C_1}{g_1}$	$\frac{C_2}{g_2}$		$\frac{g_3}{g_4}$	$\frac{g_5}{g_6}$	$\sqrt{\frac{g_5 g_1 g_2}{g_6 C_1 C_2}}$	$\frac{g_3 g_2}{g_4 C_2}$	$\frac{g_4}{g_3}\sqrt{\frac{g_5 g_1 C_2}{g_6 g_2 C_1}}$
Fig. 9.5	$\frac{C_1}{g_1}$	$\frac{C_2}{g_2}$		$\frac{g_3}{g_4}$	1	$\sqrt{\frac{g_1 g_2}{C_1 C_2}}$	$\frac{g_3 g_2}{g_4 C_2}$	$\frac{g_4}{g_3}\sqrt{\frac{g_1 C_2}{g_2 C_1}}$
Fig. 9.6	$\frac{C_1}{g_1}$	$\frac{C_2}{g_2}$		$\frac{g_3}{g_4}$	$\frac{g_3}{g_4}$	$\sqrt{\frac{g_3 g_1 g_2}{g_4 C_1 C_2}}$	$\frac{g_3 g_2}{g_4 C_2}$	$\sqrt{\frac{g_4 g_1 C_2}{g_3 g_2 C_1}}$
Fig. 9.7	$\frac{C_1}{g_1}$	$\frac{C_2}{g_2}$		1	1	$\sqrt{\frac{g_1 g_2}{C_1 C_2}}$	$\frac{g_2}{C_2}$	$\sqrt{\frac{g_1 C_2}{g_2 C_1}}$
Fig. 9.8	$\frac{C_1}{g_1}$	$\frac{C_2}{g_2}$	$\frac{g_3}{g_5}$		$\frac{g_4}{g_5}$	$\sqrt{\frac{g_4 g_1 g_2}{g_5 C_1 C_2}}$	$\frac{g_3 g_1}{g_5 C_1}$	$\frac{1}{g_3}\sqrt{\frac{g_4 g_5 g_2 C_1}{g_1 C_2}}$
Fig. 9.9	$\frac{C_1}{g_1}$	$\frac{C_2}{g_2}$	$\frac{g_3}{g_4}$		$\frac{g_3}{g_4}$	$\sqrt{\frac{g_3 g_1 g_2}{g_4 C_1 C_2}}$	$\frac{g_3 g_1}{g_4 C_1}$	$\sqrt{\frac{g_4 g_2 C_1}{g_3 g_1 C_2}}$
Fig. 9.11	$\frac{C_1}{g_1}$	$\frac{C_2}{g_2}$		$\frac{g_3}{g_2}$	1	$\sqrt{\frac{g_1 g_2}{C_1 C_2}}$	$\frac{g_3}{C_2}$	$\frac{1}{g_3}\sqrt{\frac{g_1 g_2 C_2}{C_1}}$
Fig. 9.12	$\frac{C_1}{g_1}$	$\frac{C_2}{g_2}$		$1+\frac{g_3}{g_2}$	1	$\sqrt{\frac{g_1 g_2}{C_1 C_2}}$	$\frac{g_2+g_3}{C_2}$	$\frac{1}{g_2+g_3}\sqrt{\frac{g_1 g_2 C_2}{C_1}}$

9.7.3 Tunability

Electronic tunability is available. Parameters ω_o and Q in all structures may be tuned by controlling the associated transconductances g_j through adjustment of bias voltages or currents. It is recalled from Chapter 8 that the relation of the transconductance and the bias current can be expressed as

$$g_j = kI_{Bj} \tag{9.74}$$

9.8. VERSATILE FILTER FUNCTIONS

where k is a constant equal to 19.2 at room temperature, depending on semiconductor material, etc., and I_{Bj} is the bias current. To see more clearly about how we can tune the bias current or voltage to change the filter parameters we write the expressions of the TT biquad parameters directly in terms of the bias current. For the same type OTA and $g_1 = g_2 = g$ we can derive using the expressions given in Table 9.4:

$$\omega_o = \frac{g}{\sqrt{C_1 C_2}}, \quad Q = \frac{g}{g_3}\sqrt{\frac{C_2}{C_1}}, \quad \frac{\omega_o}{Q} = \frac{g_3}{C_2} \tag{9.75}$$

Substituting Eq. (9.74) into Eq. (9.75) gives

$$\omega_o = \frac{k}{\sqrt{C_1 C_2}} I_B, \quad Q = \sqrt{\frac{C_2}{C_1} \frac{I_B}{I_{B3}}}, \quad \frac{\omega_o}{Q} = \frac{k}{C_2} I_{B3} \tag{9.76}$$

From this equation we can see that by tuning I_B we can change ω_o and Q proportionally without influencing $\frac{\omega_o}{Q}$, the bandwidth for the BP filter, which is determined by I_{B3}. Trimming I_{B3} will lead to a proportional change in $\frac{\omega_o}{Q}$ and an inversely proportional alteration in Q, without impact on ω_o which is determined by I_B. This orthogonal tunability can be very convenient for some design tasks. For example it can be used to design a BP filter with the same bandwidth for different center frequencies or the same center frequency with different bandwidths. In fully integrated filters tuning is automatic and adaptive [1, 2, 19, 41, 43, 44].

Independent tuning of ω_o and Q is very important. From the relations in Table 9.4 we can see that ω_o and Q of most structures can be independently tuned, which is indicated in Table 9.3.

9.8 Versatile Filter Functions Based on Node Current Injection

In the above we have shown how to generate filter functions from the basic OTA-C network by properly applying voltage inputs to the OTA input terminals and taking voltage outputs from circuit nodes directly. These input and output methods are very simple and easy. There is no need for any extra components, resulting in the simple structure. The input method may however cause the differential input application of some OTAs which may increase the effects of OTA input impedances. In some structures filter functions are still limited and filter design may suffer from less flexibility. The biggest problem is that we cannot control zeros independently, because all the transconductances are related to filter poles.

From this section on we will look at other input and output methods using extra input and output OTA networks. Use of additional OTAs will give more design flexibility, although the structure may be more complex. For example, arbitrary scaling of the filter gain may be achieved and various universal biquads can be readily attained.

In this section we concentrate on the technique of node current injection. This input method is to apply a voltage through an extra single-ended input OTA to some circuit node. Because the input voltage is converted into the current, which is then injected into the circuit node, we give the name of node current injection. This method will not introduce any extra circuit node and except the inherent differential input application of the g_2 OTA in Figs. 9.4 through 9.7, 9.9, and 9.12. All the other OTAs are single ended.

9.8.1 DF Structures with Node Current Injection

We redraw the circuit of Fig. 9.5, but with voltages being input to circuit nodes through g_{aj} OTAs, as shown in Fig. 9.13. With $\tau_1 = C_1/g_1$, $\tau_2 = C_2/g_2$, $k_{22} = g_3/g_4$, and $\alpha_j = g_{aj}/g_j$, j = 1, 2, $\alpha_3 = g_{a3}/g_4$, the general relations are derived by routine circuit analysis as

$$D_{DF2}(s)V_{o1} = V_{i1}\alpha_1(\tau_2 s + k_{22}) - V_{i2}\alpha_2 + V_{i3}\alpha_3 \tag{9.77}$$

$$D_{DF2}(s)V_{o2} = V_{i1}\alpha_1 + V_{i2}\alpha_2\tau_1 s - V_{i3}\alpha_3\tau_1 s \tag{9.78}$$

$$D_{DF2}(s)V_{o3} = V_{i1}\alpha_1 k_{22} + V_{i2}\alpha_2 k_{22}\tau_1 s$$
$$+ V_{i3}\alpha_3\left(\tau_1\tau_2 s^2 + 1\right) \tag{9.79}$$

$$D_{DF2}(s)(V_{o1} - V_{o3}) = V_{i1}\alpha_1\tau_2 s - V_{i2}\alpha_2(k_{22}\tau_1 s + 1)$$
$$- V_{i3}\alpha_3\tau_1\tau_2 s^2 \tag{9.80}$$

where D_{DF2} is the same as Eq. (9.44) and is given below for convenience.

$$D_{DF2}(s) = \tau_1\tau_2 s^2 + k_{22}\tau_1 s + 1 \tag{9.81}$$

It can be seen that the DF structure with node current injection in Fig. 9.13 have the LP, BP, BS, and HP characteristics. With $V_{i1} = V_{i2} = V_{i3} = V_{i123}$ the outputs from V_{o3} and $V_{o1} - V_{o3}$ will be universal in function. Take the former as an example. The universal transfer function is given by

$$\frac{V_{o3}}{V_{i123}} = \frac{\alpha_3\tau_1\tau_2 s^2 + \alpha_2 k_{22}\tau_1 s + (\alpha_3 + \alpha_1 k_{22})}{\tau_1\tau_2 s^2 + k_{22}\tau_1 s + 1} \tag{9.82}$$

For the DF structure of Fig. 9.6 with node current injection as shown in Fig. 9.14 we have the following equations with $\tau_1 = C_1/g_1$, $\tau_2 = C_2/g_2$, $k = g_3/g_4$, $\alpha_j = g_{aj}/g_j$, j = 1, 2, $\alpha_3 = g_{a3}/g_4$:

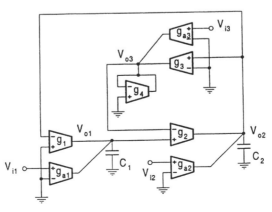

FIGURE 9.13
DF structure of Fig. 9.5 with node current injection.

9.8. VERSATILE FILTER FUNCTIONS

$$D_{DF3}(s)V_{o1} = V_{i1}\alpha_1(\tau_2 s + k) - V_{i2}\alpha_2 k - V_{i3}\alpha_3 \tau_2 s \qquad (9.83)$$

$$D_{DF3}(s)V_{o2} = V_{i1}\alpha_1 + V_{i2}\alpha_2 \tau_1 s - V_{i3}\alpha_3(\tau_1 s + 1) \qquad (9.84)$$

$$D_{DF3}(s)V_{o3} = V_{i1}\alpha_1 k + V_{i2}\alpha_2 k \tau_1 s + V_{i3}\alpha_3 \tau_1 \tau_2 s^2 \qquad (9.85)$$

where

$$D_{DF3}(s) = \tau_1 \tau_2 s^2 + k\tau_1 s + k \qquad (9.86)$$

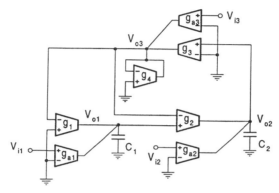

FIGURE 9.14
DF structure of Fig. 9.6 with node current injection.

Note that these inputs can support the LP, BP, and HP functions. With $V_{i1} = V_{i2} = V_{i3} = V_{i123}$ and the output taken from V_{o3} we can obtain a universal biquadratic function as

$$\frac{V_{o3}}{V_{i123}} = \frac{\alpha_3 \tau_1 \tau_2 s^2 + \alpha_2 k \tau_1 s + \alpha_1 k}{\tau_1 \tau_2 s^2 + k\tau_1 s + k} \qquad (9.87)$$

Further study of this general biquad gives, in Table 9.5, all the standard filter functions. This deserves some comments due to many interesting features. First, we can see that due to three extra input OTAs we can freely control the numerator parameters by α_j, without any influence on the denominator parameters which are entirely determined by k_{ij} and τ_j. We can thus say that the zeros and poles of the system are completely independently tunable, a very useful feature in filter design. Second, any filter characteristics can be realized, making the architecture truly universal in function. Third, no coefficient difference matching is involved, thus the architecture is insensitive to variation. All this contrasts with those structures corresponding to Eqs. (9.34), (9.61), and (9.69). A final note relates to $-\alpha_2$ and $\alpha_j = 0$ in the table. The former simply means that the input should be applied to the inverting terminal of the g_{a2} OTA, rather than the non-inverting terminal as in Fig. 9.14; while the latter implies that the g_{aj} OTA should be removed since $g_{aj} = 0$. In the remainder of the chapter we will assume this for all similar cases to avoid the inconvenience of redrawing the circuit. Thus, whenever we have a negative or zero α_j, this is accomplished through the interchange of the input terminals or removal of the g_{aj} OTA, respectively.

9.8.2 SF Structures with Node Current Injection

The KHN OTA-C biquad with the voltage inputs through extra OTAs to circuit nodes is shown in Fig. 9.15 [16]. With $\tau_1 = C_1/g_1$, $\tau_2 = C_2/g_2$, $k_{12} = g_4/g_5$, $k_{11} = g_3/g_5$, $\alpha_j = g_{aj}/g_j$, $j = 1, 2$,

Table 9.5 Filter Realizations Based on Universal Biquad Derived from Fig. 9.14

Type	Numerators	Parameter conditions	Design formulas
LP	$K_{LP}\omega_o^2$	$\alpha_3 = \alpha_2 = 0$	$K_{LP} = \alpha_1$
BP	$K_{BP}\frac{\omega_o}{Q}s$	$\alpha_3 = \alpha_1 = 0$	$K_{BP} = \alpha_2$
HP	$K_{HP}s^2$	$\alpha_2 = \alpha_1 = 0$	$K_{HP} = \alpha_3$
BS	$K_{BS}(s^2 + \omega_o^2)$	$\alpha_2 = 0, \alpha_3 = \alpha_1 = \alpha$	$K_{BS} = \alpha$
LPN	$K_{LPN}(\frac{\omega_o}{\omega_n})^2(s^2 + \omega_n^2), \omega_n > \omega_o$	$\alpha_2 = 0, \alpha_3 < \alpha_1$	$K_{LPN} = \alpha_1, \omega_n^2 = \frac{\alpha_1}{\alpha_3}\omega_o^2$
HPN	$K_{HPN}(s^2 + \omega_n^2), \omega_n < \omega_o$	$\alpha_2 = 0, \alpha_3 > \alpha_1$	$K_{HPN} = \alpha_3, \omega_n^2 = \frac{\alpha_1}{\alpha_3}\omega_o^2$
AP	$K_{AP}(s^2 - \frac{\omega_o}{Q}s + \omega_o^2)$	$\alpha_3 = -\alpha_2 = \alpha_1 = \alpha$	$K_{AP} = \alpha$

$\alpha_3 = g_{a3}/g_5$, the input and output relations are formulated as

$$D_{SF1}(s)V_{o1} = V_{i1}\alpha_1\tau_2 s - V_{i2}\alpha_2 k_{12} - V_{i3}\alpha_3\tau_2 s \tag{9.88}$$

$$D_{SF1}(s)V_{o2} = V_{i1}\alpha_1 + V_{i2}\alpha_2(\tau_1 s + k_{11}) - V_{i3}\alpha_3 \tag{9.89}$$

$$D_{SF1}(s)V_{o3} = V_{i1}\alpha_1(k_{11}\tau_2 s + k_{12}) + V_{i2}\alpha_2 k_{12}\tau_1 s$$
$$+ V_{i3}\alpha_3\tau_1\tau_2 s^2 \tag{9.90}$$

where

$$D_{SF1}(s) = \tau_1\tau_2 s^2 + k_{11}\tau_2 s + k_{12} \tag{9.91}$$

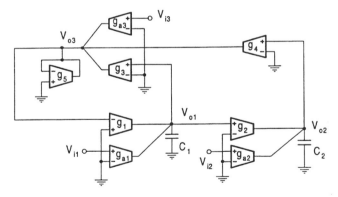

FIGURE 9.15
SF structure of Fig. 9.8 with node current injection.

Note that these inputs can support the LP, BP, and HP functions. With $V_{i1} = V_{i2} = V_{i3} = V_{i123}$ and the output taken from V_{o3} we can obtain a universal biquadratic function, given by

$$\frac{V_{o3}}{V_{i123}} = \frac{\alpha_3\tau_1\tau_2 s^2 + (\alpha_1 k_{11}\tau_2 + \alpha_2 k_{12}\tau_1)s + \alpha_1 k_{12}}{\tau_1\tau_2 s^2 + k_{11}\tau_2 s + k_{12}} \tag{9.92}$$

For the SF structure of Fig. 9.9 with node current injection as shown in Fig. 9.16, with $\tau_1 = C_1/g_1$, $\tau_2 = C_2/g_2$, $k = g_3/g_4$, $\alpha_j = g_{aj}/g_j$, j=1, 2, and $\alpha_3 = g_{a3}/g_4$ we write the equations of node voltages as

$$D_{SF2}(s)V_{o1} = V_{i1}\alpha_1\tau_2 s + V_{i2}\alpha_2 k - V_{i3}\alpha_3\tau_2 s \quad (9.93)$$

$$D_{SF2}(s)V_{o2} = -V_{i1}\alpha_1 + V_{i2}\alpha_2(\tau_1 s + k) + V_{i3}\alpha_3 \quad (9.94)$$

$$D_{SF2}(s)V_{o3} = V_{i1}\alpha_1 k(\tau_2 s + 1) - V_{i2}\alpha_2 k\tau_1 s + V_{i3}\alpha_3\tau_1\tau_2 s^2 \quad (9.95)$$

where

$$D_{SF2}(s) = \tau_1\tau_2 s^2 + k\tau_2 s + k \quad (9.96)$$

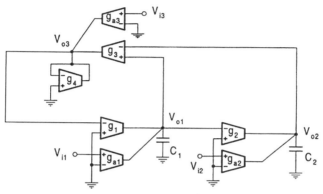

FIGURE 9.16
SF structure of Fig. 9.9 with node current injection.

From these equations, taking the output from V_{o3} and connecting V_{i1}, V_{i2}, and V_{i3} together as V_{i123}, we can obtain the following universal voltage transfer function

$$\frac{V_{o3}}{V_{i123}} = \frac{\alpha_3\tau_1\tau_2 s^2 - (\alpha_2\tau_1 - \alpha_1\tau_2)ks + \alpha_1 k}{\tau_1\tau_2 s^2 + k\tau_2 s + k} \quad (9.97)$$

Now we want to explain the generation method of universal biquads in the above from another viewpoint, which will be useful to facilitate our further investigation in the next two sections. In the above we connect all three voltage inputs together as a single voltage input. We can express this by saying that the single input voltage is distributed onto all circuit nodes by converting the voltage into currents using extra g_{aj} OTAs. This can be drawn as shown in Fig. 9.17(a), which clearly shows how to produce node input currents I_{i1}, I_{i2}, and I_{i3} from a single voltage input V_i. We intuitively call the circuit the distributor. Thus we say that we can construct universal biquads from basic structures by using the input distribution method [21, 23, 24, 25, 47].

9.9 Universal Biquads Using Output Summation Approach

As we have found, using additional OTAs can achieve many advantages. To implement arbitrary transmission zeros three programmable parameters are needed to independently tune the numerator

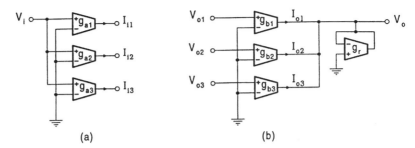

FIGURE 9.17
(a) Input distributor and (b) output combiner.

coefficients. The input distribution and output summation techniques can be utilized. As has been already demonstrated, for instance, we can use the input distributor to convert a single voltage input to weighted multiple current inputs, collect the terms in the output related to these multiple inputs with the weights to the single input, and adjust the weight coefficients to produce arbitrary transmission zeros and gains without any influence on the poles. In a similar way we may also add the summer in Fig. 9.17(b) to combine the multiple outputs for a certain input to generate any desired zeros and gains [21]–[23], [25, 47], which is the topic of this section.

9.9.1 DF-Type Universal Biquads

The summation method can be used to construct the universal biquad based on the DF realization in Fig. 9.5. From Eqs. (9.40 through 9.42) summing the three node voltages for V_{i3}, or V_{i4}, or V_{i13} (defined as before) we can obtain the respective universal biquads. For the input V_{i4} for example, summation of the LP (V_{o1}), BP (V_{o2}) and BS (V_{o3}) outputs with the weights of $\beta_1 = g_{b1}/g_r$, $\beta_2 = g_{b2}/g_r$ and $\beta_3 = g_{b3}/g_r$ gives:

$$\frac{V_o}{V_{i4}} = \frac{\beta_3 \tau_1 \tau_2 s^2 - \beta_2 \tau_1 s + (\beta_1 + \beta_3)}{\tau_1 \tau_2 s^2 + k_{22} \tau_1 s + 1} \tag{9.98}$$

The reader can verify that taking the output of $V_{o1} - V_{o3}$ as a summing element will generate more universal biquads. Note that the β_j definitions for all universal biquads are the same as those given above throughout Section 9.9. We shall therefore not repeat them in the following.

For the DF structure in Fig. 9.6, summation of the three output voltages for V_{i3}, V_{i4}, or V_{i14} can also produce corresponding universal biquads. Taking V_{i14} as an example, the LP, BP, and BS outputs from V_{o1}, V_{o2}, and V_{o3}, respectively, are summed together to give a universal transfer function:

$$\frac{V_o}{V_{i14}} = \frac{\beta_3 \tau_1 \tau_2 s^2 - \beta_2 \tau_1 s + (\beta_1 + \beta_3) k}{\tau_1 \tau_2 s^2 + k \tau_1 s + k} \tag{9.99}$$

Of course, many universal biquads can also be obtained from Fig. 9.4 using the summation technique. But considering the summation will introduce four more OTAs, so the total number of OTAs in such biquads would reach ten and thus not be very attractive.

9.9.2 SF-Type Universal Biquads

It is easy to derive a universal biquad based on the KHN OTA-C structure in Fig. 9.8. Equations reveal that summing all three output node voltages for V_{i3}, V_{i4} and V_{i5} will result in three correspond-

9.9. UNIVERSAL BIQUADS

ing universal biquads. And if the output $V_{i1} - V_{o3}$ is also taken into summation, more combinations can be produced. For instance, summation of V_{o1} (LP), V_{o2} (BP) and $V_{i1} - V_{o3}$ (HP) for V_{i1} will lead to a universal biquad as shown in Fig. 9.18, with the transfer function given by

$$\frac{V_o}{V_{i1}} = \frac{\beta_3 \tau_1 \tau_2 s^2 + \beta_2 \tau_2 s + \beta_1}{\tau_1 \tau_2 s^2 + k_{11} \tau_2 s + k_{12}} \tag{9.100}$$

Note that summing $V_{i1} - V_{o3}$ is realized by connecting V_{i1} to the non-inverting terminal of the g_{b3} OTA and V_{o3} to the inverting. This method introduces the differential input application of the g_{b3} OTA.

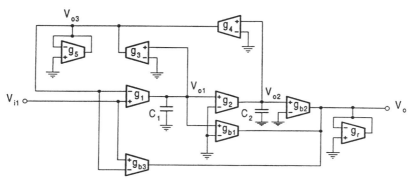

FIGURE 9.18
Universal KHN OTA-C biquad.

For the SF structure in Fig. 9.9, we can similarly generate some universal biquads. One example is the summation of V_{o1} (BP), V_{o2} (LP) and V_{o3} (HP) with V_{i3} as input, giving a universal transfer function of

$$\frac{V_o}{V_{i3}} = \frac{\beta_3 \tau_1 \tau_2 s^2 - \beta_1 \tau_2 s + \beta_2}{\tau_1 \tau_2 s^2 + k \tau_2 s + k} \tag{9.101}$$

It should be pointed out that in some cases we may need to consider the sign of β. We must emphasize that the negative β_j can be simply achieved by inputing the voltage to the non-inverting terminal of the g_{bj} OTA. This will also be implied in the next section.

9.9.3 Universal Biquads Based on Node Current Injection and Output Summation

Now we generate some universal biquads from the structures with node current inputs by using the output summation technique. We present four such biquads for the illustration purpose, based on the DF structure in Fig. 9.13, DF structure in Fig. 9.14, SF structure in Fig. 9.15, and SF structure in Fig. 9.16, respectively. All these universal biquads are obtained by means of summation of three node voltages V_{oj}, j = 1, 2, 3 with input V_{i3} using the output combiner in Fig. 9.17(b). The first one is based on the DF structure in Fig. 9.13. Summing three output voltages for V_{i3} leads to

$$\frac{V_o}{V_{i3}} = \alpha_3 \frac{\beta_3 \tau_1 \tau_2 s^2 - \beta_2 \tau_1 s + (\beta_1 + \beta_3)}{\tau_1 \tau_2 s^2 + k_{22} \tau_1 s + 1} \tag{9.102}$$

For the DF structure in Fig. 9.14 we have by summing the three node voltages

$$\frac{V_o}{V_{i3}} = \alpha_3 \frac{\beta_3 \tau_1 \tau_2 s^2 - (\beta_2 \tau_1 + \beta_1 \tau_2)s - \beta_2}{\tau_1 \tau_2 s^2 + k\tau_1 s + k} \tag{9.103}$$

For the SF structure in Fig. 9.15, the addition of the summation network leads to

$$\frac{V_o}{V_{i3}} = \alpha_3 \frac{\beta_3 \tau_1 \tau_2 s^2 - \beta_1 \tau_2 s - \beta_2}{\tau_1 \tau_2 s^2 + k_{11} \tau_2 s + k_{12}} \tag{9.104}$$

Its current transfer function version was given in [18].

Finally the SF structure in Fig. 9.16 with the summation network has

$$\frac{V_o}{V_{i3}} = \alpha_3 \frac{\beta_3 \tau_1 \tau_2 s^2 - \beta_1 \tau_2 s + \beta_2}{\tau_1 \tau_2 s^2 + k\tau_2 s + k} \tag{9.105}$$

9.9.4 Comments on Universal Biquads

Several important notes should be highlighted in the construction of universal biquads. The summation method will introduce an additional resistive node, the overall output node. This node will produce a pole at high frequencies due to parasitic capacitances, which should be dealt with carefully for HF applications. The summation method also needs one more OTA (for voltage output) than the distribution method.

Inspection of the universal biquads generated using both the distribution and summation methods indicates that some biquads may need difference matching for some particular filters, whereas others may realize them directly. The need of difference matching is a disadvantage in filter design, because of the high sensitivity to the component variation.

Note that node 3, the output node of the g_3 OTA, in all filter structures has a particular contribution to the universality and multifunctionality. The distribution-type biquads all use V_{o3} as the output and the summation-type biquads use V_{i3} or some others relevant to node 3 as the input.

To see clearly how to apply the input distributor and output summer to obtain a general biquadratic function the equations of node voltages in terms of node currents can be written, since as indicated in Fig. 9.17 the distributor and summer perform weighted voltage-to-current conversion.

Using both the distribution and summation methods can of course produce more universal biquads but will need too many OTAs and is not necessary since either the distribution or the summation can achieve sufficient numerator parameter control and fulfill the set of functions that the combination of the two can provide. However for the circuit which does not have the term of s^2 for any input and output, then both methods may be needed to obtain a universal realization. This is discussed in the next section.

9.10 Universal Biquads Based on Canonical and TT Circuits

For the canonical biquad we must include the input voltage in the combination since the basic circuit cannot offer the characteristics with term s^2. Using the input distribution and output summation methods we can obtain five universal biquads as shown in Fig. 9.19. Their transfer functions are

9.11. EFFECTS AND COMPENSATION OF OTA NONIDEALITIES

given in the order of the figures below:

$$H_a(s) = \frac{g_{z2}C_1C_2s^2 + (g_{z2}g_2C_1 - g_{z1}g_1C_2)s + (g_{z2} - g_{z0})g_1g_2}{g_r\left(C_1C_2s^2 + g_2C_1s + g_1g_2\right)} \quad (9.106)$$

$$H_b(s) = \frac{g_{z2}C_1C_2s^2 + (g_{z2}g_2C_1 - g_{z1}g_{z3}C_1)s + (g_{z2}g_1g_2 - g_{z0}g_{z3}g_2)}{g_r\left(C_1C_2s^2 + g_2C_1s + g_1g_2\right)} \quad (9.107)$$

$$H_c(s) = \frac{g_{z2}C_1C_2s^2 + (g_{z2}g_2C_1 - g_{z1}g_{z3}C_2)s + (g_{z2}g_1g_2 - g_{z1}g_{z3}g_2 - g_{z0}g_{z3}g_2)}{g_r\left(C_1C_2s^2 + g_2C_1s + g_1g_2\right)} \quad (9.108)$$

$$H_d(s) = \frac{g_{z2}C_1C_2s^2 + (g_{z2}g_2C_1 - g_{z1}g_{z3}C_2)s + (g_{z2}g_1g_2 - g_{z1}g_{z3}g_2 - g_{z0}g_{z3}g_1)}{g_r\left(C_1C_2s^2 + g_2C_1s + g_1g_2\right)} \quad (9.109)$$

$$H_e(s) = \frac{g_{z2}C_1C_2s^2 + (g_{z2}g_2C_1 - g_{z1}g_{z3}C_1)s + (g_{z2}g_1g_2 - g_{z0}g_{z3}g_1)}{g_r\left(C_1C_2s^2 + g_2C_1s + g_1g_2\right)} \quad (9.110)$$

These expressions show the variety of universal biquads that can be achieved. The structures in Fig. 9.19(a) and (b) were given in [22] and [24], respectively. As an exercise, the reader may be interested to check the equations for the structures in Fig. 9.19(c) through (e). Reference [23] has also given a complete set of universal biquads based on the TT structure. The generation method is similar to the above based on the canonical configuration.

9.11 Effects and Compensation of OTA Nonidealities

In the above we presented many two integrator loop filter structures. In this section we deal with practical frequency responses and dynamic range of these filters.

9.11.1 General Model and Equations

In Chapter 8 we discussed the OTA nonideality effects in some detail. It should be stressed that parasitics, in particular, the high frequency parasitic parameters, should be carefully considered, as OTA filters are used at high frequencies [29]–[33], [43, 44], [2, 9, 24]. In the following we assess the effects of OTA nonidealities on the TT OTA-C circuit for illustrative purpose. To do this we first define OTA nonideality symbols and then consider the circuit in Fig. 9.20.

Denote Y_{ij} and Y_{oj} as the input and output admittances of the jth OTA, respectively. They can be written as $Y_{ij} = G_{ij} + sC_{ij}$ and $Y_{oj} = G_{oj} + sC_{oj}$, where G_{ij}, C_{ij}, G_{oj} and C_{oj} are the input conductance, input capacitance, output conductance, and output capacitance, respectively. The finite bandwidth results in the transconductance frequency dependence, which can be approximately expressed as $g_j(s) \approx g_{j0}(1 - \frac{s}{\omega_{bj}})$ where g_{j0} is the nominal transconductance and ω_{bj} is the OTA bandwidth. For simplicity we will drop subscript 0 in the following.

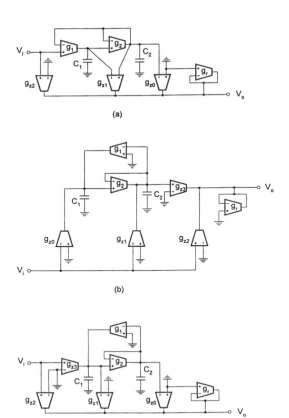

FIGURE 9.19
Universal biquads based on canonical structure.

In the ideal case, typical analysis of the circuit in Fig. 9.20 yields

$$\frac{V_{o1}}{V_i} = \frac{g_0 Y_2}{Y_1 Y_2 + g_1 g_2}, \quad \frac{V_{o2}}{V_i} = \frac{g_0 g_2}{Y_1 Y_2 + g_1 g_2} \tag{9.111}$$

From these equations we can verify that selecting $Y_1 = sC_1$ and $Y_2 = sC_2 + g_3$ leads to a lowpass filter from V_{o2}, with the transfer function given by

$$H(s) = \frac{V_{o2}}{V_i} = \frac{g_0 g_2}{C_1 C_2 s^2 + g_3 C_1 s + g_1 g_2} \tag{9.112}$$

and the pole parameters given by

$$\omega_o = \sqrt{\frac{g_1 g_2}{C_1 C_2}}, \quad Q = \frac{\sqrt{g_1 g_2 C_1 C_2}}{g_3 C_1} \tag{9.113}$$

The resulting structure is the same as that in Fig. 9.11 in Section 9.6.

When $Y_1 = sC_1 + g_4$ and $Y_2 = sC_2$ the circuit will have a bandpass function from V_{o1} and a lowpass from V_{o2}. For $Y_1 = sC_1 + g_4$ and $Y_2 = sC_2 + g_3$, a lowpass transfer function can be

9.11. EFFECTS AND COMPENSATION OF OTA NONIDEALITIES

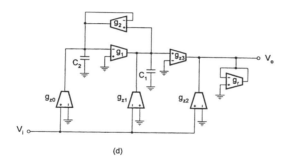

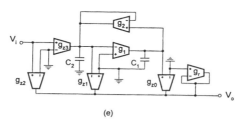

FIGURE 9.19
Universal biquads based on canonical structure (continued).

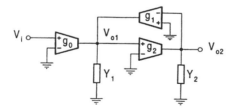

FIGURE 9.20
A circuit model for analysis of OTA nonidealities.

attained from V_{o2}, given by

$$\frac{V_{o2}}{V_i} = \frac{g_0 g_2}{C_1 C_2 s^2 + (g_3 C_1 + g_4 C_2) s + (g_1 g_2 + g_3 g_4)} \tag{9.114}$$

The corresponding circuit is displayed in Fig. 9.21.

Taking OTA finite input and output impedances and finite bandwidth into consideration, we have

$$\frac{V_{o1}}{V_i} = \frac{g_0(s) Y_2'}{Y_1' Y_2' + g_1(s) g_2(s)}, \quad \frac{V_{o2}}{V_i} = \frac{g_0(s) g_2(s)}{Y_1' Y_2' + g_1(s) g_2(s)} \tag{9.115}$$

where Y_1' and Y_2' depend on the nominal assignment of Y_1 and Y_2. For example, when $Y_1 = sC_1$ and $Y_2 = sC_2 + g_3$ and note in particular that g_3 is a grounded OTA resistor, not a discrete resistor, we have $Y_1' = sC_1 + (Y_{o0} + Y_{o1} + Y_{i2})$ and $Y_2' = sC_2 + g_3(s) + (Y_{i1} + Y_{o2} + Y_{i3} + Y_{o3})$.

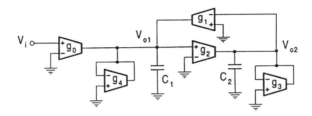

FIGURE 9.21
LP filter with complete compensation of finite OTA impedances.

9.11.2 Finite Impedance Effects and Compensation

For the structure of $Y_1 = sC_1$ and $Y_2 = sC_2 + g_3$, taking finite OTA impedances into consideration and denoting $C'_1 = C_1 + (C_{o0} + C_{o1} + C_{i2})$, $C'_2 = C_2 + (C_{i1} + C_{o2} + C_{i3} + C_{o3})$, $g'_3 = g_3 + (G_{i1} + G_{o2} + G_{i3} + G_{o3})$ and $g_p = G_{o0} + G_{o1} + G_{i2}$, we have

$$H'(s) = \frac{V_{o2}}{V_i} = \frac{g_0 g_2}{C'_1 C'_2 s^2 + \left(g'_3 C'_1 + g_p C'_2\right)s + \left(g_1 g_2 + g'_3 g_p\right)} \quad (9.116)$$

We can see that OTA nonidealities have changed all filter coefficients. For instance, the pole frequency and quality factor now become

$$\omega'_o = \sqrt{\frac{g_1 g_2 + g'_3 g_p}{C'_1 C'_2}}, \quad Q' = \frac{\sqrt{\left(g_1 g_2 + g'_3 g_p\right) C'_1 C'_2}}{g'_3 C'_1 + g_p C'_2} \quad (9.117)$$

and the magnitude at the zero frequency becomes

$$H'(0) = \frac{g_0 g_2}{g_1 g_2 + g'_3 g_p} \quad (9.118)$$

instead of $H(0) = g_0/g_1$, although $H'(\infty) = 0$. When $Y_1 = sC_1 + g_4$ and $Y_2 = sC_2$ the circuit will have similar effects from OTA parasitics.

From the above analysis we can see that all parasitic capacitances are referred to the ground due to the single-ended input of the OTA and these parasitic capacitances can be absorbed into the grounded circuit capacitances. This again confirms the conclusion that using single-ended OTAs and grounded capacitors can reduce the effects of parasitic capacitances. This is also true for the finite conductance problem. As we can see, the grounded circuit conductance (the g_3 OTA) can compensate all parasitic conductances including the finite OTA conductance. Therefore for the lowpass function if we select $Y_1 = sC_1 + g_4$ and $Y_2 = sC_2 + g_3$, that is, using one more grounded OTA resistor, all finite input and output conductances and capacitances can be absorbed.

Consider the circuit in Fig. 9.21 again. Suppose that OTAs in the circuit have finite conductances and capacitances as

$$G_{i1} = G_{i2} = G_{i3} = G_{i4} = 0, \quad G_{o0} = G_{o1} = G_{o2} = G_{o3} = G_{o4} = 0.5\mu S,$$

$$C_{i1} = C_{i2} = C_{i3} = C_{i4} = 0.04\text{pF}, \quad C_{o0} = C_{o1} = C_{o2} = C_{o3} = C_{o4} = 0.2\text{pF}$$

9.11. EFFECTS AND COMPENSATION OF OTA NONIDEALITIES

To see how absorption works, we denote and calculate

$$\Delta C_1 = C_{i2} + C_{i4} + C_{o0} + C_{o1} + C_{o4} = 0.68 \text{pF},$$

$$\Delta C_2 = C_{i1} + C_{i3} + C_{o2} + C_{o3} = 0.48 \text{pF},$$

$$\Delta g_3 = G_{i1} + G_{i3} + G_{o2} + G_{o3} = 1 \mu S,$$

$$\Delta g_4 = G_{i2} + G_{i4} + G_{o0} + G_{o1} + G_{o4} = 1.5 \mu S$$

It is recalled from Chapter 8 that the absorption approach determines the real component values by subtracting the nominal values with the increments due to nonideal OTA parameters, that is

$$C_{\text{real}} = C_{\text{nominal}} - \Delta C, \quad g_{\text{real}} = g_{\text{nominal}} - \Delta g$$

This requires that

$$C_{\text{nominal}} > \Delta C, \quad g_{\text{nominal}} > \Delta g$$

For example, the nominal value for C_1 must be much bigger than 0.68 pF and the nominal value of the g_4 transconductance must be much larger than $1.5 \mu S$. It should be noted that at very high frequencies this may not be always met. Careful design is thus needed to handle the parasitic effects.

9.11.3 Finite Bandwidth Effects and Compensation

Finite bandwidth or transconductance frequency dependence or phase shift will also affect the filter performance as shown in Eq. (9.115). Analysis can be conducted using these equations directly. The modified transfer function for $Y_1 = sC_1$ and $Y_2 = sC_2 + g_3$, for example, is derived as

$$H''(s) = \frac{\frac{g_0 g_2}{\omega_{b0} \omega_{b2}} s^2 - g_0 g_2 \left(\frac{1}{\omega_{b0}} + \frac{1}{\omega_{b2}} \right) s + g_0 g_2}{\left(C_1 C_2 - \frac{g_3 C_1}{\omega_{b3}} + \frac{g_1 g_2}{\omega_{b1} \omega_{b2}} \right) s^2 + \left[g_3 C_1 - g_1 g_2 \left(\frac{1}{\omega_{b1}} + \frac{1}{\omega_{b2}} \right) \right] s + g_1 g_2} \quad (9.119)$$

from which we can see that the frequency response is indeed changed. Putting it in a standard form of

$$H''(s) = K \frac{s^2 - \frac{\omega_z}{Q_z} s + \omega_z^2}{s^2 + \frac{\omega_o'}{Q'} s + \omega_o'^2} \quad (9.120)$$

the modified parameters can be obtained as below:

$$K = \frac{\frac{g_0 g_2}{\omega_{b0} \omega_{b2}}}{C_1 C_2 - \frac{g_3 C_1}{\omega_{b3}} + \frac{g_1 g_2}{\omega_{b1} \omega_{b2}}} \quad (9.121)$$

$$\omega_o' = \sqrt{\frac{g_1 g_2}{C_1 C_2 - \frac{g_3 C_1}{\omega_{b3}} + \frac{g_1 g_2}{\omega_{b1} \omega_{b2}}}} \quad (9.122)$$

$$Q' = \frac{\sqrt{g_1 g_2 \left(C_1 C_2 - \frac{g_3 C_1}{\omega_{b3}} + \frac{g_1 g_2}{\omega_{b1} \omega_{b2}} \right)}}{g_3 C_1 - g_1 g_2 \left(\frac{1}{\omega_{b1}} + \frac{1}{\omega_{b2}} \right)} \tag{9.123}$$

$$\omega_z = \sqrt{\omega_{b0} \omega_{b2}} \tag{9.124}$$

$$Q_z = \frac{\sqrt{\omega_{b0} \omega_{b2}}}{\omega_{b0} + \omega_{b2}} \tag{9.125}$$

Note also that $H''(\infty) = K$ is no longer zero, although $H''(0) = H(0) = g_0/g_1$.

The circuit may also be unstable if the design does not care this problem very much, since the finite bandwidth may cause the denominator coefficients to be negative and shift the poles into the right-half s plane. To ensure the stability, the following conditions with the first-order approximation should be met:

$$C_1 > \frac{g_1 g_2}{g_3} \left(\frac{1}{\omega_{b1}} + \frac{1}{\omega_{b2}} \right), \quad C_2 > \frac{g_3}{\omega_{b3}} \tag{9.126}$$

Compensation of the effects of transconductance frequency dependence is possible [9, 33, 46]. For the ideal integrator we can put a resistor in series with the capacitor [9] and this resistor can be realized using a MOSFET in the ohm region in integrated circuit design [46]. From the circuit in Fig. 9.22 we can write

$$H(s) = \frac{g_m}{sC} \frac{1 + sRC}{1 + s/\omega_b} \tag{9.127}$$

It is clear that setting

$$R = 1/\omega_b C \tag{9.128}$$

the circuit will be an ideal integrator. Doing so for all integrators, the effects of finite bandwidth in the canonical structure in Fig. 9.7, for example, can be completely compensated.

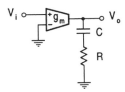

FIGURE 9.22
Passive compensation of finite bandwidth effects.

An alternative active compensation method [33] is based on the two OTAs of different transconductances, which are connected in parallel with opposite polarities, as shown in Fig. 9.23. This arrangement of the two OTAs is equivalent to a differential input OTA with the transconductance equal to the difference of the two OTAs and with reduced excess phase. The principle of compensation is now explained. The total effective transconductance can be expressed in terms of individual single-pole characteristics with the first-order approximation as

$$g_e(s) = g_1(s) - g_2(s) = g_1 \left(1 - \frac{s}{\omega_{b1}} \right) - g_2 \left(1 - \frac{s}{\omega_{b2}} \right) \tag{9.129}$$

9.11. EFFECTS AND COMPENSATION OF OTA NONIDEALITIES

where the subscripts 1 and 2 refer to OTAs 1 and 2, respectively. Equation (9.129) can also be manipulated to

$$g_e(s) = (g_1 - g_2)\left[1 - s\frac{g_1/\omega_{b1} - g_2/\omega_{b2}}{g_1 - g_2}\right] \quad (9.130)$$

It can be seen from this equation that if the condition $g_1/\omega_{b1} = g_2/\omega_{b2}$ is met, the two OTAs will behave like a single OTA having an effective transconductance $g_e = g_1 - g_2$ which is frequency independent.

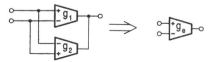

FIGURE 9.23
Active compensation of OTA finite bandwidth.

It should be noted that the first-order cancellation of excess phase is valid at frequencies much lower than ω_{b1} and ω_{b2}. The active compensation can achieve a wider tuning range but cause a reduction in the effective transconductance, both due to the difference of two transconductances. As two OTAs are used, the compensation scheme will have an increase in the power dissipation and chip area. It is, therefore, usually used to replace the OTAs whose excess phase severely affects the filter performance, e.g., the OTAs realizing integrators.

9.11.4 Selection of OTA-C Filter Structures

Filter structures may be equivalently transformed for some functions in the ideal case. For example, the canonical biquad in Fig. 9.24(a) can be equivalently converted into the circuit in Fig. 9.24(b), which is similar to the TT biquad. Notice that all two-integrator filters offer the lowpass and bandpass

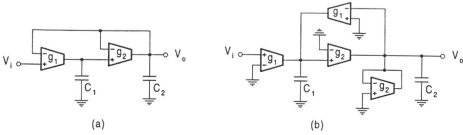

FIGURE 9.24
Illustration of equivalent structures.

functions. It can be generally said that these architectures are ideally equivalent in realizing the same LP or BP characteristic. It must be realized, however, that this happens only in the ideal situation with everything being perfect. In practice, filter performances may vary from structure to structure. For instance, the effects of OTA nonidealities on Figs. 9.24(a) and (b) will be different. For the desired function we may select the best structure with respect to practical performances among those which can ideally realize the function. This is also why we generate many filter configurations.

The illustration in Fig. 9.24 also shows how a filter using the differential input OTA can be converted into the equivalent using the single-ended input OTA, or the converse of this. Note that Fig. 9.24(a) contains two OTAs but Fig. 9.24(b) has four OTAs. It is also interesting that among all the structures presented in the chapter, only the KHN and TT biquads do not use the differential input OTA inherent in the pole network. In all the other structures the g_2 OTA has the differential input application.

9.11.5 Selection of Input and Output Methods

As we have shown, various input and output techniques may be used to realize a desired transfer function for a chosen structure. Theoretically we can inject a voltage signal through a capacitor, or input the voltage using the differential input terminals of OTAs, or apply the voltage through an extra single-ended input OTA onto the circuit node. The output can be the voltage of some circuit node or the converted node voltage using an OTA voltage amplifier. Ideally they may produce the same function required. But practically using different inputs and outputs will result in different performances. How to choose different input and output techniques may require various factors such as the effects of OTA nonidealities to be taken into account in particular applications. We have already excluded the method of applying the input voltage through the ungrounded capacitor, since it is not desirable from the viewpoint of integration and cascading. The method for inputing the voltage to the OTA input terminals is the simplest and easiest way, but may increase the parasitic effects due to the finite OTA input impedances. Imposing the voltage through an additional single-ended input OTA to the circuit node needs more OTAs although in some cases it may reduce the feedthrough effects. The output through additional OTAs may only be considered if a universal biquad is required, as the summing node of this method will have a parasitic pole at high frequencies. But generally the method of applying the input voltage and taking output voltage with additional OTAs can offer some design flexibility and ease of independent tuning. With these approaches it is also easy to form single-input and single-output universal biquads and to utilize two different sets of OTAs to generate system poles and zeros separately by using the voltage-to-current input distributor or voltage-to-current output summer, as we have shown in the chapter.

9.11.6 Dynamic Range Problem

According to the definition in Chapter 5, the dynamic range can be obtained by calculating the ratio of the maximum signal magnitude to the noise level at either the input or the output nodes of the system.

Generally, given the desired filter transfer function, the dynamic range of a filter is dependent on the dynamic ranges of the network elements, especially the active devices, and the filter network architecture. The limited dynamic range of the OTA is confined by the linear input range and noise level, which restricts the dynamic range of the filter. Several publications have dealt with noise performance analysis [34]–[37], [43] (due to OTA input voltage noise) and large-signal capability [38]–[40], [16, 24] (due to OTA nonlinearity) of OTA-C filters.

Here we want to consider the upper limit, the maximum signal level. The finite maximum magnitude of signal is mainly due to the limited linear range of the OTA as discussed in Chapter 3. For the given OTAs, to maximize the maximum input voltage, it was shown in Section 5.5.3 of Chapter 5 that the maximum values of all the OTA output voltages must be equal from structural viewpoint. For the KHN structure in Fig. 9.8 and for $Q > 1$, for example, using Eqs. (9.56 through 9.58) the maximum values of V_{o1}, V_{o2} and V_{o3} for V_{i5} can be approximated as

$$\left|\frac{V_{o2}}{V_{i5}}\right|_{max} = \frac{Q}{k_{12}}, \quad \left|\frac{V_{o1}}{V_{i5}}\right|_{max} = \frac{1}{k_{11}}, \quad \left|\frac{V_{o3}}{V_{i5}}\right|_{max} = Q \quad (9.131)$$

which are the magnitude values at ω_o.

To make the three values equal requires $k_{12} = 1$ and $k_{11} = 1/Q$. Taking this into consideration it can be seen that the gain K_{LP} of the lowpass filter must be equal to 1 and the gain K_{BP} of the

9.12. SUMMARY

bandpass filter must be equal to Q. In other words, only $K_{LP} = 1$ and $K_{BP} = Q$ can be achieved for the maximum signal operation, rather than arbitrary values. In the case of maximum signal swing, according to the above results and Eq. (9.63) we have

$$\tau_1 = \tau_2 = 1/\omega_o \tag{9.132}$$

Another example is the design for the maximum signal swing of the TT filter in Fig. 9.11. From Eqs. (9.70) and (9.71) we can see that the TT biquad using the input V_{i3}, for example, has

$$\left|\frac{V_{o1}}{V_{i3}}\right|_{max} = \sqrt{\frac{\tau_2}{\tau_1}}, \quad \left|\frac{V_{o2}}{V_{i3}}\right|_{max} = 1 \tag{9.133}$$

For the maximum signal swing design we must have $\tau_1 = \tau_2$ and for given ω_o and Q we can determine

$$\tau_1 = \tau_2 = \frac{1}{\omega_o}, \quad k_{22} = \frac{1}{Q} \tag{9.134}$$

To finish the discussion of practical design considerations we mention the loading and mismatch effects. As we have already discussed in various places, real OTAs do not have infinite output impedances. Besides the previously mentioned problems this will also cause loading effects. Thus, OTA-C circuits should be designed to drive high-impedance nodes such as the inputs of other OTAs. If the low-impedance nodes must be driven, the opamp or OTA buffer circuits must be used [2]. This should be kept in mind particularly when designing high-order filters by cascading OTA-C biquads. Only those biquads with the input voltage applied to OTA input terminals can be directly cascaded. In integrated OTA-C filter design, identical transconductances are usually used to make on-chip tuning, design, and layout easier. In practice, mismatch in transconductances can be produced due to fabrication error, which may cause performance change. Thus the sensitivity of the transfer function to the mismatch error must be small.

9.12 Summary

In this chapter we have comprehensively and systematically investigated generation and design of integrator-based second-order OTA-C filter structures. Many filter architectures have been generated. We have proved that these architectures can offer all types of filter function without use of the capacitor injection. Simultaneous outputs of different filter characteristics for some single input can be obtained. Universal biquads and realizations have been extensively studied. The filter structures have been compared and practical design considerations have been given.

The proposed first-order and second-order OTA-C filters can be cascaded or coupled to realize high-order specifications. Note that all the structures presented in the chapter are suitable for cascading, because the input is at the high-impedance input terminals of OTAs. In odd-order cascade design, while first-order sections can be used, third-order sections are also quite often utilized. Third-order OTA-C filters can be found in [25] and will be discussed in the following chapters. We will investigate high order OTA-C filter design in Chapters 10 and 11, introducing the LC ladder simulation and multiple loop feedback methods, respectively.

It is noted that the filter architectures generated in the chapter can also be derived from a general multiple loop feedback model [47]. This will be discussed in Chapter 11. Another note is that the second-order integrator loop systems have been realized using OTA-based voltage integrators and amplifiers in this chapter. In fact, we can also realize them using OTA-based current integrators and amplifiers [48, 49], which will be discussed in Chapter 12.

References

[1] Tsividis, Y.P. and Voorman, J.O., Eds., *Integrated Continuous-Time Filters: Principles, Design and Applications,* IEEE Press, 1993.

[2] Schaumann, R., Ghausi, M.S., and Laker, K.R., *Design of Analog Filters: Passive, Active RC and Switched Capacitor,* Prentice Hall, Englewood Cliffs, NJ, 1990.

[3] Laker, K.R. and Sansen, W., *Design of Analog Integrated Circuits and Systems,* McGraw-Hill, New York, 1994.

[4] Kardontchik, J.E., *Introduction to the Design of Transconductor-Capacitor Filters,* Kluwer Academic Publishers, Massachusetts, 1992.

[5] Huelsman, L.P., *Active and Passive Analog Filter Design,* McGraw-Hill, New York, 1993.

[6] Bialko, M. and Newcomb, R.W., Generation of all finite linear circuits using the integrated DVCCS, *IEEE Trans. Circuit Theory,* 18, 733–736, 1971.

[7] Contreras, R.A. and Fidler, J.K., VCT active filters, *Proc. European Conf. on Circuit Theory and Design,* 1, 361–369, 1980.

[8] Malvar, H.S., Electronically tunable active filters with operational transconductance amplifiers, *IEEE Trans. Circuits Syst.,* 29(5), 333–336, 1982.

[9] Urbaś, A. and Osiowski, J., High-frequency realization of C-OTA second-order active filters, *Proc. IEEE Intl. Symp. Circuits Syst.,* 1106–1109, 1982.

[10] Geiger, R.L. and Ferrel, J., Voltage controlled filter design using operational transconductance amplifiers, *Proc. IEEE Intl. Symp. Circuits Syst.,* 594–597, 1983.

[11] Malvar, H.S., Electronically controlled active-C filters and equalizers with operational transconductance amplifiers, *IEEE Trans. Circuits Syst.,* 31(7), 645–649, 1984.

[12] Geiger, R.L. and Sánchez-Sinencio, E., Active filter design using operational transconductance amplifiers: a tutorial, *IEEE Circuits and Devices Magazine,* 20–32, Mar. 1985.

[13] Nawrocki, R. and Klein, U., New OTA-capacitor realization of a universal biquad, *Electron. Lett.,* 22(1), 50–51, 1986.

[14] Plett, C., Copeland, M.A., and Hadway, R.A., Continuous time filters using open loop tunable transconductance amplifiers, *Proc. IEEE Intl. Symp. Circuits Syst.,* 1173–1176, 1986.

[15] Tan, M.A. and Schaumann, R., Design of a general biquadratic filter section with only transconductance and grounded capacitors, *IEEE Trans. Circuits Syst.,* 35(4), 478–480, 1988.

[16] Sánchez-Sinencio, E., Geiger, R.L., and Nevarez-Lozano, H., Generation of continuous-time two integrator loop OTA filter structures, *IEEE Trans. Circuits Syst.,* 35(8), 936–946, 1988.

[17] Ananda Mohan, P.V., Generation of OTA-C filter structures from active RC filter structures, *IEEE Trans. Circuits Syst.,* 37, 656–660, 1990.

[18] Chang, C.M. and Chen, P.C., Universal active filter with current gain using OTAs, *Intl. J. Electronics,* 71, 805–808, 1991.

[19] Loh, K.H., Hiser, D.L., Adams, W.J., and Geiger, R.L., A versatile digitally controlled continuous-time filter structure with wide range and fine resolution capability, *IEEE Trans. Circuits Syst.,* 39, 265–276, 1992.

[20] Acar, C., Anday, F., and Kuntman, H., On the realization of OTA-C filters, *Intl. J. Circuit Theory and Applications,* 21(4), 331–341, 1993.

[21] Sun, Y. and Fidler, J.K., Some design methods of OTA-C and CCII-RC active filters, *Proc. 13th IEE Saraga Colloquium on Digital and Analogue Filters and Filtering Systems,* 7/1–7/8, 1993.

[22] Sun, Y. and Fidler, J.K., Novel OTA-C realizations of biquadratic transfer functions, *Intl. J. Electron.,* 75, 333–348, 1993.

[23] Sun, Y. and Fidler, J.K., Resonator-based universal OTA-grounded capacitor filters, *Intl. J. Circuit Theory and Applications,* 23, 261–265, 1995.

[24] Sun, Y. and Fidler, J.K., Synthesis and performance analysis of a universal minimum component integrator-based IFLF OTA-grounded capacitor filter, *IEE Proc. Circuits, Devices and Systems,* 143, 107–114, 1996.

[25] Sun, Y., Jefferies, B., and Teng, J., Universal third-order OTA-C filters, *Intl. J. Electron.,* 85(5), 597–609, 1998.

[26] Nawrocki, R., Electronically tunable all-pole low-pass leapfrog ladder filter with operational transconductance amplifier, *Intl. J. Electronics,* 62(5), 667–672, 1987.

[27] Nawrocki, R., Building set for tunable component simulation filters with operational transconductance amplifiers, *Proc. Midwest Symp. Circuits and Systems,* 227–230, 1987.

[28] Nawrocki, R., Electronically controlled OTA-C filter with follow-the-leader-feedback structures, *Intl. J. Circuit Theory and Applications,* 16, 93–96, 1988.

[29] Bowron, P. and Gahir, H.M., Modelling of nonideal active devices in continuous-time OTA-C filters, *Proc. European Conf. Circuit Theory and Design,* 128–131, 1989.

[30] Nevárez-Lozano, H., Hill, J.A., and Sánchez-Sinencio, E., Frequency limitations of continuous-time OTA-C filters, *Proc. IEEE Intl. Symp. Circuits Syst.,* 2169–2172, 1988.

[31] Sun, Y. and Fidler, J.K., Performance analysis of multiple loop feedback OTA-C filters, *Proc. IEE 14th Saraga Colloquium on Digital and Analogue Filters and Filtering Systems,* 9/1–9/7, London, 1994.

[32] Nevárez-Lozano, H. and Sánchez-Sinencio, E., Minimum parasitic effects biquadratic OTA-C filter architectures, *Analog Integrated Circuits and Signal Processing,* 1, 297–319, 1991.

[33] Ramírez-Angulo, J. and Sánchez-Sinencio, E., Active compensation of operational transconductance amplifiers using partial positive feedback, *IEEE J. Solid-State Circuits,* 25(4), 1024–1028, 1990.

[34] Espinosa, G., Montecchi, F., Sánchez-Sinencio, E., and Maloberti, F., Noise performance of OTA-C Filters, *Proc. IEEE Intl. Symp. Circuits Syst.*, 2173–2176, 1988.

[35] Brambilla, A., Espinosa, G., Montecchi, F., and Sánchez-Sinencio, E., Noise optimization in operational transconductance amplifier filters, *Proc. IEEE Intl. Symp. Circuits Syst.*, 118–121, 1989.

[36] Abidi, A.A., Noise in active resonators and the available dynamic range, *IEEE Trans. Circuits Syst.*, 39, 296–299, 1992.

[37] Bowron, P. and Mezher, K.A., Noise analysis of second-order analogue active filters, *IEE Proc.-Circuits Devices Syst.*, 141, 350–356, 1994.

[38] Bowron, P., Mezher, K.A., and Muhieddine, A.A., The dynamic range of second-order continuous-time active filters, *IEEE Trans. Circuits Syst. I*, 43(5), 370–373, 1996.

[39] Groenewold, G., The design of high dynamic range continuous-time integratable bandpass filters, *IEEE Trans. Circuits Syst.*, 38, 838–852, 1991.

[40] Hiser, D.L. and Geiger, R.L., Impact of OTA nonlinearities on the performance of continuous-time OTA-C bandpass filters, *Proc. IEEE Intl. Symp. Circuits Syst.*, 1167–1170, 1990.

[41] Moulding, K.W., Quartly, J.R., Rankin, P.J., Thompson, R.S., and Wilson, G.A., Gyrator video filter IC with automatic tuning, *IEEE J. Solid-State Circuits*, SC-16, 963–968, 1980.

[42] Visocchi, P., Taylor, J., Mason, R., Betts, A., and Haigh, D., Design and evaluation of a high-precision, fully tunable OTA-C bandpass filter implemented in GaAs MESFET technology, *IEEE J. Solid-State Circuits*, 29, 840–843, 1994.

[43] Park, C.S. and Schaumann, R., Design of a 4-MHz analog integrated CMOS transconductance-C bandpass filter, *IEEE J. Solid-State Circuits*, 23(4), 987–996, 1988.

[44] Silva-Martinez, J., Steyaert, M.S.J., and Sansen, W., A 10.7-MHz 68-dB SNR CMOS continuous-time filter with on-chip automatic tuning, *IEEE J. Solid-State Circuits*, 27, 1843–1853, 1992.

[45] Kobe, M.R., Sánchez-Sinencio, E., and Ramirez-Angulo, J., OTA-C biquad-based filter silicon compiler, *Analog Integrated Circuits and Signal Processing*, 3(3), 243–258, 1993.

[46] Nedungadi, A.P. and Geiger, R.L., High-frequency voltage-controlled continuous-time low-pass filter using linearized CMOS integrators, *Electron. Lett.*, 22, 729–731, 1986.

[47] Sun, Y. and Fidler, J.K., Structure generation and design of multiple loop feedback OTA-grounded capacitor filters, *IEEE Trans. Circuits Syst., Part-I*, 43, 1–11, 1997.

[48] Sun, Y. and Fidler, J.K., Structure generation of current-mode two integrator loop dual output-OTA grounded capacitor filters, *IEEE Trans. Circuits Syst., Part II*, 43, 659–663, 1996.

[49] Sun, Y. and Fidler, J.K., Design of current-mode multiple output OTA and capacitor filters, *Intl. J. Electron.*, 81, 95–99, 1996.

[50] Kerwin, W.J., Huelsman, L.P., and Newcomb, R.W., State-variable synthesis for insensitive integrated circuit transfer functions, *IEEE J. Solid-State Circuits*, 2(3), 87–92, 1967.

REFERENCES

[51] Thomas, L.C., The biquad: part I—some practical design considerations, and The biquad: part II—a multipurpose active filtering system, *IEEE Trans. Circuit Theory,* 18(3), 350–357 and 358–361, 1971.

[52] Girling, F.E.J. and Good, E.F., Active filters: the two integrator loop, *Wireless World,* Part 7, 76–80, Feb.; Part 8, 134–139, Mar. 1970.

[53] Fidler, J.K., An active biquadratic circuit based on the two integrator loop, *Intl. J. Circuit Theory and Applications,* 4, 407–411, 1976.

[54] Ackerberg, D. and Mossberg, K., A versatile active RC building block with inherent compensation for the finite bandwidth of the amplifier, *IEEE Trans. Circuits and Systems,* 21(1), 75–78, 1974.

[55] Bruton, L.T., *RC-Active Circuits: Theory and Design,* Prentice Hall, Englewood Cliffs, NJ, 1980.

Chapter 10

OTA-C Filters Based on Ladder Simulation

10.1 Introduction

In Chapters 8 and 9 we discussed the design of low-order OTA-C filters. In this chapter we deal with high-order OTA-C filter design [1]–[42]. The most popular method for high-order filter design is the cascade method due to its modularity of structure and simplicity of design and tuning. As discussed in Chapter 5, for a given transfer function we first factorize it into low-order functions and then realize these functions using the filter structures proposed in Chapters 8 and 9. Finally we cascade the designed sections, the whole circuit giving the desired transfer function. For lowpass and highpass filters we can simply use the lowpass and highpass sections, respectively. For bandpass filters we can use either both lowpass and highpass sections, or only bandpass sections. This should be decided before factorization. The cascade method is general in that arbitrary transmission zeros, as required in equalizer design, can be realized. The principles of cascade design were established in Chapter 5, and so will not be repeated here. Suffice it to say that any of the OTA structures reported in Chapters 8 and 9 may be incorporated into such an architecture. The reader is also encouraged to refer to the relevant papers and books for OTA-C filter design examples based on this method [1]–[4], [9]–[12].

The cascade method however has a very high sensitivity to component tolerances. It has already been established that resistively terminated lossless LC filters have very low passband sensitivity. To achieve low sensitivity, OTA-C filters can thus be designed by simulating passive LC filters, as we have done for the design of opamp-based active RC filters in Chapter 6. In this chapter we investigate the simulation method for OTA-C filter design. Again, we assume the availability of design tables or appropriate computer software for the generation of LC ladder network component values, and therefore concentrate on how to simulate these passive LC ladders using only OTAs and capacitors.

Various methods for OTA-C simulation of doubly terminated passive LC ladders will be introduced in a systematic way. These can be broadly classified into three categories: component substitution, signal flow simulation, and coupled biquad realization, as discussed in Chapter 6. The first category, belonging to the topological approach, includes the inductor substitution, the Bruton transformation and the impedance/admittance block substitution. The second, being the functional or operational approach, contains the leapfrog (LF) structure and its derivatives as well as matrix methods (including the wave filter method). The third embraces the biquad-based LF structure, one of the multiple loop feedback configurations, and the follow-the-leader-feedback (FLF) structure (see Chapter 5). The component substitution methods keep the active filter structure and equations identical to those of the original passive ladder. The signal simulation method has the same equations as, but different structures from, the original ladder. The coupled biquad approach may have different equations and structures. Various practical design considerations will also be presented.

With outstanding low-sensitivity performance, ladder simulation OTA-C filters are complex in structure and difficult to tune, compared with the cascade method. While the component substitution and LF methods are most popular, the matrix decomposition and coupled biquad approaches have also been used. The coupled biquad method also has modular design properties. In general terms and also to enjoy the modular design we present the ladder simulation from the block viewpoint, leaving component-level simulation as its special cases.

We will first describe OTA-C filter design using the inductor substitution and Bruton transformation methods, followed by discussion of the admittance simulation approach. OTA-C filter design based on the signal flow simulation of passive LC ladders will then be discussed. The equivalence of the admittance substitution and signal simulation methods will also be studied. Next, the matrix methods for OTA-C filter design and coupled biquad OTA-C configurations are briefly explained. After some comments on practical OTA-C design problems, a summary of the chapter is finally presented.

10.2 Component Substitution Method

OTA-C filter design based on a passive LC ladder can be conducted by substituting resistors and inductors by OTA-C counterparts. Such an OTA-C circuit has as low a sensitivity as the passive counterpart, except for the imperfections in the realization of the active resistor and inductor and the increase in the total number of components. The Bruton method transforms the passive LC ladder into some new equivalent ladder which contains no inductors, but some new components, which are then replaced by OTA-C counterparts. The admittance block substitution method deals with each ladder arm as a whole and replaces it by the OTA-C circuit which has the same impedance or admittance function.

10.2.1 Direct Inductor Substitution

OTA-C realization based on doubly terminated passive LC ladders by direct component substitution requires the simulation of inductors and resistors. Simulation of resistors has been given in Chapter 8. For convenience we present them again in Fig. 10.1, with the input termination for ideal voltage input, general floating and grounded resistors being shown in Figs. 10.1(a), (b), and (c), respectively. All have the resistance equal to the inversion of the OTA transconductance, i.e.,

$$R = \frac{1}{g} \tag{10.1}$$

Note that the output termination resistor in the passive prototype is grounded.

Note also that the grounded resistors and input termination require a single OTA and the general floating resistor requires two OTAs. Effects of OTA nonidealities on the OTA resistors were discussed in Chapter 8.

OTA-C Inductors

Now we consider OTA-C simulation of the inductor. As discussed in Chapter 3, the OTA is most convenient for realizing the gyrator because the gyrator contains only two CCVSs. The OTA gyrator, when terminated by a capacitor, will produce a simulated OTA-C inductor with the inductance being given by

$$L = \frac{C}{g_1 g_2} \tag{10.2}$$

as depicted in Fig. 10.2(a) [7].

10.2. COMPONENT SUBSTITUTION METHOD

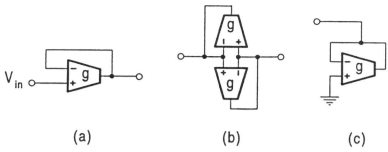

FIGURE 10.1
OTA simulation of resistors.

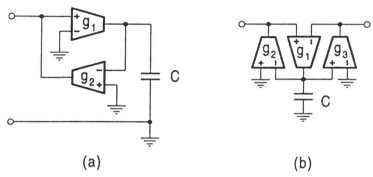

FIGURE 10.2
OTA-C simulation of inductors: (a) grounded and (b) floating.

A floating OTA-C inductor based on two gyrators connected back-to-back may be reduced to the three OTA architecture shown in Fig. 10.2(b) [8]. It can be shown that when $g_2 = g_3 = g$, the equivalent inductance is given by

$$L = \frac{C}{g_1 g} \qquad (10.3)$$

Note that a grounded inductor can be simulated using two OTAs and one capacitor. But a floating inductor needs three OTAs and one capacitor. The inductor substitution technique described above leads to a realization that has the same topology as the original passive ladder network. The difference is that each inductor is replaced by a circuit using OTAs and the capacitor.

As an example, consider the OTA-C simulation of a floating inductor of the inductance equal to 173.64 μH in a 10 MHz lowpass passive LC ladder. Choosing $g_1 = g_2 = g_3 = g = 200\mu S$, we can calculate the required capacitance, $C = g^2 L = 6.95 pF$.

Tolerance Sensitivity of Filter Function

The inductor substitution method has low sensitivity. It leaves the capacitor as it is in the original ladder and realizes the terminal resistor using a single OTA. Therefore the sensitivities to the capacitance and the transconductance of the OTA resistor are not changed. Now consider the inductor which is realized using either two or three OTAs and one capacitor. The relative sensitivities of the filter function to these OTA transconductances and the capacitance will depend on the sensitivity of the inductance to these parameters according to the following relation:

$$S_{g_i,C}^{H(s)} = S_L^{H(s)} S_{g_i,C}^L \qquad (10.4)$$

where $H(s)$ is the transfer function, and g_i and C are the related transconductances and capacitance simulating the inductor.

As can be seen from Eqs. (10.2) and (10.3), for both grounded and floating OTA-C inductors, we have $S_C^L = -S_{g_i}^L = 1$. The performance sensitivities to these parameters will therefore be the same as the original sensitivity to the passive inductance, as indicated in Eq. (10.4).

Parasitic Effects on Simulated Inductor

For the grounded inductor with $g_1 = g_2 = g$ for example, taking finite OTA impedances into consideration, we can derive the terminal admittance of the OTA-C inductor as

$$Y_{\text{in}} = G' + sC' + \frac{g^2}{G' + s(C + C')} \tag{10.5}$$

where $G' = G_i + G_o$ and $C' = C_i + C_o$. G_i, C_i and G_o, C_o are the respective input and output conductances and capacitances.

The equivalent circuit is depicted in Fig. 10.3. It consists of an inductor with a series parasitic inductor and loss resistor and parallel parasitic capacitor and loss resistor, whose values are given by

$$L = \frac{C}{g^2}, \quad L_s = \frac{C'}{g^2}, \quad R_s = \frac{G'}{g^2}, \quad C_p = C', \quad R_p = \frac{1}{G'} \tag{10.6}$$

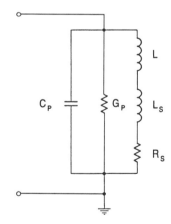

FIGURE 10.3
Equivalent circuit of practical grounded OTA-C inductor.

Now we consider only the output conductance effects. The input admittance in Eq. (10.5) becomes

$$Y_{\text{in}} = G_o + \frac{g^2}{G_o + j\omega C} \tag{10.7}$$

[if the input conductances also need to be considered, we can simply use G' to replace G_o in Eq. (10.7) to include G_i] or the input impedance with the first-order approximation is given by

$$Z_{\text{in}} = j\omega L + r_L = j\omega \frac{C}{g^2} + G_o \left[\frac{1}{g^2} + \omega^2 \left(\frac{C}{g^2} \right)^2 \right] \tag{10.8}$$

10.2. COMPONENT SUBSTITUTION METHOD

which is an ideal inductor in series with a parasitic resistor. The inductor quality factor is derived as

$$q_L = \frac{\omega L}{r_L} = \frac{1}{2}\frac{g}{G_o}\frac{2\left(\frac{\omega C}{g}\right)}{1+\left(\frac{\omega C}{g}\right)^2} \qquad (10.9)$$

The maximum value for q_L is given by $\frac{1}{2}A_{dc}$. Note that $A_{dc} = \frac{g}{G_o}$ is the maximum voltage gain at dc, set by the nonzero conductance at the output of the OTA. The inductor quality factor is therefore at most one-half the dc voltage gain of the OTAs used [17]. It is very important to reduce the output conductance in order to realize a high-quality inductor, especially for high-frequency design since the loss resistance increases (the quality factor decreases) with frequency.

The OTA may contain internal poles at high frequencies, which become a serious problem when the signals in the filter passband approach the frequencies of these undesirable poles. If the two OTAs constituting the gyrator have one dominant high-frequency pole (ω_b), then the transconductance will be frequency dependent, given by

$$g(s) = \frac{g}{1+s/\omega_b} \qquad (10.10)$$

The terminal admittance of the OTA-C inductor can be obtained with a first-order approximation as [17]

$$Y_{in} = \frac{1}{j\omega\left(C/g^2\right)} - \frac{2}{\omega_b C/g^2} \qquad (10.11)$$

and the terminal impedance with a first-order approximation can be derived either from the OTA-C inductor circuit or by inverting Y_{in} in Eq. (10.11) as

$$Z_{in} = j\omega L + r_L = j\omega\frac{C}{g^2} - 2\omega\frac{C}{g^2}\frac{\omega}{\omega_b} \qquad (10.12)$$

The quality factor of the OTA-C inductor is obtained from Eq. (10.12), given by

$$q_L = -\frac{1}{2}\frac{\omega_b}{\omega} \qquad (10.13)$$

From Eq. (10.12) it can be seen that the effect of finite bandwidth corresponds to a negative resistance in series with the inductance. According to Eq. (10.11) we can compensate for this effect by using a positive grounded resistor in parallel with the input port. This resistor can also absorb the output conductances of the related OTAs. Equation (10.12) also shows that the series parasitic loss resistance due to finite ω_b also increases with frequency.

Parasitic Effects on Filter Function

To facilitate discussion, we define the tolerance sensitivity as the sensitivity of a function to the variation in the component value and the parasitic sensitivity as the sensitivity of a function to the parasitic parameter of the component. Passive LC ladders have very low magnitude tolerance sensitivity in the passband, but the phase tolerance sensitivity may not be low. It can be shown that the parasitic impact is bigger than the tolerance influence on the magnitude response in the passband [46]. It is therefore important to reduce parasitic effects on the magnitude response in order to achieve a good passband magnitude response with the simulation method. It has also been shown in [47, 48] that the magnitude and phase parasitic sensitivities are related to the phase and magnitude tolerance sensitivities, respectively. This reveals that the phase tolerance sensitivity is also important for the magnitude performance due to the parasitic effects. There are several methods which are available for computation of tolerance sensitivity, but not much work has been done for the parasitic counterpart.

It can be shown that parasitic sensitivities can be computed based on tolerance sensitivities. Consider the network containing an inductor with loss, that is, $Z_L = sL + r_L$. The transfer function of the network can be expressed as a bilinear function of Z_L, given by

$$H(s) = \frac{A_{11} + A_{12}(sL + r_L)}{A_{21} + A_{22}(sL + r_L)} = \frac{N}{D} \tag{10.14}$$

where A_{ij} are coefficients independent of Z_L. Direct differentiation of Eq. (10.14) gives

$$\frac{\partial H}{\partial r_L} = \frac{1}{s}\frac{\partial H}{\partial L} = \frac{A_{12}A_{21} - A_{11}A_{22}}{D^2} \tag{10.15}$$

which leads to the parasitic sensitivity function as

$$P_{r_L}^H = \frac{1}{H}\frac{\partial H}{\partial r_L} = \frac{1}{sL}S_L^H \tag{10.16}$$

Writing $H = |H|e^{j\phi}$ and using Eq. (10.16) we can obtain the relative change in the magnitude due to the inductor loss resistance as

$$\frac{\Delta |H|}{|H|} = r_L P_{r_L}^{|H|} = \frac{1}{q_L}Q_L^\phi \tag{10.17}$$

where $Q_L^\phi = L\frac{\partial \phi}{\partial L}$, which is the phase tolerance sensitivity and is calculated at the nominal state, i.e., $r_L = 0$.

We can see that the magnitude change due to r_L can be calculated using the phase tolerance sensitivity with no need of any extra circuit analysis. This also reveals that the phase tolerance sensitivity is also important for the magnitude frequency response due to the existence of the parasitic loss resistance. In our case, q_L can be obtained from Eqs. (10.9) and (10.13) for the effects of the output conductance and the impact of excess phase, respectively. The total variation can be derived as

$$\frac{\Delta |H|}{|H|} = 2Q_L^\phi \left[\frac{1 + \left(\frac{\omega C}{g}\right)^2}{2\left(\frac{\omega C}{g}\right)}\frac{G_o}{g} - \frac{\omega}{\omega_b}\right] \tag{10.18}$$

It is stressed again that the parasitic effects increase with the frequency and to achieve a high-frequency performance we must reduce these parasitics.

We now consider the impact of parasitic capacitances on the filter function. From Eq. (10.5) and Fig. 10.3 we can see that the parasitic capacitances in parallel with the circuit capacitance will cause the shift in the equivalent inductance by L_s. Therefore, the following formula in Eq. (10.19) for computation of the relative change in the magnitude due to the tolerance in the inductance can be used to evaluate the effects of the parasitic capacitances.

$$\frac{\Delta |H|}{|H|} = \frac{\Delta L}{L}S_L^{|H|} \tag{10.19}$$

10.2. COMPONENT SUBSTITUTION METHOD

The change of the magnitude due to the parasitic capacitances, $C' = C_i + C_o$ in parallel with the circuit capacitance C, can be obtained from Eq. (10.19) by noting that $L = C/g^2$ and $\Delta L = L_s = C'/g^2$, given by

$$\frac{\Delta |H|}{|H|} = \frac{C'}{C} S_L^{|H|} \tag{10.20}$$

From this equation we can see that to reduce the change, the parasitic to normal capacitance ratio must be small and the magnitude tolerance sensitivity to the inductance must also be small. It is very fortunate that $S_L^{|H|}$ is indeed very small. It should be very interesting to note from the above analysis that the magnitude change due to parasitics depends on the ratios of G_o/g, C'/C, ω/ω_b and also the tolerance phase and magnitude sensitivities. To reduce the parasitic effects, these factors must be taken into consideration, especially for very high-frequency applications.

10.2.2 Application Examples of Inductor Substitution

It is well known that the inductor substitution method is most economical for simulation of highpass LC ladders, as the inductors in highpass ladders are grounded. In this section, however, two other examples are presented to illustrate how to design OTA-C filters from passive LC ladders by using the inductor substitution method introduced above and also to introduce some concepts for later use in the chapter.

OTA-C Biquad Derived from RLC Resonator Circuit

The OTA-C active resonator is the most popular biquad in practice. It can be generated by OTA-C simulation of resistors and inductors of a passive RLC resonator. It has very low sensitivity and parasitic effects.

Consider the passive RLC resonator circuit in Fig. 10.4(a). It has the driving point impedance

$$Z(s) = \frac{V_{\text{out}}}{I_{\text{in}}} = \frac{sL}{s^2 LC_1 + s\frac{L}{R} + 1} \tag{10.21}$$

Direct substitution of the OTA resistor and OTA-C inductor produces the corresponding active OTA-C resonator as shown in Fig. 10.4(b). The simulated function is derived by noting that $L = C/g_1 g_2$ and $R = 1/g_3$, given by

$$H(s) = \frac{V_{\text{out}}}{I_{\text{in}}} = \frac{sC_2}{s^2 C_1 C_2 + s g_3 C_2 + g_1 g_2} \tag{10.22}$$

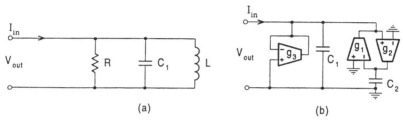

FIGURE 10.4
RLC and OTA-C resonators.

We discussed the OTA nonideality effects on the two integrator loop structure in Chapter 9. Now we analyze the effects based on the inductor simulation. Inspection of the OTA-C resonator reveals that the input and output capacitances of all the OTAs can be absorbed by the circuit capacitances C_1 and C_2. Also, the output conductance of the g_1 OTA, the input conductance of the g_2 OTA and both the input and output conductances of the g_3 OTA can be absorbed by the transconductance g_3. Therefore we only need to consider the output conductance of the g_2 OTA and the input conductance of the g_1 OTA, in addition to the frequency dependent transconductances. It is clear that we can evaluate the parasitic effects on the whole transfer function based on the general formulas derived in the previous section. But for second-order filters, parameters ω_o and Q are of more interest. The relative changes in these parameters due to the inductor parasitic losses are generally given in [47]

$$\frac{\Delta \omega_o}{\omega_o} = -\frac{1}{2Qq_L}\left(S_L^{\omega_o} + S_L^Q\right) \tag{10.23}$$

$$\frac{\Delta Q}{Q} = \frac{1}{2Qq_L}\left[\left(4Q^2 - 1\right)S_L^{\omega_o} - S_L^Q\right] \tag{10.24}$$

Again they are dependent on both tolerance sensitivities and the finite q_L. From Eq. (10.21) we can deduce that $S_L^{\omega_o} = -\frac{1}{2}$ and $S_L^Q = -\frac{1}{2}$. Equations (10.23) and (10.24) become, respectively,

$$\frac{\Delta \omega_o}{\omega_o} = \frac{1}{2Qq_L} \tag{10.25}$$

$$\frac{\Delta Q}{Q} = -\frac{1}{2Qq_L}\left(2Q^2 - 1\right) \tag{10.26}$$

The series loss resistance due to the output conductance of the g_2 OTA, G_{o2} and the input conductance of the g_1 OTA, G_{i1} is given by $r_L = (G_{o2} + G_{i1})/g_1 g_2$. The inductor quality factor can be derived as

$$q_L = \frac{\omega_o C_2}{G_{o2} + G_{i1}} \tag{10.27}$$

Substitution of this into Eqs. (10.25) and (10.26) yields the relative change of ω_o and Q. We can also see that the finite q_L has different impacts on Q and ω_o. A comparison may be made by writing

$$\frac{\Delta Q}{Q} / \frac{\Delta \omega_o}{\omega_o} = -\left(2Q^2 - 1\right) \tag{10.28}$$

which shows that the change in ω_o due to parasitic loss is smaller than that in Q. It is noted that $\Delta \omega_o / \omega_o$ being small is important as the total magnitude change depends more on it.

Finally we point out that the ideal LC resonator and its OTA-C counterpart correspond to $R = \infty$ and $g_3 = 0$, that is, the removal of the resistor and the corresponding OTA, respectively. Both lossy and ideal OTA-C resonators will be needed in the simulation of some passive LC ladders.

A Lowpass OTA-C Filter

For illustration we consider the fifth-order lowpass finite zero LC ladder in Fig. 10.5(a). We replace the input and output termination resistors by the OTA counterparts in Figs. 10.1(a) and (c), respectively, and the two floating inductors by the OTA-C equivalent in Fig. 10.2(b). The resulting OTA-C filter is displayed in Fig. 10.5(b). The component values can be determined using the formulas in Eqs. (10.1–10.3) as

$$g_1 = 1/R_1, \quad C_3' = g_2 g_3 L_3, \quad C_5' = g_4 g_5 L_5, \quad g_6 = 1/R_6 \tag{10.29}$$

10.2. COMPONENT SUBSTITUTION METHOD

where g_2 and g_3 (g_4 and g_5) can be used to produce a proper value for C'_3 (C'_5).

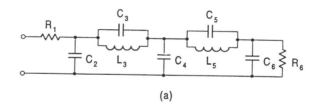

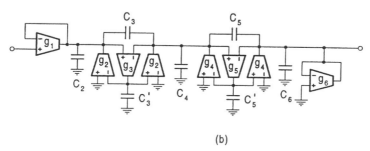

FIGURE 10.5
Fifth-order finite zero lowpass LC ladder and its OTA-C simulation.

In IC implementation, the grounded capacitor is simpler to implement in technology. The floating capacitor has substantial parasitic capacitances (about 10% of the capacitance value) from the bottom plate to the substrate, i.e., the ac ground. For grounded capacitors, the bottom plate should be connected to ground and thus the parasitic capacitances are shorted out for signal currents and play no role [3]. Figure 10.6 gives a circuit which can simulate the floating capacitor using OTAs and the grounded capacitor [14]. It can be formulated that

$$C_f = \frac{g_1 g_2}{g_3 g_4} C_g \tag{10.30}$$

Or we can say that for the given floating capacitance C_f, the grounded capacitance can be determined by

$$C_g = \frac{g_3 g_4}{g_1 g_2} C_f \tag{10.31}$$

and the value can be adjusted by the related transconductances.

It must be noted that the price paid for grounding the floating capacitor is the extra five OTAs. The increased number of OTAs may cause other problems such as extra noise and power consumption. Compared with the block substitution method we shall introduce in Section 10.3, the separate treatment of the floating capacitor and inductor in the series arm is not economical, requiring three more OTAs. Finally it should be pointed out that the convertor circuit in Fig. 10.6 has an internal node, node A, which does not have any component to ground, which we call the suspending node. The suspending node with parasitic capacitances will produce an extra pole and this parasitic pole will influence the filter responses at high frequencies.

10.2.3 Bruton Transformation and FDNR Simulation

By dividing the impedance of each branch in the passive ladder network by s the Bruton transformation converts the inductors, capacitors, and resistors of the ladder to the resistors, frequency-

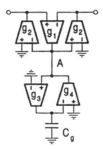

FIGURE 10.6
Conversion of floating capacitor to grounded capacitor.

dependent-negative-resistance (FDNR) components, and capacitors, respectively, as shown in Section 6.6. Thus the OTA-C realization becomes the substitution of the new set of components.

The FDNR has an impedance given by

$$Z(s) = \frac{1}{s^2 D} \tag{10.32}$$

The grounded FDNR can be implemented with five OTAs and two grounded capacitors and the floating FDNR requires six OTAs and two grounded capacitors, as shown in Figs. 10.7(a) [8, 14] and (b) [14], respectively. It can be shown that both of them have

$$D = \frac{g_1 g_2}{g_3 g_4 g_5} C_1 C_2 \tag{10.33}$$

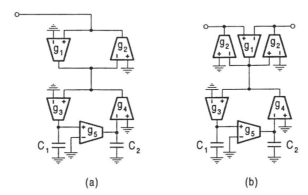

(a)   (b)

FIGURE 10.7
Realizations of grounded and floating FDNRs.

The synthesis procedure of opamp-RC filters based on the Bruton Transformation has been given in Chapter 6, but is again now summarized for OTA-C filters. For a given transfer function, first design a passive LC filter. Then use the Bruton transformation method to transform the LC filter, that is, replace the resistors R by capacitors of value $1/R$; the inductors L by resistors of value L; and the capacitors C by FDNRs whose D value is C. Next use the OTA resistors and OTA-C FDNRs to replace the corresponding resistors and FDNRs in the transformed circuit. If required, the resulting floating capacitors may also be replaced by the OTA-grounded capacitor circuit. The new element

10.2. COMPONENT SUBSTITUTION METHOD

values resulting from the substitution can be determined; taking the FDNR as an example, from Eq. (10.33) we obtain

$$C_1 C_2 = \frac{g_3 g_4 g_5}{g_1 g_2} D \qquad (10.34)$$

All the values are now electronically controllable through the related transconductances. The final OTA-C filter which realizes the given transfer characteristic is thus obtained.

The Bruton transformation method is widely used in active RC filter design. In the above we have shown that the method can also be used to design OTA-C filters. We want now to evaluate the performance of the method in OTA-C filter design and decide whether the method is suitable or not.

The Bruton Transformation does not affect the topology of the ladder circuit, nor does it affect the transfer function realized. Therefore, similar to the inductor substitution method, the FDNR substitution method can produce OTA-C filters that have as low a sensitivity as the passive circuit.

The philosophy behind the Bruton method for active RC filter design however may not be suitable for OTA-C filters. Unlike the active RC case where the transformation of inductors to resistors is clearly an advantage, in OTA-C filter design the resistors resulting from the transformation of the inductors by the Bruton method also need to be replaced. The OTA-C realization of the FDNR is clearly more complex than that of the simulated inductor, as revealed in Fig. 10.2 and Fig. 10.7. These two factors make the number of components needed for the inductor and FDNR substitution approaches much different, the total numbers of OTAs and capacitors with the FDNR method being much bigger than those required by the inductor replacement approach.

Take the third-order all-pole lowpass filter and its transformed version in Fig. 10.8 as an example. The two terminated resistors each uses one OTA and the floating inductor needs three OTAs and one capacitor. Altogether five OTAs and three capacitors are needed for the inductor substitution method. With the FDNR method, the floating resistor requires two OTAs and the two grounded FDNRs each requires five OTAs and two capacitors, which leads to the OTA and capacitor total numbers of twelve and six, respectively, with the input terminal capacitor being floating.

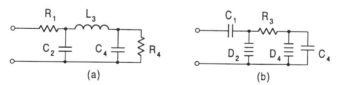

FIGURE 10.8
Third-order all-pole inductor and FDNR ladders.

Similar comparison can be conducted for the highpass filter with all zeros at the origin in Fig. 10.9. The inductor substitution approach requires six OTAs and three capacitors, while the Bruton method needs eight OTAs and four capacitors, both having a floating capacitor.

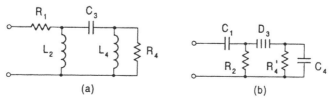

FIGURE 10.9
Third-order highpass inductor and FDNR ladders.

We should say that for both lowpass and highpass filters the FDNR method will generate additional suspending nodes which are inherent in the FDNR circuits in Fig. 10.7. In addition, the Bruton method will always have a floating capacitor at the input terminal resulting from the input termination resistor, which will also cause a dc path problem, as discussed in Chapter 6.

From the above analysis we can conclude that the Bruton transformation method may not be attractive for OTA-C filter design, although for highpass filter design the difference of the component numbers is not so great. Use of the direct inductor substitution method can also save the Bruton transformation step. A full set of OTA-C simulations of components for the inductor replacement and Bruton transformation methods with emphasis on tunability can be found in [14].

10.3 Admittance/Impedance Simulation

In Chapters 3 and 6 various opamp impedance conversion or transformation circuits and methods were introduced and applied in active RC filter design based on passive ladders. In this section we deal with the OTA-C counterparts. Generally, each arm of a ladder network is actually a one-driving-port network and may contain several components as in cases of finite zero lowpass, finite zero highpass, bandpass, and bandstop filters. Rather than dealing with individual components as previously we simulate the impedance or admittance of each arm as a whole using OTAs and capacitors. This may be achieved by direct OTA-C substitution of those admittance or impedance blocks in the prototype ladder. In the following, however, we will present a method which first converts the floating series admittance to the grounded impedance using OTAs and then simulate all grounded impedances using OTAs and capacitors.

10.3.1 General Description of the Method

To facilitate discussion, Fig. 10.10 gives a general ladder with series arm admittances and parallel arm impedances. The admittances in the series arms are floating. If the floating admittance is a complex combination of inductors and capacitors, then the individual treatment of floating inductors and floating capacitors in the arm will have redundant OTAs. We now do a simple transformation, which converts the floating admittance as a whole into a grounded impedance. The circuit for realizing this conversion is exhibited in Fig. 10.11 [8]. The relation between the floating admittance Y_f and the grounded impedance Z_g is given by

$$Y_f = g_1 g_2 Z_g \tag{10.35}$$

From the design standpoint when Y_f is given, then Z_g can be determined as

$$Z_g = \frac{1}{g_1 g_2} Y_f \tag{10.36}$$

where g_1 and g_2 can be used to scale the impedance level.

Replacing all the floating admittances by the OTA-grounded impedance circuit in Fig. 10.11, the general ladder in Fig. 10.10 can be simulated as shown in Fig. 10.12 which consists of only OTAs and grounded impedances and where for example, $Z'_3 = Y_3/g_3 g_4$. The problem left is simply to simulate all grounded impedances using OTAs and capacitors. This can be done using the inductor substitution method in Section 10.2.1. The structure in Fig. 10.12 may be simplified by using the well known fact that any two single-input OTAs with equal transconductances and opposite input polarities, whose outputs are connected to the same node, can be equivalently replaced by a single differential-input OTA with the same transconductance, for example, when $g_2 = g_4 = g_6$. This will

10.3. ADMITTANCE/IMPEDANCE SIMULATION

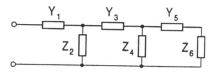

FIGURE 10.10
General admittance and impedance ladder.

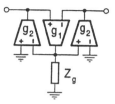

FIGURE 10.11
Conversion of floating admittance to a grounded impedance using OTAs.

be discussed in Section 10.5. Also note that the first OTA at the input end can be discarded for ideal voltage input. For very complex arms in the passive filter, multiple levels of impedance conversion and inductor substitution may be needed. But two-level simulation will suffice for most practical ladders, as can be seen in the next section.

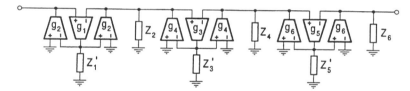

FIGURE 10.12
OTA-grounded impedance simulation of general ladder.

10.3.2 Application Examples and Comparison

To appreciate this general method we now present some typical examples. First reconsider the fifth-order finite zero lowpass LC filter in Fig. 10.5(a). There are two floating admittances which are the parallel LC resonators, needing to be treated. In general, consider the floating parallel LC resonator in Fig. 10.13(a). The floating admittance is

$$Y_f = sC + \frac{1}{sL} \tag{10.37}$$

Using the conversion circuit in Fig. 10.11 and Eq. (10.36) we have the grounded impedance

$$Z_g = s\frac{C}{g_1 g_2} + \frac{1}{sLg_1g_2} \tag{10.38}$$

which is a series LC resonator with inductance $L' = C/g_1g_2$ and capacitance $C' = g_1g_2L$. This is shown in Figs. 10.13(b) and (c) which correspond to the arrangements of the capacitor being grounded and floated, respectively, [22]. In Figs. 10.13(b) and (c) we also give the corresponding OTA-C simulations of the grounded impedance, in both cases the inductor-related capacitor having $C'' = g_3 g_4 L'$.

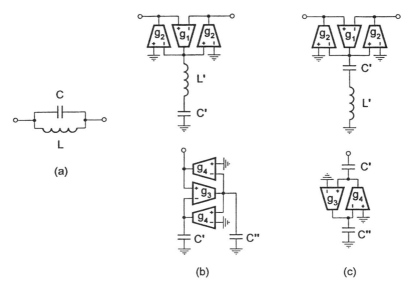

FIGURE 10.13
OTA-C simulation of parallel LC resonator in series arms.

It is evident that this is a two-step method. In both cases we can first choose g_1 or g_2 to give C' some proper value (L' is fixed accordingly) and then generate the appropriate value for C'' using g_3 or g_4. It is also very interesting to note that C'' can be tuned independently from C', which are related to the original capacitor and inductor, respectively.

Comparing the two realizations with the capacitor grounded or floating in Figs. 10.13(b) and (c) we can see that the realization with the grounded capacitor contains one more OTA than the implementation with the floating capacitor. This is obviously because the former contains the floating inductor, while the latter the grounded inductor. Also, the grounded capacitor realization has one suspending node, while the floating capacitor counterpart has two suspending nodes. The floating capacitor may be a disadvantage in IC design. Therefore (also for the reason that will follow immediately), we would not be interested in the floating capacitor realization.

Now we compare the results with those in Section 10.2.1. First look at Fig. 10.5(b) and Fig. 10.13(c). It is apparent that if the floating capacitors are allowed, the structure in Fig. 10.5(b) is more attractive than the one in Fig. 10.13(c) in terms of the number of OTAs needed (two less for each series arm) and the suspending nodes (no suspending nodes), another reason for the above negative claim about the method in Fig. 10.13(c). For grounded-capacitor-only realizations, the method in Fig. 10.13(b) is better than the method in Figs. 10.5(b) and 10.6, since the number of OTAs needed for each series arm is reduced by two.

Figure 10.14 presents the whole OTA-C circuit based on the passive LC prototype in Fig. 10.5(a) and realized using the floating admittance to grounded impedance conversion method and with grounded capacitors only. Clearly, when R, L, and C in Fig. 10.5(a) are known, the values of related components in Fig. 10.14 can be determined (with C_2, C_4, and C_6 remaining unchanged):

$$g_1 = 1/R_1, \quad C'_3 = g_2 g_3 L_3, \quad C''_3 = \frac{g_4 g_5}{g_2 g_3} C_3,$$

10.3. ADMITTANCE/IMPEDANCE SIMULATION

$$C_5' = g_6 g_7 L_5, \quad C_5'' = \frac{g_8 g_9}{g_8 g_7} C_5, \quad g_{10} = 1/R_6 \qquad (10.39)$$

It is possible to have identical transconductances (except the termination OTAs). Setting $g_i = g_j = g$, $i, j = 2, 3, \ldots, 9$ leads to $C_3' = g^2 L_3$, $C_3'' = C_3$, $C_5' = g^2 L_5$ and $C_5'' = C_5$. However, it may not be possible to have identical capacitances, unless $C_2 = C_4 = C_6$. But we can reduce the number of different values to three. For example, we can set $C_3' = C_3'' = C_2$ and obtain $g_2 g_3 = C_2/L_3$ and $g_4 g_5 = C_2^2/(C_3 L_3)$. Similarly we can make $C_5' = C_5'' = C_6$ and determine $g_6 g_7$ and $g_8 g_9$.

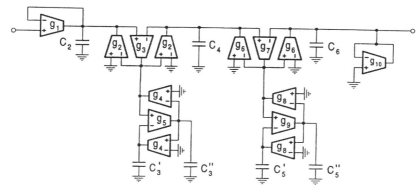

FIGURE 10.14
OTA-C simulation of finite zero lowpass ladder by admittance substitution.

Figure 10.15 gives another example of an eighth-order bandpass filter. From the passive LC ladder in Fig. 10.15(a) we can identify that

$$Y_1 = \frac{1}{R_1 + sL_1 + \frac{1}{sC_1}}, \quad Z_2 = \frac{1}{sC_2 + \frac{1}{sL_2}},$$

$$Y_3 = \frac{1}{sL_3 + \frac{1}{sC_3}}, \quad Z_4 = \frac{1}{\frac{1}{R_4} + sC_4 + \frac{1}{sL_4}} \qquad (10.40)$$

Transforming the two floating admittances into the grounded impedances and then realizing all grounded impedances by the inductor substitution method we can obtain the OTA-C circuit as shown in Fig. 10.15(b), where the corresponding component values are determined as (C_2 and C_4 are left unchanged)

$$g_3 = g_1 g_2 R_1, \quad C_1' = g_1 g_2 L_1, \quad C_1'' = \frac{g_4 g_5}{g_1 g_2} C_1, \quad C_2' = g_6 g_7 L_2,$$

$$C_3' = g_8 g_9 L_3, \quad C_3'' = \frac{g_{10} g_{11}}{g_8 g_9} C_3,$$

$$C_4' = g_{12} g_{13} L_4, \quad g_{14} = \frac{1}{R_4} \qquad (10.41)$$

We can see that through the floating admittance to grounded impedance transformation, the floating resistor, inductor, and capacitor of the series resonator in the series arm become the grounded resistor, capacitor, and inductor of the parallel resonator. This bandpass OTA-C filter architecture has a very good feature in that each circuit node has a grounded capacitor and thus can be expected to have very low parasitic capacitance effects. In fact all parasitic capacitances can be absorbed by these grounded capacitances.

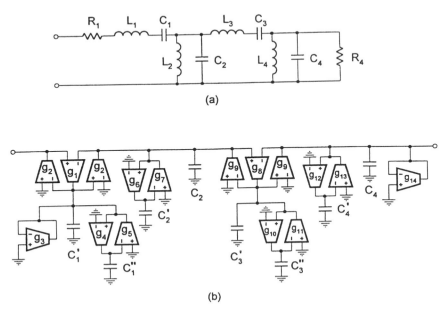

FIGURE 10.15
Bandpass RLC ladder and OTA-C simulation by admittance substitution.

10.3.3 Partial Floating Admittance Concept

Partial floating of the series arm admittance may be useful in some cases and makes the admittance substitution method more general. As shown in Fig. 10.16(a) and (b), if the floating admittance can be split into two parts $Y_j = Y_{j1} + Y_{j2}$, we can leave one part Y_{j1} where it is and simulate the other part Y_{j2} using three OTAs and the grounded impedance $Z'_{j2} = \frac{1}{g_1 g_2} Y_{j2}$. For example, if the series arm is a LC parallel resonator as shown in Fig. 10.16(c), we may leave the capacitor $Y_{j1} = sC$ unchanged and convert only the inductor $Y_{j2} = \frac{1}{sL}$ to the grounded capacitor $Z'_{j2} = \frac{1}{sLg_1g_2}$ using the OTA gyrator, as shown in Fig. 10.16(d). For the LC ladder in Fig. 10.5(a), the partial floating concept will result in the same OTA-C circuit as that in Fig. 10.5(b) which was obtained by the inductor substitution.

We can also explain the partial transformation concept in the following way. If $Y_{j1} = 0$, it will mean that we do not leave any series admittance, or any part of it, floating. Examples of this are all-pole lowpass and all-pole bandpass filters. If $Y_{j2} = 0$, it will imply that we will not convert any series admittance, or any part of it, into the grounded form. For instance, highpass filters may be dealt with in this way. The examples for both nonzero Y_{j1} and Y_{j2} may be all-pole bandstop and finite zero lowpass filters. Thus the partial concept is more comprehensive and general.

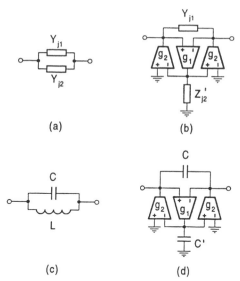

FIGURE 10.16
Illustration of partial conversion of series arm admittance.

10.4 Signal Flow Simulation and Leapfrog Structures

In the preceding sections, we have introduced the component substitution and admittance simulation methods. Another popular theory for OTA-C filter design is to simulate signal flow relations in the LC ladder circuit. As discussed for active RC filter design in Chapter 6, by this method, the circuit equations that describe the topology of the passive ladder structure are first written. Then a signal flow block diagram is drawn based on these equations. Finally the block diagram is realized using OTAs and capacitors. In the simulation of LC ladders, the original equations are of the mixed current and voltage type. We can convert these equations to their voltage-only counterparts by scaling, which is the technique that will be discussed in this section. The mixed equations can also be scaled to the current signal only. This will be dealt with in Chapter 12.

Two techniques for active signal simulation of a passive ladder exist. One is to simulate relations of series-arm currents and parallel-arm voltages, treating the respective arm as a single-port network. The other is to do component-level signal simulations, that is, to simulate relations of signals in all individual elements, for example, individual capacitor voltages and inductor currents. The first type of signal flow simulation structures are block based, since the series and parallel arms are treated as a whole, no matter how many components are there. The second method has a case-by-case feature, for the different passive LC structures the signal flow equations may be very different. Also, note that if each arm in the passive ladder structure is simply a single component, such cases including all-pole LP and zero-at-origin HP LC ladders, the block method will reduce to the component method. In the following we will therefore concentrate on the block signal simulation method for OTA-C filter design based on passive LC ladders. A systematic treatment will be given.

10.4.1 Leapfrog Simulation Structures of General Ladder

The general ladder network with series admittances and parallel impedances is shown in Fig. 10.17. The equations relating the currents flowing in the series arms, I_j, and the voltages across parallel arms, V_j, can be written as

$$I_1 = Y_1(V_{in} - V_2), \quad V_2 = Z_2(I_1 - I_3), \quad I_3 = Y_3(V_2 - V_4),$$

$$V_4 = Z_4(I_3 - I_5), \quad I_5 = Y_5(V_4 - V_6), \quad V_{out} = V_6 = Z_6 I_5 \qquad (10.42)$$

The transfer function V_{out}/V_{in} can be obtained from these equations by eliminating the intermediate variables. These equations can be represented by a signal flow diagram depicted in Fig. 10.18. Observe that the output of each block is fed back to the input of the preceding block and therefore the structure is called the leapfrog (LF) structure [53], as recalled from Chapter 6.

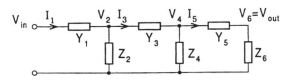

FIGURE 10.17
General admittance and impedance ladder with signals indicated.

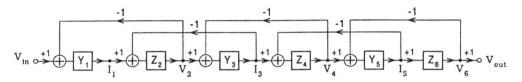

FIGURE 10.18
Leapfrog block diagram of general ladder.

In contrast with the cascaded topology, these blocks are not isolated from each other, and any change in one block affects the voltages and currents in all the other blocks. This type of coupling between the blocks makes the tuning of the whole network more difficult, but gives rise to the much lower sensitivity [49].

In active filter design the mixed current and voltage signal equations are normally converted by scaling into their counterparts with voltage signals only. Scaled by a conductance g_m, Eq. (10.42) can be written as

$$V'_1 = \frac{Y_1}{g_m}(V_{in} - V_2), \quad V_2 = g_m Z_2(V'_1 - V'_3), \quad V'_3 = \frac{Y_3}{g_m}(V_2 - V_4),$$

$$V_4 = g_m Z_4(V'_3 - V'_5), \quad V'_5 = \frac{Y_5}{g_m}(V_4 - V_6),$$

$$V_{out} = V_6 = g_m Z_6 V'_5 \qquad (10.43)$$

where $V'_j = I_j/g_m$. The Y_j/g_m and $g_m Z_i$ are voltage transfer functions. It is clear that these equations will lead to the same transfer function V_{out}/V_{in} as that from Eq. (10.42). The corresponding block diagram is shown in Fig. 10.19.

As traditionally done for opamp-RC filter design (see Chapter 6), to realize this new block diagram we can similarly synthesize the voltage summers and voltage transfer functions of Y_j/g_m and $g_m Z_i$ using OTAs and capacitors. Of course, different ladders will have different Y_j and Z_i values and the

10.4. SIGNAL FLOW SIMULATION

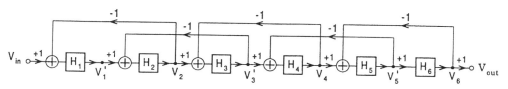

FIGURE 10.19
Scaled leapfrog block diagram of general ladder.

associated OTA-C structures thus will be different. In the following, however, we will not follow the conventional method. We will present a new, systematic, and more efficient method unique to OTA-C filters by using the feature of the OTA. This method is similar to that proposed in [22].

From Eq. (10.43) we can see that the voltage relations have a typical form of

$$U_j = H_j \left(U_{j-1} - U_{j+1} \right) \tag{10.44}$$

where U_j can be V_j or V'_j, and

$$H_j = Y_j/g_m, \quad \text{for odd } j; \quad H_j = g_m Z_j, \quad \text{for even } j \tag{10.45}$$

Equation (10.44) can be realized using an OTA with a transconductance of g_j and a grounded impedance of

$$Z'_j = H_j/g_j \tag{10.46}$$

as shown in Fig. 10.20. This is an OTA-grounded impedance section. The summation operation is simply realized by the OTA differential input. It can be verified that the voltage transfer function from the OTA input to output is equal to $g_j Z'_j = H_j$. Note that we relate the voltage transfer function H_j to the grounded impedance Z'_j. Thus the voltage transfer function realization can now become the simulation of the normal grounded impedance, which can be easily done using the inductor replacement method.

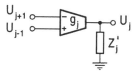

FIGURE 10.20
OTA-grounded impedance section.

Using Fig. 10.20 as a building block we can readily obtain the OTA-grounded impedance LF structure from Eq. (10.43) or Fig. 10.19, as shown in Fig. 10.21.

The grounded impedances have the values calculated by

$$Z'_1 = \frac{1}{g_1 g_m} Y_1, \quad Z'_2 = \frac{g_m}{g_2} Z_2, \quad Z'_3 = \frac{1}{g_3 g_m} Y_3, \quad Z'_4 = \frac{g_m}{g_4} Z_4,$$

$$Z'_5 = \frac{1}{g_5 g_m} Y_5, \quad Z'_6 = \frac{g_m}{g_6} Z_6 \tag{10.47}$$

From Eq. (10.47) we can see that besides the general scaling by g_m, each new grounded impedance has a separate transconductance which can be used to adjust the impedance level. We can also note

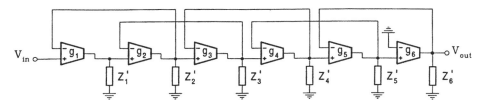

FIGURE 10.21
General LF OTA-grounded impedance realization.

that Z'_j are not the original impedances Z_j in the ladder. For the even number subscript, Z'_j is the original impedance Z_j in the parallel arm of the ladder multiplied by the ratio of g_m/g_j, while for the odd number subscript, Z'_j is the inversion of the original impedance Z_j or the admittance Y_j in the series arm divided by the product of $g_j g_m$. When $g_j = g_m = g$, we have $Z'_j = Y_j/g^2$ for odd j and $Z'_j = Z_j$ for even j. Further if $g_j = g_m = 1$, then $Z'_j = Y_j$ for odd j. Note that in many OTA-C publications, unity values of transconductances and the scaling conductance are used for simplicity.

The salient feature of the structure is that the OTA-C realization problem becomes the OTA-C realization of the grounded impedances only and the simple inductor substitution method can be conveniently used to simulate the impedance constituents. It is also noted that each OTA output node having a grounded subnetwork and the subnetwork having no extra connection with other parts of the circuit, as is the case in Fig. 10.21, can ensure the low sensitivity of the structure to the transconductance [23]. As seen from Eq. (10.47), a change in a tranconductance will change the impedance level of the grounded subnetwork connected to its output only and, according to the sensitivity property, the sensitivity of the filter function to the transconductance will be equal to the sum of all the sensitivities of the function to the constituent elements (real or simulated) in the subnetwork. Since the sensitivities to constituent passive elements are usually low or low by proper design, the sensitivity to the transconductance will also be low. This argument suits all the transconductances in Fig. 10.21. Therefore, the OTA-C filters based on Fig. 10.21 will have low sensitivity. Moreover, the original floating inductor or capacitor in the series arm can be converted into the grounded capacitor and inductor, respectively. As will be seen soon, each capacitor of the resulting OTA-C filter will have a corresponding reactive element (capacitor or inductor) in the passive counterpart, which, as demonstrated in Section 10.2, can guarantee the low sensitivity of the transfer function of the OTA-C filter to its capacitors. Thus OTA-C filters based on the method discussed above will have low sensitivity to both transconductances and capacitors.

The above method is similar to the admittance simulation method in Section 10.3 in that the floating admittances are converted into the grounded impedances using OTAs. We will prove in Section 10.5 that the OTA-C filter structures obtained using the admittance simulation method and the signal simulation method can be the same under certain conditions.

In the following we introduce some OTA-C structures derived from passive LC ladders. Since we have not put any limitations on the impedances and admittances in the general ladder, the signal simulation method is therefore suitable for arbitrary LC ladders. For simplicity, and also because of popularity, only some typical LC ladder simulations will be presented.

10.4.2 OTA-C Lowpass LF Filters

Consider the fifth-order all-pole LC ladder with termination resistors in Fig. 10.22(a). Comparing the circuit with the general ladder in Fig. 10.17 gives $Y_1 = 1/R_1$, $Z_2 = 1/sC_2$, $Y_3 = 1/sL_3$,

10.4. SIGNAL FLOW SIMULATION

$Z_4 = 1/sC_4$, $Y_5 = 1/sL_5$ and $Z_6 = \frac{1}{sC_6+1/R_6}$. The circuit equations accordingly become

$$I_1 = \frac{1}{R_1}(V_{in} - V_2), \quad V_2 = \frac{1}{sC_2}(I_1 - I_3), \quad I_3 = \frac{1}{sL_3}(V_2 - V_4),$$

$$V_4 = \frac{1}{sC_4}(I_3 - I_5), \quad I_5 = \frac{1}{sL_5}(V_4 - V_6),$$

$$V_{out} = V_6 = \frac{1}{sC_6+1/R_6} I_5 \qquad (10.48)$$

Scaling Eq. (10.48) by the factor of g_m results in voltage functions H_j given in Eq. (10.45) and realized in the way as shown in Fig. 10.20, where grounded impedances Z'_j are given in Eq. (10.46). The OTA-C filter structure is given in Fig. 10.22(b). For given R_j, C_j, and L_j, we can compute the new parameter values as

$$g'_1 = g_1 g_m R_1, \quad C'_2 = \frac{g_2}{g_m} C_2, \quad C'_3 = g_3 g_m L_3, \quad C'_4 = \frac{g_4}{g_m} C_4,$$

$$C'_5 = g_5 g_m L_5, \quad C'_6 = \frac{g_6}{g_m} C_6, \quad g'_6 = \frac{g_6}{g_m}\frac{1}{R_6} \qquad (10.49)$$

The values can be adjusted overall by g_m and individually by g_j. Two design techniques can be utilized. One is to make all transconductances identical, that is, $g_1 = g_2 = g_3 = g_4 = g_5 = g_6 = g$ with different capacitances which can be calculated from Eq. (10.49) as $C'_j = gg_m L_j$ when j is odd; $C'_j = \frac{g}{g_m} C_j$ when j is even. The other is to select the same value for all capacitances, that is, $C'_2 = C'_3 = C'_4 = C'_5 = C'_6 = C$ with different transconductances which are determined from Eq. (10.49) as $g_j = \frac{C}{g_m L_j}$, for odd j and $g_j = \frac{g_m C}{C_j}$, for even j. In many cases g_m can be chosen to be unity.

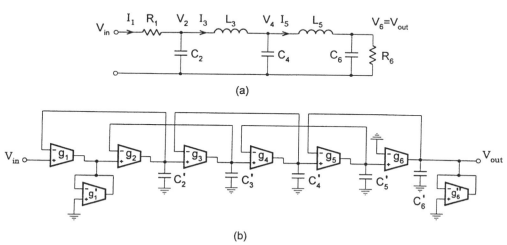

FIGURE 10.22
Fifth-order all-pole LC ladder and LF OTA-C realization.

The OTA-C filter contains only integrators and summers. Once again we see that the integrator is the basic building block in active filter design. Note that the structure requires only grounded capacitors, an advantage for integration. It is also simple, as only eight OTAs and five capacitors are needed. When $R_6 = 1/g_m$, the g'_6 termination OTA can be removed if the inverting terminal of the g_6 OTA is connected to the output. This has been used very often in the OTA-C literature such as [20]. The first paper studying the OTA-C realization of all-pole leapfrog filters based on LC ladders was published in [18], where other possibilities of terminations were also given. In the following we will first give an example to the above all-pole filter design and then look at the OTA-C architecture for the finite zero lowpass LC ladders.

Example

We design a fifth-order 1 dB-ripple 4 MHz Chebyshev filter based on the LF OTA-C simulation of the passive ladder. The fifth-order 1 dB-ripple Chebyshev lowpass filter has a normalized characteristic of (see appendix)

$$H_d(s) = \frac{0.061415}{s^5 + 0.93682s^4 + 1.68882s^3 + 0.97440s^2 + 0.58053s + 0.12283}$$

The corresponding normalized component values of the ladder in Fig. 10.22(a) are given by

$$R_1 = R_6 = 1, \quad C_2 = 2.135, \quad L_3 = 1.091,$$

$$C_4 = 3.001, \quad L_5 = 1.091, \quad C_6 = 2.135$$

We first denormalize the component values with the frequency of $f_o = 4MHz$ and the resistance of $R = 10k\Omega$ (see Chapter 2). The corresponding denormalized component values are obtained as

$$R_1 = R_6 = 10k\Omega, \quad C_2 = C_6 = 8.495pF,$$

$$L_3 = L_5 = 434.1\mu H, \quad C_4 = 11.941pF$$

Then using the formulas in Eq. (10.49) with the choice of equal transconductances

$$g_1 = g_2 = g_3 = g_4 = g_5 = g_6 = g_m = 1/R = 100\mu S$$

we obtain

$$g'_1 = g'_6 = 100\mu S, \quad C'_2 = C'_6 = 8.495pF,$$

$$C'_4 = 11.941pF, \quad C'_3 = C'_5 = 4.341pF$$

The results show that all the transconductances are equal and there are only three different capacitance values with the maximum capacitance spread (ratio) of only 2.75. Also, note that all capacitors are grounded. So this is a good design for integrated circuit implementation. Note that in this design $g_6 = g'_6$. Thus one OTA can be saved at the output end.

Now we consider a fifth-order finite zero passive LC ladder, as shown in Fig. 10.23. Similarly, identifying that

$$Y_1 = 1/R_1, \quad Z_2 = 1/sC_2, \quad Y_3 = sC_3 + \frac{1}{sL_3},$$

10.4. SIGNAL FLOW SIMULATION

$$Z_4 = 1/sC_4, \quad Y_5 = sC_5 + \frac{1}{sL_5}, \quad Z_6 = \frac{1}{sC_6 + 1/R_6} \quad (10.50)$$

and following the same design procedure we can obtain the OTA-C counterpart as shown in Fig. 10.23(b), similar to that in [20]. The difference from the all-pole type is in Y_3 and Y_5 which are a combination admittance of two components and involve two steps. Taking Y_3 as an example, we first have the corresponding grounded impedance as

$$Z_3' = \frac{Y_3}{g_3 g_m} = sL_3' + \frac{1}{sL_3 g_3 g_m} \quad (10.51)$$

where $L_3' = \frac{C_3}{g_3 g_m}$. The second term in the equation represents a capacitance of the value $C_3' = L_3 g_3 g_m$. But the first term is equivalent to an inductor. This should then be further replaced by an OTA-C inductor with $L_3' = \frac{C_3''}{g_3' g_3''}$. Combining the two steps we can also obtain C_3'' in terms of C_3. The design formulas of the OTA-C filter for all components are given below.

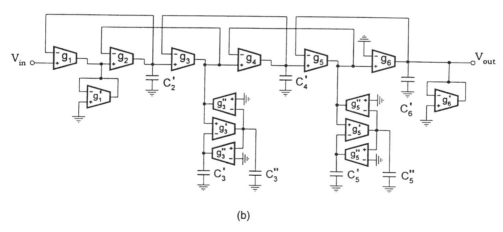

FIGURE 10.23
Fifth-order finite zero passive LC and active OTA-C LF filters.

$$g_1' = g_1 g_m R_1, \quad C_2' = \frac{g_2}{g_m} C_2, \quad C_3' = g_3 g_m L_3, \quad C_3'' = \frac{g_3' g_3''}{g_3 g_m} C_3,$$

$$C_4' = \frac{g_4}{g_m} C_4, \quad C_5' = g_5 g_m L_5, \quad C_5'' = \frac{g_5' g_5''}{g_5 g_m} C_5,$$

$$C_6' = \frac{g_6}{g_m} C_6, \quad g_6' = \frac{g_6}{g_m} \frac{1}{R_6} \quad (10.52)$$

where g_m is the scaling conductance. Similarly, using the equation the OTA-C filter can be conveniently designed to have the same transconductances or the same capacitances. If all transconductances are equal to g_m, we will have from Eq. (10.52) that $g_1' = g_m^2 R_1$, $g_6' = 1/R_6$, $C_j' = C_j$ for even j = 2, 4, 6, $C_j' = g_m^2 L_j$ for j = 3, 5 and $C_j'' = C_j$ for j = 3,5.

10.4.3 OTA-C Bandpass LF Filter Design

The complexity of the OTA-C filter based on the LF structure will depend on the number of elements in the series and shunt branches of the passive ladder circuit. The bandpass filter design may be conducted from the all-pole lowpass LC filter by applying the lowpass to bandpass frequency transformation $s \to \frac{1}{B}s + \frac{1}{sB/\omega_0^2}$, where ω_o is the center frequency and B is the bandwidth of the bandpass filter to be designed (see Chapters 2 and 6). The bandpass LC structure from the all pole lowpass prototype will typically have series resonators in series arms and parallel resonators in parallel arms. We start design from the bandpass LC ladder only and take the LF simulation of the circuit in Fig. 10.24(a) as an example. Recognizing that Y_1 is an RLC series resonator, Z_4 is a parallel RLC resonator and Z_2 and Y_3 are the ideal parallel and series LC resonators and following the same design procedure we can obtain the LF OTA-C filter structure as shown in Fig. 10.24(b). The component values can be formulated as

$$g_1''' = g_1 g_m R_1, \quad C_1' = g_1 g_m L_1, \quad C_1'' = \frac{g_1' g_1''}{g_1 g_m} C_1, \quad C_2' = \frac{g_2}{g_m} C_2,$$

$$C_2'' = \frac{g_2' g_2'' g_m}{g_2} L_2, \quad C_3' = g_3 g_m L_3, \quad C_3'' = \frac{g_3' g_3''}{g_3 g_m} C_3, \quad C_4' = \frac{g_4}{g_m} C_4,$$

$$C_4'' = \frac{g_4' g_4'' g_m}{g_4} L_4, \quad g_4''' = \frac{g_4}{g_m} \frac{1}{R_4} \quad (10.53)$$

where g_m is the scaling conductance. Further design can be carried out based on the equation.

10.4.4 Partial Floating Admittance Block Diagram and OTA-C Realization

The partial floating admittance concept can offer more flexibility in OTA-C filter realizations based on passive LC ladders, as discussed in Section 10.3.3. This concept also suits the signal simulation approach. Consider the general ladder in Fig. 10.25. We want to leave admittances Y_{j1} floating and simulate the nodal voltages and the currents flowing in admittances Y_{j2} (not the total currents in the series arms). The equations for the whole ladder in Fig. 10.25 can be written as

$$I_1 = Y_{12}(V_{in} - V_2), \quad V_2 = Z_2[(I_1 - I_3) + Y_{11}(V_{in} - V_2) - Y_{31}(V_2 - V_4)],$$

$$I_3 = Y_{32}(V_2 - V_4), \quad V_4 = Z_4[(I_3 - I_5) + Y_{31}(V_2 - V_4) - Y_{51}(V_4 - V_6)],$$

$$I_5 = Y_{52}(V_4 - V_6), \quad V_{out} = V_6 = Z_6[I_5 + Y_{51}(V_4 - V_6)] \quad (10.54)$$

Scaling Eq. (10.54) by conductance g_m and denoting $V_j' = I_j/g_m$ we can obtain

$$V_1' = \frac{Y_{12}}{g_m}(V_{in} - V_2),$$

10.4. SIGNAL FLOW SIMULATION

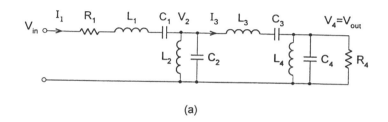

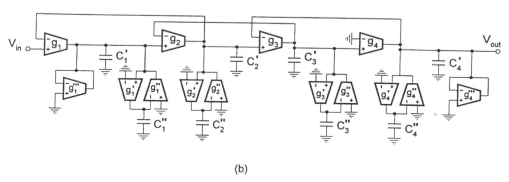

FIGURE 10.24
Eighth-order bandpass LC and LF OTA-C filter.

$$V_2 = g_m Z_2 \left[(V_1' - V_3') + \frac{Y_{11}}{g_m}(V_{in} - V_2) - \frac{Y_{31}}{g_m}(V_2 - V_4) \right],$$

$$V_3' = \frac{Y_{32}}{g_m}(V_2 - V_4), \quad V_4 = g_m Z_4 \left[(V_3' - V_5') \right.$$

$$\left. + \frac{Y_{31}}{g_m}(V_2 - V_4) - \frac{Y_{51}}{g_m}(V_4 - V_6) \right],$$

$$V_5' = \frac{Y_{52}}{g_m}(V_4 - V_6),$$

$$V_{out} = V_6 = g_m Z_6 \left[V_5' + \frac{Y_{51}}{g_m}(V_4 - V_6) \right] \tag{10.55}$$

In Fig. 10.26 we give the corresponding OTA-grounded impedance LF structure with floating admittances, where g_j is the OTA transconductance and $g_2 = g_4 = g_6$, which simulates the relations in Eq. (10.55). The design formulas for the grounded impedances Z_j', $j = 1, 2, \ldots, 6$, are the same as those in Eq. (10.47) in form, except that Y_1, Y_3 and Y_5 should be replaced by Y_{12}, Y_{32} and Y_{52}, respectively, and $g_2 = g_4 = g_6$. The design formulas for the floating admittances Y_{11}', Y_{31}' and Y_{51}' are given below:

$$Y_{11}' = \frac{g_2}{g_m} Y_{11}, \quad Y_{31}' = \frac{g_4}{g_m} Y_{31}, \quad Y_{51}' = \frac{g_6}{g_m} Y_{51} \tag{10.56}$$

To illustrate the method we consider the fifth-order finite zero LC ladder in Fig. 10.27(a). Assigning $Y_{11} = 0$, $Y_{12} = 1/R_1$, $Y_{31} = sC_3$, $Y_{32} = \frac{1}{sL_3}$, $Y_{51} = sC_5$ and $Y_{52} = \frac{1}{sL_5}$, and using the above

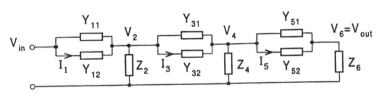

FIGURE 10.25
General ladder illustrating partial floating admittance.

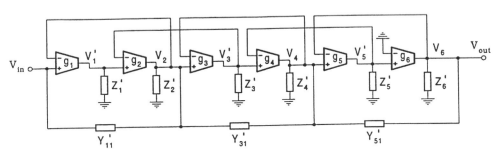

FIGURE 10.26
OTA-grounded impedance version with partial floating admittance.

method we can obtain the OTA-C filter shown in Fig. 10.27(b), similar to those in [4, 20, 21]. The values for g'_1, C'_j, and g'_6 can be calculated using Eq. (10.49). The new values for the two floating capacitances can be computed from Eq. (10.56), given by $C''_3 = \frac{g_4}{g_m} C_3$ and $C''_5 = \frac{g_6}{g_m} C_5$. Note again that there should be $g_2 = g_4 = g_6$.

For the R_1 arm we can of course assign $Y_{11} = 1/R_1$ and $Y_{12} = 0$, which means that the whole resistor is left floating. In this case the simulation structure can be simplified at the input end, requiring fewer OTAs and not having the resistive node. This consideration is in fact suitable for all the cases in which Y_1 is a resistor and is often used [21, 22, 40]. We leave this for the reader to investigate.

Note that the OTA-C simulation with floating capacitors requires fewer OTAs, but may occupy more chip area and increase parasitic effects due to floating capacitors. Also, unlike Section 10.4.1 the sensitivity to the OTA transconductance may increase due to the coupling from the floating capacitors.

10.4.5 Alternative Leapfrog Structures and OTA-C Realizations

Most active filters are composed of integrators with coupling and in most cases such coupling can be explained by feedback theory. Understanding the filter structure in this way is very beneficial, as will be seen more generally in Section 10.7 and Chapter 11. The leapfrog configuration we have discussed so far is very convenient and straightforward for description from the feedback viewpoint, as all the coupling (feedback) paths are on the upper side and impedances in their natural positions. This is especially true when feedforward techniques are used to produce transmission zeros, since thus we can have the very convenient arrangement in that all feedback is on one side and feedforward on the other, as was shown in Figs. 10.26 and 10.27 and will be seen in Chapter 11.

Many other alternative (equivalent) forms of the simulation structure of the general ladder can be obtained. The original form of the LF signal flow graph [49, 53] for the general ladder is slightly different from the one in Fig. 10.18 in that coupling appears on both upper and lower sides, as was presented in Chapters 5 and 6. Chapter 6 also drew the LF structure in a different way.

10.4. SIGNAL FLOW SIMULATION

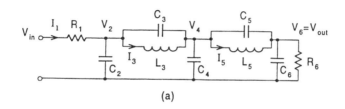

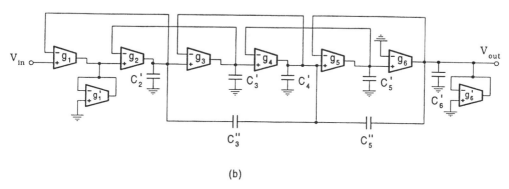

FIGURE 10.27
Fifth-order finite zero LF OTA-C filter with partial floating admittance.

In Fig. 10.28 we introduce another two alternative forms of the simulation structure of the general ladder in Fig. 10.17, all of which can be obtained by simply rearranging the LF structure in Fig. 10.18. Figure 10.28(a) can also be found in other books such as [2]. It can be easily verified that the two alternatives have the same signal relations as Eq. (10.42). We also show four interesting alternative OTA-grounded impedance realizations in Fig. 10.29. Note that we assume that the scaling conductance $g_m = 1$ and the OTAs have unity transconductances and thus the values of the grounded impedances Z'_j are equal to Y_j and Z_j for odd and even j, respectively. Again, it can be verified that they are the same as the structure in Fig. 10.21. The OTA-C simulation can be similarly conducted by further simulating the grounded impedances using the inductor substitution method.

References [19, 20] investigated the OTA-C realization of Fig. 10.28(a) and Fig. 10.29(a). Only single-input and single-output transconductors were initially used [19], usually resulting in complex filter structures with a large number of transconductors, although some very simple four transistor transconductors [45] are available which may give similar complexity at the transistor level, compared with the differential input OTA counterparts. The authors realized this and attempted to reduce the number of active devices based on the voltage and current (controlled) source shift theorem for the ladders with the capacitor loop and inductor cut-set, respectively, [20]. This can also be done by using the well known fact that any two single-input OTAs with equal transconductances and opposite input polarities, whose outputs are connected to the same node, can be equivalently replaced by a single differential-input OTA with the same transconductance [18]. The papers in [21]–[25] are mainly about OTA-C realization of the second alternative OTA-grounded impedance version in Fig. 10.29(b). The lowpass OTA-C filters in [28, 29] belong to the third type form in Fig. 10.29(c). We will later present an example for the fourth alternative OTA-grounded impedance version in Fig. 10.29(d). We must emphasize that all the LF forms are equivalent; it is just a matter of drawing. This also means that we have shown that the OTA-C structures given in the literature are all equivalent, although superficially they may look quite different. It should also be noted that all the LF structures can be seen as the cross-cascade interconnection.

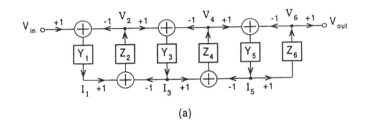

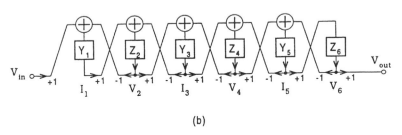

FIGURE 10.28
Alternative LF block diagrams.

The partial floating admittance concept is suitable for all the alternative LF structures, simply leaving the floating part in site and simulating the other part. Again, consider the fifth-order finite zero lowpass LC ladder filter in Fig. 10.27(a). Using the fourth alternative form in Fig. 10.29(d) and the partial floating concept we can obtain the OTA-C equivalent in Fig. 10.30, where we select the unity scaling conductance and OTA transconductances.

10.5 Equivalence of Admittance and Signal Simulation Methods

In the above we have discussed two major methods: the admittance simulation and the signal simulation. Both methods first convert, or partially convert, the admittances in the series arms into the grounded impedances and then use the inductor substitution technique to realize the grounded impedances. The only difference is that the admittance simulation method is by substitution of the floating admittance with an OTA-grounded impedance circuit, while the signal simulation approach is based on the simulation of the circuit equations with the building block of the OTA-grounded impedance voltage section. This observation suggests that there may be some relation between the two methods, and the superficial similarity between the OTA-C filters derived from the two methods seemingly also supports this. We now try to prove that under certain conditions the two methods are equivalent.

It is clear that the general proof of the equivalence must be done for the first-level simulation, since the two methods use the same technique for the second-level simulation. We come back to the general OTA-grounded impedance structure obtained by the admittance simulation method in Fig. 10.12, which is redrawn in Fig. 10.31(a) for reference. Circuit node A is associated with the outputs of two single-input and opposite-polarity OTAs, the g_2 and g_4 OTAs. If they have equal transconductance values, $g_2 = g_4 = g_m$, then the two OTAs can be combined into one differential-input OTA, the g_m OTA, in the way shown in Fig. 10.31(b). It can be shown easily that the current to the node from this OTA will be equal to the total current from the original two OTAs and the filter

10.5. EQUIVALENCE OF ADMITTANCE AND SIGNAL SIMULATION

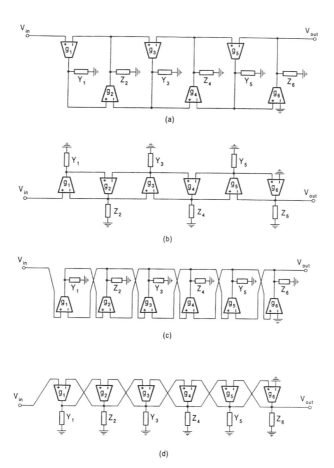

FIGURE 10.29
Alternative LF OTA-grounded impedance structures.

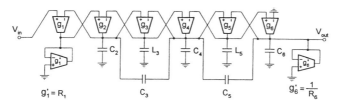

FIGURE 10.30
Finite zero LP OTA-C filter based on alternative LF structure and partial admittance conversion.

function will thus remain the same. In the same way we can do a simplification for all the other nodes. Such a manipulation at node B in Fig. 10.31(a) will require $g_4 = g_6$. Thus we can see that if all related nodes are concerned, then all the OTAs with even-number subscripts must have the same transconductance. For Fig. 10.31(a) we should have $g_2 = g_4 = g_6 = g_m$. The new structure resulting from this treatment is shown in Fig. 10.31(b), noting that the first OTA at the input end can be discarded for ideal voltage input without influence on the filter function because the ideal input voltage source can have arbitrary current. Now let us look at the signal simulation structure in Fig. 10.21. If $g_2 = g_4 = g_6 = g_m$, Fig. 10.21 will become exactly the same as Fig. 10.31(b). This means that we have proved that under the condition of $g_2 = g_4 = g_6 = g_m$, the admittance simulation structure and the signal simulation structure will be identical. From this equivalence we therefore stress that the admittance simulation and the signal simulation methods are equally useful for continuous-time integrated OTA-C filter design.

We can appreciate the above demonstration from the signal viewpoint only. In fact, the admittance simulation is the lowest-level signal simulation, the simulation of the Ohm's law equation which involves the current and voltage of the branch (series arm) only, having a local feature. The signal (LF) simulation is carried out at a high level, which is the simulation of Kirchhoff's law equations involving currents and voltages of different branches, having a global feature. When we consider the node currents for the admittance simulation configuration, that means that we also consider Kirchhoff's law relations. This is why the simplified structure obtained in this way can be equivalent to the configuration attained using the signal flow simulation method. This argument also leads us to conclude that OTA-C simulation can be based on even higher-level circuit equations such as the two-port network equations and the nodal equations, which is the topic of the next section.

We can also discuss these methods from the admittance standpoint only. The admittance simulation is the simulation of the individual admittance. In the signal simulation we can put the Kirchhoff equations in a form of admittance matrix and therefore we may say that signal simulation is a generalized admittance simulation, a simulation of the admittance matrix. When dealing with the currents of a node as in the above proof we deal equivalently with the admittances related to that node or handle some admittance matrix, no longer a single admittance. This is why we can reach the equivalence. From this discussion we can also realize that OTA-C filter design from passive LC ladders can be based on the matrix form of various equations, which will be introduced in the following section.

10.6 OTA-C Simulation of LC Ladders Using Matrix Methods

Another category of very useful but more complex methods for active filter design based on passive LC ladder simulation is based on the matrix descriptions. The first matrix method is based on the two-port transmission matrix equation [55]. The fundamental principle as discussed in Chapter 7 is the linear transformation of port variables of a network from the voltage and current domain to a new voltage domain in which active realizations are conducted. The design procedure of this approach for use in OTA-C filter design is now summarized as below:

1. Divide the passive LC ladder into component sections in cascade and write down their chain matrices.

2. Transform the voltage and current terminal variables into some new voltage variables by choosing the transformation matrices and write down the four voltage transfer functions.

10.6. OTA-C SIMULATION OF LC LADDERS

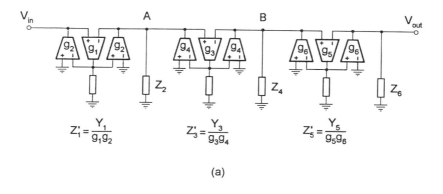

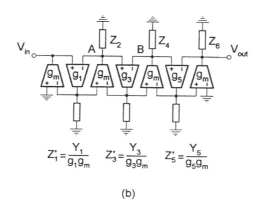

FIGURE 10.31
Admittance simulation structure (a) and its simplified equivalent (b).

3. The resulting voltage transfer functions of each section are realized using OTAs and capacitors and their building blocks.

4. Connect all the designed sections together according to the interconnection rules. If the interconnection matrix is not a 0-1 matrix for the chosen transformation matrices, then some interconnection coefficients may also need to be realized using OTAs.

This method is a combination of the cascade and simulation methods. It is general; the leapfrog structure and the wave active filter (see Chapter 7) appear as its special cases. This method can transform the original impedance/admittance ladder into many forms of structure by properly choosing the transformation matrices. In the literature only a very special case of this method has been discussed to show how to use this method to derive OTA-C filters [28, 29]. Further investigation into the application of the two-port matrix method in OTA-C filter design is needed.

Another powerful matrix method for active simulation of passive ladders has been recently proposed in [56]. This new method has been successfully used in active RC, switched-capacitor, and OTA-C filter design. The design procedure of this method for OTA-C filters is as follows [30, 31]:

1. A set of nodal equations that describe the passive prototype network is written in a matrix equation form as

$$[J] = \left([G] + s[C] + s^{-1}[\Gamma]\right)[V] \qquad (10.57)$$

where $[V]$ is a vector representing the nodal voltages, $[J]$ is a vector representing the input current sources and $[G]$, $[C]$, and $[\Gamma]$ are matrices whose elements are simple algebraic combinations of respective passive conductance, capacitance and inverted inductance values.

2. The second-order matrix equation in Eq. (10.57) is decomposed into two first-order design equations by introduction of a vector of auxiliary variables. A large number of decompositions are possible for OTA-C filters. The choice of decomposition for a particular filter design is dictated by the type of building blocks available and the nature of the desired response.

3. To form the active OTA-C filter, each row of each design equation is implemented by a first-order OTA/capacitor section.

We should stress that the two-port matrix method (due to the large number of section divisions and transformations of the passive prototype) and the nodal matrix method (due to the large number of possible decompositions of the total matrix equation of the passive prototype) can generate many useful active filter structures. The conventional simulation structures such as the LF can be derived as a special case corresponding to a particular interconnection and decomposition for the two-port and nodal methods, respectively. We must say that the two matrix methods introduced above are very useful for exploration of new simulation structures and complex design implementations. Because these methods are quite mathematically intensive and highly specialized, we will not discuss the details of the methods. The reader is strongly encouraged to refer to the relevant papers for more information.

10.7 Coupled Biquad OTA Structures

In the above we have obtained several useful OTA-C filter structures from simulation of passive LC ladders. The LF structure may be generalized in the way that the impedance/admittance sections including the associated OTAs are replaced by biquadratic filters. The coupling or feedback of the biquads can also be different. This leads to an independent and systematic study area, which is called the coupled biquad filters. As discussed in Chapter 6, in coupled biquad filter design ideal lossless active resonators are often required. Many bandpass filters in Chapters 8 and 9 can be designed to have an infinite pole Q, equivalent to an ideal active resonator.

Besides the LF structure, there are also several other coupled biquad configurations such as the follow-the-leader feedback (FLF) and inverse FLF (IFLF), which are also called multiple-loop feedback filters, as in Chapter 5. It can be proved that the FLF and IFLF are the adjoints of each other. Figure 10.32 shows the diagram of the OTA-C FLF structure, where biquads can be the OTA-C resonator shown in Section 10.2 or one of those generated in Chapters 8 and 9. This method is modular, since it is based on biquads similar to the cascade approach, and has low sensitivity due to complex interstage coupling similar to the signal flow simulation.

Similar to the active RC counterpart [2, 57, 59], denoting the input coefficient by $\alpha = g_{a0}/g_A$, the transfer function of the jth biquad by $T_j(s)$, j = 1, 2, ..., n, feedback coefficient by $F_i = g_{ai}/g_A$, i = 1, 2, ... n, and feedforward coupling coefficient by $K_i = g_{bi}/g_B$, i = 0, 1, 2, ..., n, we can write the following equations:

$$-V_0 = \alpha V_{\text{in}} + \sum_{i=1}^{n} F_i V_i \qquad (10.58)$$

10.7. COUPLED BIQUAD OTA STRUCTURES

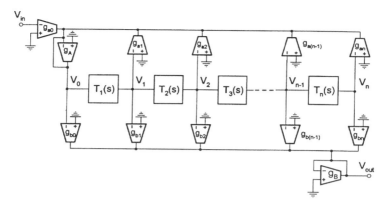

FIGURE 10.32
General OTA-C FLF structure.

$$V_i = V_0 \prod_{j=1}^{i} T_j(s), \quad i = 1, 2, \ldots, n \qquad (10.59)$$

$$V_{\text{out}} = -\sum_{i=0}^{n} K_i V_i \qquad (10.60)$$

From Eqs. (10.58) and (10.59) we can obtain

$$H_0(s) = \frac{V_0}{V_{\text{in}}} = -\alpha \frac{1}{1 + \sum_{k=1}^{n} \left[F_k \prod_{j=1}^{k} T_j(s) \right]} \qquad (10.61)$$

$$H_i(s) = \frac{V_i}{V_{\text{in}}} = -\alpha \frac{\prod_{j=1}^{i} T_j(s)}{1 + \sum_{k=1}^{n} \left[F_k \prod_{j=1}^{k} T_j(s) \right]} \qquad (10.62)$$

and

$$H_n(s) = \frac{V_n}{V_{\text{in}}} = -\alpha \frac{\prod_{j=1}^{n} T_j(s)}{1 + \sum_{k=1}^{n} \left[F_k \prod_{j=1}^{k} T_j(s) \right]} \qquad (10.63)$$

Then using Eqs. (10.60) and (10.61–10.63) we attain the overall transfer function as

$$H(s) = \frac{V_{\text{out}}}{V_{\text{in}}} = \alpha \frac{K_0 + \sum_{k=1}^{n} \left[K_k \prod_{j=1}^{k} T_j(s) \right]}{1 + \sum_{k=1}^{n} \left[F_k \prod_{j=1}^{k} T_j(s) \right]} \qquad (10.64)$$

For $T_j(s) = \tau_j s$, that is, a voltage integrator, a general nth-order transfer function can be obtained immediately. More generally T_j blocks can be OTA-C lossy integrators, first-order filters, or biquads. The first- and second-order OTA-C filters developed in Chapters 8 and 9 can be used here. In Reference [36], for example, an all-pole bandpass OTA-C FLF filter using the canonical two integrator loop bandpass biquad was designed. Very few papers on coupled biquad OTA-C filters have been published [36]–[38]. There is still much work that needs to be done to enhance the use of the method in continuous-time integrated OTA-C filter design.

10.8 Some General Practical Design Considerations

To design low-sensitivity, high-frequency, and large dynamic range OTA-C filters, many practical factors should be considered carefully. While we can use high-quality components (for example, high-frequency, low-noise, highly linear, and simple OTAs) to enhance the filter performance, in this section we will however discuss the issues from the viewpoint of filter structures and design methods.

10.8.1 Selection of Capacitors and OTAs

In some implementations, grounded capacitors are preferred, because they need less chip area and can absorb parasitic capacitances. We can convert the floating capacitor to the grounded capacitor by using extra OTAs. The trade-off between the increased number of OTAs and the use of floating capacitors however may need to be considered when choosing the structure, as the attempt of grounding capacitors may require too many extra OTAs which may balance out the advantages from grounding these capacitors. It should also be cautioned that for finite-zero LP filters, the effort of conversion will result in suspending internal nodes which are vulnerable to parasitics. In the balanced implementation as will be discussed in Chapter 12, splitting a single floating capacitor into two grounded may not reduce the total chip area, because the total capacitance of the two grounded capacitors will be four times the value of the floating capacitance.

The realizations using only single-input and single-output OTAs may require too many OTAs, although they may have some advantages of reduction in feedthrough effects. Therefore in the simulation design, the differential-input application of the OTA is very popular. Generally, any two single-input OTAs with equal transconductances and opposite input polarities, whose outputs are connected to the same node, can be equivalently replaced by a single differential-input OTA with the same transconductance. This results in simple circuitry and possibly reduced power consumption and chip area.

The simulation method can ensure that all but one, or at most two, transconductances are identical. Identical g_m values can make on-chip tuning, design, and layout much easier, since a free transconductance cell can be used throughout the circuit [3, 19]. Equal capacitance values may be achieved with different values of transconductances, which is quite popular in discrete design. But in IC implementation it seems that identical transconductances are more important as reasoned above.

In high-frequency design, the circuit capacitance is very small (the time constant C/g is small). The smallest design capacitance must be larger than the total of all parasitic capacitances on the same node for the circuit capacitance to absorb all the parasitic capacitances. For very high frequencies, this may not always be possible and the OTA parasitic capacitances are quite often used as the design capacitance, although they are less controllable. Because of the relatively large parasitic capacitances, the grounded capacitor may no longer be advantageous for very high-frequency design.

In practice, component value determination should also meet the requirements concerning dynamic range, noise, and power consumption. To optimize the filter dynamic range, for example, all integrator output swings should be made equal. The signal simulation method has the possibility of scaling the component values such that the circuits have the maximum possible dynamic range. This scaling is not normally available in the component substitution design.

10.8.2 Tolerance Sensitivity and Parasitic Effects

The simulation method can retain the low-sensitivity property of passive LC ladders, because the one-to-one correspondence of components or signals between the passive filter and active counterpart is preserved. A caution may be given to the block-based admittance substitution or signal simulation methods. In some cases when the admittance or impedance block in the general ladder is too complex, the one-to-one correspondence between the active filter and the associated passive filter may not be retained so well and thus the sensitivity of the derived active filter may not be guaranteed to be as low as that of the passive LC counterpart.

Because the parasitic sensitivity is greater than the tolerance sensitivity in passive LC ladders, it is especially important to keep the parasitic sensitivity low. To decrease the effects of the parasitic conductances and capacitances on the filter magnitude frequency response we must reduce the inductance shift and resistive loss due to these parasitics. It is very important to note that the magnitude response change due to the parasitic parameters is proportional to the phase and magnitude tolerance sensitivities to the inductance. The passive LC ladders have very low magnitude tolerance sensitivity, but the phase tolerance sensitivity may not be low, which should receive attention when dealing with the equivalent parasitic loss resistor. It should also be noted that the parasitic effects will become worse as the frequencies increase.

10.8.3 OTA Finite Impedances and Frequency-Dependent Transconductance

It is often said that OTA-C filters are especially subjected to parasitics. The capability of absorbing the parasitics is thus necessary. OTA finite impedances will cause the performance change, as we have already discussed in Section 10.2, Chapters 8 and 9. The differential input application of the OTA may cause the capacitive coupling in the circuit due to the floating finite input impedance. The resistive node that has only a grounded OTA resistor may present an extra pole by parasitic capacitances. The suspending node that does not have any grounded resistor or capacitor is vulnerable to both parasitic capacitances and resistances. In the simulation of LC ladders, most structures can well absorb OTA capacitances and thus no parasitic poles and zeros can be produced. But in some cases, for example, simulating the lowpass ladder filters of finite transmission zeros, using only grounded capacitors often causes some suspending nodes. Note in particular that in integrated circuits the design capacitance is only one or two orders of magnitude higher than, or may even well be in the similar value range of, parasitic capacitances at very high frequencies. The parasitic poles may thus be quite near the cutoff frequency [24]. The best structures should be those in which each internal circuit node has a grounded capacitor, since absorption design of parasitic capacitances can be possible.

The excess phase of the OTA will also pose a stability problem at higher frequencies. For the given OTA, to extend the working frequency of the filter, we must overcome the finite bandwidth effects. The passive compensation technique using a MOSFET resistor and the active compensation method based on the two OTAs of different transconductances, which are connected in parallel with opposite polarities, can be used as discussed in Chapter 9.

10.9 Summary

This chapter has introduced methods for design of high-order OTA-C filters based on doubly terminated passive LC ladders. Different OTA-C filter architectures have been obtained by using the component substitution, admittance simulation, signal flow simulation, and coupled biquad methods. The most outstanding advantage of this class of OTA-C filters is their low passband sensitivity performance. However, their design and tuning procedures are more complex compared with the cascade method, and more complex structures are often required. In most high-order filters with stringent requirements, the sensitivity advantage usually prevails, making it desirable to use a ladder-based method.

The inductance substitution method is one of the most popular methods used for OTA-C filter design. The Bruton Transformation method may not be suitable for OTA-C filter design because no advantages could be achieved, especially in terms of the number of components. The two-step admittance simulation method is appropriate to OTA-C filters. The LF and their alternative structures resulting from signal simulation are most widely utilized for OTA-C filter design. We have shown that the different forms of LF structures all are equivalent, thus giving the same performance. We have also generally demonstrated that the admittance substitution and signal simulation methods are equivalent in OTA-C filter design under certain conditions. These two methods both are popular in OTA-C filter design. The matrix signal flow simulation methods and coupled biquad approach are more complex, but they can realize more functions and the latter has also a modular feature.

It should be pointed out that some insights into the simulation of passive LC ladders presented in this chapter are general. They may also be suitable for other types of active filters, although they are discussed for OTA-C filter design. In Chapter 11 we will introduce another useful method for the design of high-order OTA-C filters. Current-mode OTA-C filter design based on ladder simulation will be dealt with in Chapter 12.

References

[1] Tsividis, Y.P. and Voorman, J.O., Eds., *Integrated Continuous-Time Filters: Principles, Design and Applications,* IEEE Press, 1993.

[2] Schaumann, R., Ghausi, M.S., and Laker, K.R., *Design of Analog Filters: Passive, Active RC and Switched Capacitor,* Prentice-Hall, Englewood Cliffs, NJ, 1990.

[3] Schaumann, R., Continuous-time integrated filters, in *The Circuits and Filters Handbook,* Chen, W.K., Ed., CRC Press, Boca Raton, FL, 1995.

[4] Schaumann, R. and Tan, M.A., Continuous-time filters, in *Analogue IC Design: The Current-Mode Approach,* Toumazou, C., Lidgey, F.J., and Haigh, D.G., Eds., Peter Peregrinus Ltd., London, 1990.

[5] Laker, K.R. and Sansen, W., *Design of Analog Integrated Circuits and Systems,* McGraw-Hill, New York, 1994.

[6] Kardontchik, J.E., *Introduction to the Design of Transconductor-Capacitor Filters,* Kluwer Academic Publishers, Massachusetts, 1992.

[7] Bialko, M. and Newcomb, R.W., Generation of all finite linear circuits using the integrated DVCCS, *IEEE Trans. Circuit Theory,* 18, 733–736, 1971.

[8] Geiger, R.L. and Sánchez-Sinencio, E., Active filter design using operational transconductance amplifiers: a tutorial, *IEEE Circuits and Devices Magazine,* 20–32, Mar. 1985.

[9] Loh, K.H., Hiser, D.L., Adams, W.J., and Geiger, R.L., A versatile digitally controlled continuous-time filter structure with wide range and fine resolution capability, *IEEE Trans. Circuits Syst.,* 39, 265–276, 1992.

[10] Park, S. and Schaumann, R., Design of a 4-MHz analog integrated CMOS transconductance-C bandpass filter, *IEEE J. Solid-State Circuits,* 23(4), 987–996, 1988.

[11] Silva-Martinez, J., Steyaert, M.S.J., and Sansen, W., A 10.7-MHz 68-dB SNR CMOS continuous-time filter with on-chip automatic tuning, *IEEE J. Solid-State Circuits,* 27, 1843–1853, 1992.

[12] Dehaene, W., Steyaert, M.S.J., and Sansen, W., A 50-MHz standard CMOS pulse equalizer for hard disk read channels, *IEEE J. Solid-State Circuits,* 32(7), 977–988, 1997.

[13] Sun, Y. and Fidler, J.K., Some design methods of OTA-C and CCII-RC active filters, *Proc. 13th IEE Saraga Colloquium on Digital and Analogue Filters and Filtering Systems,* 7/1–7/8, 1993.

[14] Nawrocki, R., Building set for tunable component simulation filters with operational transconductance amplifiers, *Proc. Midwest Symp. Circuits and Systems,* 227–230, 1987.

[15] Moulding, K.W., Quartly, J.R., Rankin, P.J., Thompson, R.S., and Wilson, G.A., Gyrator video filter IC with automatic tuning, *IEEE J. Solid-State Circuits,* SC-16, 963–968, 1980.

[16] Krummenacher, F. and Joehl, N., A 4-MHz CMOS continuous-time filter with on-chip automatic tuning, *IEEE J. Solid-State Circuits,* 23(3), 750–758, 1988.

[17] Wang, Y.T. and Abidi, A.A., CMOS active filter design at very high frequencies, *IEEE J. Solid-State Circuits,* 25(6), 1562–1574, 1990.

[18] Nawrocki, R., Electronically tunable all-pole low-pass leapfrog ladder filter with operational transconductance amplifier, *Intl. J. Electronics,* 62(5), 667–672, 1987.

[19] Tan, M.A. and Schaumann, R., Simulating general-parameter filters for monolithic realization with only transconductance elements and grounded capacitors, *IEEE Trans. Circuits Syst.,* 36(2), 299–307, 1989.

[20] Tan, M.A. and Schaumann, R., A reduction in the number of active components used in transconductance grounded capacitor filters, *Proc. IEEE Intl. Symp. Circuits Syst.,* 2276–2278, 1990.

[21] de Queiroz, A.C.M., Caloba, L.P., and Sánchez-Sinencio, E., Signal flow graph OTA-C integrated filters, *Proc. IEEE Intl. Symp. Circuits Syst.,* 2165–2168, 1988.

[22] Caloba, L.P., de Queiroz, A.C.M., and Sánchez-Sinencio, E., Signal-flow graph OTA-C bandpass and band-reject integrated filters, *Proc. IEEE Intl. Symp. Circuits Syst.,* 1624–1627, 1989.

[23] Caloba, L.P. and de Queiroz, A.C.M., OTA-C simulation of passive filters via embedding, *Proc. IEEE Intl. Symp. Circuits Syst.,* 1083–1086, 1989.

[24] de Queiroz, A.C.M. and Caloba, L.P., Some practical problems in OTA-C filters related with parasitic capacitances, *Proc. IEEE Intl. Symp. Circuits Syst.,* 2279–2282, 1990.

[25] de Queiroz, A.C.M. and Caloba, L.P., OTA-C filters derived from unbalanced lattice passive structures, *Proc. IEEE Intl. Symp. Circuits Syst.,* 2256–2256C, 1993.

[26] Chang, Z.Y., Haspeslagh, D., and Verfaillie, J., A highly linear CMOS G_m-C bandpass filter with on-chip frequency tuning, *IEEE J. Solid-State Circuits,* 32(3), 388–397, 1997.

[27] Gopinathan, V., Tsividis, Y.P., Tan, K.S., and Hester, R.K., Design considerations for high-frequency continuous-time filters and implementation of an antialiasing filter for digital video, *IEEE J. Solid-State Circuits,* 25(6), 1368–1378, 1990.

[28] Hwang, Y.S., Liu, S.I., Wu, D.S., and Wu, Y.P., Table-based linear transformation filters using OTA-C techniques, *Electron. Lett.,* 30, 2021–2022, 1994.

[29] Hwang, Y.S., Chiu, W., Liu, S.I., Wu, D.S., and Wu, Y.P., High-frequency linear transformation elliptic filters employing minimum number of OTAs, *Electron. Lett.,* 31, 1562–1564, 1995.

[30] Greer, N.P.J., Henderson, R.K., Li, P., and Sewell, J.I., Matrix methods for the design of transconductor ladder filters, *IEE Proc.-Circuits Devices Syst.,* 141, 89–100, 1994.

[31] Yue, L., Greer, N.P.J., and Sewell, J.I., Efficient design of ladder-based transconductor-capacitor filters and equalizers, *IEE Proc. Circuits Devices Syst.,* 142, 263–272, 1995.

[32] Sun, Y. and Fidler, J.K., OTA-C realization of general high-order transfer functions, *Electron. Lett.,* 29, 1057–1058, 1993.

[33] Sun, Y. and Fidler, J.K., Synthesis and performance analysis of a universal minimum component integrator-based IFLF OTA-grounded capacitor filter, *IEE Proceedings: Circuits, Devices and Systems,* 143, 107–114, 1996.

[34] Sun, Y. and Fidler, J.K., Structure generation and design of multiple loop feedback OTA-grounded capacitor filters, *IEEE Trans. Circuits Syst., Part-I,* 43, 1–11, 1997.

[35] Nedungadi, A.P. and Geiger, R.L., High-frequency voltage-controlled continuous-time lowpass filter using linearized CMOS integrators, *Electron. Lett.,* 22, 729–731, 1986.

[36] Nawrocki, R., Electronically controlled OTA-C filter with follow-the-leader-feedback structures, *Intl. J. Circuit Theory and Applications,* 16, 93–96, 1988.

[37] Krummenacher, F., Design considerations in high-frequency CMOS transconductance amplifier capacitor (TAC) filters, *Proc. IEEE Intl. Symp. Circuits Syst.,* 100–105, 1989.

[38] Gonuleren, A.N., Kopru, R., and Kuntman, H., Multiloop feedback bandpass OTA-C filters using quads, *Proc. Euro. Conf. Circuit Theory and Design,* 2, 607–610, Istanbul, 1995.

[39] Szczepanski, S. and Schaumann, R., Nonlinearity-induced distortion of the transfer function shape in high-order OTA-C filters, *Analog Integrated Circuits and Signal Processing,* 3(2), 143–151, 1993.

[40] Sun, Y., Jefferies, B., and Teng, J., Universal third-order OTA-C filters, *Intl. J. Electronics,* 85(5), 597–609, 1998.

[41] Sun, Y. and Fidler, J.K., Current-mode multiple loop feedback filters using dual output OTAs and grounded capacitors, *Intl. J. of Circuit Theory and Applications,* 25, 69–80, 1997.

[42] Sun, Y. and Fidler, J.K., Current-mode OTA-C realisation of arbitrary filter characteristics, *Electron. Lett.,* 32, 1181–1182, 1996.

[43] Ramirez-Angulo, J. and Sánchez-Sinencio, E., High frequency compensated current-mode ladder filters using multiple output OTAs, *IEEE Trans. Circuits Syst. - II,* 41, 581–586, 1994.

[44] Sun, Y. and Fidler, J.K., Structure generation of current-mode two integrator loop dual output-OTA grounded capacitor filters, *IEEE Trans. Circuits Syst., Part II,* 43, 659–663, 1996.

[45] Park, C.S. and Schaumann, R., A high-frequency CMOS linear transconductance element, *IEEE Trans. Circuits Syst.*, 33(11), 1132–1138, 1986.

[46] Milic, L. and Fidler, J.K., Comparison of effects of tolerance and parasitic loss in components of resistively terminated LC ladder filters, *IEE Proceedings-G*, 128(2), 87–90, 1981.

[47] Fidler, J.K., Calculation of the effects of component parasitics on circuit response, *Proc. 4th Intl. Symp. Network Theory*, 175–179, Ljubljiana, Yugoslavia, 1979.

[48] Fidler, J.K., Sensitivity assessment of parasitic effects in second-order active-filter configurations, *IEE Proc. Electronic Circuits and Systems*, 2(6), 181–185, 1978.

[49] Daryanani, G., Principles of active network synthesis, John Wiley & Sons, 1976.

[50] Chen, W.K., *Passive and Active Filters: Theory and Implementations*, John Wiley & Sons, 1986.

[51] Antoniou, A., Realization of gyrators using operational amplifiers, and their use in RC-active-network synthesis, *Proc. IEE*, 116(11), 1838–1850, 1969.

[52] Bruton, L.T., Network transfer functions using the concept of frequency-dependent negative resistance, *IEEE Trans. Circuit Theory*, 16, 406–408, 1969.

[53] Girling, F.E.J. and Good, E.F., Active filters: part 12, the leapfrog or active-ladder synthesis, *Wireless World*, 340–345, Jul., 1970.

[54] Brackett, P. and Sedra, A.S., Direct SFG simulation of LC ladder networks with applications to active filter design, *IEEE Trans. Circuits Syst.*, 23(2), 61–67, 1976.

[55] Dimopoulos, H.G. and Constantinides, A.G., Linear transformation active filters, *IEEE Trans. Circuits Syst.*, 25(10), 845–852, 1978.

[56] Ping, L., Henderson, R.K., and Sewell, J.I., A methodology for integrated ladder filter design, *IEEE Trans. Circuits Syst.*, CAS-38(8), 853–868, 1991.

[57] Tow, J., Design and evaluation of shifted companion form (follow the leader feedback) active filters, *Proc. IEEE Intl. Symp. Circuits Syst.*, 656–660, 1974.

[58] Perry, D.J., New multiple feedback active RC network, *Electron. Lett.*, 11(16), 364–368, 1975.

[59] Laker, K.R., Schaumann R., and Ghausi, M.S., Multiple-loop feedback topologies for the design of low-sensitivity active filters, *IEEE Trans. Circuits Syst.*, 26(1), 1–21, 1979.

Chapter 11

Multiple Integrator Loop Feedback OTA-C Filters

11.1 Introduction

The system performance of high-frequency continuous-time OTA-C filters depends on both the constituent components (OTAs and capacitors) and the circuit architecture (how OTAs and capacitors are connected). OTAs have been developed in different technologies such as bipolar, CMOS, BiCMOS, and GaAs, and their performances have been continuously improved, making a major contribution to enhanced continuous-time filter specifications. Many new filter architectures offering a variety of performances have been generated, and comparisons of the architectures from various viewpoints have also been made in order to find the optimum structure in one sense or another.

The structure generation, design methods, and performance evaluation of high-frequency OTA-C filters have been of considerable interest to filter designers and researchers [1]–[31]. In Chapters 8 and 9 we have discussed second-order OTA-C filters based on a single-OTA model and two integrator loop configurations, respectively. In Chapter 10 we have dealt with high-order OTA-C filter design. While the biquadratic OTA-C filters proposed in Chapters 8 and 9 can be cascaded to realize high-order specifications, Chapter 10 has introduced various approaches based on passive LC ladder simulation. No matter what the approach, the filter structures almost always consist of integrators and amplifiers as the most basic building blocks, and have feedback loops at this basic level. A general approach therefore may be developed based on the multiple loop feedback structure constructed with integrators and amplifiers. This chapter will introduce such a general multiple integrator loop feedback approach for OTA-C filter design.

As discussed in [10] and previous chapters, in the synthesis of OTA-C filters several important issues should be taken into consideration. The filter architecture should be simple and require a small number of components. To achieve the canonical or minimum component realization is of importance for both discrete and IC design because this will reduce volume, noise, parasitic effects, and power dissipation. This seems especially significant for high-order OTA-C filter design, because in OTA-C filters the transconductance gain of the OTA is used like a resistance in conventional active RC filters and the number of OTAs will therefore grow rapidly as the filter order increases. The second consideration, as we have seen, is to use grounded capacitors which can absorb parasitic capacitances and need smaller chip areas than floating ones do. To avoid producing internal nodes that are without grounded capacitors is also important, since otherwise parasitic poles will result due to OTA nonidealities and circuit parasitic capacitances. The versatility of the filter network is another practical concern; in many situations it would be ideal for the filter to provide any type of characteristic without alteration of the configuration. Furthermore, the design method and equations should be simple to use. Architectures which have appeared in the literature cannot achieve all the objectives simultaneously. For example, the ladder simulation methods are normally based on a particular passive LC prototype and can only realize zeros on the imaginary axis, thus being not general enough. Simple structures usually have floating capacitors, and grounding all capacitors will require more OTAs and produce suspending internal nodes without grounded capacitors for finite transmission zeros, besides complex design procedures and need of passive filter knowledge. The multiple integrator loop feedback method to be presented in this chapter has all the advantages.

As we have already stressed in the previous chapters, practical performance must be effectively assessed in the design. In addition to the well known sensitivity criteria, one must also note that OTAs are not ideal in both their frequency response and dynamic range. The frequency response nonidealities involve finite input and output impedances, and transconductance frequency dependence or excess phase. The dynamic range nonidealities are due to the finite linear differential input voltage and limited noise level. The former nonidealities influence the filter frequency performance and stability, while the latter restrict the filter dynamic range. Any useful design methods and filter structures should therefore have less impact from these OTA nonidealities.

This chapter will show how to generate, analyze, and design multiple integrator loop feedback filter structures using OTAs and grounded capacitors for synthesis of both transmission poles and zeros. The discussion is mainly based on the work in papers [8, 10, 12, 13]. General theory and a systematic scheme for generating all-pole filter structures is first established, with concentration on minimum component OTA-C realizations and the enumeration of canonical filter structures. Two general methods for the generation of transmission zeros are then introduced together with illustrative examples. We next formulate general sensitivity relations and analyze sensitivities for different structures. This is followed by the evaluation of dynamic range and the effects of OTA nonidealities. The chapter concludes with a brief summary.

11.2 General Design Theory of All-Pole Structures [9, 12, 28]

In this section we will generally address the multiple loop feedback method for the design of all-pole OTA-C filters.

11.2.1 Multiple Loop Feedback OTA-C Model

The basic building blocks in the construction of OTA-C filters are voltage integrators and amplifiers as shown in Figs. 11.1(a) and (b), respectively, as recalled from Chapter 9. The voltage transfer functions $H(s) = V_{out}/V_{in}$ (V_{in} is the differential input voltage to the OTA) of the integrator and amplifier are simply shown as

$$H(s) = \frac{1}{sC/g}$$

and

$$H(s) = \frac{g_1}{g_2}$$

respectively.

The general multiple integrator loop feedback OTA-C model with all capacitors being grounded to be addressed in this chapter is shown in Fig. 11.2. As depicted, this model is composed of a feedforward network consisting of n OTA-capacitor integrators connected in cascade and a feedback network that may contain OTA voltage amplifiers and/or pure wire connections.

11.2.2 System Equations and Transfer Function

To analyze the model generally, the feedback network may be described as

11.2. GENERAL DESIGN THEORY OF ALL-POLE STRUCTURES

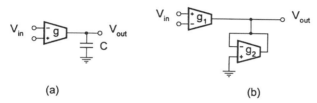

FIGURE 11.1
Voltage integrator and amplifier.

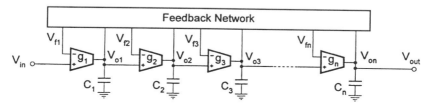

FIGURE 11.2
Multiple integrator loop feedback OTA-C model.

$$V_{fi} = \sum_{j=i}^{n} f_{ij} V_{oj}, \quad i = 1, 2, ..., n \tag{11.1}$$

where f_{ij} is the voltage feedback coefficient from the output of integrator j to the input of integrator i. This coefficient f_{ij} can be realized with an open circuit or an amplifier for the zero or nonzero values, respectively. The former means no feedback exists, while the latter suits any amount of feedback, between the ith and jth integrators. In particular, we may also realize $f_{ij} = 1$, i.e., the unity feedback by direct connection, as an alternative to using a unity gain amplifier.

Equation (11.1) can also be written in the matrix form.

$$[V_f] = [F][V_o] \tag{11.2}$$

where $[V_o] = [V_{o1}\ V_{o2} \cdots V_{on}]^t$, the output voltages of integrators, $[V_f] = [V_{f1}\ V_{f2} \cdots V_{fn}]^t$, the feedback voltages to the inverting input terminals of integrators, and $[F] = [f_{ij}]_{n \times n}$, the feedback coefficient matrix. The superscript t stands for transpose.

The currents flowing into and out of the feedback network all are zero, since they are related to the input terminals of the OTAs in the feedforward circuit or in the feedback network, which are ideally infinite impedance. Noting this and denoting time constants $\tau_j = C_j/g_j$, we can write the equations for the feedforward network by inspection

$$s\tau_1 V_{o1} = V_{in} - V_{f1}, \quad s\tau_{j+1} V_{oj+1} = V_{oj} - V_{fj+1} \tag{11.3}$$

where s is the complex frequency.

Equation (11.3) can also be condensed in a matrix form

$$[V_o] = [M(s)]^{-1} ([B]V_{in} - [V_f]) \tag{11.4}$$

where

$$[M(s)] = \begin{bmatrix} s\tau_1 & & & & \\ -1 & s\tau_2 & & & \\ & -1 & s\tau_3 & & \\ & & & \ddots & \\ & & & -1 & s\tau_n \end{bmatrix} \quad (11.5)$$

$$[B] = [1\ 0 \cdots 0]^t \quad (11.6)$$

Combining Eqs. (11.2) and (11.4) we can obtain the equation for the whole system as

$$[A(s)][V_o] = [B]V_{\text{in}} \quad (11.7)$$

where

$$[A(s)] = [M(s)] + [F] \quad (11.8)$$

Equation (11.7) establishes the relationship between the overall circuit input and the integrator outputs including the overall circuit output. Using the equation we can formulate various useful expressions for the general model. Here we first demonstrate the circuit transfer function using Eq. (11.7), while we will also refer back to this equation when coping with other problems, for example the realization of transmission zeros in Section 11.4 and sensitivity computation in Section 11.5.

Solving Eq. (11.7) yields

$$\frac{[V_o]}{V_{\text{in}}} = [A(s)]^{-1}[B] = \frac{1}{|A(s)|} \begin{bmatrix} A_{11}(s) \\ A_{12}(s) \\ \vdots \\ A_{1n}(s) \end{bmatrix} \quad (11.9)$$

where $|A(s)|$ and $A_{ij}(s)$ represent the determinant and cofactors of $[A(s)]$, respectively.

Since the overall circuit output $V_{\text{out}} = V_{on}$, from Eq. (11.9) it can be readily identified that the system transfer function is given by

$$H(s) = \frac{V_{\text{out}}}{V_{\text{in}}} = \frac{A_{1n}(s)}{|A(s)|} \quad (11.10)$$

Noting that $[F]$ is an upper triangular matrix and using Eq. (11.8) the system matrix $[A(s)]$ may be generally written as

$$[A(s)] = \begin{bmatrix} s\tau_1 + f_{11} & f_{12} & f_{13} & & f_{1n-1} & f_{1n} \\ -1 & s\tau_2 + f_{22} & f_{23} & \cdots & f_{2n-1} & f_{2n} \\ & -1 & s\tau_3 + f_{33} & & f_{3n-1} & f_{3n} \\ & & & \vdots & & \\ & & & & -1 & s\tau_n + f_{nn} \end{bmatrix} \quad (11.11)$$

11.2. GENERAL DESIGN THEORY OF ALL-POLE STRUCTURES

Based on Eq. (11.11) we can see that $A_{1n}(s) = 1$. Thus the transfer function $H(s)$ can be simplified as

$$H(s) = \frac{1}{|A(s)|} \quad (11.12)$$

11.2.3 Feedback Coefficient Matrix and Systematic Structure Generation

The feedback matrix $[F]$ is defined by Eq. (11.2) and has the property that

$$f_{ij} \begin{cases} \neq 0 & \text{if there is feedback between } V_{oj} \text{ and } V_{fi} \\ = 0 & \text{otherwise} \end{cases}$$

As can be seen from Eq. (11.2), if all the elements in a row of $[F]$, say row i, are zero, the corresponding feedback voltage V_{fi} will be zero and so is the converse. $V_{fi} = 0$ means that the inverting terminal of the OTA in the ith integrator is grounded.

Note that $[F]$ is an upper triangular matrix; f_{ij} is nonzero for all $i \leq j$. If we further suppose that no inverting integrator terminals are grounded, the feedback matrix will also have the property that each row has one and only one nonzero element, which implies that f_{nn} is always nonzero under the assumption.

As discussed before, the nonzero feedback coefficient can always be realized using an OTA voltage amplifier and the unity feedback coefficient can also be achieved by pure wire connection.

In the following by the canonical or minimum component realization we mean that for realizing unity dc gain nth-order all-pole lowpass filters, only n OTAs and n capacitors (i.e., n integrators) are required. For the general model in Fig. 11.2 the canonical realization is clearly equivalent to no components existing in the feedback network.

If all the nonzero feedback coefficients are unity and are realized with pure wire connection, there will be no OTAs in the feedback network. The whole system then has the minimum number of components. Alternatively we can say that for canonical architectures, the feedback matrix $[F] = [f_{ij}]_{n \times n}$ defined by Eq. (11.2) obviously has only zero and unit elements, and is thus unimodular, since feedback can be achieved only by direct connection.

It is apparent that there is a one-to-one correspondence between the feedback matrix $[F]$ and the circuit configuration, and different $[F]$ will give rise to different circuit structures. To show this we consider the situation that feedback is realized only by direct connection and none of the OTA inverting terminals in the integrators are grounded. According to the features of the general $[F]$ discussed above, the feedback matrix $[F]$ now becomes an upper triangular $(0, 1)$ matrix and has one and only one unit element in each row, leading to $f_{nn} = 1$. Therefore for the nth-order model there are $n!$ combinations of unit element positions in the matrix. Note that the unit element $f_{ij} = 1$ in each combination is realized by a direct connection between the negative input terminal of integrator i and the output of integrator j. Thus we have $n!$ different combinations of feedback connections, i.e., $n!$ different filter structures.

It is of particular interest that this also suggests a method for generating all possible filter architectures that are canonical and without grounded integrator inverting terminals. That is, for any given order n, we first find all $n!$ combinations of unit element positions in $[F]$. Direct connections are then made corresponding to all unit feedback coefficients in each combination; this is repeated for all $n!$ different combinations. All possible filter configurations are thus obtained, which correspond to the $n!$ different feedback connection combinations. This method is extensively studied and exemplified in Section 11.3.

11.2.4 Filter Synthesis Procedure Based on Coefficient Matching

From Eq. (11.11) we can see that the determinant $|A(s)|$ may normally be an nth-order polynomial of s. The transfer function in Eq. (11.12) may therefore have the all-pole filter characteristic. The all-pole filter structures with different feedback configurations can be generated using the method given in the preceding section. We now discuss how to design these filters to fulfill the required specification.

The general form of all-pole lowpass transfer functions may be expressed as

$$H_d(s) = \frac{A_0}{B_n s^n + B_{n-1} s^{n-1} + \cdots + B_1 s + 1} \quad (11.13)$$

To synthesize this desired transfer function $H_d(s)$ we may follow the generic procedure shown below.

Based on Eqs. (11.11) and (11.12), by expansion of $|A(s)|$ we can obtain the circuit transfer function as

$$H(s) = \frac{g_0(\tau_h, f_{ij})}{g_n(\tau_h, f_{ij}) s^n + g_{n-1}(\tau_h, f_{ij}) s^{n-1} + \cdots + g_1(\tau_h, f_{ij}) s + 1} \quad (11.14)$$

Comparing Eq. (11.13) and Eq. (11.14), to achieve the desired characteristic the following set of coefficient matching equations must be satisfied.

$$g_n(\tau_h, f_{ij}) = B_n, \quad g_{n-1}(\tau_h, f_{ij}) = B_{n-1}, \cdots, g_1(\tau_h, f_{ij}) = B_1,$$

$$g_0(\tau_h, f_{ij}) = A_0 \quad (11.15)$$

Solving Eq. (11.15) we can obtain τ_h and f_{ij}. To finish the design we can then compute the values of each C and g from τ_h and f_{ij}.

The efficient expansion of $|A(s)|$ to reach the polynomial form in s of Eq. (11.14) is the first step in the design. Some symbolic analysis techniques may be required generally to deal with $|A(s)|$ to get coefficient matching equations. However, the issue may be quite easily handled for low-order and some general high-order filters as will be shown in the next section.

The coefficient matching equations are usually nonlinear. Note that to produce the item s^k, there is at least one group of k integrators making a multiplicative contribution to the corresponding coefficient. Hence, $g_k(\tau_h, f_{ij})$ will contain at least one term of multiplication of k integration constants. In most cases a nonlinear equation solver may need to be invoked to solve the derived parameter value determination equations. In the later section we will show that the design equations of many structures can be easily solved explicitly.

To further determine each g and C there exist n degrees of freedom in the canonical realization and more than n degrees in the noncanonical. Thus the transconductances or the capacitances can be arbitrarily assigned to be identical. Taking the canonical realization as an example we may set $g_1 = g_2 = \cdots = g_n = g$ and then calculate $C_j = g\tau_j$, for any j, or let $C_1 = C_2 = \cdots = C_n = C$ and then compute $g_j = C/\tau_j$.

As can be seen from Eq. (11.8) the network performance is a function of $[F]$. Different $[F]$ matrices will lead to different transfer characteristics in Eq. (11.12). $[F]$ is also linked with filter structures and different $[F]$ matrices will correspond to different architectures. Thus the relationship between the performance and the structure is established through the feedback matrix. The significance is even more in that the generality, regularity, and systematicality of the design method is obtained from the introduction of $[F]$.

11.3 Structure Generation and Design of All-pole Filters [9, 12, 28]

In Section 11.2 we have discussed generally the proposed method for generation, analysis, and design of OTA-C filters. In this section we will investigate the application of the method. For simplicity and clarity, we concentrate on canonical filter configurations with no feedback voltages (inverting inputs of g_j OTAs) being grounded for a given order n using the method given in Section 11.2.3. Some component value determination formulas and design considerations are also presented.

11.3.1 First- and Second-Order Filters

In the simplest first-order case $[F] = f_{11}$, the general model has the transfer function

$$H(s) = 1/(\tau_1 s + f_{11})$$

The canonical structure by direct feedback connection that corresponds to $f_{11} = 1$ is given in Fig. 11.3(a).

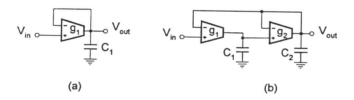

FIGURE 11.3
(a) First- and (b) second-order canonical OTA-C filter structures.

For the second-order model the general transfer function is derived using Eqs. (11.11) and (11.12) as

$$H(s) = 1/\left[\tau_1\tau_2 s^2 + (\tau_2 f_{11} + \tau_1 f_{22})s + (f_{11}f_{22} + f_{12})\right] \quad (11.16)$$

With $f_{22} = f_{12} = 1$ and $f_{11} = 0$, the canonical second-order filter is obtained as shown in Fig. 11.3(b) [4, 5], which was also discussed in terms of two integrator loop structures in Chapter 9 and the transfer function in Eq. (11.16) accordingly reduces to

$$H(s) = 1/\left(\tau_1\tau_2 s^2 + \tau_1 s + 1\right)$$

which can realize the unity dc gain ($A_0 = 1$) all-pole characteristic in Eq. (11.13) with $\tau_2 = B_2/B_1$ and $\tau_1 = B_1$.

The other combination of $[F]$ unit elements is $f_{11} = f_{22} = 1$ and $f_{12} = 0$. The corresponding filter, a cascade of two first-order canonical sections, is rejected since it cannot realize complex poles.

Having shown that the first- and second-order filters can be derived from the general model in Fig. 11.2, we now turn to a demonstration of the power of the approach in generating high-order OTA-C filter structures in the following sections.

11.3.2 Third-Order Filters

For the third-order model that is derived from Fig. 11.2 but with $n = 3$, using general $[F]$ and Eqs. (11.11) and (11.12) we formulate the general transfer function as

$$H(s) = 1/\left\{\tau_1\tau_2\tau_3 s^3 + (\tau_1\tau_2 f_{33} + \tau_1\tau_3 f_{22} + \tau_2\tau_3 f_{11})s^2\right.$$

$$+ [\tau_1(f_{22}f_{33} + f_{23}) + \tau_2 f_{11}f_{33} + \tau_3(f_{11}f_{22} + f_{12})]s$$

$$\left. + (f_{11}f_{22}f_{33} + f_{11}f_{23} + f_{12}f_{33} + f_{13})\right\} \quad (11.17)$$

As proved in Section 11.2.3, there are altogether six possible configurations. It can be verified that the last term in the denominator of Eq. (11.17) is equal to 1 for all the structures, and so $A_0 = 1$ in Eq. (11.13). The results are presented below.

When $f_{13} = f_{23} = f_{33} = 1$ and the other elements are zero, we have the structure in Fig. 11.4(a) and the circuit transfer function in Eq. (11.17) becomes

$$H(s) = 1/\left(\tau_1\tau_2\tau_3 s^3 + \tau_1\tau_2 s^2 + \tau_1 s + 1\right) \quad (11.18)$$

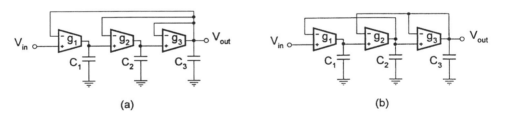

FIGURE 11.4
Third-order canonical OTA-C filters.

The parameter value equations are demonstrated as

$$\tau_1 = B_1, \quad \tau_2 = B_2/B_1, \quad \tau_3 = B_3/B_2 \quad (11.19)$$

If we select $f_{12} = f_{23} = f_{33} = 1$ and the other $f_{ij} = 0$, the filter architecture in Fig. 11.4(b) results. The corresponding transfer function and the parameter value determination formulas are derived as

$$H(s) = 1/\left(\tau_1\tau_2\tau_3 s^3 + \tau_1\tau_2 s^2 + (\tau_1 + \tau_3)s + 1\right) \quad (11.20)$$

$$\tau_3 = B_3/B_2, \quad \tau_2 = B_2/(B_1 - B_3/B_2), \quad \tau_1 = B_1 - B_3/B_2 \quad (11.21)$$

For $[F]$ with $f_{11} = f_{23} = f_{33} = 1$ and the other $f_{ij} = 0$, or $f_{12} = f_{22} = f_{33} = 1$ and the other $f_{ij} = 0$, the circuits become cascaded by a first-order and a second-order canonical sections in Fig. 11.3.

11.3. STRUCTURE GENERATION OF ALL-POLE FILTERS

The other two combinations which correspond to the $[F]$ of a unity matrix and the $[F]$ with $f_{13} = f_{22} = f_{33} = 1$ and the other $f_{ij} = 0$, respectively, do not seem practical, because the structure with a unity matrix $[F]$ is a cascade of three first-order canonical sections in Fig. 11.3, which can only realize some real poles, and for the structure corresponding to $f_{13} = f_{22} = f_{33} = 1$ and the other $f_{ij} = 0$ the solutions of component values for the Butterworth and Chebyshev approximations are not real positive numbers, which has been numerically verified. Therefore they are rejected.

11.3.3 Fourth-Order Filters

For the fourth-order general model of Fig. 11.2, again from Eqs. (11.11) and (11.12), the general transfer function can be written with some tedious manipulation as

$$H(s) = 1/\Big\{(\tau_1\tau_2\tau_3\tau_4)s^4$$

$$+ (\tau_1\tau_2\tau_3 f_{44} + \tau_1\tau_2\tau_4 f_{33} + \tau_1\tau_3\tau_4 f_{22} + \tau_2\tau_3\tau_4 f_{11})s^3$$

$$+ [\tau_1\tau_2(f_{33}f_{44} + f_{34}) + \tau_1\tau_3 f_{22}f_{44} + \tau_1\tau_4(f_{22}f_{33} + f_{23})$$

$$+ \tau_2\tau_3 f_{11}f_{44} + \tau_2\tau_4 f_{11}f_{33} + \tau_3\tau_4(f_{11}f_{22} + f_{12})]s^2$$

$$+ [\tau_1(f_{22}f_{33}f_{44} + f_{22}f_{34} + f_{23}f_{44} + f_{24}) + \tau_2(f_{11}f_{33}f_{44} + f_{11}f_{34})$$

$$+\tau_3(f_{11}f_{22}f_{44} + f_{12}f_{44}) + \tau_4(f_{11}f_{22}f_{33} + f_{11}f_{23} + f_{12}f_{33} + f_{13})]s$$

$$+ (f_{11}f_{22}f_{33}f_{44} + f_{11}f_{22}f_{34} + f_{11}f_{44}f_{23}$$

$$+f_{12}f_{33}f_{44} + f_{11}f_{24} + f_{13}f_{44} + f_{12}f_{34} + f_{14})\Big\} \qquad (11.22)$$

For any particular $[F]$ we can easily draw the associated structure, obtain the corresponding transfer function, and calculate the component values. There are altogether 24 combinations of possible filter configurations according to the discussion in Section 11.2.3. There are ten practical structures, five of which are shown in Fig. 11.5, together with the corresponding $[F]$s, transfer functions, and some design formulas below. Note that in each case the f_{ij}s that are not written out are treated as zero and the realization of the unity dc gain all-pole lowpass characteristic in Eq. (11.13) with $A_0 = 1$ is dealt with.

<u>Fig. 11.5(a)</u>: $f_{12} = f_{23} = f_{34} = f_{44} = 1$

$$H(s) = 1/\Big[\tau_1\tau_2\tau_3\tau_4 s^4 + \tau_1\tau_2\tau_3 s^3 + (\tau_1\tau_2 + \tau_1\tau_4 + \tau_3\tau_4)s^2 + (\tau_1 + \tau_3)s + 1\Big]$$

$$\tau_4 = B_4/B_3, \quad \tau_3 = B_3/B, \quad \tau_2 = B/(B_1 - B_3/B),$$

$$\tau_1 = B_1 - B_3/B, \quad B = B_2 - B_1 B_4/B_3$$

Fig. 11.5(b): $f_{13} = f_{23} = f_{34} = f_{44} = 1$

$$H(s) = 1/\left[\tau_1\tau_2\tau_3\tau_4 s^4 + \tau_1\tau_2\tau_3 s^3 + (\tau_1\tau_2 + \tau_1\tau_4) s^2 + (\tau_1 + \tau_4) s + 1\right]$$

$$\tau_4 = B_4/B_3, \quad \tau_3 = B_3/\left[B_2 - (B_1 - B_4/B_3) B_4/B_3\right],$$

$$\tau_2 = B_2/(B_1 - B_4/B_3) - B_4/B_3, \quad \tau_1 = B_1 - B_4/B_3$$

Fig. 11.5(c): $f_{13} = f_{24} = f_{34} = f_{44} = 1$

$$H(s) = 1/\left[\tau_1\tau_2\tau_3\tau_4 s^4 + \tau_1\tau_2\tau_3 s^3 + \tau_1\tau_2 s^2 + (\tau_1 + \tau_4) s + 1\right]$$

$$\tau_4 = B_4/B_3, \tau_3 = B_3/B_2, \tau_2 = B_2/(B_1 - B_4/B_3), \tau_1 = B_1 - B_4/B_3$$

Fig. 11.5(d): $f_{14} = f_{23} = f_{34} = f_{44} = 1$

$$H(s) = 1/\left[\tau_1\tau_2\tau_3\tau_4 s^4 + \tau_1\tau_2\tau_3 s^3 + (\tau_1\tau_2 + \tau_1\tau_4) s^2 + \tau_1 s + 1\right]$$

$$\tau_4 = B_4/B_3, \tau_3 = B_3/(B_2 - B_1 B_4/B_3), \tau_2 = B_2/B_1 - B_4/B_3, \tau_1 = B_1$$

Fig. 11.5(e): $f_{14} = f_{24} = f_{34} = f_{44} = 1$

$$H(s) = 1/\left[\tau_1\tau_2\tau_3\tau_4 s^4 + \tau_1\tau_2\tau_3 s^3 + \tau_1\tau_2 s^2 + \tau_1 s + 1\right]$$

$$\tau_1 = B_1, \tau_2 = B_2/B_1, \tau_3 = B_3/B_2, \tau_4 = B_4/B_3$$

It is observed that the circuits in Figs. 11.5(a)–(e) can be easily designed using the attached formulas. The other five practical structures correspond to $f_{12} = f_{22} = f_{34} = f_{44} = 1$, $f_{12} = f_{24} = f_{34} = f_{44} = 1$, $f_{13} = f_{22} = f_{34} = f_{44} = 1$, $f_{14} = f_{22} = f_{34} = f_{44} = 1$, $f_{14} = f_{23} = f_{33} = f_{44} = 1$, respectively.

In addition to the 10 configurations presented above there are another 14 possible structures. These 14 configurations, however, have been found not suitable for realizing the Butterworth and Chebyshev approximation filters; 10 of them are a cascade of canonical sections of either 4 first-orders, or 2 first-orders and 1 second-order, or 1 first-order and 1 third-order, which cannot realize 2 pairs of complex poles, while the other 4 (corresponding to $f_{12} = f_{24} = f_{33} = f_{44} = 1$, $f_{13} = f_{24} = f_{33} = f_{44} = 1$, $f_{14} = f_{22} = f_{33} = f_{44} = 1$, and $f_{14} = f_{24} = f_{33} = f_{44} = 1$, respectively) have no practical solutions of the associated design equations.

11.3. STRUCTURE GENERATION OF ALL-POLE FILTERS

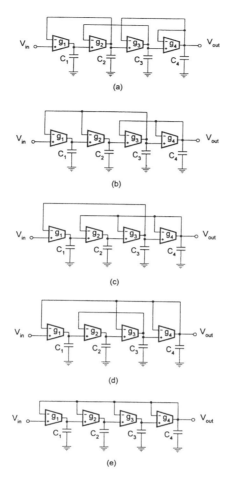

FIGURE 11.5
Fourth-order all-pole canonical OTA-C filter structures.

11.3.4 Design Examples of Fourth-Order Filters

Numerical design examples for the five structures in Fig. 11.5 are now presented. For the 4th-order Butterworth lowpass filter the normalized transfer function is (see Chapter 2 and Appendix A)

$$H_d(s) = \frac{1}{s^4 + 2.61313 s^3 + 3.41421 s^2 + 2.61313 s + 1}$$

We use the five structures given in Fig. 11.5 to realize this characteristic. Identifying that $B_4 = 1$, $B_3 = B_1 = 2.61313$ and $B_2 = 3.41421$, the parameter values of the structures are calculated from the individual coefficient matching equations by using the formulated explicit solutions, which are given in Table 11.1.

The realization of the fourth-order 455 kHz unity gain Butterworth filter using the structure in Fig. 11.5(a) is now further considered. The equal transconductance design is adopted with the transconductance value being $g = 30.1 \mu S$. The normalized capacitances are calculated from τ_j in

Table 11.1 Parameter Values for Normalized 4th-Order Butterworth Filter

Circuit	τ_1	τ_2	τ_3	τ_4
Fig. 11.5(a)	1.53073	1.57716	1.08239	0.382683
Fig. 11.5(b)	2.23044	1.14805	1.02049	0.382683
Fig. 11.5(c)	2.23044	1.53073	0.765367	0.382683
Fig. 11.5(d)	2.61313	0.92388	1.08239	0.382683
Fig. 11.5(e)	2.61313	1.30656	0.765367	0.382683

Table 11.1 as

$$C_1 = 46.1\mu F, \quad C_2 = 47.5\mu F, \quad C_3 = 32.6\mu F, \quad C_4 = 11.5\mu F$$

For the cutoff frequency 455 kHz, frequency denormalization leads to the nominal circuit capacitances

$$C_1 = 16.1 pF, \quad C_2 = 16.6 pF, \quad C_3 = 11.4 pF, \quad C_4 = 4.0 pF$$

The Chebyshev lowpass filters can also be realized. For instance, the transfer function of the 1dB ripple, frequency-normalized Chebyshev filter with unity dc gain is obtained as

$$H_d(s) = \frac{1}{3.62808s^4 + 3.45688s^3 + 5.27496s^2 + 2.69429s + 1}$$

The circuit parameter values for realizing this characteristic using the five canonical configurations in Fig. 11.5 are listed in Table 11.2.

Table 11.2 Parameter Values for 4th-Order 1dB Ripple Chebyshev Filter

Circuit	τ_1	τ_2	τ_3	τ_4
Fig. 11.5(a)	1.28172	1.90934	1.41256	1.04953
Fig. 11.5(b)	1.64476	2.15761	0.974114	1.04953
Fig. 11.5(c)	1.64476	3.20713	0.655337	1.04953
Fig. 11.5(d)	2.69429	0.908307	1.41256	1.04953
Fig. 11.5(e)	2.69429	1.95783	0.655337	1.04953

11.3.5 General nth-Order Architectures

The above examples reveal that using the proposed method we may systematically generate a large number of practical all-pole filter structures. General nth-order architectures can also be thus derived. Rather than trying to exhaustively enumerate all $n!$ possible general structures due to the complexity of the problem, we present some typical ones for illustrative purposes.

11.3.5.1 General IFLF Configuration

If $[F]$ is chosen so that the elements in the last column all are unity and the other elements of the matrix are zero, then the circuit has the inverse follow-the-leader feedback (IFLF) structure as

11.3. STRUCTURE GENERATION OF ALL-POLE FILTERS

shown in Fig. 11.6. System matrix $[A(s)]$ becomes

$$[A(s)] = \begin{bmatrix} s\tau_1 & 0 & & 0 & 1 \\ -1 & s\tau_2 & & 0 & 1 \\ & & \cdots & & \\ 0 & 0 & & s\tau_{n-1} & 1 \\ 0 & 0 & & -1 & s\tau_n + 1 \end{bmatrix} \quad (11.23)$$

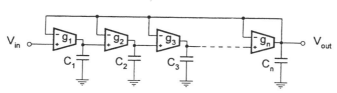

FIGURE 11.6
General IFLF structure.

The third-order IFLF filter has been given in Fig. 11.4(a) and fourth-order counterpart is the circuit in Fig. 11.5(e). We now derive the general explicit design formulas for the nth-order IFLF circuit. By expansion of $[A(s)]$ using the last column we obtain

$$|A(s)| = \tau_1 \tau_2 \cdots \tau_n s^n + \tau_1 \tau_2 \cdots \tau_{n-1} s^{n-1} + \cdots + \tau_1 \tau_2 s^2 + \tau_1 s + 1 \quad (11.24)$$

Comparing it with Eq. (11.13) gives the design equations

$$\prod_{i=1}^{j} \tau_i = B_j \quad j = 1, 2, \ldots, n \quad (11.25)$$

By simple manipulation of Eq. (11.25) we obtain

$$\tau_1 = B_1, \quad \tau_i = B_i / B_{i-1} \quad (11.26)$$

The IFLF structure can also be derived by using the signal flow graph method [7].

11.3.5.2 General LF Configuration

If the choice is made of $f_{i(i+1)} = 1$, $i = 1, 2, \ldots, n - 1$, $f_{nn} = 1$, and all the other f_{ij}s are zero, then the leapfrog (LF) configuration results as shown in Fig. 11.7. The derivative third- and fourth-order counterparts have been exhibited in Fig. 11.4(b) and Fig. 11.5(a), respectively.

The transfer function may be obtained explicitly in an iterative way and the analytic expressions for determining parameter values may also be attainable from the coefficient matching equations. To appreciate this we write the system matrix

$$[A(s)] = \begin{bmatrix} s\tau_1 & 1 & & & & \\ -1 & s\tau_2 & 1 & \cdots & & \\ & -1 & s\tau_3 & & & \\ & & & \cdots & & \\ & & & & s\tau_{n-1} & 1 \\ & & & & -1 & s\tau_n + 1 \end{bmatrix} \quad (11.27)$$

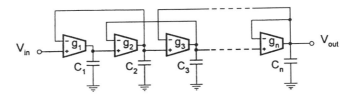

FIGURE 11.7
General LF configuration.

from which we can obtain $|A(s)|$ as

$$|A(s)| = (s\tau_n + 1) A_{nn}(s) + A_{(n-1)n}(s) = s\tau_1 A_{11}(s) + A_{12}(s) \tag{11.28}$$

where $A_{jn}(s)$ can be formulated in an iterative way as

$$A_{1n}(s) = 1, \quad A_{2n}(s) = \tau_1 s,$$

$$A_{jn}(s) = s\tau_{j-1} A_{(j-1)n}(s) + A_{(j-2)n}(s) \tag{11.29}$$

and $A_{1j}(s)$ can be determined by

$$A_{1n}(s) = 1, \quad A_{1(n-1)}(s) = s\tau_n + 1,$$

$$A_{1j}(s) = s\tau_{j+1} A_{1(j+1)}(s) + A_{1(j+2)} \tag{11.30}$$

For any order, using the above iterative formulas we can derive the corresponding transfer function $H(s) = V_{\text{out}}/V_{\text{in}}$. We can verify this for the third- and fourth-order LF filters discussed previously. We now take the fifth-order filter as another example. Using Eqs. (11.28–11.30) we have

$$|A(s)| = \tau_1\tau_2\tau_3\tau_4\tau_5 s^5 + \tau_1\tau_2\tau_3\tau_4 s^4 + (\tau_1\tau_2\tau_3 + \tau_1\tau_2\tau_5 + \tau_1\tau_4\tau_5 + \tau_3\tau_4\tau_5) s^3$$

$$+ (\tau_1\tau_2 + \tau_1\tau_4 + \tau_3\tau_4) s^2 + (\tau_1 + \tau_3 + \tau_5) s + 1 \tag{11.31}$$

To realize the fifth-order transfer function in Eq. (11.13), the pole parameters, τ_j are determined as

$$\tau_5 = \frac{B_5}{B_4}, \quad \tau_4 = \frac{B_4}{B_3 - B_2\tau_5}, \quad \tau_3 = \frac{B_3 - B_2\tau_5}{B_2 - (B_1 - \tau_5)\tau_4},$$

$$\tau_2 = \frac{B_2 - (B_1 - \tau_5)\tau_4}{B_1 - \tau_3 - \tau_5}, \quad \tau_1 = B_1 - \tau_3 - \tau_5 \tag{11.32}$$

11.3.6 Other Types of Realization

It should be noticed that the explicit expressions of the transfer function of, for example, the third- and fourth-order models given in Eq. (11.17) and Eq. (11.22) are general; they are actually suitable for any realizations of feedback coefficients.

In the above we have considered canonical structures with integrator inverting input terminals being ungrounded. Based on the discussion we may conveniently further comment on the other types of realization structure.

First consider the noncanonical realizations with the inverting terminals of integrators remaining ungrounded. If some or all feedback coefficients are realized with OTA voltage amplifiers, many more structures may be obtained. For example, for the second-order two integrator loop feedback biquads based on the general model in Fig. 11.2, we may also choose $f_{22} = g_{b3}/g_{b4}$, $f_{12} = 1$, or $f_{22} = 1$, $f_{12} = g_{b5}/g_{b6}$, or $f_{22} = g_{b3}/g_{b4}$, $f_{12} = g_{b5}/g_{b6}$, all with $f_{11} = 0$. (The corresponding biquadratic structures have been investigated in Chapter 9.)

If grounded integrator inverting terminals are further allowed, more structure varieties may be obtained. For example, if we select $f_{1j} \neq 0$, $j = 1, 2, ..., n$ and the other $f_{ij} = 0$, with f_{1j} being realized by OTA voltage amplifiers, then the general FLF structure can be obtained. Note that in this structure, $V_{fi} = 0$ for all $i = 2, 3, ..., n$.

Note that the noncanonical synthesis produces some non-integrating nodes (for example, if f_{ij} is realized with an OTA voltage amplifier, there will be a node without any circuit capacitance connected, which is the output node of the amplifier, also the inverting input terminal of the ith integrator). The nonideal OTA capacitances and circuit parasitic capacitances associated with the node may thus influence the high-frequency performance, producing an unwanted pole. More OTAs will also cause other problems, as will be discussed later in the chapter.

11.4 Generation and Synthesis of Transmission Zeros

In the above sections we concentrated on the generation of all-pole filters. In this section we address the issue of implementing the transmission zeros, that is, the synthesis of the general transfer function

$$H_d(s) = \frac{A_n s^n + A_{n-1} s^{n-1} + \cdots + A_1 s + A_0}{B_n s^n + B_{n-1} s^{n-1} + \cdots + B_1 s + 1} \tag{11.33}$$

Note that this is a universal expression, since any characteristics of any order and any type can be derived from the expression. In the extreme case that the nth-order system has n zeros and is without lack of any terms of s, the system will have $2n + 1$ independent coefficients. Clearly for the universal realization, the minimal number of capacitors is n, the order of the filter; while the minimum number of OTAs equals $2n + 1$, that is, the number (n) of integrators plus the number ($n + 1$) of coefficients of the numerator of the transfer function. The latter $n + 1$ OTAs guarantee that the numerator coefficients are controllable separately from each other, which ensures that any special numerators can be achieved by selecting these transconductances, and separately from the denominator which is determined by the n integrators, thus keeping the natural modes unchanged when the numerator is adjusted.

For a given input to a node, different nodes may support different types of output characteristic, while for a fixed output node the output function may vary as the input node changes. Therefore by altering input and output nodes we may realize some transmission zeros. For example noncanonical second-order filters may support a variety of filtering functions such as the lowpass, bandpass, highpass and bandstop characteristics in this way, as has been shown in Chapter 9.

More generally, for a given input we may combine the different node outputs with a summation OTA network to give the overall circuit output, or for a fixed output distribute the overall input onto different circuit nodes using an OTA distribution network. A general transfer function can thus be obtained. Then by properly selecting the summation or distribution weights for respective cases one may attain any filter characteristics. In the following we will formulate design equations of the two methods and illustrate the realization of various numerator characteristics using different architectures.

11.4.1 Output Summation of OTA Network [12]

Here we connect a summation OTA network to the circuit in Fig. 11.2, as shown in Fig. 11.8. Denoting $H_j(s) = V_{oj}/V_{in}$ and $\alpha_j = g_{aj}/g_r$ we derive

$$H(s) = V_{out}/V_{in} = \alpha_0 + \sum_{j=1}^{n} \alpha_j H_j(s) \tag{11.34}$$

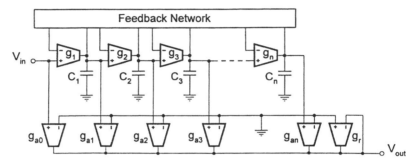

FIGURE 11.8
OTA-C filter model with output summation OTA network.

Using the results given in Section 11.2.2 [Eq. (11.9)] we know that

$$H_j(s) = A_{1j}(s)/|A(s)| \tag{11.35}$$

Substituting Eq. (11.35) into Eq. (11.34) we have the circuit transfer function

$$H(s) = \alpha_0 + \frac{1}{|A(s)|} \sum_{j=1}^{n} \alpha_j A_{1j}(s) \tag{11.36}$$

The overall transfer function in Eq. (11.36) may have the general form of Eq. (11.33) with reference to matrix $[A(s)]$ in Eq. (11.11) and the transmission zeros may be controlled arbitrarily by transconductances g_{aj} through weights α_j.

11.4.2 Input Distribution of OTA Network [12]

In this approach, the voltage signal is applied to circuit nodes by an input OTA network as shown in Fig. 11.9. In this way finite transmission zeros can also be achieved. Exactly the same formulation process as that in Section 11.2.2 can be followed to derive the design equations for this case. All the relations in Eqs. (11.1), (11.2), (11.4), (11.5), (11.7), and (11.8) apply here, with only one exception that instead of $[B] = [1 \ 0 \cdots 0]^t$ of Eq. (11.6) in Section 11.2.2, now

$$[B] = [\beta_1 \ \beta_2 \cdots \beta_n]^t \tag{11.37}$$

11.4. GENERATION AND SYNTHESIS OF TRANSMISSION ZEROS

where $\beta_j = g_{aj}/g_j$, $j = 1, 2, \ldots, n$, since Eq. (11.3) in Section 11.2.2 correspondingly becomes for the present case

$$s\tau_1 V_{o1} = \beta_1 V_{in} - V_{f1}, \quad s\tau_{j+1} V_{o(j+1)} = \beta_{j+1} V_{in} + (V_{oj} - V_{f(j+1)}) \quad (11.38)$$

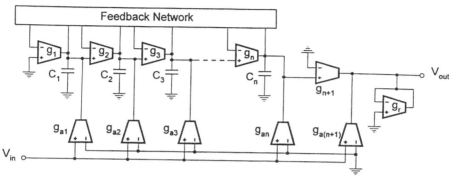

FIGURE 11.9
OTA-C filter model with input distribution OTA network.

This exception is clearly due to the change of input form; in Fig. 11.2 the input voltage is applied only to the first integrator output node and this is realized directly through the OTA in the integrator, while in the present case the input voltage is distributed to all the integrator output nodes and this is accomplished by extra OTAs.

First solving Eq. (11.7) for $[V_o]$, then substituting Eq. (11.37) and finally observing $V_{out} = \gamma(\beta_{n+1} V_{in} + V_{on})$, where $\gamma = g_{n+1}/g_r$ and $\beta_{n+1} = g_{a(n+1)}/g_{n+1}$, we can formulate that

$$H(s) = \frac{V_{out}}{V_{in}} = \gamma \left[\beta_{n+1} + \frac{1}{|A(s)|} \sum_{j=1}^{n} \beta_j A_{jn}(s) \right] \quad (11.39)$$

Equation (11.39) also indicates the possibility of arbitrary transmission zero realization by adjusting β_j, that is, g_{aj}.

Note that the distribution method actually involves the superposition theorem, since the responses corresponding to the different resulting node inputs are superposed at the output node. This method can therefore also be understood in the way that the different node inputs are collected with weights into a single input.

It is also noted that if the maximum order in the numerator is required to be $n - 1$, then we can remove the $g_{a(n+1)}$ OTA, and the g_{n+1} and g_r OTAs for $\gamma = 1$ and simply output the voltage V_{on} directly in the distribution case (this leads to an advantage that the resistive summing node that will have effects of the parasitics at very high frequencies is avoided), while for the summation the g_{a0} OTA should be deleted. It is also of interest to note that when the transadmittance functions are required, we can eliminate the g_r OTA in both input distribution and output summation configurations.

In the next two sections, we will illustrate the general output summation and input distribution methods for arbitrary transmission zero realization developed above. We will investigate third-order OTA-C structures and general nth-order IFLF and LF configurations. Note that universal OTA-C biquads can also be derived based on the output summation and input distribution methods, which have already been discussed in Chapter 9.

11.4.3 Universal and Special Third-Order OTA-C Filters [13]

The general third-order transfer function can be expressed as

$$H_d(s) = \frac{A_3 s^3 + A_2 s^2 + A_1 s + A_0}{B_3 s^3 + B_2 s^2 + B_1 s + 1} \qquad (11.40)$$

Four third-order universal OTA-C structures are given in Fig. 11.10, which consist of the canonical IFLF and output summation network, canonical IFLF and input distribution network, canonical LF and output summation network, and canonical LF and input distribution network. Deriving the circuit transfer functions of the structures and comparing them with the desired characteristic in Eq. (11.40) we can obtain the design formulas for filter parameters.

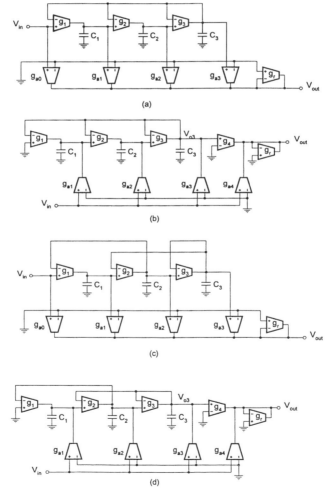

FIGURE 11.10
Four universal third-order OTA-C filters.

11.4. GENERATION AND SYNTHESIS OF TRANSMISSION ZEROS

11.4.3.1 IFLF and Output Summation Structure in Fig. 11.10(a)

Denote that $\tau_j = C_j/g_j$ and $\alpha_j = g_{aj}/g_r$. Using Eq. (11.36) the voltage transfer function $H(s) = V_{\text{out}}/V_{\text{in}}$ of the structure is derived as

$$H(s) = \frac{\alpha_0 \tau_1 \tau_2 \tau_3 s^3 + (\alpha_0 \tau_1 \tau_2 + \alpha_1 \tau_2 \tau_3) s^2 + (\alpha_0 \tau_1 + \alpha_1 \tau_2 + \alpha_2 \tau_3) s + (\alpha_0 + \alpha_1 + \alpha_2 + \alpha_3)}{\tau_1 \tau_2 \tau_3 s^3 + \tau_1 \tau_2 s^2 + \tau_1 s + 1} \quad (11.41)$$

Comparison of Eqs. (11.41) and (11.40) leads to the following design formulas:

$$\tau_1 = B_1, \quad \tau_2 = B_2/B_1, \quad \tau_3 = B_3/B_2 \quad (11.42)$$

$$\alpha_0 = A_3/B_3, \quad \alpha_1 = (A_2 - \alpha_0 B_2) B_1/B_3 ,$$

$$\alpha_2 = [A_1 - \alpha_0 B_1 - \alpha_1 B_2/B_1)] B_2/B_3 ,$$

$$\alpha_3 = A_0 - (\alpha_0 + \alpha_1 + \alpha_2) \quad (11.43)$$

11.4.3.2 IFLF and Input Distribution Structure in Fig. 11.10(b)

For the filter in Fig. 11.10(b), with $\beta_j = g_{aj}/g_j$ and $\gamma = g_4/g_r$ and using Eq. (11.39) we derive the voltage transfer function

$$H(s) = \gamma \frac{\beta_4 \tau_1 \tau_2 \tau_3 s^3 + (\beta_4 + \beta_3) \tau_1 \tau_2 s^2 + (\beta_4 + \beta_2) \tau_1 s + (\beta_4 + \beta_1)}{\tau_1 \tau_2 \tau_3 s^3 + \tau_1 \tau_2 s^2 + \tau_1 s + 1} \quad (11.44)$$

The τs can be determined using Eq. (11.42) and the βs are obtained as follows (set $\gamma = 1$):

$$\beta_4 = A_3/B_3, \quad \beta_3 = A_2/B_2 - \beta_4 ,$$

$$\beta_2 = A_1/B_1 - \beta_4, \quad \beta_1 = A_0 - \beta_4 \quad (11.45)$$

11.4.3.3 LF and Output Summation Structure in Fig. 11.10(c)

The voltage transfer function of the structure in Fig. 11.10(c) is derived as

$$H(s) = \frac{\alpha_0 \tau_1 \tau_2 \tau_3 s^3 + (\alpha_0 \tau_1 \tau_2 + \alpha_1 \tau_2 \tau_3) s^2 + [\alpha_0 (\tau_1 + \tau_3) + \alpha_1 \tau_2 + \alpha_2 \tau_3] s + (\alpha_0 + \alpha_1 + \alpha_2 + \alpha_3)}{\tau_1 \tau_2 \tau_3 s^3 + \tau_1 \tau_2 s^2 + (\tau_1 + \tau_3) s + 1} \quad (11.46)$$

The parameters can be determined using the equations below

$$\tau_1 = B_1 - B_3/B_2, \quad \tau_2 = B_2/(B_1 - B_3/B_2), \quad \tau_3 = B_3/B_2 \quad (11.47)$$

$$\alpha_0 = A_3/B_3, \quad \alpha_1 = (A_2 - \alpha_0 B_2)(B_1 - B_3/B_2)/B_3 ,$$

$$\alpha_2 = [A_1 - \alpha_0 B_1 - \alpha_1 B_2/(B_1 - B_3/B_2)] B_2/B_3 ,$$

$$\alpha_3 = A_0 - (\alpha_0 + \alpha_1 + \alpha_2) \quad (11.48)$$

11.4.3.4 LF and Input Distribution Structure in Fig. 11.10(d)

The voltage transfer function of Fig. 11.10(d) is demonstrated as

$$H(s) = \gamma \frac{\beta_4 \tau_1 \tau_2 \tau_3 s^3 + (\beta_4 + \beta_3) \tau_1 \tau_2 s^2 + [\beta_4(\tau_1 + \tau_3) + \beta_2 \tau_1] s + (\beta_4 + \beta_3 + \beta_1)}{\tau_1 \tau_2 \tau_3 s^3 + \tau_1 \tau_2 s^2 + (\tau_1 + \tau_3) s + 1} \quad (11.49)$$

Taking $\gamma = 1$ we have the following design formulas [τs are calculated using Eq. (11.47)]

$$\beta_4 = A_3/B_3, \quad \beta_3 = A_2/B_2 - \beta_4, \quad \beta_2 = (A_1 - \beta_4 B_1)/(B_1 - B_3/B_2),$$

$$\beta_1 = A_0 - (\beta_4 + \beta_3) \quad (11.50)$$

11.4.3.5 Realization of Special Characteristics

If the maximum order in the numerator is required to be $n - 1 = 2$ and $\gamma = 1$ is selected, then g_{a4}, g_4 and g_r OTAs all can be removed and the output voltage V_{o3} of the third integrator can be directly used as the filter output in the distribution structures, while for the summation type, the g_{a0} OTA should be deleted. In any case, if the calculated $\alpha_j = 0$ or $\beta_j = 0$, then the corresponding g_{aj} OTA should be eliminated. Thus, for some specific characteristics the filter structures can be simplified. For example, using Figs. 11.10(a)–(d) to realize the desired numerator with $A_3 = A_1 = 0$ we will have, respectively,

$$\alpha_0 = 0, \quad \alpha_1 = A_2 B_1/B_3, \quad \alpha_2 = -A_2 (B_2/B_3)^2,$$

$$\alpha_3 = A_0 - (\alpha_1 + \alpha_2) \quad (11.51)$$

$$\gamma = 1, \quad \beta_4 = 0, \quad \beta_3 = A_2/B_2, \quad \beta_2 = 0, \quad \beta_1 = A_0 \quad (11.52)$$

$$\alpha_0 = 0, \quad \alpha_1 = (B_1 - B_3/B_2) A_2/B_3, \quad \alpha_2 = -A_2 (B_2/B_3)^2,$$

$$\alpha_3 = A_0 - (\alpha_1 + \alpha_2) \quad (11.53)$$

$$\gamma = 1, \quad \beta_4 = 0, \quad \beta_3 = A_2/B_2,$$

$$\beta_2 = 0, \quad \beta_1 = A_0 - A_2/B_2 \quad (11.54)$$

Note that using the above formulas we may have negative values for some α_j and β_j. This simply means the need for the interchange of the two input terminals of the associated g_{aj} OTA. Figs. 11.11(a)–(d) show the resulting structures corresponding to Eqs. (11.51–11.54).

11.4.3.6 Design of Elliptic Filters

We now show how to use the filter structures above to realize an elliptic lowpass filter of

$$H_d(s) = \frac{0.588358 s^2 + 1}{1.67029 s^3 + 1.41856 s^2 + 1.91391 s + 1}$$

11.4. GENERATION AND SYNTHESIS OF TRANSMISSION ZEROS

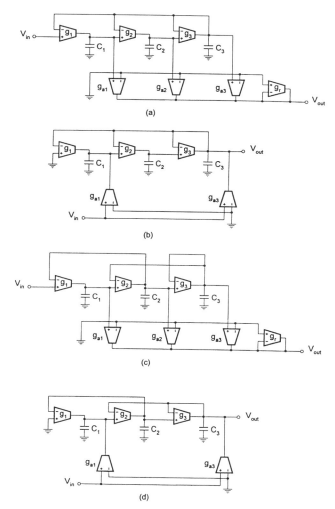

FIGURE 11.11
Four elliptic third-order OTA-C filters.

Using Eq. (11.42) we can determine the IFLF pole parameters as

$$\tau_1 = 1.91391, \quad \tau_2 = 0.741184, \quad \tau_3 = 1.17745$$

and the LF pole parameters are given by Eq. (11.47) as

$$\tau_1 = 0.736455, \quad \tau_2 = 1.926199, \quad \tau_3 = 1.17745$$

The zero parameters for all the structures are computed using Eqs. (11.51–11.54), given in Table 11.3.
Suppose that the cutoff frequency is required to be 500 kHz. We further determine the component values of the structure in Fig. 11.11(b). We choose the normalized values of capacitances as

$$C_1 = 42.5 \mu F, \quad C_2 = 30.3 \mu F, \quad C_3 = 48.2 \mu F$$

Table 11.3 Parameter Values of Third-Order Elliptic OTA-C Filter

Fig. 11.11	α_0	α_1	α_2	α_3
(a)	0	0.674173	-0.424378	0.750205
(c)	0	0.259416	-0.424378	1.16496
Fig. 11.11	β_4	β_3	β_2	β_1
(b)	0	0.414757	0	1
(d)	0	0.414757	0	0.585243

and compute
$$g_{a1} = g_1 = 22.2\mu S, \quad g_2 = g_3 = 40.9\mu S, \quad g_{a3} = 17.0\mu S$$

using the data in Table 11.3. Denormalization with the frequency of 500 kHz we have the real capacitances as
$$C_1 = 13.5pF, \quad C_2 = 9.7pF, \quad C_3 = 15.3pF$$

11.4.4 General nth-Order OTA-C Filters

We now try to further establish the explicit design equations for universal IFLF and LF architectures which are suitable for any values of the order n.

11.4.4.1 Universal IFLF Architectures [8, 10]

As an illustration, for the canonical IFLF structure ($f_{jn} = 1$, $j = 1, 2, ..., n$ and the other $f_{ij} = 0$) with the output summation OTA network [8], using Eq. (11.23) it can be demonstrated that

$$|A(s)| = \tau_1 \tau_2 \cdots \tau_n s^n + \tau_1 \tau_2 \cdots \tau_{n-1} s^{n-1}$$
$$+ \cdots + \tau_1 \tau_2 s^2 + \tau_1 s + 1 \tag{11.55}$$

$$A_{1j}(s) = \tau_{j+1} \tau_{j+2} \cdots \tau_n s^{n-j} + \tau_{j+1} \tau_{j+2} \cdots \tau_{n-1} s^{n-j-1}$$
$$+ \cdots + \tau_{j+1} \tau_{j+2} s^2 + \tau_{j+1} s + 1 \tag{11.56}$$

where $j = 1, 2, 3, ..., n - 1$

Substitution of relations (11.55) and (11.56) into Eq. (11.36) yields the general circuit transfer function and comparing this function with that in Eq. (11.33) we have the design equations

$$B_j = \prod_{i=1}^{j} \tau_i, \quad (j = 1, 2, \cdots, n) \tag{11.57}$$

and

$$A_{n-j} = \sum_{i=0}^{j} \left(\alpha_i \prod_{h=i+1}^{n-j+i} \tau_h \right), \quad (j = 0, 1, 2, \ldots, n-1),$$

11.4. GENERATION AND SYNTHESIS OF TRANSMISSION ZEROS

$$A_0 = \sum_{i=0}^{n} \alpha_i \tag{11.58}$$

From the design viewpoint, if the transfer characteristic of Eq. (11.33) is desired the circuit parameters must then be determined in terms of coefficients A_j and B_j from Eqs. (11.57) and (11.58). With $B_0 = 1$ it is easy to demonstrate that

$$\tau_j = B_j / B_{j-1}, \quad (j = 1, 2, 3, \ldots, n) \tag{11.59}$$

and

$$A_{n-j} = \sum_{i=0}^{j} \left(\frac{B_{n-j+i}}{B_i} \right) \alpha_i, \quad (j = 0, 1, 2, \ldots, n) \tag{11.60}$$

Equation (11.59) can be directly used for calculation of integration constants τ_j. From Eq. (11.60) the iterative computation formulas of summation weights α_j can also be obtained, given by

$$\alpha_0 = \frac{A_n}{B_n}, \quad \alpha_j = \frac{B_j}{B_n} \left[A_{n-j} - \sum_{i=0}^{j-1} \left(\frac{B_{n-j+i}}{B_i} \right) \alpha_i \right],$$

$$(j = 1, 2, 3, \ldots, n) \tag{11.61}$$

The parameter value determination formulas in Eqs. (11.59) and (11.61) apply to any order realizations, including second-order. It can be observed from numerator coefficient expressions (11.58) or (11.60) that the circuit may realize any special transfer functions, since we can enable any coefficient A_i of the numerator to be any value including zero by properly choosing the values and signs of g_{aj}, $j \leq n - i$. On the other hand, for any required zeros, that is any values of A_i, we can easily compute the associated parameters α_j by means of Eq. (11.61). If the calculated α_j is negative, we can simply interchange the two input terminals of the related OTA with g_{aj}. If the computed value of α_j is zero, then the g_{aj} OTA should be removed.

Now we consider the canonical IFLF structure with the distribution network [10, 28]. Using Eq. (11.23) we formulate

$$A_{1n}(s) = 1, \quad A_{jn} = \tau_1 \tau_2 \tau_3 \cdots \tau_{j-1} s^{j-1}, \quad j = 2, 3, 4, \ldots, n \tag{11.62}$$

and $|A(s)|$ is given by Eq. (11.55).
Combining Eq. (11.62) with Eq. (11.39) we have the circuit transfer function

$$H(s) = \frac{\gamma}{|A(s)|} \left[\beta_{n+1} \prod_{i=1}^{n} \tau_i s^n + (\beta_{n+1} + \beta_n) \prod_{i=1}^{n-1} \tau_i s^{n-1} + \cdots \right.$$

$$\left. + (\beta_{n+1} + \beta_2) \tau_1 s + (\beta_{n+1} + \beta_1) \right] \tag{11.63}$$

Comparing Eq. (11.63) with Eq. (11.33) when $\gamma = 1$ and noting that the τ_j are calculated using Eq. (11.59) we get

$$\beta_{n+1} = A_n / B_n, \quad \beta_1 = A_0 - A_n / B_n,$$

$$\beta_j = A_{j-1}/B_{j-1} - A_n/B_n, \quad j = 2, 3, \ldots, n \qquad (11.64)$$

Any filters may be realized through adjusting distribution weights β_j, that is, the associated g_{aj}.

In odd-order elliptic filter realizations, $A_n = 0$ can be achieved by removing the g_{a0} and $g_{a(n+1)}$ OTAs for the output summation and input distribution methods, respectively. For other zero A_j realizations, the input distribution method is advantageous when compared to the output summation technique in that the former does not require any component difference matching or equality constraints. For instance, if a zero coefficient is required, from Eq. (11.60) we may see that some restriction on the relationship between α_j values will be needed for the output summation approach to make a coefficient equal to zero. However, inspection of Eq. (11.63) indicates that a zero coefficient, say $A_j = 0$ can be achieved simply by setting β_{j+1} to zero, that is, eliminating the OTA with $g_{a(j+1)}$ for the input distribution method.

11.4.4.2 Universal LF Architectures

In Section 11.3.5 we have proved that $|A(s)|$, $A_{jn}(s)$, and $A_{1j}(s)$ of the general LF structure can be obtained in an iterative way. Substituting these into Eqs. (11.36) and (11.39) we can obtain the circuit transfer functions for the output summation and input distribution methods of transmission zero realization. Taking the fifth-order structure as an example, we have determined $|A(s)|$ and τ_j in Eqs. (11.31) and (11.32). Now for the output summation type, the numerator of the transfer function is formulated using Eq. (11.36) as

$$N(s) = \alpha_0 \tau_1 \tau_2 \tau_3 \tau_4 \tau_5 s^5 + (\alpha_0 \tau_1 \tau_2 \tau_3 \tau_4 + \alpha_1 \tau_2 \tau_3 \tau_4 \tau_5) s^4$$

$$+ [\alpha_0 (\tau_1 \tau_2 \tau_3 + \tau_1 \tau_2 \tau_5 + \tau_1 \tau_4 \tau_5 + \tau_3 \tau_4 \tau_5) + \alpha_1 \tau_2 \tau_3 \tau_4 + \alpha_2 \tau_3 \tau_4 \tau_5] s^3$$

$$+ [\alpha_0 (\tau_1 \tau_2 + \tau_1 \tau_4 + \tau_3 \tau_4) + \alpha_1 (\tau_2 \tau_3 + \tau_2 \tau_5 + \tau_4 \tau_5) + \alpha_2 \tau_3 \tau_4 + \alpha_3 \tau_4 \tau_5] s^2$$

$$+ [\alpha_0 (\tau_1 + \tau_3 + \tau_5) + \alpha_1 (\tau_2 + \tau_4) + \alpha_2 (\tau_3 + \tau_5) + \alpha_3 \tau_4 + \alpha_4 \tau_5] s$$

$$+ (\alpha_0 + \alpha_1 + \alpha_2 + \alpha_3 + \alpha_4 + \alpha_5) \qquad (11.65)$$

and for the input distribution structure, using Eq. (11.39) we have the numerator of the transfer function as ($\gamma = 1$)

$$N(s) = \beta_6 \tau_1 \tau_2 \tau_3 \tau_4 \tau_5 s^5 + (\beta_6 + \beta_5) \tau_1 \tau_2 \tau_3 \tau_4 s^4$$

$$+ [\beta_6 (\tau_1 \tau_2 \tau_3 + \tau_1 \tau_2 \tau_5 + \tau_1 \tau_4 \tau_5 + \tau_3 \tau_4 \tau_5) + \beta_4 \tau_1 \tau_2 \tau_3] s^3$$

$$+ [(\beta_6 + \beta_5)(\tau_1 \tau_2 + \tau_1 \tau_4 + \tau_3 \tau_4) + \beta_3 \tau_1 \tau_2] s^2 \qquad (11.66)$$

$$+ [\beta_6 (\tau_1 + \tau_3 + \tau_5) + \beta_4 (\tau_1 + \tau_3) + \beta_2 \tau_1] s + (\beta_6 + \beta_5 + \beta_3 + \beta_1)$$

To realize the general fifth-order function in Eq. (11.33), the pole parameters, τ_j in the denominator of the filter transfer function as shown in Eq. (11.32) can be determined using Eq. (11.33). The numerator parameters, α_j for the summation type and β_j for the distribution can then be easily determined from Eqs. (11.65) and (11.66) with comparison with Eq. (11.33) in an iterative way, respectively.

11.5. GENERAL FORMULATION OF SENSITIVITY ANALYSIS

Figure 11.12 shows a special fifth-order lowpass filter realizing

$$H_d(s) = \frac{A_4 s^4 + A_2 s^2 + A_0}{B_5 s^5 + B_4 s^4 + B_3 s^3 + B_2 s^2 + B_1 s + 1} \quad (11.67)$$

(the numerator coefficients of $A_j = 0$, $j = 1, 3, 5$), which is obtainable from the general input distribution structure by removing the g_{a6} OTA ($\beta_6 = 0$), replacing the g_6 and g_r OTAs by a direct connection ($\gamma = 1$), and removing the g_{a2} and g_{a4} OTAs ($\beta_2 = \beta_4 = 0$).

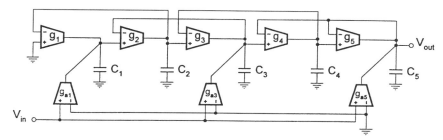

FIGURE 11.12
Lowpass LF-input distribution filter that can realize imaginary- and real-axis zeros.

With τ_j calculated using Eq. (11.32), the numerator parameters of the filter can be determined from Eq. (11.66) compared with Eq. (11.67) as

$$\beta_5 = A_4/B_4, \quad \beta_3 = (A_2 - \beta_5 B_2)/\tau_1 \tau_2, \quad \beta_1 = A_0 - (\beta_3 + \beta_5) \quad (11.68)$$

The reader is encouraged to further realize a particular elliptic function given in the appendix, with a denormalization frequency of 10 MHz.

11.5 General Formulation of Sensitivity Analysis [12, 28]

In the preceding sections we have discussed many interesting OTA-C structures of the multiple integrator loop configuration. In the design of active filters sensitivity is one of the most important criteria in assessing the filter quality, as has been mentioned on several occasions in this book. This section focuses on sensitivity analysis of all-pole OTA-C filters. Instead of calculating the sensitivity of individual structures generated we will present a general approach.

Since after calculation of the τ_j and f_{ij} sensitivities we can easily further compute the gs and Cs sensitivities using the relations $\tau_j = C_j/g_j$ and $f_{ij} = g_{ij1}/g_{ij2}$, in the following we deal with only the sensitivities to τ_j and f_{ij}.

11.5.1 General Sensitivity Relations

To formulate sensitivity functions we differentiate $[V_o]/V_{\text{in}}$ in Eq. (11.9) with respect to circuit parameter x using the well known inverse matrix differentiation formula and obtain the derivative of $[V_o]/V_{\text{in}}$ as

$$\frac{\partial ([V_o]/V_{\text{in}})}{\partial x} = -[A(s)]^{-1} \frac{\partial [A(s)]}{\partial x} [A(s)]^{-1} [B] \quad (11.69)$$

where $[A(s)]$ and $[B]$ were shown in Eqs. (11.8) and (11.6), respectively.

When $x = \tau_j$, since $[F]$ is independent of τ_j, using Eqs. (11.8) and (11.5), or just Eq. (11.11) we have

$$\frac{\partial [A(s)]}{\partial \tau_j} = \frac{\partial [M(s)]}{\partial \tau_j} = \begin{bmatrix} \vdots \\ \cdots s \cdots \\ \vdots \end{bmatrix} \begin{matrix} jth \\ \\ jth \\ \\ \end{matrix} \quad (11.70)$$

Substituting Eq. (11.70) into Eq. (11.69), together with Eq. (11.6) yields

$$\frac{\partial ([V_o]/V_{in})}{\partial \tau_j} = -\frac{s}{|A(s)|^2} \begin{bmatrix} A_{j1}(s)A_{1j}(s) \\ A_{j2}(s)A_{1j}(s) \\ \vdots \\ A_{jn}(s)A_{1j}(s) \end{bmatrix} \quad (11.71)$$

From Eq. (11.71) and noting that the output voltage is the last element in vector $[V_o]$ we can identify that

$$\frac{\partial H(s)}{\partial \tau_j} = -s \frac{A_{jn}(s)A_{1j}(s)}{|A(s)|^2} \quad (11.72)$$

Thus using Eq. (11.72) and Eq. (11.12), the sensitivities of the transfer function $H(s)$ with respect to integration constants τ_j can be readily obtained, given by

$$S_{\tau_j}^{H(s)} = \frac{\tau_j}{H(s)} \frac{\partial H(s)}{\partial \tau_j} = -s\tau_j \frac{A_{jn}(s)A_{1j}(s)}{|A(s)|} \quad (11.73)$$

Considering that $A_{1n}(s) = 1$, from Eq. (11.73) we can also write the simplified sensitivity relations for $j = 1$ and $j = n$, given by

$$S_{\tau_1}^{H(s)} = -s\tau_1 \frac{A_{11}(s)}{|A(s)|}, \quad S_{\tau_n}^{H(s)} = -s\tau_n \frac{A_{nn}(s)}{|A(s)|} \quad$$

Next we consider the transfer function sensitivities to feedback coefficients f_{ij}. Using Eq. (11.8) or Eq. (11.11) and considering that $[M(s)]$ is not related to f_{ij} we derive

$$\frac{\partial [A(s)]}{\partial f_{ij}} = \frac{\partial [F]}{\partial f_{ij}} = \begin{bmatrix} \vdots \\ \cdots 1 \cdots \\ \vdots \end{bmatrix} \begin{matrix} jth \\ \\ ith \\ \\ \end{matrix} \quad (11.74)$$

Then substituting Eq. (11.74) into Eq. (11.69) (now $x = f_{ij}$) and incorporating Eq. (11.6) we can obtain that

$$\frac{\partial ([V_o]/V_{in})}{\partial f_{ij}} = -\frac{1}{|A(s)|^2} \begin{bmatrix} A_{i1}(s)A_{1j}(s) \\ A_{i2}(s)A_{1j}(s) \\ \vdots \\ A_{in}(s)A_{1j}(s) \end{bmatrix} \quad (11.75)$$

11.5. GENERAL FORMULATION OF SENSITIVITY ANALYSIS

From Eq. (11.75) we can identify

$$\frac{\partial H(s)}{\partial f_{ij}} = -\frac{A_{in}(s) A_{1j}(s)}{|A(s)|^2} \tag{11.76}$$

Using Eq. (11.76) and the sensitivity definition, plus Eq. (11.12) we can prove the relative sensitivity functions as

$$S_{f_{ij}}^{H(s)} = -f_{ij} \frac{A_{in}(s) A_{1j}(s)}{|A(s)|}, \quad i \le j \tag{11.77}$$

When considering $A_{1n}(s) = 1$, we may simplify Eq. (11.77) for the cases of $i = 1$, $j = n$, and both $i = 1$ and $j = n$ as shown below

$$S_{f_{1j}}^{H(s)} = -f_{1j} \frac{A_{1j}(s)}{|A(s)|}, \quad S_{f_{in}}^{H(s)} = -f_{in} \frac{A_{in}(s)}{|A(s)|}, \quad S_{f_{1n}}^{H(s)} = -\frac{f_{1n}}{|A(s)|}$$

Using Eq. (11.73) and Eq. (11.77) we can also readily demonstrate the following relation

$$S_{f_{jj}}^{H(s)} / S_{\tau_j}^{H(s)} = f_{jj} / s\tau_j \tag{11.78}$$

Equation (11.78) can also be obtained from the sensitivity definition and the structural feature of matrix $[A(s)]$ in Eq. (11.11). The relation may be used to compute the sensitivity to f_{jj} when the sensitivity to τ_j is known, or vice versa.

From the sensitivity functions developed above, we may easily obtain the magnitude and the phase sensitivities of $H(j\omega)$, since they are the real and imaginary parts of $S_x^{H(j\omega)}$ (x is τ_i or f_{ij}), respectively. That is, with $|H(j\omega)|$ and $\phi(\omega)$ being the magnitude and phase frequency responses, respectively, we have

$$S_x^{|H(j\omega)|} = Re\left\{S_x^{H(j\omega)}\right\}, \quad Q_x^{\phi(\omega)} = x \frac{\partial \phi(\omega)}{\partial x} = Im\left\{S_x^{H(j\omega)}\right\} \tag{11.79}$$

The Schoeffler's measure introduced in Section 5.3 [32] can also be readily calculated by

$$S = \sum_{i=1}^{n} \left|S_{\tau_i}^{H(j\omega)}\right|^2 + \sum_{i=1}^{n} \sum_{h=1, h \ge i}^{n} \left|S_{f_{ih}}^{H(j\omega)}\right|^2 \tag{11.80}$$

Note that the formulation is general. Using these formulas, we can calculate sensivities of specific structures without knowing their transfer functions.

11.5.2 Sensitivities of Different Filter Structures

The first example involves the sensitivity analysis of third-order structures. The general τ_j sensitivity functions of the third-order structures are derived using Eq. (11.73) as

$$S_{\tau_1}^{H(s)} = -H(s)\left[\tau_1 \tau_2 \tau_3 s^3 + (\tau_1 \tau_2 f_{33} + \tau_1 \tau_3 f_{22}) s^2 + \tau_1 (f_{22} f_{33} + f_{23}) s\right],$$

$$S_{\tau_2}^{H(s)} = -H(s)\left[\tau_1 \tau_2 \tau_3 s^3 + (\tau_1 \tau_2 f_{33} + \tau_2 \tau_3 f_{11}) s^2 + \tau_2 f_{11} f_{33} s\right],$$

$$S_{\tau_3}^{H(s)} = -H(s)\left[\tau_1\tau_2\tau_3 s^3 + (\tau_1\tau_3 f_{22} + \tau_2\tau_3 f_{11})s^2 + \tau_3(f_{11}f_{22} + f_{12})s\right] \quad (11.81)$$

where $H(s)$ has been given in Eq. (11.17) of Section 11.3.2.

Two special canonical cases are given below, when realizing the unity dc gain characteristic ($A_0 = 1$) in Eq. (11.13). The sensitivities for $f_{13} = f_{23} = f_{33} = 1$ and the other $f_{ij} = 0$ in $[F]$, i.e., the configuration in Fig. 11.4(a), are calculated with substitution of Eq. (11.19), given by

$$S_{\tau_h}^{H(s)} = -\left(\sum_{j=h}^{3} B_j s^j\right) \bigg/ \left(\sum_{j=1}^{3} B_j s^j + 1\right), \quad h = 1, 2, 3 \quad (11.82)$$

For $[F]$ with $f_{12} = f_{23} = f_{33} = 1$ and the other $f_{ij} = 0$, i.e., the structure in Fig. 11.4(b), with incorporation of Eq. (11.21) the sensitivities in Eq. (11.81) accordingly reduce to

$$S_{\tau_1}^{H(s)} = -\left[B_3 s^3 + B_2 s^2 + (B_1 - B_3/B_2)s\right]/D_d(s),$$

$$S_{\tau_2}^{H(s)} = -\left(B_3 s^3 + B_2 s^2\right)/D_d(s),$$

$$S_{\tau_3}^{H(s)} = -\left[B_3 s^3 + (B_3/B_2)s\right]/D_d(s),$$

$$D_d(s) = B_3 s^3 + B_2 s^2 + B_1 s + 1 \quad (11.83)$$

To illustrate the sensitivity computation method formulated above we consider the third-order unity dc gain all-pole Butterworth characteristic

$$H_d(s) = \frac{1}{s^3 + 2s^2 + 2s + 1}$$

Identifying $B_1 = 2$, $B_2 = 2$ and $B_3 = 1$ and substituting them into the expressions in Eq. (11.82), then utilizing Eq. (11.80) (where the second part is now zero) the Schoeffler's multi-parameter sensitivity for the structure in Fig. 11.4(a) is computed as

$$S_1 = \frac{3\omega^6 + 4\omega^4 + 4\omega^2}{1 + \omega^6} \quad (11.84)$$

Similarly, using Eqs. (11.83) and (11.80) we also obtain the Schoeffler's sensitivity for the structure in Fig. 11.4(b), given by

$$S_2 = \frac{3\omega^6 + 4\omega^4 + 2.5\omega^2}{1 + \omega^6} \quad (11.85)$$

Comparing the two results in Eqs. (11.84) and (11.85) we have

$$S_1 - S_2 = \frac{1.5\omega^2}{1 + \omega^6} \quad (11.86)$$

which clearly shows that the structure in Fig. 11.4(a) is more sensitive than the circuit in Fig. 11.4(b) at all (both passband and stopband) frequencies. So in terms of sensitivity the latter architecture is better than the former.

It is noted that for some general nth-order structures, by calculating the relevant cofactors of $[A(s)]$ we can also attain the sensitivity functions. For example, the sensitivity functions of the general all-pole IFLF configuration in Fig. 11.6 are given by

$$S_{\tau_h}^{H(s)} = -\frac{\sum_{j=h}^{n} \left(\prod_{i=1}^{j} \tau_i\right) s^j}{\sum_{j=1}^{n} \left(\prod_{i=1}^{j} \tau_i\right) s^j + 1} \tag{11.87}$$

Substituting the design equations results in

$$S_{\tau_h}^{H(s)} = -\left(\sum_{j=h}^{n} B_j s^j\right) / \left(\sum_{j=1}^{n} B_j s^j + 1\right) \quad h = 1, 2, ..., n \tag{11.88}$$

11.6 Determination of Maximum Signal Magnitude

In the following we present a method for determining the maximum input voltage $|V_{in}|_{max}$ of the filter when the maximum linear differential input voltages of the OTAs are given [10]. Noise performance analysis can be found elsewhere [11], [22]–[25].

We denote the maximum linear input voltage of the g_j OTA with V_{Tj}, and the voltage across the input terminals of the g_j OTA in the filter with V_{dj}. The relation in Eq. (11.89) must be met for all the related OTAs to ensure their operation in their respective linear regions.

$$|V_{dj}| \leq V_{Tj} \tag{11.89}$$

Using $H_j(s)$ to represent the transfer function of signal voltages from the filter input to the differential input of the g_j OTA, defined as

$$V_{dj} = H_j(s) V_{in} \tag{11.90}$$

and substituting Eq. (11.90) into Eq. (11.89) we have

$$|V_{in}| \leq V_{Tj} / |H_j(j\omega)|_{max} \tag{11.91}$$

From Eq. (11.91) we can see that for the given V_{Tj} of OTAs, by finding the maximum values of magnitude of concerned signal transfer functions we can obtain the maximum input voltage of the filter as

$$|V_{in}|_{max} = \min \left\{ V_{Tj} / |H_j(j\omega)|_{max} : \text{ for all related } j \right\} \tag{11.92}$$

For the third-order elliptic filter in Fig. 11.13, which is redrawn from Fig. 11.11(b) for convenience, the maximum allowable input voltage of the filter is given by

$$|V_{in}|_{max} = \min \left\{ \frac{V_{Ta1}}{|H_{a1}(j\omega)|_{max}}, \frac{V_{Ta3}}{|H_{a3}(j\omega)|_{max}}, \right.$$

$$\left.\frac{V_{T1}}{|H_1(j\omega)|_{\max}}, \frac{V_{T2}}{|H_2(j\omega)|_{\max}}, \frac{V_{T3}}{|H_3(j\omega)|_{\max}}\right\} \quad (11.93)$$

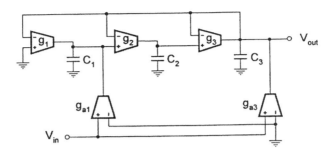

FIGURE 11.13
Third-order elliptic IFLF and input distribution filter structure.

It is easy to see that $H_{a1}(s) = H_{a3}(s) = 1$ and $H_1(s) = H(s)$ that has been already obtained in Section 11.4.3 and is rewritten below for convenience. $H_2(s)$ and $H_3(s)$ can also be formulated from Fig. 11.13 and are given below.

$$H_1(s) = \left(\beta_3 \tau_2 \tau_1 s^2 + \beta_1\right)/D(s),$$

$$H_2(s) = \left[(\beta_1 \tau_2 \tau_3 - \beta_3 \tau_2 \tau_1) s^2 + (\beta_1 \tau_2 - \beta_3 \tau_2) s\right]/D(s),$$

$$H_3(s) = \left[-\beta_3 \tau_2 \tau_1 s^2 + (\beta_1 \tau_3 - \beta_3 \tau_1) s - \beta_3\right]/D(s),$$

$$D(s) = \tau_3 \tau_2 \tau_1 s^3 + \tau_2 \tau_1 s^2 + \tau_1 s + 1 \quad (11.94)$$

When realizing the characteristic $H_d(s)$ in Eq. (11.40) with $A_3 = A_1 = 0$ using $H_1(s)$ with relations in Eqs. (11.42) and (11.52), the associated functions in the above become, respectively,

$$H_1(s) = \left(A_2 s^2 + A_0\right)/D_d(s),$$

$$H_2(s) = \left[(A_0 B_3/B_1 - A_2) s^2 + (A_0 B_2/B_1 - A_2/B_1) s\right]/D_d(s),$$

$$H_3(s) = \left[-A_2 s^2 + (A_0 B_3/B_2 - A_2 B_1/B_2) s - A_2/B_2\right]/D_d(s),$$

$$D_d(s) = B_3 s^3 + B_2 s^2 + B_1 s + 1 \quad (11.95)$$

For the normalized third-order elliptic lowpass filter with $A_2 = 0.588358$, $A_0 = 1$, $B_3 = 1.67029$, $B_2 = 1.41856$, $B_1 = 1.91391$ as shown in Section 11.4.3, using Eqs. (11.95) we can draw $|H_1(j\omega)|$, $|H_2(j\omega)|$ and $|H_3(j\omega)|$. From these graphs we can identify the maximum values of these magnitudes as

$$|H_1(j\omega)|_{\max} = 1, \quad |H_2(j\omega)|_{\max} = 1.07, \quad |H_3(j\omega)|_{\max} = 0.87$$

which appear at the normalized frequencies of 0.9, 1, and 1, respectively. Thus, together with $|H_{a1}(j\omega)|_{\max} = |H_{a3}(j\omega)|_{\max} = 1$ we obtain the maximum input voltage of the filter as

$$|V_{\text{in}}|_{\max} = \min\{V_{Ta1}, V_{Ta3}, V_{T1}, 0.93 V_{T2}, 1.149 V_{T3}\}$$

It is thus obvious that V_{T2} is crucial for making $|V_{\text{in}}|_{\max}$ as high as possible. In the event that all threshold voltages are identical and equal to V_T, we will have $|V_{\text{in}}|_{\max} = 0.93 V_T$.

11.7 Effects of OTA Frequency Response Nonidealities [10]

As discussed previously, OTA frequency nonidealities including finite input and output impedances and frequency dependence of the transconductance will cause deterioration of the filter performances. It should be noted that the input resistance of the CMOS OTA is usually very large and as a result it can be ignored in analysis. However the input resistance of the bipolar OTA must be taken into consideration, since it is comparable with the output resistance. Here our analysis is conducted for CMOS OTA-C filters. We can generally obtain the equivalent circuit of the filter incorporating the nonidealities and parasitics and formulate the real transfer function. The filter frequency performances can then be analyzed. Consider the third-order elliptic filter in Fig. 11.13, whose design in the ideal case was given in Section 11.4.3 and dynamic range was considered in Section 11.6.

For the convenience of comparison, the ideal transfer function of the circuit is repeated again as

$$H(s) = \frac{\beta_3 \tau_2 \tau_1 s^2 + \beta_1}{\tau_3 \tau_2 \tau_1 s^3 + \tau_2 \tau_1 s^2 + \tau_1 s + 1} \tag{11.96}$$

For the normalized elliptic characteristic with the cutoff frequency of 500 kHz, it has been determined in Section 11.4.3 that $g_{a1} = g_1 = 22.2 \mu S$, $g_2 = g_3 = 40.9 \mu S$, $g_{a3} = 17.0 \mu S$, $C_1 = 13.5 pF$, $C_2 = 9.7 pF$, $C_3 = 15.3 pF$.

Let us use, for example, C_{i2}, C_{o2}, G_{o2} and $g_2(s)$ to represent the input capacitance, output capacitance, output conductance, and frequency-dependent transconductance of the g_2 OTA, respectively, and C_{p2} denotes the parasitic capacitance of node 2. We also introduce such symbols as $C_1' = C_1 + C_{o1} + C_{oa1} + C_{p1}$, $C_2' = C_2 + C_{o2} + C_{p2}$, $C_3' = C_3 + C_{o3} + C_{oa3} + C_{i1} + C_{p3}$, $G_1 = G_{o1} + G_{oa1}$, $G_2 = G_{o2}$ and $G_3 = G_{o3} + G_{oa3}$. The OTA transconductance frequency dependence can be modeled in various ways, such as the one pole, two pole, one pole and one zero, two pole and one zero, and excess phase or phase shift. All the models can be reasonably simplified as $g_j(s) = g_j(1 - s/\omega_{bj})$, as mentioned in the previous chapters.

For example, the practical nonideal parameters of the CMOS OTAs with the above-calculated transconductance values are given by [21]

$$C_{i1} = C_{i2} = C_{i3} = C_d = 0.0385 pF,$$

$$C_{o1} = C_{o2} = C_{o3} = C_{oa1} = C_{oa3} = C_o = 0.52 pF,$$

$$G_{o1} = G_{oa1} = 56 nS, \quad G_{o2} = G_{o3} = 954 nS, \quad G_{oa3} = 39 nS,$$

$$1/\omega_{b1} = 1/\omega_{ba1} = 31.1 ns, \quad 1/\omega_{b2} = 1/\omega_{b3} = 2.57 ns, \quad 1/\omega_{ba3} = 33 ns$$

The common-mode input capacitances (of the same value for all the OTAs) are treated equivalently as the parasitic node capacitances of $C_{p1} = C_{p2} = 0.0502 pF$ and $C_{p3} = 0.1506 pF$.

When both OTA finite impedances and transconductance frequency dependence are taken into consideration the formulas will become very complicated. We therefore present only a first-order approximation of the effects.

$$H'(s) = \frac{A'_3 s^3 + A'_2 s^2 + A'_1 s + A'_0}{B'_3 s^3 + B'_2 s^2 + B'_1 s + B'_0} \quad (11.97)$$

where

$$A'_3 = -\frac{\beta_3}{\tau_3} \frac{C'_1}{C_1} \frac{C'_2}{C_2} \frac{1}{\omega_{ba3}}$$

$$A'_2 = \frac{\beta_3}{\tau_3} \frac{C'_1}{C_1} \frac{C'_2}{C_2} + \frac{\beta_3}{\tau_3} \frac{C'_1}{C_1} \frac{C_{i3}}{C_2} + \frac{\beta_1}{\tau_1} \frac{C'_2}{C_2} \frac{C_{i2}}{C_3} + \frac{\beta_3}{\tau_3} \frac{C'_2}{C_2} \frac{C_{i2}}{C_1}$$

$$A'_1 = \frac{\beta_1}{\tau_2 \tau_1} \frac{C_{i3}}{C_3} + \frac{\beta_3}{\tau_3} \frac{C'_2}{C_2} \frac{G_1}{C_1} + \frac{\beta_3}{\tau_3} \frac{C'_1}{C_1} \frac{G_2}{C_2} - \frac{\beta_1}{\tau_3 \tau_2 \tau_1} \left(\frac{1}{\omega_{ba1}} + \frac{1}{\omega_{b2}} + \frac{1}{\omega_{b3}} \right)$$

$$A'_0 = \frac{\beta_1}{\tau_3 \tau_2 \tau_1}$$

$$B'_3 = \frac{C'_1}{C_1} \frac{C'_2}{C_2} \frac{C'_3}{C_3} + \frac{C'_1}{C_1} \frac{C'_2}{C_2} \frac{C_{i2}}{C_3} + \frac{C'_2}{C_2} \frac{C'_3}{C_3} \frac{C_{i2}}{C_1} + \frac{C'_1}{C_1} \frac{C'_2}{C_2} \frac{C_{i3}}{C_3}$$

$$+ \frac{C'_1}{C_1} \frac{C'_3}{C_3} \frac{C_{i3}}{C_2} - \frac{1}{\tau_3} \frac{C'_1}{C_1} \frac{C'_2}{C_2} \frac{1}{\omega_{b3}}$$

$$B'_2 = \frac{1}{\tau_3} \frac{C'_1}{C_1} \frac{C'_2}{C_2} + \frac{1}{\tau_2} \frac{C'_1}{C_1} \frac{C_{i3}}{C_3} + \frac{1}{\tau_1} \frac{C'_2}{C_2} \frac{C_{i2}}{C_3} + \frac{1}{\tau_3} \frac{C'_2}{C_2} \frac{C_{i2}}{C_1} + \frac{G_3}{C_3} \frac{C'_1}{C_1} \frac{C'_2}{C_2}$$

$$+ \frac{G_2}{C_2} \frac{C'_1}{C_1} \frac{C'_3}{C_3} + \frac{G_1}{C_1} \frac{C'_2}{C_2} \frac{C'_3}{C_3} - \frac{1}{\tau_3 \tau_2} \frac{C'_1}{C_1} \left(\frac{1}{\omega_{b2}} + \frac{1}{\omega_{b3}} \right)$$

$$B'_1 = \frac{1}{\tau_3 \tau_2} \frac{C'_1}{C_1} + \frac{1}{\tau_2 \tau_1} \frac{C_{i3}}{C_3} + \frac{1}{\tau_3} \frac{C'_2}{C_2} \frac{G_1}{C_1} + \frac{1}{\tau_3} \frac{C'_1}{C_1} \frac{G_2}{C_2}$$

$$- \frac{1}{\tau_3 \tau_2 \tau_1} \left(\frac{1}{\omega_{b1}} + \frac{1}{\omega_{b2}} + \frac{1}{\omega_{b3}} \right)$$

$$B'_0 = \frac{1}{\tau_3 \tau_2 \tau_1} + \frac{1}{\tau_3 \tau_2} \frac{G_1}{C_1}$$

The filter performance is affected at all frequencies. Equation (11.97) and its coefficients indicate that filter frequency characteristics will perform differently from the ideal case of Eq. (11.96). In addition to the change of all coefficients, the numerator is increased in order $(A'_3 s^3)$ and has an unexpected term $(A'_1 s)$. The system poles and zeros will be accordingly altered. In particular, the zeros will vary from the ideal two imaginary to three general zeros.

Note that the effect of the parasitic capacitances C_{oj}, C_{pj}, and C_{i1} causes the element value shift. This effect may be eliminated by absorbing these capacitive parasitics into the circuit capacitances. However we must consider that such compensation design is by reducing the circuit capacitances to accommodate the parasitic capacitances. Therefore, comparatively, if the parasitic capacitances are too large, the circuit capacitances will become too small and the parasitic part will thus become dominant. The parasitics-induced element shift can also be effectively reduced by tuning.

It is important to note that each node in the circuit has a grounded capacitor and thus every node parasitic capacitance can be absorbed into the corresponding circuit capacitance and no parasitic poles will be produced.

The OTA input capacitances C_{i2} and C_{i3} and output conductances G_{oj} can have a major impact. The influences due to these nonidealities are mutually dependent as can be seen from the coefficient expressions of Eq. (11.97). However, equations for A'_0 and B'_0 do show that the finite output resistances mainly influence the low frequency response, and the input capacitances primarily affect the high-frequency characteristic as revealed by the expressions of A'_2 and B'_3. It can also be seen that the OTA transconductance frequency dependence degrades the high-frequency characteristic more than the low-frequency characteristic.

Generally the differential input application of OTAs causes feedthrough effects or unexpected signal paths due to finite input capacitances like C_{i2} and C_{i3} [20]. To overcome this problem single-input OTAs may be applied, since as recalled from Chapter 9, structures with differential-input OTAs can be equivalently converted into structures with single-ended input OTAs by splitting a differential-input OTA into two single-ended input OTAs of the same transconductance but opposite polarity. However, this may need double the number of OTAs compared with the differential input applications of OTAs, if the type of OTA remains the same. The large number of active devices will lead to other problems, for example, the increase of the equivalent node conductances G_j and the circuit capacitance spread. Also the overall effect of OTA excess phase will be accordingly increased. This is also unfavorable for power consumption and the chip area. These trade-offs between the two applications must be accounted for in some situations, especially in the design of high-order filters.

The reader is encouraged to conduct (Spice or Matlab) simulation using the data given above to observe the deviation of the nonideal performance from the ideal one to confirm the theoretical analysis of the effects of OTA nonidealities in the above. The PSpice result can be found in [10].

For high-order OTA-C filter design we have so far introduced the cascade (biquads in Chapters 8 and 9), ladder simulation (Chapter 10) and multiple integrator loop feedback (this chapter) approaches. To this point, the reader may also be interested to compare these three methods, in details, generally or using some typical realizations. For this consideration, we mention that reference [13] has shown that the multiple integrator loop feedback structures may be advantageous over the cascade and ladder simulation architectures in terms of generality of function, simplicity of structure, and insensitivity to OTA nonidealities.

11.8 Summary

A general multiple loop feedback approach for the realization of OTA-C filters has been proposed. The systematic generation, analysis, and design of different filter configurations have been addressed with emphasis on canonical structures. We have formulated general relations for all-pole and arbitrary transmission zero realizations and exemplified the theory extensively. The method described in the chapter has the following advantages: (a) it is systematic and general due to the introduction of the feedback matrix [F] and the relationship between [F] and the feedback connection (some well known filter configurations are simply special cases of the approach); (b) a variety of new structures with different performances can be generated, with both canonical and noncanonical realizations being available; (c) all capacitors are grounded and canonical realization can guarantee that all internal nodes have a grounded capacitor; (d) it is also flexible in assigning element values and in various cases simple explicit design formulas are applicable.

The essence of the method is the establishment of the relationship between the filter structure and the feedback matrix, which makes systematic structure generation and general analysis and design equation formulation possible. Using the one-to-one correspondence between the feedback connection matrix and the circuit configuration one can deal with any particular applications based on these general equations. For example, if the circuit topology is known, we may write the feedback matrix [F] and analyze the filtering characteristic and sensitivity performance. For the desired transfer function, on the other hand, an [F], that is, a circuit structure, may be defined to realize the transfer function.

In the chapter, we have demonstrated the general expressions for sensitivity computation and results for different structures have been given. We have also analyzed the effects of OTA nonidealities, which embrace finite impedances, phase shift, and nonlinearity. The OTA impedances and excess phase cause a shift of filter frequency characteristics from ideal ones. In particular they change the pole positions and hence pose a potential risk of instability, while OTA nonlinearity and noise limit the filter dynamic range.

In the chapter we have mainly dealt with canonical filter structures. Noncanonical realization with OTA amplifiers realizing general feedback coefficients can provide some design flexibility and result in more architectures. They will also cause problems due to the greater number of OTAs needed and resistive nodes introduced, such as poor frequency performance, the large chip area, and power consumption. Trade-offs between the feedthrough effects (parasitic zeros) and the problems related to the large number of OTAs must also be considered when deciding whether to exploit the differential input or single input OTAs, as the differential input application gives rise to unwanted signal paths, whereas the single input application results in the increase in the number of OTAs.

The general unbalanced models can be converted into the balanced equivalents by using differential four-input and two-output OTAs in integrators and mirroring the feedback network in the upper part to the lower part. This will be discussed in the next chapter. Note that the OTA-C filters in this chapter are based on voltage integrators and voltage feedback. A general multiple current-integrator loop current-feedback model can be similarly established using dual output OTAs (DO-OTA) and capacitors, which will also be studied in Chapter 12.

From the viewpoint of the whole book, we have so far discussed opamp-RC filters (Chapters 4–7) and OTA-C filters (Chapters 8–11). In recent years many other new high-performance structures and design methods of continuous-time active filters have been proposed and widely used in practice. In Chapter 12 we shall introduce these new topics, commercially used or in the research front, to satisfy different needs of the readers.

References

[1] Tsividis, Y.P. and Voorman, J.O., Eds., *Integrated continuous-time filters: principles, design and applications,* IEEE Press, 1993.

REFERENCES

[2] Schaumann, R., Ghausi, M.S., and Laker, K.R., *Design of Analog Filters: Passive, Active RC and Switched Capacitor*, Prentice-Hall, Englewood Cliffs, NJ, 1990.

[3] Laker, K.R. and Sansen, W., *Design of Analog Integrated Circuits and Systems*, McGraw-Hill, New York, 1994.

[4] Urbaś, A. and Osiowski, J., High-frequency realization of C-OTA second-order active filters, *Proc. IEEE Int. Symp. Circuits Syst.*, 1106–1109, 1982.

[5] Sánchez-Sinencio, E., Geiger, R.L., and Nevarez-Lozano, H., Generation of continuous-time two integrator loop OTA filter structures, *IEEE Trans. Circuits Syst.*, 35(8), 936–946, 1988.

[6] Sun, Y. and Fidler, J.K., Novel OTA-C realizations of biquadratic transfer functions, *Int. J. Electronics*, 75, 333–348, 1993.

[7] Guo, W., Liu, J., and Yang, S., The realization of high-order OTA-C filter, *Int. J. Electronics*, 65(6), 1153–1157, 1988.

[8] Sun, Y. and Fidler, J.K., OTA-C realization of general high-order transfer functions, *Electronics Letters*, 29, 1057–1058, 1993.

[9] Sun, Y. and Fidler, J.K., Minimum component multiple integrator loop OTA-grounded capacitor all-pole filters, *Proc. IEEE Midwest Symp. on Circuits and Systems*, 983–986, Lafayette, USA, 1994.

[10] Sun, Y. and Fidler, J.K., Synthesis and performance analysis of a universal minimum component integrator-based IFLF OTA-grounded capacitor filter, *IEE Proceedings: Circuits, Devices and Systems*, 143, 107–114, 1996.

[11] Sun, Y. and Fidler, J.K., Performance analysis of multiple loop feedback OTA-C filters, *Proc. IEE 14th Saraga Colloquium on Digital and Analogue Filters and Filtering Systems*, 9/1–9/7, London, 1994.

[12] Sun, Y. and Fidler, J.K., Structure generation and design of multiple loop feedback OTA-grounded capacitor filters, *IEEE Trans. Circuits Syst., Part-I*, 43, 1–11, 1997.

[13] Sun, Y., Jefferies, B., and Teng, J., Universal third-order OTA-C filters, *Int. J. Electronics*, 85(5), 597–609, 1998.

[14] Nedungadi, A.P. and Geiger, R.L., High-frequency voltage-controlled continuous-time low-pass filter using linearized CMOS integrators, *Electronics Letters*, 22, 729–731, 1986.

[15] Chiang, D.H. and Schaumann, R., A CMOS fully-balanced continuous time IFLF filter design for read/wire channels, *Proc. IEEE Int. Symp. Circuits Syst.*, 1, 167–170, 1996.

[16] Sun, Y. and Fidler, J.K., Some design methods of OTA-C and CCII-RC active filters, *Proc. 13th IEE Saraga Colloquium on Digital and Analogue Filters and Filtering Systems*, 7/1–7/8, 1993.

[17] Nawrocki, R., Electronically tunable all-pole low-pass leapfrog ladder filter with operational transconductance amplifier, *Int. J. Electronics*, 62(5), 667–672, 1987.

[18] Nawrocki, R., Electronically controlled OTA-C filter with follow-the-leader-feedback structures, *Int. J. Circuit Theory and Applications*, 16, 93–96, 1988.

[19] Krummenacher, F., Design considerations in high-frequency CMOS transconductance amplifier capacitor (TAC) filters, *Proc. IEEE Int. Symp. Circuits Syst.,* 100–105, 1989.

[20] Nevárez-Lozano, H., Hill, J.A., and Sánchez-Sinencio, E., Frequency limitations of continuous-time OTA-C filters, *Proc. IEEE Int. Symp. Circuits Syst.,* 2169–2172, 1988.

[21] Nevárez-Lozano, H. and Sánchez-Sinencio, E., Minimum parasitic effects biquadratic OTA-C filter architectures, *Analog Integrated Circuits and Signal Processing,* 1, 297–319, 1991.

[22] Espinosa, G., Montecchi, F., Sánchez-Sinencio, E., and Maloberti, F., Noise performance of OTA-C Filters, *Proc. IEEE Int. Symp. Circuits Syst.,* 2173–2176, 1988.

[23] Abidi, A.A., Noise in active resonators and the available dynamic range, *IEEE Trans. Circuits Syst.,* 39, 296–299, 1992.

[24] Groenewold, G., The design of high dynamic range continuous-time integratable bandpass filters, *IEEE Trans. Circuits Syst.,* 38, 838–852, 1991.

[25] Park, C.S. and Schaumann, R., Design of a 4-MHz analog integrated CMOS transconductance-C bandpass filter, *IEEE J. Solid-State Circuits,* 23(4), 987–996, 1988.

[26] Hiser, D.L. and Geiger, R.L., Impact of OTA nonlinearities on the performance of continuous-time OTA-C bandpass filters, *Proc. IEEE Int. Symp. Circuits Syst.,* 1167–1170, 1990.

[27] Szczepanski, S. and Schaumann, R., Nonlinearity-induced distortion of the transfer function shape in high-order OTA-C filters, *Analog Integrated Circuits and Signal Processing,* 3(2), 143–151, 1993.

[28] Sun, Y., *Analysis and synthesis of impedance matching networks and transconductance amplifier filters,* Ph.D. Thesis, University of York, UK, 1996.

[29] Sun, Y. and Fidler, J.K., Structure generation of current-mode two integrator loop dual output-OTA grounded capacitor filters, *IEEE Trans. Circuits Syst., Part II,* 43, 659–663, 1996.

[30] Sun, Y. and Fidler, J.K., Current-mode multiple loop feedback filters using dual output OTAs and grounded capacitors, *Int. J. of Circuit Theory and Applications,* 25, 69–80, 1997.

[31] Sun, Y. and Fidler, J.K., Current-mode OTA-C realisation of arbitrary filter characteristics, *Electronics Letters,* 32, 1181–1182, 1996.

[32] Schoeffler, J.D., The synthesis of minimum sensitivity networks, *IEEE Trans. Circuit Theory,* 11, 271–276, 1964.

[33] Snelgrove, W.M. and Sedra, A.S., Synthesis and analysis of state-space active filters using intermediate transfer functions, *IEEE Trans. Circuits Syst.,* CAS-33(3), 287–301, 1986.

[34] Roberts, G.W. and Sedra, A.S., A generalization of intermediate transfer function analysis to arbitrary linear networks, *Proc. IEEE ISCAS,* 1059–1062, 1989.

[35] Sedra, A.S. and Brackett, P.O., *Filter Theory and Design: Active and Passive,* Matrix Publishers, 1978.

[36] Heinlein, W.E. and Holmes, W.H., *Active Filters for Integrated Circuits: Fundamentals and Design Methods,* R. Oldenbourg Verlag Gmbh, Munchen, 1974.

[37] Moschytz, G.S., *Linear Integrated Networks: Design,* Van Nostrand Reinhold, NJ, 1975.

REFERENCES

[38] Tow, J., Design and evaluation of shifted companion form (follow the leader feedback) active filters, *Proc. IEEE Int. Symp. Circuits Syst.,* 656–660, 1974.

[39] Perry, D.J., New multiple feedback active RC network, *Electronics Letters,* 11(16), 364–368, 1975.

[40] Laker, K.R. and Ghausi, M.S., Synthesis of a low-sensitivity multiloop feedback active RC filter, *IEEE Trans. Circuits Syst.,* 21(2), 252–259, 1974.

[41] Laker, K.R., Schaumann, R., and Ghausi, M.S., Multiple-loop feedback topologies for the design of low-sensitivity active filters, *IEEE Trans. Circuits Syst.,* 26(1), 1–21, 1979.

[42] Girling, F.E.J. and Good, E.F., Active filters: part 12, the leapfrog or active-ladder synthesis, *Wireless World,* July, 340–345, 1970.

Chapter 12

Current-Mode Filters and Other Architectures

12.1 Introduction

Continuous-time active filter design has developed very rapidly over the last few years [1]–[60]. In the previous chapters we have basically concentrated on two main types: opamp-based active-RC filters and OTA-based active-C filters. We have also confined ourselves to the differential-input and single-output opamp and OTA. There have also been many other advanced design methods which have been utilized in practice. Multiple-input and multiple-output opamps and OTAs as well as other new active devices such as current conveyors have also been available. This chapter will introduce these new design approaches and filter structures.

Recent advances in analog integrated circuits and signal processing have shown that the current-mode approach is superior to the voltage-mode in terms of its wide bandwidth, high speed, low voltage and power, large dynamic range, and simplicity in circuit structure [7, 8]. Current-mode signal processing techniques have also been widely used, in particular, in high performance filtering applications [12]–[25], [35]–[49]. For example, in the simulation of passive LC ladders, rather than converting mixed voltage and current equations to the voltage-only counterparts, designers now scale the mixed equations to the current-only ones [19, 20, 41, 42]. Even in the realization of a voltage transfer function, it is now preferred to cascade a transadmittance function, a current transfer function, and a transimpedance function section to benefit from the current-mode approach, in contrast with the conventional active RC design that uses only voltage sections.

The OTA-C filter structures we have discussed in the previous chapters are mainly based on voltage integrators, voltage amplifiers, and voltage feedback. They are very convenient for voltage signal processing and voltage description with voltage inputs to OTA input terminals and voltage outputs from circuit nodes. For current signal processing and description we can expect that circuits based on current integrators, current amplifiers, and current feedback should be straightforward, with current inputs to circuit nodes and current outputs from OTA output terminals. Although single-output OTAs can be used to construct individual current integrators and current amplifiers, they cannot readily provide local or overall current outputs or current feedback. Therefore dual- or multiple-output OTAs (DO-OTA or MO-OTA) [10, 11] are needed for current-mode signal processing.

The current-mode DO-OTA filters were first studied in [12, 13], where a number of current-mode single or two DO-OTA filters and a current-mode Tow-Thomas or resonator type two integrator loop filter were developed. These filters can be implemented using only DO-OTAs and capacitors by replacing the resistor with the DO-OTA simulation [10]. Recently, a large number of papers on current-mode continuous-time integrated OTA filters have been published [14]–[25]. This chapter will investigate current-mode DO-OTA-C filters. A comprehensive set of first- and second-order filter structures are generated based on a current-mode single DO-OTA and five-admittance model in Section 12.2. Current-mode DO-OTA-C biquadratic architectures of two integrator loop configurations are generated in Section 12.3. Current-mode DO-OTA-C filter design based on passive ladder simulation is discussed in Section 12.4. We also deal with generation and synthesis of current-mode multiple integrator loop DO-OTA-C filters in Section 12.5.

The current-mode DO-OTA-C filters in Sections 12.2 through 12.5 correspond to the voltage-mode OTA-C filters in Chapters 8 through 11, respectively. Having those well-known advantages of voltage-mode OTA-C filters such as simplicity in structure, programmability, suitability for high frequencies, and full integration, the current-mode DO-OTA-C filters can also directly process the current signals with no need of any additional conversion components. More important is that the current-mode DO-OTA-C structures may have different, possibly better, performances such as distortion and noise than the voltage-mode counterparts, although they may have the same sensitivity [8]. The network transpose [9] and adjoint circuit [8] methods may be used to convert voltage-mode circuits to current-mode counterparts, but for the reader's convenience (to avoid the need for prior knowledge of voltage-mode OTA-C filters and some other circuit theories relating the transpose and adjoint concepts) we choose to present this chapter in a relatively independent way, rather than from the viewpoint of converting the voltage-mode OTA-C filters in Chapters 8–11 to the current-mode DO-OTA-C counterparts.

Besides the current-mode DO-OTA-C filters, many other popular continuous time filter structures such as the balanced opamp-RC and OTA-C configurations, MOSFET-C filters, OTA-C-opamp structures and active filters using current conveyors are also introduced in this chapter, which is the subject of Section 12.6. Section 12.7 summarizes the whole chapter.

12.2 Current-Mode Filters Based on Single DO-OTA Model

In Chapter 8 we discussed filter structures using a single-OTA model with voltage input, output, and feedback. Single-OTA filters are cheap to build, consume less power, and have better noise performance. Furthermore, OTA-C filters can be easily obtained from the single-OTA counterparts by OTA simulation of the resistor. In this section we introduce a current-mode model using a single DO-OTA, with current input, output, and feedback and generate the corresponding current-mode filters using a similar method.

12.2.1 General Model and Filter Architecture Generation

The symbol of the dual-output OTA (DO-OTA) is shown in Fig. 12.1(a) and its ideal equivalent circuit is given in Fig. 12.1(b). The circuit of the DO-OTA can be a simple differential amplifier with two outputs, an OTA with addition of a current mirror to its output (to provide multiple current outputs), or a series connection of two OTAs with the same transconductances and opposite polarities. The DO-OTA can be conveniently used to construct various current-mode filters.

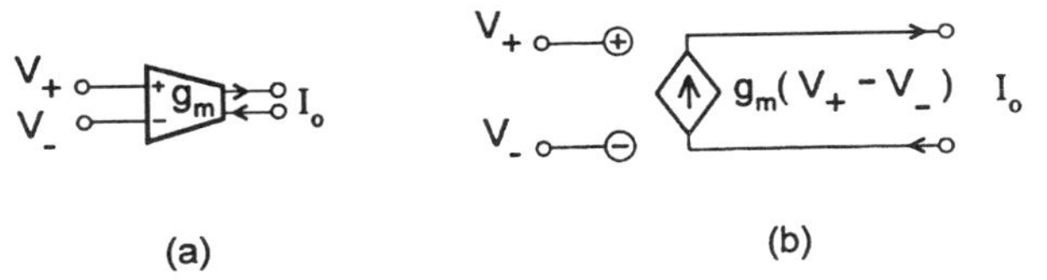

FIGURE 12.1
Symbol and equivalent circuit of DO-OTA.

Consider the general model in Fig. 12.2 [14], which consists of one DO-OTA and five admittances, with current input, current output, and current feedback. The current transfer function, $H(s) = I_o/I_i$

12.2. CURRENT-MODE FILTERS BASED ON DO-OTA MODEL

can be shown as

$$H(s) = \frac{g_m Y_2 Y_4}{Y_1 Y_2 Y_4 + Y_1 Y_2 Y_5 + Y_1 Y_3 Y_4 + Y_1 Y_3 Y_5 + Y_1 Y_4 Y_5 + Y_2 Y_3 Y_4 + Y_2 Y_3 Y_5 + Y_2 Y_4 Y_5 + g_m Y_2 Y_4} \quad (12.1)$$

which is the same in form as the general expression for the voltage-mode counterpart in Chapter 8. Thus similar design techniques may be used. Sensitivity performances will be the same, which will be shown in Section 12.2.3.

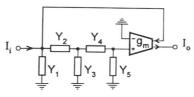

FIGURE 12.2
Current-mode model with one DO-OTA and five admittances.

Different filter structures and characteristics can be realized using the general circuit model and transfer function. This can be done by assigning different components to Y_j and checking the corresponding transfer functions in Eq. (12.1). For example, Y_j can be a resistor ($Y_j = g_j$), a capacitor ($Y_j = sC_j$), an open circuit ($Y_j = 0$), or a short circuit ($Y_j = \infty$). Both first-order and second-order filter structures can be obtained. In the following we will present a set of filter structures based on this model.

12.2.1.1 First-Order Filter Structures

For first-order filters five admittances are too much. We thus set $Y_4 = \infty$ and $Y_5 = 0$ and choose the other three admittances. Selecting $Y_1 = sC_1$, $Y_2 = \infty$, and $Y_3 = 0$ gives rise to the simplest structure as shown in Fig. 12.3(a), which has a lowpass filter function given by

$$H(s) = \frac{g_m}{sC_1 + g_m} \quad (12.2)$$

with the dc gain equal to unity and the cutoff frequency equal to g_m/C_1.

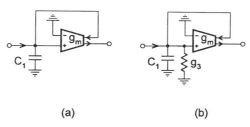

(a)   (b)

FIGURE 12.3
Two simple first-order configurations.

The circuit in Fig. 12.3(b), corresponding to $Y_1 = sC_1$, $Y_2 = \infty$ and $Y_3 = g_3$ has the lowpass characteristic as

$$H(s) = \frac{g_m}{sC_1 + (g_3 + g_m)} \quad (12.3)$$

In Fig. 12.4 we present a set of first-order filters with one capacitor and two resistors. It is first verified that when choosing $Y_1 = sC_1$, $Y_2 = g_2$ and $Y_3 = g_3$, the general model produces a lowpass filter, that is

$$H(s) = \frac{g_m g_2}{s(g_2 + g_3)C_1 + g_2(g_3 + g_m)} \qquad (12.4)$$

The circuit is shown in Fig. 12.4(a).

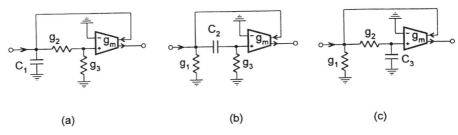

(a)   (b)   (c)

FIGURE 12.4
First-order configurations with two resistors.

Then consider the circuit in Fig. 12.4(b), which is obtained by setting $Y_1 = g_1$, $Y_2 = sC_2$, and $Y_3 = g_3$. It is found that a highpass filter is derived whose current transfer function is given by

$$H(s) = \frac{s g_m C_2}{s(g_1 + g_3 + g_m)C_2 + g_1 g_3} \qquad (12.5)$$

with the gain at the infinity frequency being $g_m/(g_1 + g_3 + g_m)$ and the cutoff frequency equal to $g_1 g_3/[(g_1 + g_3 + g_m)C_2]$.

Finally, if Y_1 and Y_2 are resistors and Y_3 a capacitor, then $H(s)$ is of the lowpass characteristic. The circuit is presented in Fig. 12.4(c) and the current transfer function is given below.

$$H(s) = \frac{g_m g_2}{s(g_1 + g_2)C_3 + g_2(g_1 + g_m)} \qquad (12.6)$$

12.2.1.2 Second-Order Filter Architectures

Suppose that each admittance is realized with one component and two and only two capacitors are used. Exhaustive search by trying all different combinations of components shows that a total of 13 different second-order structures can be derived: one highpass (HP), four bandpass (BP), and three lowpass (LP) filters with three resistors; two bandpass and two lowpass filters with two resistors; as well as one lowpass filter with one resistor. The combinations of components for the 13 structures are presented in Table 12.1. The corresponding configurations and transfer functions can be readily derived from the general model in Fig. 12.2 and the general expression in Eq. (12.1). The transfer functions of the three resistor filters in Table 12.1 are the same as the respective counterparts in Section 8.7 of Chapter 8. We will show LP and BP filters with one or two resistors in Section 12.2.3.

12.2.2 Passive Resistor and Active Resistor

Many interesting second-order filters using a single DO-OTA, two capacitors, and different number of resistors have been developed. The resistors can be passive resistors, as in discrete design. They can also be active resistors as in integrated circuits. The active resistor can be realized by a DO-OTA connected in the way as shown in Fig. 12.5 [10]. When both terminals A and B are floating, it

12.2. CURRENT-MODE FILTERS BASED ON DO-OTA MODEL

Table 12.1 Generation of Second-Order Filter Structures Based on Model in Fig. 12.2

Type	Components				
	Y_1	Y_2	Y_3	Y_4	Y_5
General					
$LP1$ [12]	sC_1	g_2	sC_3	∞	0
$LP2$	sC_1	g_2	sC_3	∞	g_5
$LP3$	g_1	∞	sC_3	g_4	sC_5
$LP4$	sC_1	g_2	sC_3	g_4	g_5
$LP5$	g_1	g_2	sC_3	g_4	sC_5
$LP6$	sC_1	g_2	g_3	g_4	sC_5
$BP1$	g_1	sC_2	sC_3	∞	g_5
$BP2$	g_1	∞	sC_3	sC_4	g_5
$BP3$	g_1	sC_2	sC_3	g_4	g_5
$BP4$ [14]	g_1	g_2	sC_3	sC_4	g_5
$BP5$	g_1	sC_2	g_3	g_4	sC_5
$BP6$	sC_1	g_2	g_3	sC_4	g_5
HP [14]	g_1	sC_2	g_3	sC_4	g_5

is a floating resistor and when terminal B is grounded, it is a grounded resistor. In both cases the conductance of the resistor is equal to the transconductance of the DO-OTA. Filters whose resistors all are active resistors will comprise only DO-OTAs and capacitors and thus are DO-OTA-C filters.

FIGURE 12.5
DO-OTA simulation of resistor.

The transfer functions of the DO-OTA-C filters with active resistors are the same as those of the single DO-OTA counterparts with passive resistors and their sensitivity performances are thus also the same due to one to one correspondence between the passive resistor and active resistor.

The differences are that the passive resistor filters may have low power consumption, noise, and parasitic effects, while the active resistor filters are suitable for full integration and their tunability is improved. The number of DO-OTAs in the DO-OTA-C filters is equal to the number of resistors plus one, since only one DO-OTA is required for the simulation of both grounded and floating resistors.

In the following discussion we show some second-order filter architectures derived from the general model in Fig. 12.2 based on components in Table 12.1. We will not distinguish the passive and active resistors and simply say the resistor, unless otherwise stated. Therefore, symbol g_j will mean the conductance of a passive resistor or the transconductance of the active resistor.

12.2.3 Design of Second-Order Filters

The simplest second-order lowpass filter with $Y_1 = sC_1$, $Y_2 = g_2$, $Y_3 = sC_3$, $Y_4 = \infty$, $Y_5 = 0$ of Fig. 12.1 is shown in Fig. 12.6 [12], which has only one resistor. The current transfer function in Eq. (12.1) becomes

$$H(s) = \frac{g_m g_2}{s^2 C_1 C_3 + s g_2 (C_1 + C_3) + g_m g_2} \tag{12.7}$$

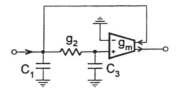

FIGURE 12.6
Simplest second-order lowpass filter.

As mentioned before, the standard form of the lowpass characteristic is normally written as

$$H_d(s) = \frac{K\omega_o^2}{s^2 + \frac{\omega_o}{Q}s + \omega_o^2} \tag{12.8}$$

Comparison of Eqs. (12.7) and (12.8) indicates that the dc gain of the filter, K, is unity and

$$\omega_o = \sqrt{\frac{g_m g_2}{C_1 C_3}}, \quad Q = \sqrt{\frac{g_m}{g_2}} \frac{\sqrt{C_1 C_3}}{C_1 + C_3} \tag{12.9}$$

For convenience of design and also from the viewpoint of cost we set $C_1 = C_3 = C$. This permits the development of simple design formulas for the component values, given by

$$C_1 = C_3 = C, \quad g_2 = \frac{\omega_o C}{2Q}, \quad g_m = 2Q\omega_o C \tag{12.10}$$

where C can be arbitrarily assigned. The sensitivities are found to be:

$$S_{g_m}^{\omega_o} = S_{g_2}^{\omega_o} = -S_{C_1}^{\omega_o} = -S_{C_3}^{\omega_o} = 1/2 \tag{12.11}$$

$$S_{g_m}^{Q} = -S_{g_2}^{Q} = \tfrac{1}{2}, \quad -S_{C_1}^{Q} = S_{C_3}^{Q} = \tfrac{1}{2}\frac{C_1 - C_3}{C_1 + C_3} = 0 \tag{12.12}$$

and these results indicate superior sensitivity performance. Note that setting $C_1 = C_3$ leads not only to practical convenience, but also to a decrease in the sensitivity of the filter to deviations in the capacitor design values. It is therefore clear from the above discussion that like the voltage-mode counterpart in Chapter 8 the current-mode DO-OTA lowpass filter has a very simple structure, minimum component count, very simple design formulas, and extremely low sensitivity.

We now consider lowpass and bandpass filters with two resistors. The lowpass filter with $Y_1 = sC_1$, $Y_2 = g_2$, $Y_3 = sC_3$, $Y_4 = \infty$, $Y_5 = g_5$ is depicted in Fig. 12.7(a). Its current transfer function is derived as

$$H(s) = \frac{g_m g_2}{s^2 C_1 C_3 + s\left[(g_2 + g_5)C_1 + g_2 C_3\right] + g_2(g_m + g_5)} \tag{12.13}$$

Comparing the transfer function in Eq. (12.13) with the desired function in Eq. (12.8) and selecting $C_1 = C_3 = C$ and $g_2 = g_5 = g$ yields the following formulas:

$$g = \frac{\omega_o C}{3Q}, \quad g_m = 3Q\omega_o C\left(1 - \frac{1}{9Q^2}\right), \quad K = 1 - \frac{1}{9Q^2} \tag{12.14}$$

12.2. CURRENT-MODE FILTERS BASED ON DO-OTA MODEL

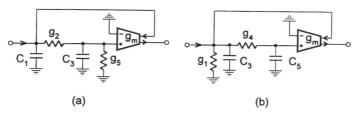

FIGURE 12.7
Lowpass filters with two resistors.

The sensitivities can be derived as

$$S_{C_1}^{\omega_o} = S_{C_3}^{\omega_o} = -S_{g_2}^{\omega_o} = -1/2, \quad S_{g_5}^{\omega_o} = 1/18Q^2,$$

$$S_{g_m}^{\omega_o} = 0.5\left(1 - 1/9Q^2\right) \tag{12.15}$$

$$S_{C_1}^Q = -S_{C_3}^Q = S_{g_2}^Q = -1/6, \quad S_{g_5}^Q = -1/3 + 1/18Q^2,$$

$$S_{g_m}^Q = S_{g_m}^{\omega_o} \tag{12.16}$$

$$S_{C_1}^K = S_{C_3}^K = S_{g_2}^K = 0, \quad -S_{g_5}^K = S_{g_m}^K = 1/9Q^2 \tag{12.17}$$

It can be seen from these results that the structure in Fig. 12.7(a) has very low sensitivity.

The second lowpass filter which corresponds to $Y_1 = g_1$, $Y_2 = \infty$, $Y_3 = sC_3$, $Y_4 = g_4$, $Y_5 = sC_5$, is shown in Fig. 12.7(b). It has the transfer function

$$H(s) = \frac{g_m g_4}{s^2 C_3 C_5 + s\left[g_4 C_3 + (g_1 + g_4)C_5\right] + g_4(g_m + g_1)} \tag{12.18}$$

Note that this lowpass filter is similar to the one discussed above.

The bandpass filter with two resistors, corresponding to $Y_1 = g_1$, $Y_2 = sC_2$, $Y_3 = sC_3$, $Y_4 = \infty$, $Y_5 = g_5$ is shown in Fig. 12.8(a). The current transfer function is derived as

$$H(s) = \frac{s g_m C_2}{s^2 C_2 C_3 + s\left[(g_1 + g_5 + g_m)C_2 + g_1 C_3\right] + g_1 g_5} \tag{12.19}$$

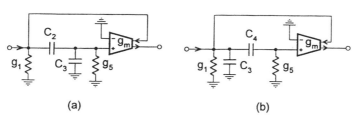

FIGURE 12.8
Bandpass filters with two resistors.

Comparing Eq. (12.19) with the ideal bandpass characteristic

$$H_d(s) = \frac{K\frac{\omega_o}{Q}s}{s^2 + \frac{\omega_o}{Q}s + \omega_o^2} \qquad (12.20)$$

leads to the following design equation with $C_2 = C_3 = C$ and $g_1 = g_5 = g$

$$g = \omega_o C, \quad g_m = \frac{\omega_o C}{Q}(1 - 3Q), \quad K = 1 - 3Q \qquad (12.21)$$

For practical Q values, g_m is negative. Note that $g_m < 0$ simply means the interchange of the DO-OTA output terminals.

The sensitivities of the filter are formulated as

$$S_{C_2}^{\omega_o} = S_{C_3}^{\omega_o} = -S_{g_1}^{\omega_o} = -S_{g_5}^{\omega_o} = -1/2, \quad S_{g_m}^{\omega_o} = 0 \qquad (12.22)$$

$$-S_{C_2}^Q = S_{C_3}^Q = S_{g_5}^Q = 1/2 - Q, \quad S_{g_1}^Q = 1/2 - 2Q,$$

$$S_{g_m}^Q = -1 + 3Q \qquad (12.23)$$

$$S_{C_2}^K = -S_{C_3}^K = -S_{g_5}^K = Q, \quad S_{g_1}^K = -2Q, \quad S_{g_m}^K = 3Q \qquad (12.24)$$

From the sensitivity results, it can be observed that the design using the circuit in Fig. 12.8(a) has low ω_o sensitivity. However, the Q and K sensitivities display a modest Q dependence, although this is no problem for low Q design. The realization of large Q with the DO-OTA output terminals interchanged may cause an increase in the Q and K sensitivities. But considering that it is the ω_o sensitivity that contributes more to the response deviation and the K sensitivity is less important, the designs are still useful for not very large Q, since the ω_o sensitivities are extremely low.

The second bandpass filter with two resistors is associated with $Y_1 = g_1$, $Y_2 = \infty$, $Y_3 = sC_3$, $Y_4 = sC_4$, $Y_5 = g_5$, as shown in Fig. 12.8(b). Its transfer function is given by

$$H(s) = \frac{sg_m C_4}{s^2 C_3 C_4 + s[g_5 C_3 + (g_1 + g_5 + g_m)C_4] + g_1 g_5} \qquad (12.25)$$

This filter function is similar to that of the above bandpass filter.

12.2.4 Effects of DO-OTA Nonidealities

The practical DO-OTA has finite input and output conductances and capacitances. It also has nonzero phase shift (finite bandwidth). Using $Y_{mi} = G_{mi} + sC_{mi}$, $Y_{mo} = G_{mo} + sC_{mo}$ and $g_m(s) = \frac{g_{m0}}{1+s/\omega_{mb}} \approx g_{m0}(1 - s/\omega_{mb})$ to represent the finite input admittance, output admittance, and transconductance frequency dependence of the g_m DO-OTA, the general model in Fig. 12.2 taking these nonidealities into consideration will have a changed current transfer function. The output short-circuit current transfer function is given by

$$H'(s) = \frac{I_o}{I_i} = \left[1 + \frac{Y_{mo}}{g_m(s)}H_v(s)\right]H_i(s) \qquad (12.26)$$

12.2. CURRENT-MODE FILTERS BASED ON DO-OTA MODEL

where

$$H_i(s) = g_m(s)Y_2Y_4/[(Y_1 + Y_{mo})Y_2Y_4 + (Y_1 + Y_{mo})Y_2(Y_5 + Y_{mi})$$

$$+ (Y_1 + Y_{mo})Y_3Y_4 + (Y_1 + Y_{mo})Y_3(Y_5 + Y_{mi})$$

$$+ (Y_1 + Y_{mo})Y_4(Y_5 + Y_{mi}) + Y_2Y_3Y_4 + Y_2Y_3(Y_5 + Y_{mi})$$

$$+ Y_2Y_4(Y_5 + Y_{mi}) + g_m(s)Y_2Y_4] \qquad (12.27)$$

$$H_v(s) = 1 + \frac{Y_3}{Y_2} + \frac{Y_2 + Y_3 + Y_4}{Y_2Y_4}(Y_5 + Y_{mi}) \qquad (12.28)$$

Using these relations we can analyze the effects of the DO-OTA nonidealities on filter performance for all structures by substituting corresponding Y_j components in Table 12.1. Of course, we can also evaluate the effects by directly considering individual filter structures. We are not going into the detail here, but the results are, in general, that the nonideal excess phase will have a Q-enhancement effect, the input and output conductances introduce losses, causing reduction of the pole and zero quality factors and the low-frequency gain. The DO-OTA capacitances degrade the high-frequency characteristic of the filter. They may produce parasitic poles for some structures. We should also note that the output capacitance of the dual-output OTA generates parasitic zeros as seen from Eq. (12.26), compared with the input capacitance of the differential input OTA in voltage-mode OTA-C circuits, which produce extra zeros in Chapter 8.

In fully integrated filter design the resistor is simulated by a DO-OTA as shown in Fig. 12.5. The nonidealities of such DO-OTAs will also have an impact on the filter performance. Taking the differential input and output admittances Y_{Ri} and Y_{Ro} and frequency-dependent transconductance $g_R(s)$ into account, the DO-OTA resistor will have an equivalent admittance given by

$$Y_R = g_R(s) + Y_{Ri} + Y_{Ro}$$

$$= (g_{R0} + G_{Ri} + G_{Ro}) + s(C_{Ri} + C_{Ro} - g_{R0}/\omega_{Rb}) \qquad (12.29)$$

This reveals that the resistor conductance is changed and also there appears a nonideal capacitor with the resistor. If the common-mode input and output admittances are also taken into consideration, then a π equivalent circuit will result for the floating resistor, where the series arm floating admittance is given by Eq. (12.29) and the two parallel arm grounded admittances have the same value, given by the sum of the common-mode input and output admittances. For the grounded resistor, the equivalent circuit will be a changed grounded admittance given by the admittance in Eq. (12.29) plus the sum of the common-mode input and output admittances.

For IC filter design the effects of the g_m OTA imperfections can still be assessed using Eqs. (12.26)–(12.28). As discussed for the voltage-mode circuits in Chapter 8, to systematically evaluate the effects of the nonidealities of the resistor DO-OTAs on different filter architectures, we can simply replace resistor conductances in individual transfer functions or the corresponding Y_j in Eqs. (12.26)–(12.28) by associated modified admittances. We can also substitute the nonideal equivalent circuit of the resistor into the filter circuit and analyze the effects of the nonidealities of the resistor simulation DO-OTAs.

12.3 Current-Mode Two Integrator Loop DO-OTA-C Filters

In Chapter 9 we have extensively investigated OTA-C biquadratic filters of the two integrator loop configuration based on voltage integrators, voltage amplifiers, and voltage feedback networks. Two integrator loop configurations have been widely employed in practical applications. In this section we realize the two integrator loop structures from the current-mode viewpoint. The resulting DO-OTA-C filters are based on current integrators, current amplifiers, and current feedback.

12.3.1 Basic Building Blocks and First-Order Filters [15]

The basic DO-OTA based current-mode building blocks are indicated in Fig. 12.9, where Figs. 12.9(a), (b), and (c) are the current integrator, amplifier, and summer, respectively. The current summer is realized simply with the circuit node.

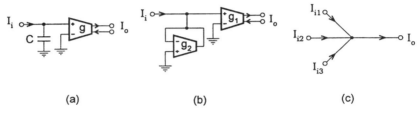

FIGURE 12.9
Basic current-mode DO-OTA-C building blocks.

First-order current-mode filters are given in Fig. 12.10. Figure 12.10(a) has the lowpass function

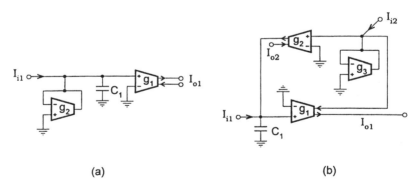

FIGURE 12.10
First-order current-mode DO-OTA-C configurations.

$$I_{o1} = I_{i1}/(\tau_1 s + k_1) \tag{12.30}$$

where $\tau_1 = C_1/g_1$ and $k_1 = g_2/g_1$. For the feedback circuit in Fig. 12.10(b), with $\tau_1 = C_1/g_1$ and $k_1 = g_2/g_3$ we have the current transfer functions as

$$I_{o1} = (I_{i1} - k_1 I_{i2})/(\tau_1 s + k_1),$$

12.3. CURRENT-MODE TWO INTEGRATOR LOOP FILTERS

$$I_{o2} = (k_1 I_{i1} + k_1 \tau_1 s I_{i2})/(\tau_1 s + k_1) \quad (12.31)$$

The two relations in Eq. (12.31) show that the filter in Fig. 12.10(b) can fulfill lowpass and highpass specifications. The circuits in Figs. 12.10(a) and (b) with I_{o1} and I_{i1} are also called lossy integrators. The two simple lossy integrators given in Fig. 12.3 are also useful.

12.3.2 Current-Mode DO-OTA-C Configurations with Arbitrary k_{ij} [15]

Two integrator loop configurations in Fig. 9.3.1 of Chapter 9 will be recalled. Using the basic current building blocks in Fig. 12.9 and the first-order current circuits in Figs. 12.3 and 12.10 we can also realize the two integrator loop architectures in the current domain. Because grounded capacitors have some advantages, and for second-order filters two capacitors are sufficient, we will use two grounded capacitors in all the current-mode DO-OTA-C realizations in this section.

The current-mode architectures corresponding to Fig. 9.3 are obtained as illustrated in Fig. 12.11. Figure 12.11(a) uses six DO-OTAs and the structure in Fig. 12.11(b) has five DO-OTAs. The pole parameter relations including ω_o, ω_o/Q and Q for the two realizations are displayed in Table 12.2. The zeros of different required characteristics can be realized using different inputs and outputs.

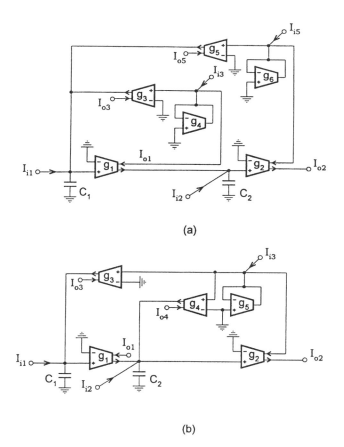

FIGURE 12.11
Second-order current-mode DO-OTA-C filters with arbitrary k_{ij}.

Table 12.2 Parameter Relations of General Current-Mode DO-OTA-C Realizations

Circuits	τ_1	τ_2	k_{11}	k_{22}	k_{12}	ω_o	$\frac{\omega_o}{Q}$	Q
Fig.12.11(a)	$\frac{C_1}{g_1}$	$\frac{C_2}{g_2}$	$\frac{g_3}{g_4}$		$\frac{g_5}{g_6}$	$\sqrt{\frac{g_5g_3g_2}{g_6C_1C_2}}$	$\frac{g_3g_1}{g_4C_1}$	$\frac{g_4}{g_3}\sqrt{\frac{g_5g_2C_1}{g_6g_1C_2}}$
Fig.12.11(b)	$\frac{C_1}{g_1}$	$\frac{C_2}{g_2}$		$\frac{g_4}{g_5}$	$\frac{g_3}{g_5}$	$\sqrt{\frac{g_3g_1g_2}{g_5C_1C_2}}$	$\frac{g_4g_2}{g_5C_2}$	$\frac{1}{g_4}\sqrt{\frac{g_3g_5g_1C_2}{g_2C_1}}$

To show how current signals are processed by these filters we apply current inputs to circuit nodes and take current outputs from DO-OTA output terminals. For the circuit in Fig. 12.11(a) we formulate the current transfer functions as

$$\left.\begin{aligned}
D_1(s)I_{o1} &= \tau_2 s I_{i1} - k_{12} I_{i2} + k_{11}\tau_2 s I_{i3} + k_{12}\tau_2 s I_{i5} \\
D_1(s)I_{o2} &= I_{i1} + (\tau_1 s + k_{11}) I_{i2} + k_{11} I_{i3} + k_{12} I_{i5} \\
D_1(s)I_{o3} &= -k_{11}\tau_2 s I_{i1} + k_{11}k_{12} I_{i2} + k_{11}\left(\tau_1\tau_2 s^2 + k_{12}\right) I_{i3} \\
&\quad - k_{11}k_{12}\tau_2 s I_{i5} \\
D_1(s)(I_{o3}+I_{o5}) &= -(k_{11}\tau_2 s + k_{12}) I_{i1} - k_{12}\tau_1 s I_{i2} \\
&\quad + k_{11}\tau_1\tau_2 s^2 I_{i3} + k_{12}\tau_1\tau_2 s^2 I_{i5} \\
D_1(s) &= \tau_1\tau_2 s^2 + k_{11}\tau_2 s + k_{12} = \tau_1\tau_2\left(s^2 + \frac{\omega_o}{Q}s + \omega_o^2\right)
\end{aligned}\right\} \quad (12.32)$$

Looking at the expressions in Eq. (12.32), especially those of I_{o3} and $I_{o3} + I_{o5}$ and noting in particular the contributions of I_{i3} we can see that the architecture in Fig. 12.11(a) can perform LP, BP, HP, and BS filtering functions. It is noted that the g_1 DO-OTA in Fig. 12.11(a) may be designed to have another output terminal, if necessary, to give a direct I_{o1} output.

The current relations of Fig. 12.11(b) are derived as

$$\left.\begin{aligned}
D_2(s)I_{o1} &= (\tau_2 s + k_{22}) I_{i1} - k_{12} I_{i2} + k_{12}\tau_2 s I_{i3} \\
D_2(s)I_{o2} &= I_{i1} + \tau_1 s I_{i2} + (k_{22}\tau_1 s + k_{12}) I_{i3} \\
D_2(s)I_{o3} &= -k_{12} I_{i1} - k_{12}\tau_1 s I_{i2} + k_{12}\tau_1\tau_2 s^2 I_{i3} \\
D_2(s)I_{o4} &= -k_{22} I_{i1} - k_{22}\tau_1 s I_{i2} + k_{22}\tau_1\tau_2 s^2 I_{i3} \\
D_2(s) &= \tau_1\tau_2 s^2 + k_{22}\tau_1 s + k_{12} = \tau_1\tau_2\left(s^2 + \frac{\omega_o}{Q}s + \omega_o^2\right)
\end{aligned}\right\} \quad (12.33)$$

Equation (12.33) indicates that the structure in Fig. 12.11(b) supports LP, BP, and HP filters, the output from I_{o3} or I_{o4} with I_{i3} making the HP contribution to the multifunctionality.

12.3.3 Current-Mode DO-OTA-C Biquadratic Architectures with $k_{12} = k_{jj}$

Consider the four DO-OTA circuit in the Fig. 12.12(a) [16], with $k_{12} = k_{11} = k = g_3/g_4$ and $\tau_j = C_j/g_j$, we can derive the current transfer relations as shown below:

$$\left.\begin{aligned}
D_1(s)I_{o1} &= \tau_2 s I_{i1} - k I_{i2} + k\tau_2 s I_{i3} \\
D_1(s)I_{o2} &= I_{i1} + (\tau_1 s + k) I_{i2} + k I_{i3} \\
D_1(s)I_{o3} &= -k(\tau_2 s + 1) I_{i1} - k\tau_1 s I_{i2} + k\tau_1\tau_2 s^2 I_{i3} \\
D_1(s) &= \tau_1\tau_2 s^2 + k\tau_2 s + k
\end{aligned}\right\} \quad (12.34)$$

From Eq. (12.34), we can see that the filter configuration in Fig. 12.12(a) offers the LP, BP, and HP characteristics.

12.3. CURRENT-MODE TWO INTEGRATOR LOOP FILTERS

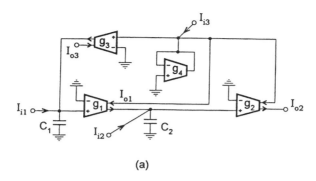

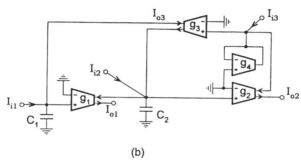

FIGURE 12.12
Current-mode DO-OTA-C filters with (a) $k_{12} = k_{11}$ and (b) $k_{12} = k_{22}$.

The four DO-OTA filter structure with $k_{12} = k_{22} = k = g_3/g_4$ and $\tau_j = C_j/g_j$ is shown in Fig. 12.12(b). We can formulate

$$\left.\begin{aligned} D_2(s)I_{o1} &= (s\tau_2 + k)I_{i1} + kI_{i2} - sk\tau_2 I_{i3} \\ D_2(s)I_{o2} &= -I_{i1} + s\tau_1 I_{i2} + (s\tau_1 + 1)kI_{i3} \\ D_2(s)I_{o3} &= kI_{i1} - sk\tau_1 I_{i2} + s^2 k\tau_1\tau_2 I_{i3} \\ D_2(s) &= \tau_1\tau_2 s^2 + k\tau_1 s + k \end{aligned}\right\} \quad (12.35)$$

Thus the circuit has the LP, BP, and HP functions. To output the HP function the g_3 OTA may need another output.

12.3.4 Current-Mode DO-OTA-C Biquadratic Architectures with $k_{12} = 1$[15]

The filter circuit with $k_{12} = 1$ is shown in Fig. 12.13(a), which has four DO-OTAs. The equations of the circuit reduce to

$$\left.\begin{aligned} D_1(s)I_{o1} &= \tau_2 s I_{i1} - I_{i2} + k_{11}\tau_2 s I_{i3} \\ D_1(s)I_{o2} &= I_{i1} + (\tau_1 s + k_{11})I_{i2} + k_{11}I_{i3} \\ D_1(s)I_{o3} &= -k_{11}\tau_2 s I_{i1} + k_{11}I_{i2} + k_{11}\left(\tau_1\tau_2 s^2 + 1\right)I_{i3} \\ D_1(s)(I_{o3} - I_{o2}) &= -(k_{11}\tau_2 s + 1)I_{i1} - \tau_1 s I_{i2} + k_{11}\tau_1\tau_2 s^2 I_{i3} \\ D_1(s) &= \tau_1\tau_2 s^2 + k_{11}\tau_2 s + 1 \end{aligned}\right\} \quad (12.36)$$

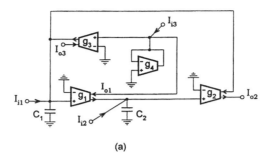

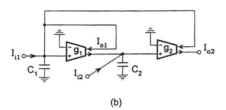

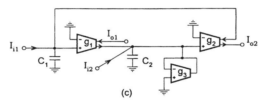

FIGURE 12.13
Current-mode DO-OTA-C biquads with $k_{12} = 1$.

The parameter relations are exhibited in Table 12.3. Figure 12.13(a) and Eq. (12.36) show that this circuit saves two DO-OTAs and realizes LP, BP, HP, and BS filters, compared with the general circuit in Fig. 12.11(a). Note that I_{i3} plays a special role in achieving the multifunctionality and the output $I_{o3} - I_{o2}$ can be realized by simply connecting the two corresponding terminals together due to their respective directions.

A two-DO-OTA realization is associated with the feedback coefficients $k_{11} = k_{12} = 1$ as shown in Fig. 12.13(b). In this case all the feedback paths reduce to simple pure wire connections, thus resulting in the simplest or canonical structure with the LP and BP functions given by Eq. (12.37)

$$D_1(s)I_{o1} = \tau_2 s I_{i1} - I_{i2}, \, D_1(s)I_{o2} = I_{i1} + (\tau_1 s + 1) I_{i2},$$

$$D_1(s) = \tau_1 \tau_2 s^2 + \tau_2 s + 1 \tag{12.37}$$

The special three DO-OTA realization which corresponds to $k_{12} = 1$ is given in Fig. 12.13(c) [13]. The parameter relations of the special realization are presented in Table 12.3 and the associated characteristic is written as

$$D_2(s)I_{o1} = (\tau_2 s + k_{22}) I_{i1} - I_{i2}, \, D_2(s)I_{o2} = I_{i1} + \tau_1 s I_{i2},$$

$$D_2(s) = \tau_1 \tau_2 s^2 + k_{22} \tau_2 s + 1 \tag{12.38}$$

12.3. CURRENT-MODE TWO INTEGRATOR LOOP FILTERS

Table 12.3 Parameter Relations of Special Current-Mode DO-OTA-C Realizations

Circuits	τ_1	τ_2	k_{11}	k_{22}	k_{12}	ω_o	$\dfrac{\omega_o}{Q}$	Q
Fig. 12.13(a)	$\dfrac{C_1}{g_1}$	$\dfrac{C_2}{g_2}$	$\dfrac{g_3}{g_4}$		1	$\sqrt{\dfrac{g_1 g_2}{C_1 C_2}}$	$\dfrac{g_3 g_1}{g_4 C_1}$	$\dfrac{g_4}{g_3}\sqrt{\dfrac{g_2 C_1}{g_1 C_2}}$
Fig. 12.13(b)	$\dfrac{C_1}{g_1}$	$\dfrac{C_2}{g_2}$	1		1	$\sqrt{\dfrac{g_1 g_2}{C_1 C_2}}$	$\dfrac{g_1}{C_1}$	$\sqrt{\dfrac{g_2 C_1}{g_1 C_2}}$
Fig. 12.13(c)	$\dfrac{C_1}{g_1}$	$\dfrac{C_2}{g_2}$		$\dfrac{g_3}{g_2}$	1	$\sqrt{\dfrac{g_1 g_2}{C_1 C_2}}$	$\dfrac{g_3}{C_2}$	$\dfrac{1}{g_3}\sqrt{\dfrac{g_1 g_2 C_2}{C_1}}$

12.3.5 DO-OTA Nonideality Effects

In this section we evaluate the effects of DO-OTA nonidealities including the finite input and output impedances as well as the transconductance frequency dependence.

Note that all OTAs in the generated architectures are used with one of the input terminals grounded. For circuits in Figs. 12.13(b) and 12.13(c) where there are capacitors on all nodes, DO-OTA input capacitances and parasitic node capacitances can be compensated by absorption design. In Figs. 12.11(a), (b), Figs. 12.12(a), (b) and Fig. 12.13(a), the input capacitances of the DO-OTAs in the feedforward path and the parasitic node capacitances on the nodes with capacitors can be absorbed by the corresponding circuit capacitances. However, the input capacitances of the DO-OTAs and the parasitic node capacitances related to the resistive nodes (from the nodes to ground there is only a DO-OTA resistor) in the feedback path(s) in these circuits will cause an increase in the order of system transfer functions, a change influencing the filter characteristics in very high-frequency designs.

The most influential parasitics are perhaps the DO-OTA output capacitances and conductances. Note that the DO-OTA may be realized by using the differential output transconductance amplifier. In this case the finite output capacitance and conductance of the DO-OTA will degrade the filter performance, causing parasitic zeros, especially the output impedance of the g_1 DO-OTA in the configurations in Figs. 12.11(a), 12.12(a), 12.13(a), 12.13(b), and the g_3 DO-OTA in Fig. 12.12(b). The DO-OTA may also be implemented by using a single-ended output transconductance amplifier followed by a current mirror or by connecting two OTAs of the same transconductances but different polarities in series. In these cases the finite output impedances may be modeled as connected from each output terminal to ground and thus the DO-OTA output capacitances related to the capacitive nodes in all the architectures in Figs. 12.11–12.13 may be absorbed by circuit capacitances. However, the finite DO-OTA output capacitances associated with the resistive nodes in Figs. 12.11(a), (b), Fig. 12.12(a), (b), and Fig. 12.13(a) will give rise to parasitic system poles. Also, any parasitic floating capacitances and resistances between the two output terminals of the DO-OTAs may deteriorate filter performance.

12.3.6 Universal Current-Mode DO-OTA-C Filters

From the formulated current transfer expressions in Eqs. (12.32)–(12.38) it is seen that the proposed filters in Figs. 12.11–12.13 all have a multifunction feature, supporting more than two functions of LP, BP, HP, and BS at different input or output positions. Inspecting the different structures in Figs. 12.11–12.13 we can also see that the structures contains two to six DO-OTAs.

As we discussed in Chapter 9, the general biquadratic characteristic can also be realized by adding more OTAs to the basic structures. For example, the input distribution OTA network in Fig. 12.14(a) can be connected to the two integrator loop filter structures to produce universal biquads. Note that the current distributor converts a single current input to weighted multiple current inputs, $I_{ij} = \alpha_j I_{in}$ where $\alpha_j = g_{aj}/g_r$ are the distribution coefficients. The coefficients can be selected to produce arbitrary transmission zeros and gain without any influence on the poles. Note, in particular, that the sign of α_j can be easily changed by interchanging the input terminals of the g_{aj} OTA, if needed.

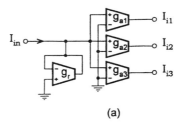

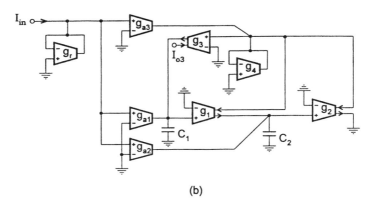

FIGURE 12.14
Input distribution network and universal biquad example.

For example, connecting the distribution network in Fig. 12.14(a) to the circuit in Fig. 12.11(b) and taking output from I_{o3}, we can obtain from Eq. (12.33) the general current transfer function

$$H(s) = \frac{I_{o3}}{I_{in}} = k_{12} \frac{\alpha_3 \tau_1 \tau_2 s^2 - \alpha_2 \tau_1 s - \alpha_1}{\tau_1 \tau_2 s^2 + k_{22} \tau_1 s + k_{12}} \quad (12.39)$$

which can support a variety of filter functions such as the LP, BP, HP, notch, and allpass characteristics.

For Fig. 12.12(a) with the distribution network and with output from I_{o3}, from Eq. (12.34) we can derive

$$H(s) = \frac{I_{o3}}{I_{in}} = k \frac{\alpha_3 \tau_1 \tau_2 s^2 - (\alpha_2 \tau_1 + \alpha_1 \tau_2) s - \alpha_1}{\tau_1 \tau_2 s^2 + k \tau_2 s + k} \quad (12.40)$$

As an example of the distribution method we give the resulting universal biquad in Fig. 12.14(b).

The general transfer function for Fig. 12.12(b) with the distribution network and output from I_{o3} is obtained from Eq. (12.35) as

$$H(s) = \frac{I_{o3}}{I_{in}} = k \frac{\alpha_3 \tau_1 \tau_2 s^2 - \alpha_2 \tau_1 s + \alpha_1}{\tau_1 \tau_2 s^2 + k \tau_1 s + k} \quad (12.41)$$

The circuit in Fig. 12.13(a), to which the distribution network is connected also has a general

12.3. CURRENT-MODE TWO INTEGRATOR LOOP FILTERS

function from I_{o3}, which can be derived from Eq. (12.36) and is given by

$$H(s) = \frac{I_{o3}}{I_{in}} = k_{11} \frac{\alpha_3 \tau_1 \tau_2 s^2 - \alpha_1 \tau_2 s + (\alpha_2 + \alpha_3)}{\tau_1 \tau_2 s^2 + k_{11} \tau_2 s + 1} \tag{12.42}$$

We can also add a weighted current summer to the two integrator loop filters to combine relevant current outputs for a certain input to generate any desired zeros. This can be done in two ways: one is to directly sum the relevant output currents, each of the weights consisting of two-OTA current amplifiers [16]; the other is to sum the related output currents via node voltages, which requires just one extra OTA for each summed current. The latter method requires just half the number of OTAs needed by the former and is thus used below. The summation network is shown in Fig. 12.15(a).

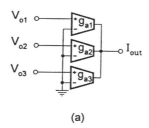

(a)

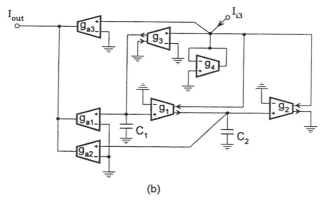

(b)

FIGURE 12.15
Output summation network and universal biquad example.

For the structures in Figs. 12.11(b), 12.12(a), 12.12(b), and 12.13(a), from Eqs. (12.33)–(12.36) it can be observed that summing I_{o1}, I_{o2} and I_{o3} with weights, with I_{i3} being as input we will have universal transfer functions. Thus connecting the summation OTA network in Fig. 12.15(a) to the circuits in, for example, Figs. 12.11(b), 12.12(a), 12.12(b), and 12.13(a) will result in four corresponding universal biquads. The universal biquad based on Fig. 12.12(a) and the summation method is representatively shown in Fig. 12.15(b). With $\beta_j = g_{aj}/g_j$ we attain the associated current transfer functions $H(s) = I_{out}/I_{i3}$ of these universal biquads, given by, respectively,

$$H(s) = \frac{\beta_3 k_{12} \tau_1 \tau_2 s^2 + (\beta_2 k_{22} \tau_1 + \beta_1 k_{12} \tau_2) s + \beta_2 k_{12}}{\tau_1 \tau_2 s^2 + k_{22} \tau_1 s + k_{12}} \tag{12.43}$$

$$H(s) = k \frac{\beta_3 \tau_1 \tau_2 s^2 + \beta_1 \tau_2 s + \beta_2}{\tau_1 \tau_2 s^2 + k \tau_2 s + k} \tag{12.44}$$

$$H(s) = k\frac{\beta_3\tau_1\tau_2 s^2 + (\beta_2\tau_1 - \beta_1\tau_2)s + \beta_2}{\tau_1\tau_2 s^2 + k\tau_1 s + k} \quad (12.45)$$

$$H(s) = k_{11}\frac{\beta_3\tau_1\tau_2 s^2 + \beta_1\tau_2 s + (\beta_3 + \beta_2)}{\tau_1\tau_2 s^2 + k_{11}\tau_2 s + 1} \quad (12.46)$$

Finally, two universal biquads based on the canonical structure of Fig. 12.13(b) are obtained with a distribution network and a summation network, as shown in Figs. 12.16(a) and (b), respectively. With $\alpha_j = g_{aj}/g_r$, j=0, 1, 2 and $\tau_j = C_j/g_j$, the current transfer function of Fig. 12.16(a) is derived

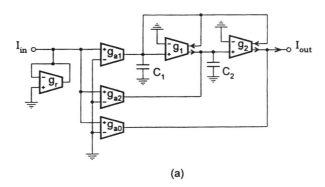

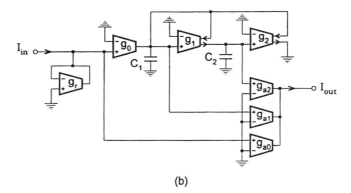

FIGURE 12.16
Universal current-mode DO-OTA-C biquads based on canonical structure.

as

$$H(s) = \frac{\alpha_0\tau_1\tau_2 s^2 + (\alpha_0\tau_2 + \alpha_2\tau_1)s + (\alpha_0 + \alpha_2 + \alpha_1)}{\tau_1\tau_2 s^2 + \tau_2 s + 1} \quad (12.47)$$

With $\beta_j = g_{aj}/g_j$ and $\tau_j = C_j/g_j$ we derive the general transfer function of Fig. 12.16(b) as

$$H(s) = \gamma\frac{\beta_0\tau_1\tau_2 s^2 + (\beta_0 + \beta_1)\tau_2 s + (\beta_0 + \beta_2)}{\tau_1\tau_2 s^2 + \tau_2 s + 1} \quad (12.48)$$

These universal structures have the input summed in order to achieve the second-order term s^2. This is different from the universal biquads obtained above which do not need to include the input in the summation since the outputs in the basic structures contain s^2 terms.

12.4 Current-Mode DO-OTA-C Ladder Simulation Filters

Active filters derived from passive LC ladders have very low sensitivity. We have already discussed opamp-RC and OTA-C filter design based on LC ladder simulation in Chapter 6 and Chapter 10, respectively. In the simulation of passive LC ladders, the original equations are of the mixed current and voltage type. In opamp-RC and OTA-C realizations we convert these equations to voltage-only counterparts by scaling. In this section we convert these mixed equations to the current-only equivalents [19, 20, 41, 42] and realize the corresponding current signal flow diagram using current-mode DO-OTA-C building blocks. Discussion will be in parallel with that in Chapter 10 for ease of understanding and comparison.

12.4.1 Leapfrog Simulation Structures of General Ladder

The general ladder network with series admittances and parallel impedances is recalled from Chapter 10 and shown in Fig. 12.17. The equations relating the currents flowing in the series arms,

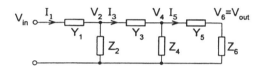

FIGURE 12.17
General admittance and impedance ladder with signals indicated.

I_j, and the voltages across the parallel arms, V_j, are also recalled and repeated below.

$$I_1 = Y_1(V_{in} - V_2), \quad V_2 = Z_2(I_1 - I_3), \quad I_3 = Y_3(V_2 - V_4),$$

$$V_4 = Z_4(I_3 - I_5), \quad I_5 = Y_5(V_4 - V_6), \quad V_{out} = V_6 = Z_6 I_5 \quad (12.49)$$

The transfer function V_{out}/V_{in} can be obtained from these equations by eliminating the intermediate variables.

The mixed current and voltage signal equations are now converted by scaling into the counterparts with current signals only. Scaled by a conductance g_m, Eq. (12.49) can be written as

$$I_1 = \frac{Y_1}{g_m}(I'_{in} - I'_2), \quad I'_2 = g_m Z_2(I_1 - I_3), \quad I_3 = \frac{Y_3}{g_m}(I'_2 - I'_4),$$

$$I'_4 = g_m Z_4(I_3 - I_5), \quad I_5 = \frac{Y_5}{g_m}(I'_4 - I'_6),$$

$$I'_{out} = I'_6 = g_m Z_6 I_5 \quad (12.50)$$

where $I'_j = g_m V_j$. The Y_j/g_m and $g_m Z_i$ are current transfer functions. It is clear that this equation will lead to the same transfer function $I'_{out}/I'_{in} = V_{out}/V_{in}$ as that from Eq. (12.49). The corresponding leapfrog (LF) signal flow diagram is shown in Fig. 12.18.

We can synthesize the current summers and current transfer functions of Y_j/g_m and $g_m Z_i$ using DO-OTAs and grounded impedances. We prefer using grounded impedances because grounded

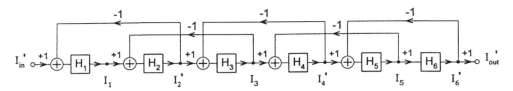

FIGURE 12.18
Current-mode leapfrog block diagram of general ladder.

capacitors and grounded DO-OTA resistors can absorb some DO-OTA finite input and output capacitances and conductances, respectively, thus reducing parasitic effects. From Eq. (12.50) we can see that the current relations have a typical form of

$$J_j = H_j \left(J_{j-1} - J_{j+1} \right) \tag{12.51}$$

where J_j can be I_j or I'_j, and

$$H_j = Y_j/g_m, \quad \text{for odd } j; \quad H_j = g_m Z_j, \quad \text{for even } j \tag{12.52}$$

Equation (12.51) can be realized using a DO-OTA with a transconductance of g_j and a grounded impedance of

$$Z'_j = H_j/g_j \tag{12.53}$$

as shown in Fig. 12.19. This is a DO-OTA-grounded impedance section. The summation operation is simply realized by the circuit node. It can be verified that the current transfer function from the OTA input to output is equal to $g_j Z'_j = H_j$. Note that we relate the current transfer function H_j to the grounded impedance Z'_j. Thus the current transfer function realization can now become the simulation of the normal grounded impedance.

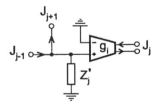

FIGURE 12.19
Current-mode DO-OTA-grounded impedance section.

Using Fig. 12.19 as a building block we can readily obtain the OTA-grounded impedance LF structure from Eq. (12.50) or Fig. 12.18, as shown in Fig. 12.20. From Eqs. (12.52), (12.53) we can show the grounded impedances have the values, given by

$$Z'_1 = \tfrac{1}{g_1 g_m} Y_1, \quad Z'_2 = \tfrac{g_m}{g_2} Z_2, \quad Z'_3 = \tfrac{1}{g_3 g_m} Y_3, \quad Z'_4 = \tfrac{g_m}{g_4} Z_4,$$

$$Z'_5 = \tfrac{1}{g_5 g_m} Y_5, \quad Z'_6 = \tfrac{g_m}{g_6} Z_6 \tag{12.54}$$

which are the same as the corresponding formulas in Chapter 10.

12.4. CURRENT-MODE LADDER SIMULATION FILTERS

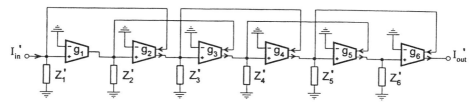

FIGURE 12.20
General current-mode LF DO-OTA-grounded impedance realization.

From Eq. (12.54) we can see that besides the general scaling by g_m, each new grounded impedance has a separate transconductance which can be used to adjust the impedance level. We can also note that Z'_j are not the original impedances Z_j in the ladder. For the even number subscript, Z'_j is the original impedance Z_j in the parallel arm of the ladder multiplied by the ratio of g_m/g_j, while for the odd number subscript, Z'_j is the inversion of the original impedance Z_j or the admittance Y_j in the series arm divided by the product of $g_j g_m$. When $g_j = g_m = g$, we have $Z'_j = Y_j/g^2$ for odd j and $Z'_j = Z_j$ for even j. Further if $g_j = g_m = 1$, then $Z'_j = Y_j$ for odd j.

From the above discussion we can see that for any concrete passive LC ladder, the DO-OTA-C realization problem becomes the DO-OTA-C realization of the grounded impedances and the simple inductor substitution method can be conveniently used to simulate the impedance constituents. In the following we introduce DO-OTA-C structures derived from different passive LC ladders using the above general method.

We mention here that if voltage input and output are preferred, at the input end an OTA with transconductance g_m can be used to convert input voltage V_{in} to input current I'_{in} and a grounded resistor simulated by an OTA with transconductance g_m can be connected to the output end to convert output current I'_{out} to output voltage V_{out}.

12.4.2 Current-Mode DO-OTA-C Lowpass LF Filters

Consider the fifth-order all-pole LC ladder with termination resistors in Fig. 12.21(a). Comparing the circuit with the general ladder in Fig. 12.17 gives $Y_1 = 1/R_1$, $Z_2 = 1/sC_2$, $Y_3 = 1/sL_3$, $Z_4 = 1/sC_4$, $Y_5 = 1/sL_5$ and $Z_6 = 1/(sC_6 + 1/R_6)$. The circuit equations accordingly become

$$I_1 = \frac{1}{R_1}(V_{in} - V_2), \quad V_2 = \frac{1}{sC_2}(I_1 - I_3), \quad I_3 = \frac{1}{sL_3}(V_2 - V_4),$$

$$V_4 = \frac{1}{sC_4}(I_3 - I_5), \quad I_5 = \frac{1}{sL_5}(V_4 - V_6),$$

$$V_{out} = V_6 = \frac{1}{sC_6 + 1/R_6} I_5 \qquad (12.55)$$

Scaling Eq. (12.55) by the factor of g_m results in current functions H_j given in Eq. (12.52) and realized in the way as shown in Fig. 12.19, where grounded impedances Z'_j are given in Eq. (12.53). The DO-OTA-C filter structure is given in Fig. 12.21(b). For given R_j, C_j, and L_j, we can compute the new parameter values as

$$g'_1 = g_1 g_m R_1, \quad C'_2 = \frac{g_2}{g_m} C_2, \quad C'_3 = g_3 g_m L_3, \quad C'_4 = \frac{g_4}{g_m} C_4,$$

$$C'_5 = g_5 g_m L_5, \quad C'_6 = \frac{g_6}{g_m} C_6, \quad g'_6 = \frac{g_6}{g_m}\frac{1}{R_6} \qquad (12.56)$$

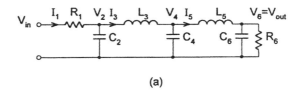

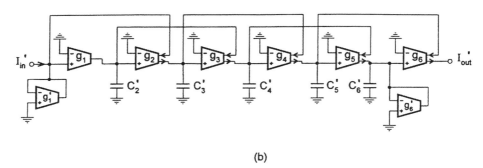

(b)

FIGURE 12.21
Fifth-order all-pole LC ladders and LF DO-OTA-C realization.

The values can be adjusted overall by g_m and individually by g_j. Two design techniques can be utilized. One is to make all transconductances identical, that is, $g_1 = g_2 = g_3 = g_4 = g_5 = g_6 = g$ with different capacitances which can be calculated from Eq. (12.56) as $C'_j = g g_m L_j$ when j is odd; $C'_j = C_j g/g_m$ when j is even. The other is to select the same value for all capacitances, that is, $C'_2 = C'_3 = C'_4 = C'_5 = C'_6 = C$ with different transconductances which are determined from Eq. (12.56) as $g_j = C/g_m L_j$, for odd j and $g_j = g_m C/C_j$, for even j. In many cases g_m can be chosen to be unity. The filter structure consists of only current integrators and summers. When $R_6 = 1/g_m$, the g'_6 termination OTA can be removed if the inverting terminal of the g_6 OTA is connected to the output.

Now we consider a fifth-order finite zero passive LC ladder, as shown in Fig. 12.22(a). Similarly, identifying that

$$Y_1 = \tfrac{1}{R_1}, \ Z_2 = \tfrac{1}{sC_2}, \ Y_3 = sC_3 + \tfrac{1}{sL_3}, \ Z_4 = \tfrac{1}{sC_4},$$

$$Y_5 = sC_5 + \tfrac{1}{sL_5}, \ Z_6 = \tfrac{1}{sC_6 + 1/R_6} \tag{12.57}$$

and following the same design procedure we can obtain the DO-OTA-C counterpart as shown in Fig. 12.22(b). The difference from the all-pole type is in Y_3 and Y_5 which are a combination admittance of two components and involve two steps. Taking Y_3 as an example, we first have the corresponding grounded impedance as

$$Z'_3 = \frac{Y_3}{g_3 g_m} = sL'_3 + \frac{1}{sL_3 g_3 g_m} \tag{12.58}$$

where $L'_3 = C_3/g_3 g_m$. The second term in the equation represents a capacitance of the value $C'_3 = L_3 g_3 g_m$. But the first term is equivalent to an inductor. This should then be further replaced by an OTA-C inductor with $L'_3 = C''_3/g'_3 g''_3$. Combining the two steps we can also obtain C''_3 in

12.4. CURRENT-MODE LADDER SIMULATION FILTERS

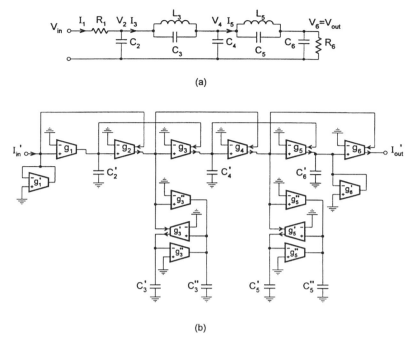

FIGURE 12.22
Fifth-order finite zero passive LC and active DO-OTA-C LF filters.

terms of C_3. The design formulas of the DO-OTA-C filter for all components are given below:

$$g'_1 = g_1 g_m R_1, \quad C'_2 = \frac{g_2}{g_m} C_2, \quad C'_3 = g_3 g_m L_3, \quad C''_3 = \frac{g'_3 g''_3}{g_3 g_m} C_3 ,$$

$$C'_4 = \frac{g_4}{g_m} C_4, \quad C'_5 = g_5 g_m L_5, \quad C''_5 = \frac{g'_5 g''_5}{g_5 g_m} C_5 ,$$

$$C'_6 = \frac{g_6}{g_m} C_6, \quad g'_6 = \frac{g_6}{g_m} \frac{1}{R_6} \qquad (12.59)$$

where g_m is the scaling conductance. Similarly, using the equation the DO-OTA-C filter can be conveniently designed to have the same transconductances or the same capacitances. Note that we can also use the form of the grounded inductor and floating capacitor, as in Chapter 10.

12.4.3 Current-Mode DO-OTA-C Bandpass LF Filter Design

The complexity of the DO-OTA-C filter based on the LF structure will depend on the number of elements in the series and shunt branches of the passive ladder circuit. The bandpass LC structure typically has series resonators in series arms and parallel resonators in parallel arms. Consider the bandpass LC filter in Fig. 12.23(a). Recognizing that Y_1 is a RLC series resonator, Z_4 is a parallel RLC resonator and Z_2 and Y_3 are the ideal parallel and series LC resonators and following the same design procedure we can obtain the LF DO-OTA-C filter structure as shown in Fig. 12.23(b). The component values can be formulated as

$$g'''_1 = g_1 g_m R_1, \quad C'_1 = g_1 g_m L_1, \quad C''_1 = \frac{g'_1 g''_1}{g_1 g_m} C_1 ,$$

12. CURRENT-MODE FILTERS AND OTHER ARCHITECTURES

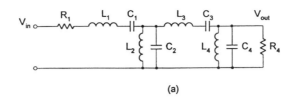

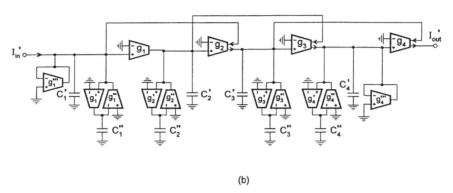

FIGURE 12.23
Eighth-order bandpass LC and LF DO-OTA-C filter.

$$C'_2 = \frac{g_2}{g_m} C_2, \quad C''_2 = \frac{g'_2 g''_2 g_m}{g_2} L_3,$$

$$C'_3 = g_3 g_m L_3, \quad C''_3 = \frac{g'_3 g''_3}{g_3 g_m} C_3, \quad C'_4 = \frac{g_4}{g_m} C_4,$$

$$C''_4 = \frac{g'_4 g''_4 g_m}{g_4} L_4, \quad g'''_4 = \frac{g_4}{g_m} \frac{1}{R_4} \tag{12.60}$$

where g_m is the scaling conductance. Further design can be carried out based on the equation.

12.4.4 Alternative Current-Mode Leapfrog DO-OTA-C Structure

One of the most outstanding features of the basic LF configuration we have studied is that the circuit can be very straightforwardly and conveniently explained with feedback theory. An alternative form of the simulation structure of the general ladder in Fig. 12.17 is given in Fig. 12.24, which can be obtained by simply rearranging the LF structure in Fig. 12.18. The corresponding OTA-grounded impedance version is shown in Fig. 12.25. Note that we assume that the scaling conductance $g_m = 1$ and the OTAs have unity transconductances and thus the values of the grounded impedances Z'_j are equal to Y_j and Z_j for odd and even j, respectively. The OTA-C simulation can then be similarly conducted by further simulating the grounded impedances using the inductor substitution method. The fifth-order lowpass LC filter in Fig. 12.21(a) is simulated in this way and its DO-OTA-C version is shown in Fig. 12.26.

12.5. CURRENT-MODE MULTIPLE LOOP FEEDBACK DO-OTA-C FILTERS

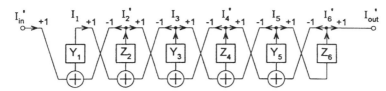

FIGURE 12.24
Alternative current-mode LF block diagram.

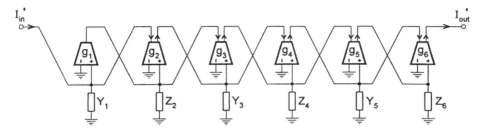

FIGURE 12.25
Alternative current-mode DO-OTA-grounded impedance LF structure.

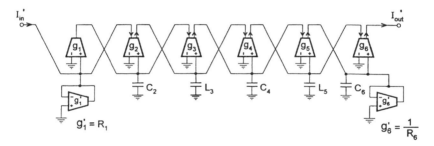

FIGURE 12.26
Alternative current-mode DO-OTA-C LF structure based on all-pole LC ladder.

12.5 Current-Mode Multiple Loop Feedback DO-OTA-C Filters

Voltage-mode multiple loop feedback OTA-C filters were studied in Chapter 11. We now investigate the design of current-mode multiple loop feedback DO-OTA-C filters. The method used is similar to that proposed in Chapter 11, but here in the current domain. Note that multiple loop feedback filter structures have many attractive features such as realizability of arbitrary transmission zeros, use of only grounded capacitors, versatility in the type of structures and requirement of fewer components. Reference [22] has shown that to realize the same transfer function, the multiple loop feedback method may need a smaller number of OTAs and have less parasitic effects than the cascade and ladder simulation methods.

12.5.1 Design of All-Pole Filters [23]

The general model of current-mode multiple loop feedback DO-OTA-C filters is shown in Fig. 12.27. It is similar to the model in Chapter 11, but differs from that in that it consists of current integrators and the current feedback network that may contain DO-OTA based current amplifiers. The circuit equations of the current-mode model can be established following the same procedure in Chapter 11. Denote the current integration constant of the jth integrator as $\tau_j = C_j/g_j$, and the current feedback coefficient from the jth integrator output to the ith integrator input as f_{ij}. We first describe the current feedback network by

$$[I_f] = [F][I_o] \tag{12.61}$$

where $[I_o] = [I_{o1}\ I_{o2} \cdots I_{on}]^t$, the output currents of integrators, $[I_f] = [I_{f1}\ I_{f2} \cdots I_{fn}]^t$, the feedback currents and $[F] = [f_{ij}]_{n \times n}$, the feedback matrix.

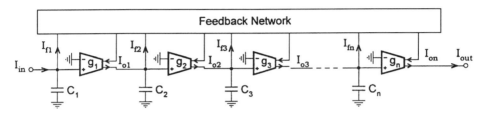

FIGURE 12.27
Current-mode multiple loop feedback DO-OTA-C model.

Then the current feedforward part is determined by

$$[M(s)][I_o] = [B]I_{in} - [I_f] \tag{12.62}$$

where

$$[M(s)] = \begin{bmatrix} s\tau_1 & & & \\ -1 & s\tau_2 & & \\ & & \ddots & \\ & & -1 & s\tau_n \end{bmatrix} \tag{12.63}$$

$$[B] = [1\ 0 \cdots 0]^t \tag{12.64}$$

The equation for the whole current-mode DO-OTA-C system is therefore derived by combining Eqs. (12.61) and (12.62) as

$$[A(s)][I_o] = [B]I_{in} \tag{12.65}$$

where

$$[A(s)] = [M(s)] + [F] = \begin{bmatrix} s\tau_1 + f_{11} & f_{12} & & f_{1n} \\ -1 & s\tau_2 + f_{22} & & f_{2n} \\ & & \ddots & \\ & & -1 & s\tau_n + f_{nn} \end{bmatrix} \tag{12.66}$$

From Eq. (12.65) we can obtain all the integrator current outputs

$$[I_o] = [A(s)]^{-1}[B]I_{in} \tag{12.67}$$

12.5. CURRENT-MODE MULTIPLE LOOP FEEDBACK FILTERS

Using Eq. (12.66) and Eq. (12.64) we derive the overall current transfer function from Eq. (12.67)

$$H(s) = I_{out}/I_{in} = I_{on}/I_{in} = 1/|A(s)| \tag{12.68}$$

where $|A(s)|$ is the determinant of matrix $[A(s)]$.

In feedback coefficient matrix $[F]$ the element $f_{ij} \neq 0$, if there exists feedback between I_{fi} and I_{oj}; otherwise, $f_{ij} = 0$. The nonzero current coefficient f_{ij} can be realized as $f_{ij} = g_{ij1}/g_{ij2}$ or simply as a direct connection if $f_{ij} = 1$. As in the voltage-mode case in Chapter 11, according to the one-to-one correspondence between feedback matrix $[F]$ and filter architectures for a given order, the generation of filter structures can be accomplished by finding all combinations of $[F]$ nonzero elements. (If the output current of any integrator is fedback to some circuit node, then $[F]$ has one and only one nonzero element in each column. And therefore for the general nth-order there are $n!$ possible combinations.) We then have all different feedback connections, i.e., all filter structures. In the minimum component or canonical realization, namely, one DO-OTA and one capacitor for one pole, to realize the unity dc gain nth-order all-pole characteristic only n DO-OTAs and n capacitors are needed in the whole system. This clearly requires that there should not be any circuit components in the feedback network. Therefore feedback can be achieved only by direct connections and the values of all nonzero elements in $[F]$ become unity.

To exemplify the general theory and show how to realize the desired all-pole filter characteristic:

$$H_d(s) = 1/\left(B_n s^n + B_{n-1} s^{n-1} + \cdots + B_1 s + 1\right) \tag{12.69}$$

we discuss the design of canonical fourth-order current-mode DO-OTA-C filters. The $[F]$ matrices, corresponding transfer functions $H(s)$ and simple design formulas for the realization of Eq. (12.69) of five cases are shown below and their structures are given in Fig. 12.28.

<u>Structure (a)</u>: $f_{11} = f_{12} = f_{23} = f_{34} = 1$,

$H(s) = 1/[\tau_1 \tau_2 \tau_3 \tau_4 s^4 + \tau_2 \tau_3 \tau_4 s^3 + (\tau_3 \tau_4 + \tau_1 \tau_4 + \tau_1 \tau_2) s^2 + (\tau_2 + \tau_4) s + 1]$

$\tau_1 = B_4/B_3$, $\tau_2 = B_3/B$, $\tau_3 = B/(B_1 - B_3/B)$, $\tau_4 = B_1 - B_3/B$, $B = B_2 - B_1 B_4/B_3$

<u>Structure (b)</u>: $f_{11} = f_{12} = f_{23} = f_{24} = 1$,

$H(s) = 1/[\tau_1 \tau_2 \tau_3 \tau_4 s^4 + \tau_2 \tau_3 \tau_4 s^3 + (\tau_3 \tau_4 + \tau_1 \tau_4) s^2 + (\tau_1 + \tau_4) s + 1]$

$\tau_1 = B_4/B_3$, $\tau_2 = B_3/[B_2 - (B_1 - B_4/B_3) B_4/B_3]$, $\tau_3 = B_2/(B_1 - B_4/B_3) - B_4/B_3$,
$\tau_4 = B_1 - B_4/B_3$

<u>Structure (c)</u>: $f_{11} = f_{12} = f_{13} = f_{24} = 1$,

$H(s) = 1/[\tau_1 \tau_2 \tau_3 \tau_4 s^4 + \tau_2 \tau_3 \tau_4 s^3 + \tau_3 \tau_4 s^2 + (\tau_1 + \tau_4) s + 1]$

$\tau_1 = B_4/B_3$, $\tau_2 = B_3/B_2$, $\tau_3 = B_2/(B_1 - B_4/B_3)$, $\tau_4 = B_1 - B_4/B_3$

<u>Structure (d)</u>: $f_{11} = f_{12} = f_{14} = f_{23} = 1$,

$H(s) = 1/[\tau_1 \tau_2 \tau_3 \tau_4 s^4 + \tau_2 \tau_3 \tau_4 s^3 + (\tau_3 \tau_4 + \tau_1 \tau_4) s^2 + \tau_4 s + 1]$

$\tau_1 = B_4/B_3$, $\tau_2 = B_3/(B_2 - B_1 B_4/B_3)$, $\tau_3 = B_2/B_1 - B_4/B_3$, $\tau_4 = B_1$

<u>Structure (e)</u>: $f_{11} = f_{12} = f_{13} = f_{14} = 1$,

$H(s) = 1/(\tau_1 \tau_2 \tau_3 \tau_4 s^4 + \tau_2 \tau_3 \tau_4 s^3 + \tau_3 \tau_4 s^2 + \tau_4 s + 1)$

$\tau_4 = B_1$, $\tau_3 = B_2/B_1$, $\tau_2 = B_3/B_2$, $\tau_1 = B_4/B_3$

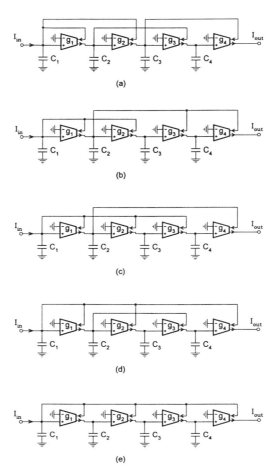

FIGURE 12.28
Fourth-order current-mode DO-OTA-C filters.

A numerical example of the fourth-order filter design for realizing the normalized Butterworth approximation with the desired function

$$H_d(s) = \frac{1}{s^4 + 2.61313s^3 + 3.41421s^2 + 2.61313s + 1}$$

is now given. Table 12.4 presents the parameter values of all the canonical fourth-order filters in Fig. 12.28, which are obtained by using the explicit design formulas given in the above.

The explicit expression of the general current transfer function of the fourth-order model with the general $[F]$ can also be derived, which is the same in form as that in Chapter 11 [23]. Some typical general nth-order current-mode DO-OTA-C architectures may include the canonical FLF structure corresponding to $f_{1i} = 1$, for $i = 1, 2, ..., n$ and the canonical LF configuration associated with $f_{11} = 1$, $f_{ij} = 1$ for $j = i + 1$ [23].

12.5. CURRENT-MODE MULTIPLE LOOP FEEDBACK FILTERS

Table 12.4 Parameter Values for Fourth-Order Butterworth Filter

Circuit	τ_4	τ_3	τ_2	τ_1
Fig. 12.28(a)	1.53073	1.57716	1.08239	0.382683
Fig. 12.28(b)	2.23044	1.14805	1.02049	0.382683
Fig. 12.28(c)	2.23044	1.53073	0.765367	0.382683
Fig. 12.28(d)	2.61313	0.92388	1.08239	0.382683
Fig. 12.28(e)	2.61313	1.30656	0.765367	0.382683

12.5.2 Realization of Transmission Zeros

The transfer function with arbitrary transmission zeros of

$$H_d(s) = \frac{A_n s^n + A_{n-1} s^{n-1} + \cdots + A_1 s + A_0}{B_n s^n + B_{n-1} s^{n-1} + \cdots + B_1 s + 1} \tag{12.70}$$

can be realized by using the input distribution and output summation methods. Although the theory is similar to that in Chapter 11, we want to repeat the formulation process to show how the circuits work in current domain.

12.5.2.1 Multiple Loop Feedback with Input Distribution

The multiple current-integrator loop current-feedback model with a current input distribution network is shown in Fig. 12.29. The current distribution coefficient to the input of the jth integrator is denoted as $\alpha_j = g_{aj}/g_r$ ($\alpha_0 = g_{a0}/g_r$ is the direct transmission coefficient from the overall input to the overall output). We can establish the equation relating the output currents I_{oj} of integrators to

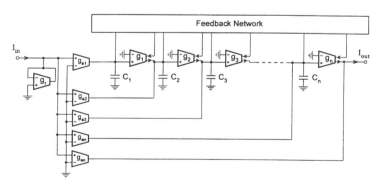

FIGURE 12.29
Current-mode multiple loop feedback and input distribution model.

the overall input current I_{in}, which is the same as Eq. (12.65) except now

$$[B] = [\alpha_1 \; \alpha_2 \cdots \alpha_n]^t \tag{12.71}$$

From the circuit we can see that the overall output current can be expressed in terms of the overall input current and the output current of the nth integrator as

$$I_{out} = \alpha_0 I_{in} + I_{on} \tag{12.72}$$

Solving I_{on} from Eq. (12.65) with $[B]$ in Eq. (12.71) leads to

$$I_{on} = \frac{1}{|A(s)|} \sum_{j=1}^{n} \alpha_j A_{jn}(s) I_{in} \qquad (12.73)$$

and substituting Eq. (12.73) into Eq. (12.72) we derive the circuit transfer function as

$$H(s) = \frac{I_{out}}{I_{in}} = \alpha_0 + \frac{1}{|A(s)|} \sum_{j=1}^{n} \alpha_j A_{jn}(s) \qquad (12.74)$$

where $A_{ij}(s)$ represent cofactors of matrix $[A(s)]$.

The transfer function in Eq. (12.74) may have the general form of Eq. (12.70) with reference to matrix $[A(s)]$ in Eq. (12.66). The system poles are determined by τ_j and f_{ij} and the transmission zeros may be controlled arbitrarily by transconductances g_{aj} through weights α_j.

12.5.2.2 Multiple Loop Feedback with Output Summation

Look at Fig. 12.30 which is a basic multiple integrator loop feedback configuration with an output summation OTA network, compared with Fig. 12.29. The input current is applied only to the input node of the first integrator. The overall output is a weighted sum of all output currents of integrators and the overall input current.

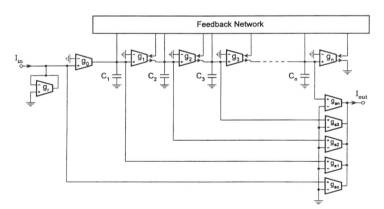

FIGURE 12.30
Current-mode multiple loop feedback and output summation model.

With $\gamma = g_0/g_r$, from Fig. 12.30 we can also obtain Eq. (12.65), but with

$$[B] = [\gamma \; 0 \cdots 0]^t \qquad (12.75)$$

From the circuit we can observe that $I_{oaj} = I_{o(j-1)} g_{aj}/sC_j$, and $I_{oj} = I_{o(j-1)} g_j/sC_j$, which give with $\beta_0 = g_{a0}/g_0$ and $\beta_j = g_{aj}/g_j$

$$I_{oaj} = \beta_j I_{oj} \qquad (12.76)$$

12.5. CURRENT-MODE MULTIPLE LOOP FEEDBACK FILTERS

Since the overall output current is the sum of output currents of all the g_{aj} OTAs, we can obtain

$$I_{\text{out}} = \beta_0 I_{o0} + \sum_{j=1}^{n} \beta_j I_{oj} \tag{12.77}$$

Note that $I_{o0} = \gamma I_{\text{in}}$. Solving Eq. (12.65) with $[B]$ in Eq. (12.75) yields

$$I_{oj} = \gamma \frac{A_{1j}(s)}{|A(s)|} I_{\text{in}} \tag{12.78}$$

From Eqs. (12.77) and (12.78) we can therefore obtain the general current transfer function as

$$H(s) = \frac{I_{\text{out}}}{I_{\text{in}}} = \gamma \left[\beta_0 + \frac{1}{|A(s)|} \sum_{j=1}^{n} \beta_j A_{1j}(s) \right] \tag{12.79}$$

Design can be carried out based on the equation for the summation type current-mode DO-OTA-C filters.

12.5.2.3 Filter Structures and Design Formulas

Many filter structures and their design formulas can be derived based on the above discussion. We can obtain explicit design formulas for general nth-order architectures with the distribution and summation networks [24]. For simplicity, in the following we illustrate four third-order structures and present their design formulas for the synthesis of elliptic functions [i.e., $n = 3$ and $A_3 = A_1 = 0$ in Eq. (12.70)].

The four structures are given in Fig. 12.31 [22]. The configuration with the FLF (that is, $f_{11} = f_{12} = f_{13} = 1$) and the input distribution network in Fig. 12.31(a) has the current transfer function as

$$H(s) = \frac{\alpha_3 \tau_1 \tau_2 s^2 + (\alpha_3 \tau_2 + \alpha_2 \tau_1) s + (\alpha_3 + \alpha_2 + \alpha_1)}{\tau_1 \tau_2 \tau_3 s^3 + \tau_2 \tau_3 s^2 + \tau_3 s + 1} \tag{12.80}$$

The parameter value equations for the denominator and numerator are formulated by comparing Eq. (12.80) with Eq. (12.70) for $n = 3$ and $A_3 = A_1 = 0$ as

$$\tau_3 = B_1, \quad \tau_2 = B_2/B_1, \quad \tau_1 = B_3/B_2 \tag{12.81}$$

$$\alpha_3 = A_2 B_1/B_3, \quad \alpha_2 = -\alpha_3 B_2^2/B_1 B_3,$$

$$\alpha_1 = A_0 - \alpha_3 - \alpha_2 \tag{12.82}$$

The current transfer function of the FLF and summation structure in Fig. 12.31(b) is derived as

$$H(s) = \gamma \frac{\beta_1 \tau_2 \tau_3 s^2 + \beta_3}{\tau_1 \tau_2 \tau_3 s^3 + \tau_2 \tau_3 s^2 + \tau_3 s + 1} \tag{12.83}$$

The parameter value equations for the numerator, with $\gamma = 1$, are demonstrated as

$$\beta_1 = A_2/B_2, \quad \beta_3 = A_0 \tag{12.84}$$

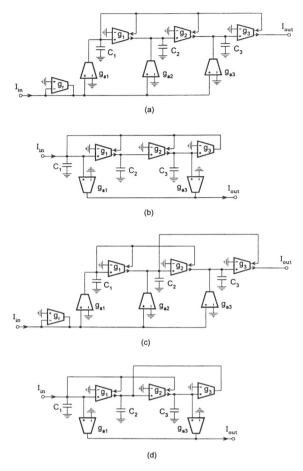

FIGURE 12.31
Third-order elliptic current-mode FLF and LF DO-OTA-C structures.

The denominator parameters are calculated using Eq. (12.81).

The circuit with the LF architecture ($f_{11} = f_{12} = f_{23} = 1$) and the input distribution network is shown in Fig. 12.31(c). It has the current transfer function as

$$H(s) = \frac{\alpha_3 \tau_1 \tau_2 s^2 + (\alpha_3 \tau_2 + \alpha_2 \tau_1) s + (\alpha_3 + \alpha_2 + \alpha_1)}{\tau_1 \tau_2 \tau_3 s^3 + \tau_2 \tau_3 s^2 + (\tau_1 + \tau_3) s + 1} \quad (12.85)$$

and the design formulas are given by

$$\tau_1 = B_3/B_2, \quad \tau_2 = B_2/(B_1 - B_3/B_2),$$

$$\tau_3 = B_1 - B_3/B_2 \quad (12.86)$$

$$\alpha_3 = A_2 \frac{B_1 - B_3/B_2}{B_3}, \quad \alpha_2 = -\alpha_3 \frac{B_2^2}{B_3(B_1 - B_3/B_2)},$$

12.6. OTHER CONTINUOUS-TIME FILTER STRUCTURES

$$\alpha_1 = A_0 - \alpha_3 - \alpha_2 \qquad (12.87)$$

For the LF and summation combination structure in Fig. 12.31(d) we have the current transfer function as

$$H(s) = \gamma \frac{\beta_1 \tau_2 \tau_3 s^2 + (\beta_1 + \beta_3)}{\tau_1 \tau_2 \tau_3 s^3 + \tau_2 \tau_3 s^2 + (\tau_1 + \tau_3) s + 1} \qquad (12.88)$$

and with $\gamma = 1$, the design formulas are derived as

$$\beta_1 = A_2/B_2, \quad \beta_3 = A_0 - \beta_1 \qquad (12.89)$$

The denominator parameters are given in Eq. (12.86).

The minus sign in α_2 in Eqs. (12.82) and (12.87) simply means the interchange of the associated g_{a2} OTA input terminals. The zero values of α_j or β_j, if any, imply that the corresponding g_{aj} OTAs should be removed. It should also be noted that to realize this particular kind of filters the distribution method [Figs. 12.31(a) and 12.31(c)] requires seven OTAs and difference matching, while the summation approach [Figs. 12.31(b) and 12.31(d)] needs five OTAs and no difference matching. Hence the summation approach is preferable. This is in contrast with the voltage-mode design in Chapter 11, where the distribution method is advantageous.

For the normalized third-order elliptic lowpass filter with

$$H_d(s) = \frac{0.588358s^2 + 1}{1.67029s^3 + 1.41856s^2 + 1.91391s + 1} \qquad (12.90)$$

The parameter values for pole realizations are calculated as $\tau_1 = 1.17745$, $\tau_2 = 0.741181$, $\tau_3 = 1.91391$ for the FLF structure and $\tau_1 = 1.17745$, $\tau_2 = 1.92618$, $\tau_3 = 0.736459$ for the LF structure. The parameter values for zero realizations are computed as $\alpha_3 = 0.674176$, $\alpha_2 = -0.424379$, $\alpha_1 = 0.750203$ for the FLF and distribution architecture, $\alpha_3 = 0.259418$, $\alpha_2 = -0.424379$, $\alpha_1 = 1.16496$ for the LF and distribution structure, $\beta_1 = 0.414758$, $\beta_3 = 1$ for the FLF and summation architecture, and $\beta_1 = 0.414758$, $\beta_3 = 0.585242$ for the LF and summation configuration.

12.6 Other Continuous-Time Filter Structures

In this section we briefly overview other popular continuous-time active filter architectures and design techniques. This includes balanced opamp-RC and OTA-C structures, MOSFET-C and OTA-C-opamp filters, and filters using current conveyors.

12.6.1 Balanced Opamp-RC and OTA-C Structures

The opamp-RC and OTA-C (voltage- and current-mode) filters we have discussed so far all are single ended. In this section we discuss active filter design based on balanced architectures.

Balanced structures are most widely utilized in continuous-time integrated filter design [1]–[4], [26]–[30]. This is because balanced structures can increase the common-mode rejection ratio, eliminate the even-order harmonic distortion components and reduce the effects of power supply noise. Balanced configurations can be obtained from single-ended structures. The single-ended to balanced conversion can be generally achieved by first mirroring the whole single-ended circuit at ground (duplicating all the components and changing the terminal polarities of all mirrored active elements) and then combining each original amplifier and its mirrored counterpart into a balanced differential input-differential output device with inverting-noninverting gains. Note that the resulting balanced version will have an added benefit of 3dB improvement in dynamic range. Because signals of both polarities are now available, sign inverters with unity gain can be readily realized by a simple crossing of wires and thereby save the components in question.

These general rules for converting single-ended structures to balanced differential counterparts are suitable for opamp-RC filters in Chapters 4–7, OTA-C filters in Chapters 8–11 and current-mode DO-OTA-C filters in Sections 12.2–12.5. For example, the process of converting a single-ended opamp-RC integrator into a balanced one is shown in Fig. 12.32. The balanced OTA-C integrator is

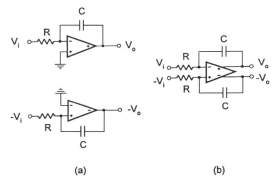

FIGURE 12.32
Balanced opamp-RC integrator.

shown in Fig. 12.33(a), which can be derived from the single-ended prototype in Fig. 12.33(b) using the conversion method.

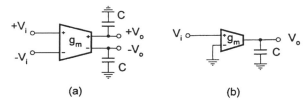

FIGURE 12.33
Balanced OTA-C integrator.

We should say that balanced filter structures can also be generated without having single-ended equivalents. Note that compared with the single-ended circuit the balanced equivalent requires twice the number of passive components and active components with balanced differential inputs and differential outputs often consist of more complicated circuitry than their single-ended counterparts. It is also noted that in integrated filters where the balanced version is used, the whole IC filter is presumably customer designed [2]. Thus although balanced differential opamps and OTAs may not be commercially available, this will not pose a major problem. More discussion on balanced filter structures can be found in [2, 4, 26, 28].

12.6.2 MOSFET-C Filters

As is well known, automatic electronic tuning is crucial for fully integrated filters to compensate the drifts of element values and filter performances due to component tolerance, device nonideality, parasitic effects, temperature, environment and aging. Integration of analog circuits in MOS technology is also driven towards a single chip implementation of mixed circuits and systems because digital circuits are integrated in MOS technology.

In conventional active RC filters, the resistor is the problem; it has a very limited range of values (normally $R \leq 40k\Omega$ without use of special processing techniques and resistances beyond the limit will be physically too large) and is not electronically tunable.

As is well known, a MOSFET can be used as a voltage-controlled resistor biased in the ohmic region, with the resistance being adjustable by the bias gate voltage. It is therefore obvious that using the MOSFET to replace the resistor in active RC filters can meet the two requirements and the resulting filters are called the MOSFET-C filters.

MOSFET-C filters are usually implemented using a balanced structure [26]–[30]. Figure 12.34 shows a balanced MOSFET-C integrator. The balanced active RC prototype is also given in the figure for comparison. MOSFET-C filters can be similarly constructed by replacing resistors in active RC filters by their corresponding MOSFETs.

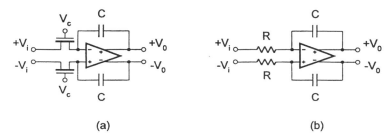

FIGURE 12.34
Balanced MOSFET-C integrator.

As in active RC filter design, we must consider the opamp nonideality effects including the frequency limitation and compensation techniques may be needed [29]. A new and important problem peculiar to MOSFET-C filters is the nonideality of the MOSFET. The MOSFET is in nature a nonlinear resistor and we must reduce or eliminate the MOSFET nonlinearity. This is also why MOSFET-C filters must have balanced structures (a single-ended configuration results in high nonlinearities and distortion, but balanced structures can eliminate the even-function nonlinearity). In the literature, a special approach for a complete cancellation of MOSFET nonlinearity has been proposed [30], which is shown in Fig. 12.35.

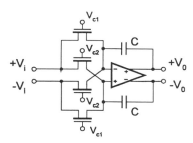

FIGURE 12.35
Modified MOSFET-C integrator.

We must stress that the MOSFET-C approach is one of the most popular methods in continuous-time integrated filter design, which is second to only the OTA-C method. The major advantage of the MOSFET-C approach is that the well developed active RC design methods can be directly used and the constituent components (opamps, MOSFETs and capacitors) all are standard and available in most IC design libraries. Like conventional active RC filters, this approach however remains suffering from the limited frequency range due to the use of opamps and the inherent nonlinearities of MOSFET resistors that must be canceled. As we mentioned before, OTAs require much simpler circuitry than opamps (for example, the first stage of the opamp) and have a very high-frequency response and an electronically tunable transconductance. The OTA-C filtering approach dominates in high-frequency integrated continuous-time filter design and has been successfully implemented in different IC technologies including CMOS. The reader may refer to references [1]–[4], [26]–[32] for details of MOSFET-C filter design.

12.6.3 OTA-C-Opamp Filter Design

Since an OTA has a simple circuitry suitable for MOS integration and has a transconductance tunable by the bias voltage or current, we can also use the OTA to replace the resistor in opamp-RC filters, resulting in the OTA-C-opamp filters, which can resolve the problems due to the use of resistors. More often, OTA-C-opamp filters are developed to solve the parasitic capacitance problem in OTA-C filters [32]. That is, inserting an opamp into the OTA-C integrator to reduce the parasitic effects by the virtual ground property of the opamp. Figure 12.36(a) shows a single-ended integrator which consists of an OTA and an opamp with feedback from a Miller capacitor [34]. The corresponding balanced structure is given in Fig. 12.36(b). Similar to the compensation technique for the OTA-C integrator introduced in Chapter 9 we can insert a small MOSFET resistor in series with the Miller capacitor to compensate the nonideal frequency characteristics of the OTA and opamp.

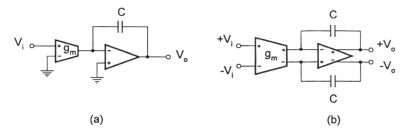

FIGURE 12.36
OTA-C-opamp integrators.

The generation of OTA-C-opamp filter structures is simple, either by replacing the resistor in opamp-RC filters by the OTA or by substituting the OTA-C-opamp integrator for the OTA-C integrator in OTA-C filters.

In some high-frequency OTA-C filters where a simple single-stage OTA is normally used, the dc gain of the OTA is very low and usually parasitic capacitances are high. Therefore, normally a very wide range of tuning is required to compensate the nonideal effects such as parasitic capacitances and poor output impedances [33]. The OTA-C-opamp technique may present an alternative to the OTA-C method since OTA-C-opamp filters are insensitive to parasitic capacitances due to the virtual ground of the high gain opamp input and have high dc gain due to the two stage arrangement of integrators. However, the OTA-C-opamp method suffers from the frequency limitation imposed by the opamp and the large power consumption and chip area due to more active devices used (one extra opamp for each integrator). Also, considering that OTA-C filters which may be seen as a result of replacement of both opamps and resistors in opamp-RC filters usually have the structures that have a grounded capacitor on each node which can be used to absorb OTA parasitic capacitances, as has been seen from Chapters 8–11, the OTA-C-opamp approach may no longer be attractive.

We can also have current-mode DO-OTA-C-opamp filter structures. Figure 12.37 gives a current-mode integrator which uses an opamp, a capacitor, and a DO-OTA. Using this integrator to replace

12.6. OTHER CONTINUOUS-TIME FILTER STRUCTURES

the DO-OTA-C integrator in current-mode DO-OTA-C filters in Sections 12.2–12.5 we can easily generate corresponding current-mode DO-OTA-C-opamp filters.

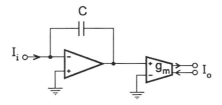

FIGURE 12.37
Current-mode DO-OTA-C-opamp integrator.

12.6.4 Active Filters Using Current Conveyors

The second-generation current conveyor (CCII) which was first proposed in 1970 [35] has recently been widely used in high performance analog signal processing circuits including active filters [36]–[38] due to its mixed voltage and current functions, wide bandwidth and large dynamic range. In Chapter 3 a very brief introduction was given. Now we discuss applications of the CCII in active filter design.

It will be recalled from Chapter 3 that the CCII is defined as [35]

$$\begin{bmatrix} I_y \\ V_x \\ I_z \end{bmatrix} = \begin{bmatrix} 0 & 0 & 0 \\ 1 & 0 & 0 \\ 0 & \pm 1 & 0 \end{bmatrix} \begin{bmatrix} V_y \\ I_x \\ V_z \end{bmatrix}$$

Normally the CCIIs with the positive and negative polarities are represented by the CCII+ and CCII−, respectively.

The CCII is a mixed voltage and current active building block. With terminal Z grounded the CCII can be used as a voltage follower from terminal Y to terminal X. With terminal Y grounded the CCII can also be used as a current follower from terminal X to terminal Z. What is more, the CCII together with an external resistor connected from terminal X to ground can be used as a transconductance amplifier with the transconductance being equal to the reciprocal of the resistance, with voltage input to terminal Y and current output from terminal Z. Ideally the CCII transconductance amplifier has the characteristic of $I_z = \pm \frac{1}{R} V_y$.

Because of its versatility of function the CCII can be used to construct many interesting filter structures. The building blocks such as the voltage and current integrators and the simulated inductor are illustrated in Fig. 12.38(a), (b), and (c), respectively, [35]. For simplicity the ground connection in the CCII symbol in Chapter 3 is dropped. The equivalent inductance of the simulated inductor is given by $L = R_1 R_2 C$.

A two-CCII active RC biquad is displayed in Fig. 12.39. This circuit may be considered to be either a cross-connection of an ideal integrator and a lossy integrator or a parallel connection of an active inductor, a capacitor, and a resistor. The voltage outputs from the circuit nodes of the biquad with grounded capacitors ideally give the lowpass and bandpass filters for the voltage input through R_3, given by

$$H_{LP}(s) = \frac{V_{oLP}}{V_{in}} = \frac{\frac{1}{R_2 R_3 C_1 C_2}}{s^2 + \frac{1}{R_3 C_1} s + \frac{1}{R_1 R_2 C_1 C_2}} \quad (12.91)$$

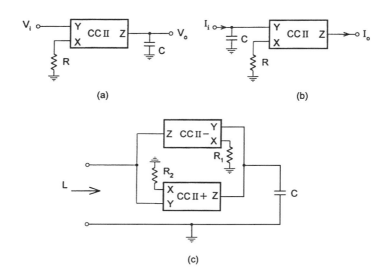

FIGURE 12.38
CCII-based building blocks.

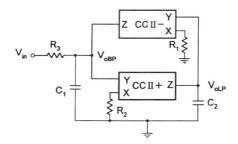

FIGURE 12.39
Two-CCII biquad.

$$H_{BP}(s) = \frac{V_{oBP}}{V_{in}} = \frac{\frac{1}{R_3 C_1} s}{s^2 + \frac{1}{R_3 C_1} s + \frac{1}{R_1 R_2 C_1 C_2}} \quad (12.92)$$

To give another example we consider the CCII-RC structure using one CCII in Fig. 12.40. Routine circuit analysis yields

$$H_{LP}(s) = \frac{V_{\text{out}}}{V_{iLP}} = \frac{g_1 g_2}{s^2 C_1 C_2 + s g_1 (C_1 + C_2) + g_1 g_2} \quad (12.93)$$

$$H_{BP}(s) = \frac{V_{\text{out}}}{V_{iBP}} = \frac{s g_1 C_2}{s^2 C_1 C_2 + s g_1 (C_1 + C_2) + g_1 g_2} \quad (12.94)$$

indicating that the circuit can offer the lowpass and bandpass functions. The reader should check that the ω_o and Q sensitivities are very low.

In practice, CCII-RC filter design must also take the effects of CCII nonidealities into consideration. These nonidealities include finite impedances at terminals Y, X, and Z and the error or frequency dependence of voltage and current following characteristics (from terminals Y to X and

12.7. SUMMARY 425

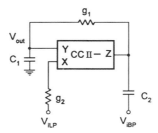

FIGURE 12.40
Single-CCII biquad.

terminals X to Z, respectively). They will adversely affect the filter frequency performance in various ways. Detailed discussion and compensation methods can be found in [39, 40].

12.6.5 Log-Domain, Current Amplifier, and Integrated-RLC Filters

These filtering techniques are very active research topics these days. We will not discuss them in details in this book, but give references for the reader who may be interested to know them. Log-domain filters (designed based on the log operation of the transistor nonlinear characteristic) have been investigated in [45]–[49]. They are internally nonlinear externally linear current-mode filters. References [41]–[44] are concerned with current amplifier filters, linear current-mode filters. Publications on new integrated RLC filters are given in [50]–[53], the interest in which has been aroused by recent surge in wireless communications.

12.7 Summary

The design of current-mode DO-OTA-C filters has been studied. First- and second-order current-mode filters have been derived using a single DO-OTA and five-admittance model. They are suitable for both discrete and IC implementations. Current-mode two-integrator loop DO-OTA-C architectures have been obtained and the range of filter functions which are supported by the various architectures has been shown. The filter performances including the effects of DO-OTA nonidealities have also been analyzed. Current-mode DO-OTA-C filters based on leapfrog simulation of passive LC ladders have been discussed. Systematic generation and design of current-mode multiple integrator loop feedback DO-OTA-C filters have been studied.

Instead of synthesizing the filters based on voltage building blocks such as voltage integrators and voltage amplifiers using single output OTAs, all the realizations in Sections 12.2–12.5 are based on current building blocks such as current integrators and current amplifiers incorporating DO-OTAs. The current-mode DO-OTA-C filter structures derived in this chapter, together with those voltage-mode OTA-C counterparts in Chapters 8–11 complementarily form a complete set of transconductance amplifier and capacitor filter architectures. This will no doubt give the filter designer more opportunities to choose the best structure for their requirements.

In this chapter we have also introduced various other successful techniques and structures for continuous-time filter design such as balanced opamp-RC and OTA-C structures, MOSFET-C configurations, OTA-C-opamp filters, and active filters using CCIIs.

We stress that continuous-time integrated filter design for high-frequency applications has progressed tremendously and remains an active topic. In this book, we have concentrated on the conventional well-established opamp-RC filters and the predominant OTA-C filters that have been developed more recently. We have also discussed other popular methods such as the current-mode DO-OTA-C and MOSFET-C approaches. The discussion in this book has been mainly on filter design methods, structure generation and performance analysis in the block level. Some important issues such as solid-state (transistor-level) implementation, automatic (on-chip) tuning, and computer-aided design of continuous-time filters have not been discussed in details. The reader may be interested to refer to relevant publications for solid-state implementation [1, 4, 5, 7, 28], automatic tuning [54]–[57] and filter CAD [58]–[60].

It is also stressed that in high-frequency continuous-time integrated filter design, the study of filter structures and synthesis methods is as important as solid-state implementation, because the performance of a filter depends on the constituent components, the connection of the components, and the principles based on which the structure is derived, together with the IC technology, fabrication techniques, and packaging methods used. This presents the variety of areas to which filter researchers and designers can be devoted to improve and enhance filter performances for advanced applications. At the end of the book we hope that we have armed the reader with sufficient background to explore further topics in this exciting and challenging area.

References

[1] Tsividis, Y.P. and Voorman, J.O., Eds., *Integrated continuous-time filters: principles, design and applications,* IEEE Press, 1993.

[2] Schaumann, R., Ghausi, M.S., and Laker, K.R., *Design of analog filters: passive, active RC and switched capacitor,* Prentice-Hall, Englewood Cliffs, NJ, 1990.

[3] Schaumann, R., Continuous-time integrated filters, in *The Circuits and Filters Handbook,* W. K. Chen, Ed., CRC Press, Boca Raton, FL, 1995.

[4] Laker, K.R. and Sansen, W., *Design of analog integrated circuits and systems,* McGraw-Hill, New York, 1994.

[5] Johns, D.A. and Martin, K., *Analog integrated circuit design,* John Wiley & Sons, New York, 1997.

[6] Ismail, M. and Fiez, T., Eds., *Analog VLSI: signal and information processing,* McGraw-Hill, New York, 1994.

[7] Toumazou, C., Lidgey, F.J., and Haigh, D.G., Eds., *Analogue IC design: the current-mode approach,* Peter Peregrinus, London, 1990.

[8] Roberts, G.W. and Sedra, A.S., All current-mode frequency selective circuits, *Electronics Letters,* 25, 759–761, 1989.

[9] Bhattacharyya, B.B. and Swamy, M.N.S., Network transposition and its application in synthesis, *IEEE Trans. Circuits Syst.,* 18, 394–397, 1971.

[10] den Brinker, C.S. and Gosling, W., The development of the voltage-to-current transactor (VCT), *Microelectronics Journal,* 8, 9–18, 1977.

[11] Szczepanski, S., Jakusz, J., and Schaumann, R., A linear fully balanced CMOS OTA for VHF filtering applications, *IEEE Trans. Circuits Syst. -II,* 44, 174–187, 1997.

[12] Fidler, J.K., Mack, R.J., and Noakes, P.D., Active filters incorporating the voltage-to-current transactor, *Microelectronics Journal,* 8, 19–22, 1977.

[13] Contreras, R.A. and Fidler, J.K., VCT active filters, *Proc. European Conf. on Circuit Theory and Design,* 1, 361–369, 1980.

[14] Al-Hashimi, B.M. and Fidler, J.K., A novel VCT-based active filter configuration, *Proc. 6th Int. Symp. Networks, Systems and Signal Processing,* Yugoslavia, June, 1989.

[15] Sun, Y. and Fidler, J.K., Structure generation of current-mode two integrator loop dual output-OTA grounded capacitor filters, *IEEE Trans. Circuits Syst., Part II,* 43, 659–663, 1996.

[16] Sun, Y. and Fidler, J.K., Design of current-mode multiple output OTA and capacitor filters, *Int. J. of Electronics,* 81, 95–99, 1996.

[17] Ramirez-Angulo, J., Robinson, M., and Sánchez-Sinencio, E., Current-mode continuous-time filters: two design approaches, *IEEE Trans. Circuits Syst. - II,* 39, 337–341, 1992.

[18] Nawrocki, R. and Klein, U., New OTA-capacitor realization of a universal biquad, *Electronics Letters,* 22(1), 50–51, 1986.

[19] Haigh, D.G., Taylor, J.T., and Singh, B., Continuous-time and switched capacitor monolithic filters based on current and charge simulation, *IEE Proc. Part G,* 137, 147–155, 1990

[20] Ramirez-Angulo, J. and Sánchez-Sinencio, E., High frequency compensated current-mode ladder filters using multiple output OTAs, *IEEE Trans. Circuits Syst. - II ,* 41, 581–586, 1994.

[21] Tingleff, J. and Toumazou, C., Integrated current mode wave active filters based on lossy integrators, *IEEE Trans. Circuits Syst.,* 42, 237–244, 1995.

[22] Sun, Y., Jefferies, B., and Teng, J., Universal third-order OTA-C filters, *Int. J. Electronics,* 85(5), 597-609, 1998.

[23] Sun, Y. and Fidler, J.K., Current-mode multiple loop feedback filters using dual output OTAs and grounded capacitors, *Int. J. of Circuit Theory and Applications,* 25, 69–80, 1997.

[24] Sun, Y. and Fidler, J.K., Current-mode OTA-C realisation of arbitrary filter characteristics, *Electronics Letters,* 32, 1181–1182, 1996.

[25] Sun, Y. and Fidler, J.K., Some design methods of OTA-C and CCII-RC active filters, *Proc. IEE Saraga Colloq. on Filters,* 7/1–7/8, London, 1993.

[26] Voorman, J.O., Van Bezooijen, A., and Ramaho, N., On balanced integrated filters, in *Integrated continuous time filters,* Tsividis, Y.P. and Voorman, J.O., Eds., IEEE Press, 1993.

[27] Banu, M. and Tsividis, Y.P., Fully integrated active RC filters in MOS technology, *IEEE J. Sloid-State Circuits,* 18(6), 644–651, 1983.

[28] Tsividis, Y.P., Banu, M., and Khoury, J., Continuous-time MOSFET-C filters in VLSI, *IEEE Trans. Circuits Syst.,* 33(2), 125–140, 1986.

[29] Khoury, J.M. and Tsividis, Y., Analysis and compensation of high-frequency effects in integrated MOSFET-C continuous-time filters, *IEEE Trans. Circuits Syst.,* 34(8), 862–875, 1987.

[30] Czarnul, Z., Modification of Banu-Tsividis continuous time integrator structure, *IEEE Trans. Circuits Syst.,* CAS-33(7), 714–716, 1996.

[31] Ismail, M., Smith, S.V., and Beale, R.G., A new MOSFET-C universal filter structure for VLSI, *IEEE J. Solid-State Circuits,* 23(1), 183–194, 1988.

[32] Tsividis, Y.P., Integrated continuous-time filter design—an overview, *IEEE J. Solid-State Circuits,* 29(3), 166–186, 1994.

[33] Shoaei, O. and Snelgrove, W.M., A wide-range tunable 25 MHz-110 MHz BiCMOS continuous-time filter, *Proc. IEEE ISCAS,* 317–320, 1996.

[34] Franco, S., Use transconductance amplifiers to make programmable active filters, *Electronic Design,* Sept, 98–101, 1976.

[35] Sedra, A.S. and Smith, K.C., A second-generation current conveyor and its applications, *IEEE Trans. Circuit Theory,* CT-17, 132–134, 1970.

[36] Sedra, A.S., Roberts, G.W., and Gohh, F., The current conveyor: history, progress and new results, *IEE Proceedings-G,* 137(2), 78–87, 1990.

[37] Sun, Y. and Fidler, J.K., Versatile active biquad based on second-generation current conveyors, *Int. J. Electronics,* 76(1), 91–98, 1994.

[38] Toumazou, C. and Lidgey, F.J., Universal active filter using current conveyors, *Electron. Lett.,* 22, 662–664, 1986.

[39] Fabre, A., Saaid, O., and Barthelemy, H., On the frequency limitations of the circuits based on second generation current conveyors, *Analog Integrated Circuits and Signal Processing,* 7, 113–129, 1995.

[40] Sun, Y. and Fidler, J.K., Analysis of current conveyor error effects in signal processing circuits, *Int. J. Cir. Theor. Appl.,* 24, 479–487, 1996.

[41] Lee, S.S., Zele, R.H., Allstot, D.J., and Liang, G., CMOS continuous-time current-mode filters for high-frequency applications, *IEEE J. Solid-State Circuits,* 28, 323–329, 1993.

[42] Zele, R.H. and Allstot, D.J., Low-power CMOS continuous-time filters, *IEEE J. Solid-State Circuits,* 31(2), 157–168, 1996.

[43] Smith, S.L. and Sanchez-Sinencio, E., Low voltage integrators for high frequency CMOS filters using current mode techniques, *IEEE Trans. Circuits Syst. II,* 43(1), 39–48, 1996.

[44] Wu, C.Y. and Hsu, H.S., The design of CMOS continuous-time VHF current and voltage-mode lowpass filters with Q-enhancement circuits, *IEEE J. Solid-State Circuits,* 31(5), 614–624, 1996.

[45] Tsividis, Y.P., Externally linear, time-invariant systems and their application to compounding signal processing, *IEEE Trans. Circuits Syst. II,* 44, 65–85, 1997.

[46] Adams, R., Filtering in the log domain, Presented at 63rd AES Conf., New York, May 1979, Reprint 1470.

[47] Frey, D., Log-domain filtering: an approach to current-mode filtering, *IEE Proc. -G,* 140(6), 406–416, 1993.

[48] Perry, D. and Roberts, G.W., The design of log-domain filters based on the operational simulation of LC ladders, *IEEE Trans. Circuits Syst. II,* 43(11), 763–774, 1996.

[49] Frey, D., Exponential state space filters: a generic current mode design strategy, *IEEE Trans. Circuits Syst. I,* 43(1), 34–42, 1996.

[50] Nguyen, N.M. and Meyer, R.G., Si IC compatible inductors and LC passive filters, *IEEE J. Solid-State Circuits,* 25, 1028–1030, 1990.

[51] Pipilos, S., Tsividis, Y., Fenk, J., and Papananos, Y., A Si 1.8 GHz RLC filter with tunable center frequency and quality factor, *IEEE J. of Solid-State Circuits,* 31(10), 1517–1525, 1996.

[52] Duncan, R., Martin, K., and Sedra, A.S., A Q-enhanced active-RLC bandpass filter, *IEEE Trans. Circuits Syst. II,* 44(5), 341–347, 1997.

[53] Kuhn, W.B., Stephenson, F.W., and Elshabini-Riad, A., A 200 MHz CMOS Q-enhanced LC bandpass filter, *IEEE J. Solid-State Circuits,* 31(8), 1112–1122, 1996.

[54] Kozma, K.A., John, D.J., and Sedra, A.S., Automatic tuning of continuous-time integrated filters using an adaptive filter technique, *IEEE Trans. Circuits Syst.,* 38, 1241–1248, 1991.

[55] Kozma, K.A., John, D.A., and Sedra, A.S., Tuning of continuous-time filters in the presence of parasitic poles, *IEEE Trans. Circuits Syst. - I,* 40(1), 13–20, 1993.

[56] Tsividis, Y.P., Self-tuned filters, *Electronics Letters,* 17, 406–407, 1981.

[57] Schaumann, R. and Tan, M.A., The problem of on-chip automatic tuning in continuous-time integrated filters, *Proc. IEEE ISCAS,* 106–109, 1989.

[58] Kobe, M.R., Sánchez-Sinencio, E., and Ramirez-Angulo, J., OTA-C biquad-based filter silicon compiler, *Analog Integrated Circuits and Signal Processing,* 3(3), 243–258, 1993.

[59] Henderson, R.K., Ping, Li, and Sewell, J.I., Analog integrated filter compilation, *Analog Integrated Circuits and Signal Processing,* 3, 217–228, 1993.

[60] Ouslis, C. and Sedra, A.S., Designing custom filters, *IEEE Circuits Dev. Mag.,* 29–37, May, 1995.

Appendix A
A Sample of Filter Functions

TABLE A.1
Butterworth Polynomials, $B_n(s)$ in Expanded and in Factored Form

n	Butterworth Polynomials
1	$s + 1$
2	$s^2 + 1.4142s + 1$
3	$s^3 + 2s^2 + 2s + 1 = (s+1)(s^2 + s + 1)$
4	$s^4 + 2.613s^3 + 3.414s^2 + 2.613s + 1 = (s^2 + 0.765s + 1)(s^2 + 1.848s + 1)$
5	$s^5 + 3.2361s^4 + 5.2361s^3 + 5.2361s^2 + 3.2361s + 1 = (s + 1.0000)(s^2 + 0.618s + 1)(s^2 + 1.618s + 1)$
6	$s^6 + 3.8637s^5 + 7.4641s^4 + 9.1416s^3 + 7.4641s^2 + 3.8637s + 1 = (s^2 + 0.5176s + 1)(s^2 + 1.4142s + 1)(s^2 + 1.9318s + 1)$
7	$s^7 + 4.4940s^6 + 10.0978s^5 + 14.5918s^4 + 14.5918s^3 + 10.0978s^2 + 4.4940s + 1 = (s + 1.0000)(s^2 + 0.445s + 1)(s^2 + 1.247s + 1)(s^2 + 1.802s + 1)$
8	$s^8 + 5.1258s^7 + 13.1317s^6 + 21.8462s^5 + 25.6884s^4 + 21.8462s^3 + 13.1371s^2 + 5.1258s + 1 = (s^2 + 0.3902s + 1)(s^2 + 1.1112s + 1)(s^2 + 1.663s + 1)(s^2 + 1.9616s + 1)$
9	$s^9 + 5.7588s^8 + 16.5817s^7 + 31.1634s^6 + 41.9864s^5 + 41.9864s^4 + 31.1634s^3 + 16.5817s^2 + 5.7588s + 1$ $= (s + 1.0000)(s^2 + 0.3474s + 1)(s^2 + 1.0000s + 1)(s^2 + 1.532s + 1)(s^2 + 1.8794s + 1)$
10	$s^{10} + 6.3925s^9 + 20.4317s^8 + 42.8021s^7 + 64.8824s^6 + 74.2334s^5 + 64.8824s^4 + 42.8021s^3 + 20.4317s^2 + 6.3925s + 1$ $= (s^2 + 0.3128s + 1)(s^2 + 0.908s + 1)(s^2 + 1.4142s + 1)(s^2 + 1.782s + 1)(s^2 + 1.9754s + 1)$

TABLE A.2
Denominator Polynomials in Expanded and in Factored Forms for Chebyshev Filters of Odd Order

n	Polynomials
	$A_{max} = 0.1$ dB ($\varepsilon = 0.15262$)
1	$s + 6.55220$
3	$s^3 + 1.93881s^2 + 2.62950s + 1.63805 = (s^2 + 0.96941s + 1.68975)(s + 0.96941)$
5	$s^5 + 1.74396s^4 + 2.77070s^3 + 2.39696s^2 + 1.43556s + 0.40951 = (s^2 + 0.33307s + 1.19494)(s^2 + 0.87198s + 0.63592)(s + 0.53891)$
7	$s^7 + 1.69322s^6 + 3.18350s^5 + 3.16925s^4 + 2.70514s^3 + 1.48293s^2 + 0.56179s + 0.10238 = (s^2 + 0.16768s + 1.09245)(s^2 + 0.46983s + 0.75322)(s^2 + 0.67893s + 0.33022)(s + 0.37678)$
9	$s^9 + 1.67270s^8 + 3.64896s^7 + 3.96385s^6 + 4.19161s^5 + 2.93387s^4 + 1.73412s^3 + 0.69421s^2 + 0.19176s + 0.025595 = (s^2 + 0.10088s + 1.05421)(s^2 + 0.29046s + 0.83437)(s^2 + 0.44501s + 0.49754)(s^2 + 0.54589s + 0.20135)(s + 0.29046)$
	$A_{max} = 0.5$ dB ($\varepsilon = 0.34931$)
1	$s + 2.86278$
3	$s^3 + 1.25291s^2 + 1.53490s + 0.71569 = (s^2 + 0.62646s + 1.14245)(s + 0.62646)$
5	$s^5 + 1.17249s^4 + 1.93738s^3 + 1.30958s^2 + 0.75252s + 0.17892 = (s^2 + 0.22393s + 1.03578)(s^2 + 0.58625s + 0.47677)(s + 0.36232)$
7	$s^7 + 1.15122s^6 + 2.41265s^5 + 1.86941s^4 + 1.64790s^3 + 0.75565s^2 + 0.28207s + 0.04473 = (s^2 + 0.11401s + 1.01611)(s^2 + 0.31944s + 0.67688)(s^2 + 0.46160s + 0.25388)(s + 0.25617)$
9	$s^9 + 1.14257s^8 + 2.90273s^7 + 2.42933s^6 + 2.78150s^5 + 1.61139s^4 + 0.98362s^3 + 0.34082s^2 + 0.09412s + 0.01118 = (s^2 + 0.06891s + 1.00921)(s^2 + 0.19841s + 0.78937)(s^2 + 0.30398s + 0.45254)(s^2 + 0.37288s + 0.15634)(s + 0.19841)$
	$A_{max} = 1$ dB ($\varepsilon = 0.50885$)
1	$s + 1.96523$
3	$s^3 + 0.73782s^2 + 1.02219s + 0.32689 = (s^2 + 0.36891s + 0.88610)(s + 0.36891)$
5	$s^5 + 0.70646s^4 + 1.49954s^3 + 0.69348s^2 + 0.45935s + 0.08172 = (s^2 + 0.13492s + 0.95217)(s^2 + 0.35323s + 0.39315)(s + 0.21831)$

TABLE A.2
Denominator Polynomials in Expanded and in Factored Forms for Chebyshev Filters of Odd Order (*continued*)

n	Polynomials
7	$s^7 + 0.69809s^6 + 1.99367s^5 + 1.03955s^4 + 1.14460s^3 + 0.38264s^2 + 0.16613s + 0.02043 = (s^2 + 0.06913s + 0.97462)(s^2 + 0.19371s + 0.63539)(s^2 + 0.27991s + 0.21239)(s + 0.15534)$
9	$s^9 + 0.69468s^8 + 2.49129s^7 + 1.38375s^6 + 2.07675s^5 + 0.85687s^4 + 0.64447s^3 + 0.16845s^2 + 0.05438s + 0.00511 = (s^2 + 0.04189s + 0.98440)(s^2 + 0.12063s + 0.76455)(s^2 + 0.18482s + 0.42773)(s^2 + 0.22671s + 0.13153)(s + 0.12063)$

$A_{max} = 2$ dB ($\varepsilon = 0.76478$)

n	Polynomials
1	$s + 1.30756$
3	$s^3 + 0.73782s^2 + 1.02219s + 0.32689 = (s^2 + 0.36891s + 0.88610)(s + 0.36891)$
5	$s^5 + 0.70646s^4 + 1.49954s^3 + 0.69348s^2 + 0.45935s + 0.08172 = (s^2 + 0.13492s + 0.95217)(s^2 + 0.35323s + 0.39315)(s + 0.21831)$
7	$s^7 + 0.69809s^6 + 1.99367s^5 + 1.03955s^4 + 1.14460s^3 + 0.38264s^2 + 0.16613s + 0.02043 = (s^2 + 0.06913s + 0.97462)(s^2 + 0.19371s + 0.63539)(s^2 + 0.27991s + 0.21239)(s + 0.15534)$
9	$s^9 + 0.69468s^8 + 2.49129s^7 + 1.38375s^6 + 2.07675s^5 + 0.85687s^4 + 0.64447s^3 + 0.16845s^2 + 0.05438s + 0.00511 = (s^2 + 0.04189s + 0.98440)(s^2 + 0.12063s + 0.76455)(s^2 + 0.18482s + 0.42773)(s^2 + 0.22671s + 0.13153)(s + 0.12063)$

$A_{max} = 3$ dB ($\varepsilon = 0.99763$)

n	Polynomials
1	$s + 1.00238$
3	$s^3 + 0.59724s^2 + 0.92835s + 0.25059 = (s^2 + 0.29862s + 0.83917)(s + 0.29862)$
5	$s^5 + 0.57450s^4 + 1.41503s^3 + 0.54894s^2 + 0.40797s + 0.06265 = (s^2 + 0.10972s + 0.93603)(s^2 + 0.28725s + 0.37701)(s + 0.17753)$
7	$s^7 + 0.56842s^6 + 1.91155s^5 + 0.83144s^4 + 1.05185s^3 + 0.30002s^2 + 0.14615s + 0.01566 = (s^2 + 0.05629s + 0.96648)(s^2 + 0.15773s + 0.62726)(s^2 + 0.22792s + 0.20425)(s + 0.12649)$
9	$s^9 + 0.56594s^8 + 2.41014s^7 + 1.11232s^6 + 1.94386s^5 + 0.67893s^4 + 0.58351s^3 + 0.13139s^2 + 0.04759s + 0.00392 = (s^2 + 0.03413s + 0.97950)(s^2 + 0.09827s + 0.75966)(s^2 + 0.15057s + 0.42283)(s^2 + 0.18470s + 0.12664)(s + 0.09827)$

TABLE A.3
Elliptic Approximation Functions for $A_{max} = 0.5$ dB

n	A_{min}	Numerator constant K	Numerator of $F(s)$	Denominator of $F(s)$
(a) $\Omega_s = 1.5$				
2	8.3	0.38540	$s^2 + 3.92705$	$s^2 + 1.03153s + 1.60319$
3	21.9	0.31410	$s^2 + 2.80601$	$(s^2 + 0.45286s + 1.14917)(s + 0.766952)$
4	36.3	0.015397	$(s^2 + 2.53555)(s^2 + 12.09931)$	$(s^2 + 0.25496s + 1.06044)(s^2 + 0.92001s + 0.47183)$
5	50.6	0.019197	$(s^2 + 2.42551)(s^2 + 5.43764)$	$(s^2 + 0.16346s + 1.03189)(s^2 + 0.57023s + 0.57601)(s + 0.42597)$
(b) $\Omega_s = 2.0$				
2	13.9	0.20133	$s^2 + 7.4641$	$s^2 + 1.24504s + 1.59179$
3	31.2	0.15424	$s^2 + 5.15321$	$(s^2 + 0.53787s + 1.14849)(s + 0.69212)$
4	48.6	0.0036987	$(s^2 + 4.59326)(s^2 + 24.22720)$	$(s^2 + 0.30116s + 1.06258)(s^2 + 0.88456s + 0.41032)$
5	66.1	0.0046205	$(s^2 + 4.36495)(s^2 + 10.56773)$	$(s^2 + 0.19255s + 1.03402)(s^2 + 0.58054s + 0.52500)(s + 0.392612)$
(c) $\Omega_s = 3.0$				
2	21.5	0.083974	$s^2 + 17.48528$	$s^2 + 1.35715s + 1.55532$
3	42.8	0.063211	$s^2 + 11.82781$	$(s^2 + 0.58942s + 1.14559)(s + 0.65263)$
4	64.1	0.00062046	$(s^2 + 10.4554)(s^2 + 58.471)$	$(s^2 + 0.32979s + 1.063281)(s^2 + 0.86258s + 0.37787)$
5	85.5	0.00077547	$(s^2 + 9.8955)(s^2 + 25.0769)$	$(s^2 + 0.21066s + 1.0351)(s^2 + 0.58441s + 0.496388)(s + 0.37452)$

Index

A

Absorption, of OTA finite input and output impedances, 253, 254, 298, 299
Active compensation, 146
Active elements, 75-106
Active filter realization
   by element simulation, 188
   by functional simulation, 195
   leap-frog method, 196
   using cascade of biquads, 155-162
   using cascade of biquartics, 171-180
   current conveyors, 423-425
   using FDNR, 193
   using primary resonators, 167, 168
   single OTA, 231-264
Active compensation, of OTA finite bandwidth, 300, 301
Activity, 31
Admittances
   combining with single DO-OTA, 389
   combining with single OTA, 232, 241, 246, 258
   simulation of floating with OTAs and grounded impedance, 321
Admittance/impedance simulation of ladders using OTA-C method, 320-325
Åkerberg-Mossberg biquad, 146, 147
Allpass biquad, 129
Allpass filter, 58
   OTA-C realization, 277, 278
All-pole function, 39-47, 54-58
All-pole lowpass filters, realization of using
   ladder simulation current-mode DO-OTA-C structures, 407, 408, 411
   ladder simulation OTA-C architectures, 328, 329, 330
   multiple loop feedback current-mode DO-OTA-C method, 411-415
   multiple loop feedback OTA-C structures, 350-363
Amplifier
   differential, 84
   inverting voltage, 98
   noninverting voltage, 98, 101
   operational, 84
   operational transconductance, 100
   unity-gain, 100
Amplifiers, using OTAs
   voltage, 270
   current, 396
Amplitude equalization, 52
Amplitude equalizers, 6, 52
Analysis, 7-10
Antoniou GIC, 91
Approximation, 7, 35-73
Attenuation
   in dB, 20
   in Nepers, 20

B

Bach's circuit, 141
Balanced filter structures, 419-422
Bandpass filter, 5, 121
Bandpass filters, based on ladder simulation,
   OTA-C, 323, 324, 332, 333
   current-mode DO-OTA-C, 409, 410
Bandstop filter, 6
Bandwidth, 63, 100, 104
Bandwidth, effects of OTA finite, *see* effects of OTA nonidealities
Bessel polynomials, 55
Bessel-Thomson filter functions, 54-58
Biquad
   Åkerberg-Mossberg, 146, 147
   allpass, 129, 138
   all-purpose, 144
   bandpass, 121
   bandstop, 126-129
   coupled OTA, 340, 341
   current conveyor, 423-425
   current-mode, 390-395, 397-404
   Deliyannis, 121
   highpass, 120
   KHN, 143
   lowpass, 116
   OTA-C, 274-303
   Sallen-Key, 116, 120
   single OTA, 235-264
   three-opamp, 141
   Tow-Thomas, 144
   two-opamp, 136
   universal current-mode, 401-404
   universal OTA-C, 287-295
Biquad magnitude characteristics
   allpass, 121

bandpass, 121
bandstop, 121
highpass, 121
lowpass, 121
Biquadratic function, alternative form of, 273, 274
Biquads; see also Biquad
   cascade of, 155-162
   coupled, 200
   two integrator-loop, 141
Biquartic sections, 171
Brackett P. O., 220
Bridged-TT network, 135
Bruton, L. T., 194
Bruton transformation, OTA-C ladder simulation
    using, 317-320
Building blocks
   current-mode, 396
   OTA-C, 270, 271
   using current conveyors, 424
Butterworth filter, 39
Butterworth lowpass filters, designing based on
    multiple loop feedback
   current-mode DO-OTA-C configurations, 414, 415
   OTA-C structures, 359, 360
Butterworth polynomial, 42, 432

C

Canonic biquads, 115
Canonical OTA-C biquad, 280
Canonical OTA-C realization, of high-order filters, 353
Capacitor, simulation of floating using OTAs and
    grounded capacitor, 317, 318
Cascade connection, 16
Cascade design, 309
   OTA-C and current-mode sections,
   see first-order filters, second-order filters and third-
    order filters
Cascade filter, 155-162
Cascade of biquads, 155-162
Cascade of biquartics, 171
Cascade sequence, 158
Cauer filter, see Elliptic filter
Causality, 35
Chebyshev filter, 42-45, 433, 434
Chebyshev lowpass filters, designing with
   ladder simulation OTA-C structure, 330
   multiple loop feedback OTA-C configurations, 360
Chebyshev polynomial, 42
Chebyshev rational functions, 42-45
Circuit analysis, 7
Classification of SABs, 115
Compensation of biquads, 146
Compensation, of OTA finite bandwidth, 300, 301
Component substitution approach, OTA-C simulations
    of LC ladders, 310-325
Complementary transfer functions, 223
Complementary transformation, 116
Complex impedance scaling, 193
Connection of two-ports
   cascade, 16

cross-cascade, 213
parallel-parallel, 15
parallel-series, 16
series-parallel, 16
series-series, 14
Constantinides A. G., 205
Continuous-time filter, 3
Continuous-time filter functions, 19
Converters, impedance, 76-81
Coupled biquads, 200
Coupled biquad OTA-C structures, 340, 341
Cross-cascade connection, 213
Current conveyor, 104
   as current follower, 423
   as voltage follower, 423
   as transconductance element, 423
   building blocks using, 423-424
   filters using, 423-425
Current-mode filters, 387-419
   based on DO-OTA-C ladder simulation, 405-411
   multiple integrator loop feedback DO-OTA-C, 411-
    419
   two integrator loop DO-OTA-C, 396-404
   using single DO-OTA, 388-395
Curve approximation, 7
Cutoff frequency, 5

D

Daryanani G., 124
DC bias current, of OTA, 249
Delay characteristics, 52-59
   of allpass functions, 58
   of Bessel-Thomson filters, 54-57
   of Pade, 58
Degree of freedom, 354
Delay denormalization, 64
Delay equalization, 59
Delay equalizer, 59
Delay, group, 21, 52-53
Delay, phase, 53
Deliyannis biquad, 121-126
Denormalization, 60, 64
Differential-input OTA, vs. single-input OTA, 301, 342,
    381
Differential amplifier, 84
Distributed networks, 1
DO-OTA-C filters, see current-mode filters
DO-OTA-RC filters, current-mode using single DO-
    OTA, 388-395
Doubly-terminated lossless twoport, 64, 183, 206
Driving point function, 20
Dual output OTA (DO-OTA), symbol and equivalent
    circuit , 388
Dynamic range, analysis of , 302, 303, 377-379
Dynamic range definition, 152
Dynamic range of LF filters, 201

E

Economical WAF, 217-220

Index

Element transformation, 65-68
Elliptic filter, 47-49
   current-mode multiple loop feedback DO-OTA-C structures, 419
   lowpass characteristic of, 48
   multiple loop feedback OTA-C architectures, 368-370
Enhanced negative feedback, 115
Enhancement, in quality factor Q due to OTA finite bandwidth, 251, 252
Enhanced positive feedback, 115
Equalizer, 6, 52, 59
Equi-ripple filter approximation, 42
Equivalence of admittance and signal simulation methods in OTA-C filter design, 336, 338, 339
Equivalence of OTA-C structures, 301
Excess phase, 145
   of OTA, *see* frequency-dependent transconductance

F

FDNR, 80
   OTA-C realization of, 317, 318
Feedback network, matrix description of, 351, 353, 412
Fettweis A., 205
Filters
   active, 3, 31
   Bessel-Thomson, 54-58
   Butterworth, 39-42
   Cauer, 47-49
   characterization, 1-4
   Chebyshev, 42-45
   continuous-time, 3
   current-mode DO-OTA-C, 387-419
   elliptic, 47-49
   finite, 3
   fundamentals, 1-33
   inverse Chebyshev, 45, 46
   linear, 2
   lumped, 1
   maximally-flat delay, 54-57
   MOSFET-C, 420-422
   OTA-C, 254-257, 269-303, 309-344, 349-382
   OTA-C-opamp, 422, 423
   OTA-RC, 231-264
   passive, 3, 29, 30
   time-invariant, 3
   using current conveyors, 423-425
First-order active filter, 108-110
   current-mode, 389, 390, 396
   OTA-C, 271
   OTA-RC, 233, 234
Fleischer, P. E., 124
Floating admittance, simulation using OTAs and grounded impedance, 320, 321
Floating capacitor
   disadvantages of, 277, 317
   simulation of using OTAs and grounded capacitor, 317, 318
Follow-the-leader feedback, (FLF) 168-171
   coupled biquad, 340, 341

current-mode DO-OTA-C, 414, 418
Frequency scaling, *see* Frequency transformation
Frequency response, 21
Frequency transformation, 59-64
Frequency-dependent negative resistor, *see* FDNR
Frequency-dependent transconductance, 249, 250
Friend biquad, 126-129
Function
   Bessel-Thomson filter, 54-57
   Butterworth filter, 39-42, 432
   Chebyshev filter, 42-45, 433, 434
   delay, 53
   elliptic filter, 47-49, 435
   inverse Chebyshev, 45, 46
   magnitude, 24, 38
   nonminimum-phase, 53
   permitted, 35
   Papoulis, 47
   phase, 21, 52, 53, 54
   positive real, 29

G

Gain distribution, 159
Gain sensitivity product, 114
Generalized impedance converter, *see* GIC
Generalized impedance inverter, 81
GIC, 76
   Antoniou, 91
   using opamps, 91
GIC biquads, 136-139
G_m-C filters, *see* OTA-C filters
Gorski-Popiel, J., 191
Grounded capacitors, vs. floating capacitors, 317, 342
Group delay, 21, 52, 53
Gyration conductances, 185
Gyrator, 81, 185
Gyrator imperfections, 186
Gyrator realizations, 90, 102, 103
Gyrator realization using OTAs, 90

H

Highpass biquad, 120
High-order filters, current-mode, 405-419, *see also* cascade design
High-order filters, OTA-C using
   cascade method, 309,
   LC ladder simulation, 309-344
   multiple loop feedback approach, 349-382
Highpass filter, 5
Hybrid parameters, 14
Huelsman, L. P., 143
Hurwitz polynomial, 29

I

Ideal controlled-sources, 75, 76
Ideal filter characteristics, 5
Ideal transformer, 78
Ideal transmission of signal, 4

IFLF (inverse FLF) configurations, OTA-C, 360, 361, 370-372
Immittance, 79
Impedance converter
    generalized, see GIC
    Riordan, 90
Impedance inverter, see Gyrator
Impedance level, 67
Impedance scaling, 70
Impedance transformation, 76, 77
Impulse response, 22
Inductance simulation, 79, 82, 224
    using OTA-C, 310, 311
    using current conveyors, 424
Inductor substitution method, OTA-C filters based on ladder simulation using, 310-317
Input methods in OTA-C filters, selection of, 302
Integration, 86, 102
Integrators
    current-mode DO-OTA-C, 396
    current-mode DO-OTA-C-opamp, 423
    MOSFET-C, 421
    OTA-C, 270, 271
    OTA-C-opamp, 422
    using current conveyor, 424
    noninverting, 146
Inverse Chebyshev, 45, 46
Inverting amplifier, 99
Inverting integrator, 86
Inverting voltage amplifier, 99, 101

K

Kerwin-Huelsman-Newcomb biquad, see KHN biquad
Key, E. L., 117, 120
KHN biquad, 143
KHN OTA-C biquad, 281, 282, 289, 290, 292, 293

L

Ladders, simulation of
    based on OTAs and capacitors, 309-344
    using current-mode method, 405-411
Ladder simulation structures with partial floating admittances, 324, 325, 332-335
LC ladder, doubly-terminated, 64, 183, 206
LC ladders, OTA-C simulation using
    admittance/impedance simulation method, 320-325
    Bruton transformation, 317-320
    inductor substitution, 310-317
    matrix methods, 338-340
    signal flow simulation, 325-337
LC ladder simulation, 184-202
Leapfrog (LF) filter architectures
    ladder-based current-mode and OTA-C, see signal flow simulations
    multiple loop feedback current-mode DO-OTA-C, 414, 418
    multiple loop feedback OTA-C, 361, 362, 372, 373
Leap-frog method, 195-201

Leap-frog topology, 163
LHP, 20
Linear filter, 2
Linear phase, 53
Linear range, effects of limited OTA input, 377, 378, 379
Linear transformation active filters, 224-229
LLF networks, 10
Loading effects, in OTA-C filters, 303
Loss function, 20
Lossless twoport, 184; see also LC ladder
Lossy integrators, OTA-C, 271
Low sensitivity of passive filters, 184
Lowpass biquad, 117
Lowpass characteristics
    Butterworth, 39-42
    Chebyshev, 42-45
    elliptic-function, 47-49
    equal-ripple in passband, 42-45
    equal-ripple in both pass and stop bands, 47-49
    equal-ripple in stopband, 45-46
    maximally flat, 39-42
Lowpass finite zero filters, based on ladder simulation,
    current-mode DO-OTA-C, 408, 409
    OTA-C, 316, 317, 322, 323, 330-332, 334, 335, 337
Lowpass-to-bandpass transformation, 62
    element replacement, 66
Lowpass-to-bandstop transformation, 63
    element replacement, 66
Lowpass-to-highpass transformation, 61
    element replacement, 66
Lowpass-to-lowpass transformation, 60
Lueder E., 157
Lumped network, 1

M

Magnitude function, 24, 38
Matching active transformer, 221, 222
Matrix
    A, see transmission
    chain, see transmission
    matrix conversion, 13
    transmission, 14
Matrix methods, of LC filter simulation for OTA-C filter design, 338-340
Maximally flat magnitude characteristic, 39-42
Maximally-flat delay filter, 54-58
Maximum power transfer, 184
Maximum signal level, determination, 302, 303, 377-379
Mismatch, in transconductances, 303
Modified transmission matrix, 206, 225
Moschytz G., 157
MOSFET-C filters, 420-422
Mossberg, K., 146, 147
Multiparameter sensitivity, 153
Multiple integrator loop feedback
    current-mode DO-OTA-C filters, 411-419
    OTA-C filters, 350-382
Multiple integrator loop feedback OTA-C filters

Index   441

all-pole canonical structures, 355-363
  determination of maximum signal magnitude, 377-379
  general model and formulation, 350-353
  OTA nonideality effects, 379-381
  sensitivity analysis, 373-377
  structure generation, 353
  structures of realising transmission zeros, 363-373
  synthesis procedure, 354
Multiple-loop feedback filters, 151, 162-171

N

Negative-impedance converter, 79, 88, 89
Negative-impedance inverter, 82
Negative resistance, 82
  parasitic associated with OTA inductor, 313
Network function, 20
Network
  active, 31
  distribution OTA, 292, 365, 402, 415
  feedback, 351, 412
  passive, 29, 30
  summation OTA, 292, 364, 403, 416
Network parameters
  a parameters, 14
  conversion table, 13
  G parameters, 12
  H parameters, 12
  Y parameters, 12
  Z parameters, 11
Network transfer function, 19, 20
Newcomb, R. W., 143
Nodal analysis, 7
Nonideal parameters of OTA, 249, 250; see effects of OTA nonidealities
Noninverting integrator, 146
Noninverting voltage amplifier, 98, 99
Nonlinearity, cancellation of MOSFET, 421
Nonminimum-phase functions, 53
Norton's theorem, 9

O

One-port network, 10
Opamp
  ideal, 84
  practical, 93
  single-pole model, 94, 98
  virtual ground, 85
Operational amplifier, see opamp
Operational transconductance amplifier, see OTA
Operations using opamps, 85–87
Operations using OTAs, 101, 102
Orchard, 184
Orchard-Wilson gyrator, 90
OTA
  ideal, 100
  practical, 103
  typical parameter values of, 231, 232
OTA-C filters

based on ladder simulation, 309-344
  derived from single OTA filters, 254-258
  multiple integrator loop feedback, 349-382
  two integrator loop, 269-304
OTA-C-opamp filters, 422-423
OTA nonidealities, effects of, 249-258, 295-301, 312-315, 379-381
OTA-RC filters, see single OTA filters
Output methods in OTA-C filters, selection of, 302
Output voltages, equalisation of the maxima of, 302
Overshoot, 24

P

Pade approximation, 58, 59
Paley-Wiener criterion, 36
Papoulis approximation, 47
Parameters, see Network parameters
Parasitic components, due to OTA nonidealities
  associated with OTA-C inductor, 312, 313
  associated with OTA resistor, 255-258, 395
  effects on OTA-C filters by inductor substitution, 313-316
Parasitics, sensitivity to 313, 314
Parasitic effects, relation with phase sensitivity, 314
Passband, 5
Passband ripple, 45
Passband variation, see Passband ripple
Passive compensation, of OTA finite bandwidth, 300
Passive filters, 3, 29, 30
  simulation of, 183-202
Passivity criteria, 29
Permitted functions, 35
Phase equalization, see Delay equalization
Phase function, 21, 52-54
Phase delay and shift, of OTA, see frequency-dependent transconductance
PIC, 79
Pole, 20
Pole sensitivity, 113
Pole-zero pairing, 156
Polynomial
  Bessel, 54-57
  Butterworth, 42, 432
  Chebyshev, 42-45, 433, 434
  Hurwitz, 29
Positive-impedance converter, 79
Positive-impedance inverter, see Gyrator
Positive real function, 29
Potential instability, 27
Power complementary functions, 216, 223
Power, maximum, 184
PRB, 165, 167, 168, 170
Predistortion, 71
Primary-resonator-block, see PRB

Q

Quality factor, 111
Q enhancement, see Selectivity enhancement
Q sensitivity, 113

R

RC-CR transformation, 69, 70
Realization of active elements
    using opamps, 87-91
Realization of biquadratics
    using opamps, 114-147
    using SABs, 114-136
Realization of biquartic sections, 175
Realization of
    first order functions, 108-110
Realization of high-order functions
    analogue computer methods, 165
    selection criteria, 151, 152
Reciprocity theorem, 32
Reflection coefficient, 49
Relative change, in filter parameters due to OTA nonidealities, 251-253, 314-316
Relative sensitivity, 112
Residue condition, 29
Resistors, OTA simulation circuits of 254, 255, 390, 391
Resonator, OTA-C simulation of
    grounded parallel LCR, 315
    floating parallel LC, 321, 322
Responses of 2nd-order functions, 121
RHP, 20
Riordan, 90
Ripple, passband, 45
Rise time, 24-26
Root sensitivity, 113
Routh-Hurwitz, 30

S

SAB
    allpass, 129, 138
    bandpass, 121
    highpass, 120
    lowpass, 116
    lowpass and highpass notch, 126-129
    using Twin-T RC networks, 134
    with a positive real zero, 132
Sallen-Key biquads, 116, 120
Sampled-data filter, 3
Saraga, W., 120
Scaling
    conductance, 326, 405
    frequency, see Frequency transformation
    impedance, 70
Scattering parameters, 207
Schoeffler's sensitivity measure, 153
Second-order filters
    current-mode DO-OTA-C, 397-404
    current-mode single DO-OTA, 390-395
    OTA-C, 254-258, 274-303
    single OTA, 235-264
    using current conveyors, 423-425
Second-order filters, design examples of
    bandpass, single OTA, 243
    bandstop, OTA-C, 276-278
    highpass, single OTA, 260

KHN OTA-C biquad, 282
lowpass, single OTA, 235, 236
Selectivity enhancement, 139
Sensitivity
    analysis of single OTA filters, 236-240, 242, 260, 261
    CBR filters, 178
    computation of current-mode single DO-OTA filters, 392-394
    measures, 111-113
    multiparameter, 153
    multiple integrator loop feedback OTA-C filters, 373-377
    parasitic, calculated based on normal sensitivity, 314-316
    pole semi-relative, 113
    relative, 112
    Schoeffler measure of , 375
    semirelative phase, 113
    WAF, 220, 221
    worst-case, 153
Sequencing of biquads, see Cascade sequence
Settling time, 24
Shifted-companion-form, 166
Signal flow simulation, of LC ladders using
    current-mode method, 405-411
    OTAs and capacitors, 325-337
Simulation of LC ladders, 184
Simulation of LC ladders using
    coupled biquads, 200
    FDNR, 193-195
    functional approach, 195-201
    GIC, 190-193
    gyrator-C, 185, 188
    leap-frog, 195-201
    linear transformation, 224-229
    signal-flow-graph, 229
    topological approach, 185
    wave active filters, 205-224
Single-ended circuits, vs. balanced ones, 419-421
Single OTA filters, 231-264
    derived from five admittance model, 258-264
    derived from four admittance models, 241-248
    derived from three admittance model, 232-240
    effects of OTA nonidealities on, 249-254, 255-258
    OTA-C filters derived from, 254-258
Slew-rate, 97
Specified curve, 7
Stability, 26-28, 36
Stability and Passivity, 31
Standard deviation, 153
Stopband, 5
Stopband edges, 37
Strictly stable, 26, 37
Subtraction of voltages, 86, 101
Summation of voltages, 85, 101
Summed-feedback topology, 162-164
Summer circuits, using OTAs, 270
Supercapacitor, 80, 92, 194
Superinductor, 80
Superposition, 2
Symmetrical notch, 131

Index   443

Synthesis, filter, 7
Synthesis of OTA-C filters, 349
Systems matrix, 352
Szentirmai, G., 200

T

Third-order filters
   current-mode, 417-419
   OTA-C, 356, 357, 366-370
Thomas, L. C., 144
Thomson filter, see Bessel-Thomson filters
Three-opamp biquads, 141
Time, 24
Time-domain stability, 26
Time-invariant filter, 3
Tolerance, 111, 153
Tow, J., 144
Tow-Thomas biquad, 144
   OTA-C, 284
Transadmittance, 20
Transconductance-C filters, see OTA-C filters
Transfer functions, 17, 18
Transformation
   complementary, 116
   frequency, 59-64
   lowpass-to-bandpass, 62
   lowpass-to-bandstop, 63
   lowpass-to-highpass, 61
   lowpass-to-lowpass, 60
   RC-CR, 69, 70
Transformer, ideal, 78
Transient response, 22, 23
Transimpedance, 20
Transition band, 5
Transmission matrix, 14
Transmission zeros, 165, 168
Transmission zeros, realization of using
   current-mode DO-OTA-C method, 401-404, 415-419
   OTA-C structures, 287-295, 363-373
Tuning, of OTA filters, 249, 286, 287
Twin-T RC networks, 135
Two-integrator biquads, 141
Two integrator loop
   configurations, 272

   current-mode filters, 397-404
   OTA-C filters, 274-303
Two integrator loop OTA-C filters, 274-304
   comparison of , 285-287
   distributed-feedback (DF), 274-280
   dynamic range of, 302, 303
   effects and compensation of OTA nonidealities on, 295-302
   summed-feedback (SF), 280-283
   universal, 287-295
   using lossy integrators, 283-285
Two-opamp biquads, 136
Twoport, lossless, see Lossless twoport
Two-port
   active, 31
   lossless, see Lossless twoport
   network, 11
Two-port interconnections, 14–16

U

Unstable, potentially, 32

V

Virtual earth, 85
Voltage addition, 85
Voltage-controlled resistors, 255, 391, 421
Voltage-mode filters, vs. current-mode filters, 387, 388
Voltage subtraction, 86
Voltage variable resistor, 101

W

WAF, 205-224
Wave active, 208–212
Wave active filters, see WAF
Wave simulation of inductance, 224
Wave variables, 205
Worst case sensitivity, 153

Z

Zeros, transmission, see Transmission zeros